**River Ecology**

# River Ecology

## Science and Management for a Changing World

**Michael A. Mallin**
*University of North Carolina Wilmington, USA*

# OXFORD
### UNIVERSITY PRESS

Great Clarendon Street, Oxford, OX2 6DP,
United Kingdom

Oxford University Press is a department of the University of Oxford.
It furthers the University's objective of excellence in research, scholarship,
and education by publishing worldwide. Oxford is a registered trade mark of
Oxford University Press in the UK and in certain other countries

Published in the United States of America by Oxford University Press
198 Madison Avenue, New York, NY 10016, United States of America

British Library Cataloguing in Publication Data
Data available

Library of Congress Control Number: 2022951025

ISBN 978–0–19–954951–1
ISBN 978–0–19–954952–8 (pbk.)

DOI: 10.1093/oso/9780199549511.001.0001

Printed and bound by
CPI Group (UK) Ltd, Croydon, CR0 4YY

# Contents

# About the Author

Dr. Michael Mallin is currently Research Professor at the University of North Carolina Wilmington, Center for Marine Science. He holds a BS in Botany from Ohio University, an MS in freshwater ecology from the University of Florida, and a PhD in marine and estuarine biology from the University of North Carolina Chapel Hill. Dr. Mallin was an Aldo Leopold Environmental Leadership Fellow and is an elected Fellow of the American Association for the Advancement of Science. He served as President of the Southeastern Estuarine Research Society from 2004 to 2006 and as President of the North Carolina Lake Management Society from 2018 to 2019. He has worked in a variety of ecosystems including freshwater streams, rivers, tidal creeks, large estuaries, reservoirs, lakes, and the coastal ocean.

# Introduction

This book is an outgrowth of a graduate course, River Ecology, that I devised in 2000 at the University of North Carolina Wilmington. I had been teaching a graduate class called Estuarine Biology, and my colleague Larry Cahoon had been teaching undergraduate Limnology, but there were no courses that dealt with streams or rivers, despite Wilmington being located at the terminus of the largest and most industrialized river in North Carolina. While there have been excellent books published on stream ecology, primarily regarding the benthos, a wholistic approach that includes the plethora of modern insults we continue to inflict upon rivers was not available, so that was my inspiration.

A little bit about my background may help at the outset. During my MS degree program at the University of Florida I conducted plankton research in the highly eutrophic Oklawaha chain of lakes in central Florida. Later in my PhD program at the University of North Carolina Chapel Hill I conducted phytoplankton, zooplankton, and trophic studies and nutrient research on a highly eutrophic estuary (the Neuse River estuary) as well as on the coastal ocean. Previous to my PhD while in the nonacademic world I had worked as a limnologist on rivers and reservoirs for a large public utility company, and even before graduate school I had worked in a wastewater treatment plant, both as an operator and as a laboratory analyst and stream sampler. My research at the University of North Carolina Wilmington has focused (at various times) on nutrient pollution, eutrophication and algal blooms, fecal microbial pollution, the impact of urbanization on streams and rivers, human sewage stream pollution, the impacts of industrial livestock production, the intersection of hurricanes, human

development, and river ecology, and, more happily, stream restoration.

The book differs from other stream ecology-oriented books (referred to within for their many helpful contributions) in that the major assumption is that there are few, if any "pristine" river ecosystems remaining on Planet Earth (not even in the tropics anymore), and thus major effort is put toward introducing and explaining the many physical, biological, chemical, and political anthropogenic insults we humans have had on river ecosystems. Basic stream and river ecology and theory are of course covered. However, strong emphasis is devoted to river nutrient pollution and eutrophication, river fragmentation, the power industry and its multiple impacts on rivers and reservoirs, chemical pollution, industrialized animal production and its discharges, the ongoing introductions of plant and animal invasive species, human watershed development impacts and subsequent stream restoration, and floods, hurricanes, and of course climate change.

This book also features chapters devoted to largely ignored or forgotten stream and river study components including blackwater streams and rivers, tidal creek ecology, and reservoir limnology. In addition, human wastewater treatment is explained, a subject I was never taught in the courses I took on stream ecology, limnology, biological oceanography, or even microbial ecology: I had to take a course in environmental engineering to learn its basics. Yet, poor sewage treatment had a major impact on early stream ecology and sanitary engineers had to deal with massive stream sewage pollution—a problem that remains in many ecosystems to this very day. Finally, because your author considers river ecosystems in a wholistic fashion,

*River Ecology*. Michael A. Mallin, Oxford University Press. © Michael A. Mallin (2023). DOI: 10.1093/oso/9780199549511.003.0001

the estuarine component is, as such, recognized and emphasized in several chapters. Sampling and analysis methods are introduced where appropriate, but this is not a "methods" book. Within I direct the reader to well-known detailed methods manuals when discussing certain topics. The book is primarily aimed at graduate and upper-level undergraduate students, but it is my hope that professional researchers, planners, environmental engineers, agency professionals, and nongovernmental organization (NGO) personnel will also find it of use.

For very helpful chapter reviews I thank Dr. JoAnn Burkholder, Dr. Larry Cahoon, Dr. John Beaver, Dr. Mike Burchell, Dr. Scott Ensign, Dr. Mike O'Driscoll, Dr. Denise Sanger, Dr. Sandra Shumway, and Dr. Steve Skrabal. Additional assistance, discussion, and input was contributed by Dr. Eric Bolen, Dr. Tom Kwak, Dr. James Merritt, Dr. Sam Mozley, Kyle Rachels, and Dr. Fred Scharf. I thank Melissa D. Smith of the University of North Carolina Wilmington Center for Marine Sciences for her excellent artwork, improving upon my own figures, and producing drawings from my photographs. For photographic contributions I thank Amy Grogan, Tom Lankford, and Matthew McIver of the University of North Carolina Wilmington; JoAnn Burkholder, Stacie Flood, and Elle Hannon of North Carolina State University; John Beaver of BSA Environmental Services, Inc.; Kemp Burdette of Cape Fear River Watch; Mark Vander Borgh, Dan Wiltsie, and Rob Emens of the North Carolina Department of Environmental Quality; Walker Golder of Audubon; Dennis Lemly of Wake Forest University; Fritz Rhode of the National Marine Fisheries Service; Denise Sanger of the South Carolina Department of Natural Resources; and Wendy Stanton of the US Fish & Wildlife Service.

Finally, I wish to thank my editor Ian Sherman of Oxford University Press for his infinite patience and help, and other Oxford team members including Charlie Bath, Katie Lakina, Bethany Kershaw, and Lucy Nash.

# CHAPTER 1

# The Physical Nature of River Ecosystems

Streams and rivers can be considered to be the veins and arteries of a continent. As well as being complex systems in themselves, they are corridors along which there is a constant multidirectional movement of living and nonliving materials. Tides cause movement of seawater and its constituents from the ocean upriver to varying degrees, but in the end the net movement of water of all major rivers is seaward. Along this path to the sea, rivers carry inorganic material such as suspended sediments, organic but nonliving material such as plant detritus, living organic material such as phytoplankton and zooplankton, dissolved materials including carbon, nitrogen, phosphorus, and silicate, and numerous toxic pollutants. There is also very active use of these corridors by aquatic, terrestrial, and avian life. Along with the multitude of organisms that permanently inhabit rivers, many species of fish and invertebrates use rivers for extensive journeys as a part of their life cycle. Anadromous fish such as salmon swim upstream from the ocean to spawn in freshwater streams, catadromous eels swim downstream to spawn in the ocean, and various crustacean meroplankton hitch rides on the tides to move between upper and lower estuaries during their life cycles. The shoreline regions are heavily used as trails for mammals such as deer and raccoons, and birds use the river corridor for movement, hunting, and nesting. The interfaces between the water, the air, riparian forests and their canopies, forests, and marshes (called **ecotones**) are areas of intense animal activity.

Streams drain an area of the landscape that is defined by elevation and slope. The area that a given stream or river drains is called its **drainage basin**, or **catchment**. The area between catchments is called a topographic divide (Gordon et al. 2004); bear in mind that some catchments will also receive water from underground sources from outside the catchment, as well as surface flows. Catchments can be very small (such as an isolated gully draining into a small brook) or vast (such as the area draining into the Amazon River and its tributaries). Major rivers accept the water and associated water-borne materials from hundreds to thousands of individual stream catchments to form very large catchments. Numerous physical, meteorological, and anthropogenic factors affect the quantity and quality of the water that drains from a given catchment; thus, let us define **watershed** as the area that is drained by a given stream and its tributaries, plus its human and animal populations, land use, and the local meteorology affecting the stream. Additionally, when one takes a watershed approach to stream study, the vertical component to a watershed must be included, that is the impact of trees and other vegetation on local hydrology and supply of organic matter to the collector streams. Some watersheds receive greater amounts of polluted atmospheric deposition than others, depending upon location and prevailing winds. As will become clear in this book, what happens in the headwaters of river system can have a major impact on the chemistry and biology of the lower river and estuary, even hundreds of miles away. That is why it is essential to consider stream systems from a watershed perspective to fully understand their structure and function—this is the ecosystem perspective. A good way to begin this investigation is to categorize the contributing streams within a watershed.

*River Ecology*. Michael A. Mallin, Oxford University Press. © Michael A. Mallin (2023). DOI: 10.1093/oso/9780199549511.003.0002

## 1.1 Stream Order

Streams and rivers come in numerous sizes, and these ecosystems function somewhat differently as such. Therefore it is helpful to classify streams based on their scale to aid in their study. A convenient way to classify streams is by the Strahler (1952) stream order scheme (Fig. 1.1). Using this classification, any stream ecologist, as well as some limnologists, fisheries biologists, and estuarine ecologists form an instant mental image of the system being described. In this scenario, the smallest permanent (perennial) streams are called first-order streams (in the USA these streams are also called blue-line streams from their designation on US Geological Survey maps). When two first-order streams join, a second-order stream is formed. It requires the joining of two second-order streams to form a third-order stream, and so on. Addition of a lower-order stream does not change the order of a higher-order stream. Medium-sized rivers are 5th to 6th order, the Mississippi River is thought to be a 10th-order stream, and the Amazon River is believed to be a 12th-order stream. The vast majority of total stream length in a watershed is comprised of first- to third-order streams, although they collectively comprise only a minor portion of total stream surface area in the watershed. These are often wadeable streams, while in 4th order and higher the depth and current become too strong for such activity. Some basic tenets of drainage system geometry are that as stream order increases, the number of direct tributary streams decrease, average stream length increases, lateral catchment area decreases, and average slope decreases (Gordon et al. 2004).

So what feeds a first-order stream? In some cases the stream may originate from a spring that flows year-round (Figs. 1.1; 1.2a). More commonly, first-order streams are fed by intermittent streams, which only contain water periodically, depending upon connectivity to the groundwater table, rainfall, season, local snowmelt, and evaporation and plant transpiration rates (Fig. 1.2a). Ephemeral streams (Fig. 1.3a) are basically natural ditches that are fed solely by rainfall runoff (Gordon et al. 2004). Note that "temporary" waterways have ecological value themselves, as seed banks for plants, egg banks for invertebrates, biological corridors for organisms, and areas for organic materials processing (Steward et al. 2012). The divide between intermittent and ephemeral streams is often difficult to determine. Keep in mind that land developers will also straighten and expand natural intermittent and ephemeral streams to enhance their drainage capacities. Because human use in one way or another impacts most stream systems, drainage ditches collecting excess water from agricultural fields or urban areas also frequently feed first-order streams (Fig. 1.3b). Collectively, this upper watershed area of springs, intermittent and ephemeral streams, and drainage ditches is referred to as the **headwaters** (Fig. 1.1).

The lowest-order streams have the narrowest width and greatest slope and most turbulent

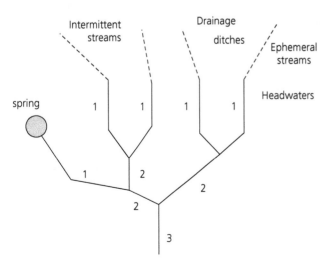

**Fig. 1.1** Stylized rendering of the headwaters region and the Strahler (1952) stream order scheme. (Figure by the author.)

(a)    (b)

**Fig. 1.2**  Figure 1.2a (left) Headwaters spring in Horseshoe Bend National Military Park, Alabama. Figure 1.2b (right) intermittent stream in northern Ohio. (Photos by the author.)

(a)    (b)

**Fig. 1.3**  Figure 1.3a (left) Ephemeral stream draining forest in central North Carolina. Figure 1.3b (right) Runoff ditch in headwaters of Bradley Creek, coastal North Carolina. (Photos by the author.)

waters. As the stream order increases seaward, this gradient is characterized by decreasing slope, increasing stream width and depth, higher mean water temperature, and less short-term fluctuation in water flow (i.e. smaller stream discharge can increase greatly during a storm). Going downstream, in most cases the character of the streambed will change as well. Lower-order streams have rougher, more deeply incised beds due to in-stream erosion, in middle-order streams the bed is smoother and wider, and in high-order rivers the bed is broad and smooth, often with a deep channel within the stream bottom that is called a **thalweg**. The sediment composition of lower-order streams (< 3rd order) is a mixture of inorganic materials of broad size range, from large boulders to very small clay particles. These lower-order streams along the stream continuum can be considered the erosional section, because rainstorms and flash floods erode particles from the bottom and banks that can be transported far downstream. Higher-order streams (> 4th order) are considered depositional areas because small particles eroded from upstream are deposited here. In higher-order streams a **floodplain** develops over which the river (if undammed) will spread out during periods of high flow. The river gradient extends seaward under the salinity concentration is 0.5 ppt, which is considered to be the head of the estuary. In mountainous areas the floodplain can be extremely compressed, with

(a)

(b)

**Fig. 1.4** Figure 1.4a (left) Constrained reach along river in Oregon. Figure 1.4b (right) Unconstrained reach of the lower Cape Fear River on the North Carolina coastal plain, showing large meander. (Photos by the author.)

slopes in close proximity to the stream. Stream stretches in such areas are called **constrained reaches** (Fig. 1.4a), which Gregory et al. (1991) define as having a valley width less than twice that of the channel width. At the other end of the spectrum are **unconstrained reaches**, low slope areas where an extensive floodplain is present (Fig. 1.4b).

How then are streams formed and maintained? Consider a rainstorm in a woodland valley catchment. When rainfall strikes the ground, the water has one of these fates: evaporate from the ground or pavement; be transpired from the vegetation back to the atmosphere; infiltrate into the ground; or flow overland downslope until it reaches a body of water (Fig. 1.5).

## 1.2 Evapotranspiration

The amount of rainfall that is evaporated from surfaces plus that transpired from vegetation is called **evapotranspiration**, which increases with higher air temperatures, wind speed, and lower relative humidity, and the amount of actively growing vegetation. High air temperatures increase direct evaporation from surfaces, and actively growing plants absorb water through their roots and transpire some of it through their stomata back into the atmosphere. Thus, in temperate regions evapotranspiration is high in late spring, summer, and early fall, but lowest during the winter months. The seasonal difference can be considerable. For example, in North Carolina, a relatively warm state in the southeastern

USA, rates of evaporation alone in July are approximately four times those of December (Robinson 2005). Regional evapotranspiration rates, in conjunction with local rainfall and snowmelt patterns, are also integral in determining when flood stages of rivers are most likely to occur year by year.

## 1.3 Infiltration and Groundwater

The amount of water that infiltrates into the ground depends on how wet the ground is to begin with (its saturation), the type and thickness of the soil, and the depth to underlying bedrock. Water that infiltrates into soil moves downward until it intercepts and enters the **water table**, the uppermost zone of saturated soils (Fig. 1.5). There are often multiple zones of saturated soils above each other in a region; these are called **aquifers**. Aquifers can be made of many types of material: unconsolidated sand and gravel, or even consolidated rocks including sandstone, limestone, dolomite, granite, and basalt (Alley et al. 1999). Within aquifers the pores or open spaces between unconsolidated particles are filled with water; with more consolidated rocky material the cracks between are filled with water. The upper aquifer is unconfined by an impermeable layer and, as mentioned, is topped by the water table. The height of the water table can fluctuate: in drought periods it will lower; in summer it will also lower due to increased evaporation and transpiration; and in winter it is likely to be at its highest through lack of evaporation and transpiration. Infiltration from

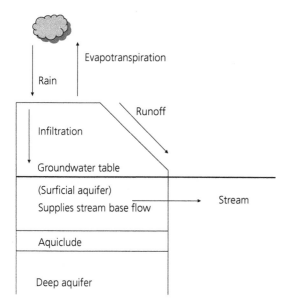

**Fig. 1.5** The three fates of rainwater after it comes to Earth, and movement of infiltrated water in a typical upland drainage to a stream. (Figure by the author.)

the surface down into the water table is commonly called **groundwater recharge**.

The infiltration (or percolation) rate depends largely upon soil permeability and is lower with fine soils and higher with coarse soils. For instance, infiltration rates are low in a clay field (clays encompass the finest soil particles), higher in a plowed field, and higher yet in a sandy field (sand grains are large and coarse particles—picture a sandy beach). The ability for a given soil or rocky material to transmit ground water is called **hydraulic conductivity**. Below the unconfined upper aquifer there will be an impermeable confining layer of dense material such as clay; these layers are sometimes called aquitards or aquicludes. These impermeable layers are what confine aquifers. In some regions there can be several aquifers separated by impermeable or semipermeable aquitards. They do not all have to be freshwater either. Barrier islands along the coasts of continents may have a freshwater aquifer perched above a brackish water aquifer. In some areas where an aquifer is held under pressure by confining layers, a spring may occur, pumping water upwards out of the aquifer; these are called **artesian** aquifers and springs (Keeney 1986). A seep is a surface area

that is wet from aquifer contact but not actively pumping. Infiltrated water in the aquifer nearest the surface may eventually work its way down to a lower aquifer, depending upon the permeability of the aquitard and the hydraulic head. Or, it may move laterally toward a stream, a process sometimes called subsurface drainage. It then enters the stream through the bed or through the banks and is critical in maintaining the stream's **base flow**, which is the portion of the stream water that is derived from groundwater inputs rather than overland flow or direct rainfall. The amount that groundwater contributes to average stream flow is approximately 40–50% in small streams (Alley et al. 1999) but varies widely. Besides streams and rivers, groundwater of course contributes to lakes and wetlands. Water movement through the soil is far slower than surface runoff; 30 cm/day is considered a high rate of movement (Alley et al. 1999).

As noted above, stream discharge is a result of subsurface-derived base flow, along with surface runoff that enters the stream. If the surface water table is higher than the channel bottom, water infiltrates into the channel and it is considered a **gaining** stream. Perennial streams are typically gaining streams. If the channel bed is above the water table then the stream is a **losing** stream, fed by overland runoff (intermittent streams are losing streams). Some streams may have both gaining reaches and losing reaches, depending upon the local geology and hydrology. It must be noted that in arid regions such as the US southwest, most streams are losing streams due to depth to the water table and only fill during rare rainstorms (Hupp and Overcamp 1996; Dodds 2002). Hydrologists can determine gaining or losing reaches by temperature changes—groundwater has a constant temperature, while runoff is subject to sunlight-driven water temperature fluctuations—and also changes in hydraulic head and discharge can be used to determine gaining or losing reaches. In both gaining and losing streams there is interaction between the streambed and the subsurface material below it. This zone where surface water and groundwater mix and interact is referred to as the **hyporheic** zone, and in this zone water moves up or down, along with any compounds that are dissolved in the water.

## 1.4 Overland Runoff

The area where the landscape meets the water is called the **riparian zone** (Fig. 1.6a, b). This term has been used to refer to as narrow an area as the water's edge, and it has been expanded to encompass as large an area outward as the limits of flooding and upward into the canopy of streamside vegetation (Gregory et al. 1991). Thus, it is an operationally defined term; regardless, it is a zone of intense interaction between the land and water where exchanges of numerous materials occur, including organic matter large and small, nutrients, chemical pollutants, and materials eroded from the landscape. Back to our fates of rain in a watershed; if the ground is already saturated and can accept no more rainwater, that water which is not evapotranspired flows over land as surface **runoff**, moving downslope through a riparian zone until it enters a stream (or in some cases a lake or wetland). Water flowing across the landscape is also known as **sheet flow** (Fig. 1.7a). Runoff is more intense with excessive rainfall, poorly drained soils, bedrock that is near the soil surface, and lower amounts of vegetation. Surface runoff will flow overland until a temporary or permanent stream is encountered or in some cases formed by runoff. In a low-order stream the volume of water flow, known as **stream discharge**, will fluctuate rapidly due to runoff inputs and a small catchment. Whereas discharge of a low-order stream is highly dependent upon an individual rain event, such an event will have little effect upon a higher-order stream or river's discharge. A high-order river responds to a multitude of inputs from lower-order streams distributed throughout the river drainage basin. In a small stream, peak discharge rapidly follows the rainfall, but peak discharge in a higher-order stream will be delayed,

**Fig. 1.6**  Figure 1.6a (left) Rocky riparian zone of Tennessee stream. Figure 1.6b (right) Broad, swampy riparian zone of a North Carolina coastal plain stream in winter. (Photos by the author.)

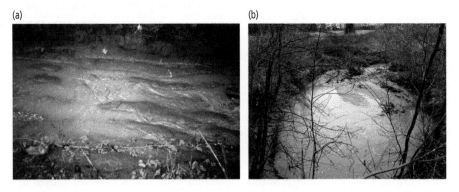

**Fig. 1.7**  Figure 1.7a (left) Sheet flow of runoff across disturbed ground in North Carolina Piedmont. Figure 1.7b (right) Excessive sediment loading from upstream construction causing high turbidity in Crabtree Creek, Raleigh, North Carolina. (Photos by the author.)

sometimes by many days after a rain, depending upon the size of the drainage network. In headwaters areas impacted by snow, a large spring melting event will lead to a peak discharge and occasionally flooding.

Overland runoff shapes and carves the landscape. It erodes loose soils and washes them downslope, where they enter a stream, lake, reservoir, or estuary as **suspended sediments** (Fig. 1.7b). Inorganic suspended sediments are also called **alluvium** (Cole 1994). Erosion will occur on rockier terrain; some formations made of sedimentary materials are softer and more amenable to erosion, while harder materials such as granite are highly resistant to erosion. Runoff (and infiltrated water) can leach into karst soils, which are calcium carbonate, chemically eroding these soils and leaving sinkholes and even extensive subsurface caverns and underground rivers, such as in the Mammoth Cave region in Kentucky. On the surface, eroded sediments that enter streams become incorporated into a dynamic that both creates and destroys habitat for aquatic plants and animals. Lighter suspended sediments, such as clays, will create a brown cloudiness or murkiness within the receiving stream that is called **turbidity** (Fig. 1.7b).

Runoff is increased, often dramatically, by physical alteration of the natural watershed vegetation, especially by clearcuts for forestry or development. Clearcuts remove the anchoring function of roots and rhizomes, as well as removing the transpiration function of the forest. The presence of trees protects soil from erosion by intercepting rainfall with leaves, branches, trunks, and epiphytes. Watershed physical alterations include crop agriculture and livestock pasturing. Crop agriculture removes the natural vegetation, disturbs the soil, and leaves crop monocultures. With livestock pasturing, vegetation is of course grazed to lower levels; in addition, trampling of streamside or streambed areas by hooves causes considerable land disruption as well. Changes in the landscape for increased human development (urbanization) such as road building, bridge building, land clearing, and pavement installation increase evapotranspiration, remove vegetative buffering, and exacerbate runoff; this is detailed in Chapter 15. In less-developed nations, another factor contributing to deforestation includes the need for firewood to cook and heat homes, in addition to agriculture and rural development. Finally, another human-caused landscape change in watersheds comes from warfare, where intense bombing removes the vegetation and greatly increases runoff (Lacombe et al. 2010).

Deforestation, regardless of cause, leads to flooding and water pollution following catastrophic weather events such as hurricanes, typhoons, and earthquakes, and in poor nations the lack of proper sanitary infrastructure complicates matters greatly. Flooding greatly increases water pollution and causes people in poverty to drink from contaminated sources. Such pollution and landscape changes associated with deforestation have exacerbated the spread of human diseases caused by bacteria, viruses, and protozoans (Morens et al. 2004; Patz et al. 2004; Enserink 2011). Well-known current examples of the ecological and health consequences of deforestation are seen in Haiti, Tibet, and Nepal.

## 1.5 Water Movement

The physical and hydrological structures of streams within a watershed change dramatically as one moves downstream from the headwaters region through middle-order streams to higher-orders rivers and into the estuary. We can refer to this seaward change as the river continuum, which is discussed in depth in Chapter 6). The streambed is shaped by a multitude of factors including runoff and the amount of base flow. These factors, along with stream order (i.e. the amount of water contributed by tributary streams) and slope determine stream **discharge**. Stream discharge, often referred to as Q, is usually given as cubic meters per second ($m^3/s$). Discharge is computed as flow rate V (m/s) multiplied by the area of the water column A ($m^2$), which itself is determined as stream width (m) multiplied by depth (m) at a given point. River or stream discharge is computed at numerous locations throughout the USA by either the US Geological Survey or state organizations and can be measured at weirs of known dimensions. Stream discharge is often represented on a hydrograph, which is a graph of water discharge per unit time (day, month, year, etc.). Such a graph is especially useful in depicting the difference in stream or

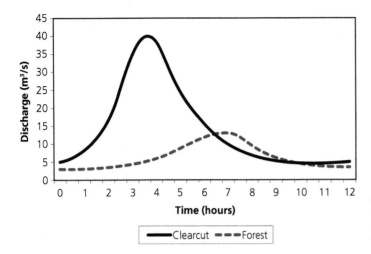

**Fig. 1.8** An illustration of how post-rain event hydrographs might differ in depicting differences in a forested stream's discharge with and without a watershed clearcut. (Figure by the author.)

ditch discharge when a major change in catchment landscape occurs, such as a clearcut (Fig. 1.8). As mentioned earlier, with the trees and their evapo-transpiration function gone, overland runoff drastically increases and the receiving stream discharge rapidly increases during a rainstorm.

The amount and variability of stream discharge erodes the streambed and banks, moves material, and creates and destroys physical structure and habitats within the stream. Most of the erosion occurs during periods of maximal discharge; the term "effective discharge" is defined as the discharge transporting the most sediment over time (see Gordon et al. 2004). Over hundreds to thousands of years, maximum discharge changes along with changing climate and land use. Physical evidence of previous river-formed land alterations can be seen as **terraces** and **alluvial fans** along the present floodplains of some rivers (Hupp and Osterkamp 1996); these alterations have been caused by long-term tectonic, volcanic, or glacial changes (Gregory et al. 1991).

Within a stream there are two basic types of flow, without an abrupt demarcation between the two in nature. **Laminar** flow consists of direct movement downstream, i.e. water layers moving in tandem downstream. In reality, true laminar flow is rare in nature, but laminar flow is approached by deep, slow-moving coastal plain streams with little in-channel roughness to cause turbulence. **Turbulent** flow is movement downstream only on average,

with water masses in the channel moving at different speeds and causing frictional forces. Turbulence, and the eddies and waves created, is caused by many factors, including different temperatures between water layers, different directional movement of adjoining water layers, wind over the surface, drag created by contact with the bottom and shoreline, and any object intruding into the water column such as rocks, woody debris, and man-made structures. As moving water nears the bottom sediments there is a boundary layer, where the frictional resistance created causes the lowest water layers to slow relative to upper layers. This has implications to benthic animal habitat, as they can more easily withstand the slower currents without being dislodged into the stream flow. There is some resistance created by contact with the air as well, so the surface water layer moves slightly slower than the layer below it (Gordon et al. 2004).

As mentioned, during and immediately following a rain event, drainage ditches and ephemeral and intermittent streams will fill with runoff and discharge will rapidly increase in headwaters streams. In lowland areas with a high water table, there will be a further increase in the water table height, so that groundwater discharge into stream channels increases further due to increased vertical head applied to the saturated zone. The result is localized high discharge that will cause in-stream erosion, move bottom sediments of a variety of sizes, and reshape animal habitat in these small streams. When discharge is sufficiently high so that the elevation of the water reaches the level of the floodplain, the stream is considered to be at **bankfull** stage. Moving downstream along the continuum, discharge response to rain events will be muted and delayed, since the actual catchment size increases greatly and all tributaries may not receive the same amounts of rainfall. In high-order rivers, localized rainfall will have little impact on discharge, but major discharge increases can be expected sometime after large rain events occur in headwaters regions. In some cases this may be a matter of weeks. For example, in the Cape Fear River, the largest river in North Carolina, peak discharge response as measured at a lock and dam in the 5th-order river best responded to rain events that occurred two to four weeks earlier in the headwaters region, located 240 km upstream of the riverine station (Mallin et al. 1999). In such scenarios, rainfall over a large headwaters region increases discharge in numerous low-order streams, which combine as flow goes seaward, creating much greater discharge in the higher-order streams until the estuary is reached near the ocean.

To summarize, in low-order streams there is typically steeper slope, rapid flow, a lot of channel contact by stream water, turbulent water movement, and erosion. There is a very narrow floodplain, if one is present at all. A rainstorm will rapidly translate to a higher discharge due to the small-sized catchment feeding the stream. The sediments often consist of a broad variety of size ranges, particularly in headwaters characterized by mountain or foothills regions. Exceptions to this occur in coastal plain or prairie streams, which have rather gentle slopes even in headwaters areas and well-sorted, rather than mixed sediments. As the continuum progresses seaward, the slope moderates and the stream widens and accepts inputs from tributaries. In high-order rivers, the water is deeper and less turbulent, with much less channel contact, slope is low, and there is much deposition of suspended particles. These rivers often feature broad floodplains.

Floods have a major impact on stream systems and are discussed more in Chapter 17. The regional pattern of annual flooding depends on several factors. In the southeastern USA, flooding occurs in winter and early spring due to rains combined with low evapotranspiration caused by lack of active plant growth. In the northeast and midwest regions of the USA, flooding occurs in early to mid-spring due to snowmelt, while snowmelt in the US and Canadian Rocky Mountains leads to early summer flooding. Flooding in Florida and northern Mexico occurs in autumn due to tropical storms, and flooding along the west coast of the USA occurs in autumn and winter due to rains. River discharge is also subject to macroscale forces, such as oceanic El Niño events (Savidge and Cahoon 2002). Note that climate change is already affecting flood timing. An analysis by Blöschl et al. (2017) showed that over the past 50 years there has been a shift to earlier spring flooding in northeastern Europe due to earlier spring snowmelt resulting from increased warming.

## 1.6 Stream Structure

In smaller streams (again with the exception of sand-bottomed, slow-flowing coastal plain streams) it is common to see **riffles**, which are shallow areas of rapid water, usually flowing over stones or gravel (Fig. 1.9a), where erosion of small particles occurs; riffles are common in lower- to middle-order streams. Also in such streams one can find **pools**, which are deeper areas of slow flow, where deposition of small particles is common in lower- to middle-order streams (Fig. 1.9b). In rivers with gravel beds, a riffle-pool sequence tends to repeat at intervals of roughly five to seven channel widths. They change location and move downstream during storms. In lower- to middle-order streams, one most commonly finds waterfalls in steep areas, and in less steep areas **cascades**, which are areas of rapid drop in elevation of the streambed that are characterized by turbulent waters (Fig. 1.10a). In middle-order streams, **runs** are a common feature (Fig. 1.10b), which are areas of steady flow (not quiet water). A more nebulous term is **reach**, which can range for 10–1000 m in length and are generally determined by the geomorphic features such as underlying bedrock and alluvial material (Gregory et al. 1991; Allan 2004) and can contain the various channel units discussed here. Riffles and pools also occur in middle-order streams.

(a)    (b)

**Fig. 1.9** Figure 1.9a (left) Riffles in a stream in Stowe, Vermont. Figure 1.9b (right) Pool in a Tennessee mountain stream. (Photos by the author.)

(a)    (b)

**Fig. 1.10** Figure 1.10a (left) Cascade in western North Carolina. Figure 1.10b (right) Run in Tuckasegee River, North Carolina. (Photos by the author.)

In high-order rivers (usually 5th order or greater) the slope is lower and the stream is sinuous and defined by **meanders**, broad curves in the river's path over the floodplain. The degree of a river's sinuosity can be determined by a sinuosity index, commonly computed as dividing the length of a river reach as measured along a channel by the length of that reach as measured along the valley (Mueller 1968). The index can range from 1.0, a straight stream, with those with a sinuosity index < 1.5 considered generally straight channel reaches. River reaches with a sinuosity index of 1.5 or greater are considered meandering, ranging up to about 4.0, which is highly intricate meandering (Leopold and Wolman 1957; Gordon et al. 2004). Over time, and in response to major storms, meanders can become cut off from the river, leaving an **oxbow**, which retains

water and functions as a lake (these are common and easily seen from an airplane in the lower Mississippi region; Fig. 1.11a). In some rivers, the channel may be **braided**, where flow passes through multiple smaller, often temporary channels rather than a single major semipermanent channel; this occurs mainly in lowland areas where much sediment has been deposited by variable flows over time (Cole 1994). **Anastomozing** channels are relatively permanent multiple channels where stable, vegetated islands are present (Gordon et al. 2004). Middle-order and higher-order streams also feature **bars**—areas of deposition in the region of the channel where flow is slow, such as along the outside of a meander (Fig. 1.11b). Riffles, pools, falls, and cascades are sometimes referred to as **channel units**. Large woody materials deposited within the stream

(a)                                        (b)

**Fig. 1.11** Figure 1.11a (left) Aerial view of lower river showing meanders and oxbows. Figure 1.11b (right) Bar on inner meander and snags in channel of the Cuyahoga River, northern Ohio. (Photos by the author.)

at least temporarily, such as large tree branches, are called **snags** (for obvious reasons—these are hazards to boaters going too fast!—Fig. 1.11b). As demonstrated in Chapter 8, snags are also important habitats for benthic organisms and key feeding areas for fish. Snags and other smaller microhabitats within a stream such as individual boulders or small bars are considered **channel subunits**; these features are physically less than a channel width in size (Gregory et al. 1991). In timbering regions such as the Pacific Northwest, it is common to see whole trees, or logs produced from trees, in rivers and estuaries and deposited on marine beaches. In middle- and high-order streams, large woody material that has been moved and deposited by the current can help shape the streambed, causing damming and pool formation or leading to bar and even island formation (Hupp and Osterkamp 1996). Where woody vegetation has had time to become established in situ along a bar, island formation or expansion can occur due to accretion of sediment material (Hupp and Osterkamp 1996). Off-channel stream and river features include backwaters and side channels to the main channel.

As a river becomes an estuary and nears the ocean, over time **avulsions** may occur, where the lower river shifts its channel to the ocean; this also occurs at the foot of mountain valleys; note the lower Mississippi River has avulsed about every 1400 years (Pearce 2021). With these basic stream features in mind, let us examine sediment inputs, movement, and impacts in more detail.

## 1.7 Stream Sediments

The sediments of a streambed cover a broad variety of shapes and sizes. The type of sediment is critical to stream biota in that sediment particles serve as substrata for microscopic and macrophytic plants and bacteria, as feeding areas and hiding places for many invertebrates, and as feeding areas and nesting materials for stream fish. Sediment sizes can be classified using the Wentworth (1922) scale (Table 1.1).

The distribution of these sediment classes depends first on where the stream is located geographically; i.e. a stream draining a mountain catchment may have larger-sized boulders and cobbles in abundance (Fig. 1.12a), whereas a stream in sandhills or a lowland coastal region may consist largely of sand (Fig. 1.12b). Current speed and turbulence is also integral to the distribution of sediment particles within a stream system. In

**Table 1.1** Sediment size classification based on Wentworth (1922).

| Class | Size range (particle diameter) |
|---|---|
| Boulder | > 256 mm |
| Cobble | 64–256 mm |
| Pebble | 32–64 mm |
| Gravel | 2–32 mm |
| Sand | 0.063–2.0 mm |
| Silt | 0.0039–0.063 mm |
| Clay | < 0.0039 mm |

(There are subdivisions within the major size terms as well. Note that particles smaller than clays are colloidal.)

(a)                                                          (b)

**Fig. 1.12** Figure 1.12a (left) Boulder and cobble habitat in a New Hampshire mountain stream. Figure 1.12b (right) Well-sorted sands form the bottom of a coastal plain stream in North Carolina; such streams do not have riffles per se, but patterns in the sand locally called ripples. (Photos by the author.)

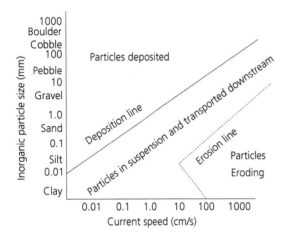

**Fig. 1.13** Relationship between current speed and ability to erode and suspend sediments of various size classes. (Modified from Hjulström [1939] with permission.)

general, faster currents are required to erode and transport larger-sized particles from a streambed, and when the current slows, these particles are deposited (Fig. 1.13). However, the finest-sized particles (clay < 0.0039 mm) require higher flows to erode from the streambed because their very small size leads to smaller frictional forces (Gordon et al. 2004). For example, a 10-mm gravel particle will settle out at any current speed up to 100 cm/s.

The size of a sediment particle is also critical to downstream movement of associated materials. The smaller a particle is, the larger is the surface to volume ratio. Thus, very small particles have (per unit volume) more sites on the surface for materials to physically and chemically attach to. Clay particles are the most reactive of sediment particles, especially clays. Examples of materials that readily attach to clay particles include nutrients such as ammonium and phosphate, metals, and bacteria, including pathogens.

## 1.8 Suspended Sediment Transport Issues

**Suspended solids** are an important water quality parameter. This term is sometimes referred to as suspended sediments, which is more specific, referring to eroded particles. Suspended solids are measured as total suspended solids (TSS)—it is the amount of organic and inorganic materials suspended in a water column and measured gravimetrically by dry weight (in mg/L). In a nutshell, a known volume of water is filtered through a preweighed 0.45-μm filter, dried in an oven, and reweighed. The percentage of organic material can then be determined by combusting in a muffle furnace at high temperature and reweighing the ash. Ash-free dry mass is TSS minus the ashed material. In some locations, government regulatory agencies will have TSS standards for water bodies. The source areas for inorganic suspended solids are channel areas (streambed and bank erosion) within the stream and its tributaries, and soils in the watershed washed into the stream or its tributaries during land-disrupting activities such as mining, forestry, agriculture, road building, bridge building, and upland commercial or housing development (Waters 1995; Wood and Armitage 1997). Other suspended solids can originate from organic materials such as leaf litter and twig particles, atmospheric deposition of various particles, and the stormwater runoff of natural and anthropogenic particulates from urban areas (Wood and Armitage 1997; Mallin et al. 2009). Additionally, some suspended particles are actually aggregates of smaller particles that have become bound together by electrochemical attraction or biologically based bonding involving bacteria and organic substances (Walling and Webb 1992; Carlough 1994). Once in the stream, the suspended solid load alternately settles, erodes, or is entrained, depending upon current speeds. Some moves downstream along the bottom as **bedload**, rolling and tumbling, while some is entrained into the current by turbulence and current. Under normal current flows, sediments are deposited in areas where water velocity slows, like pools, beds of rooted vegetation, backwaters, in the lee of large boulders and cobbles, and of course on the convex side of river meanders (Wood and Armitage 1997). By far the largest portion of annual suspended solid transport occurs during large storms (Olsen et al. 1982; Walling and Webb 1992).

Suspended solids contribute to **turbidity**, which is a measure of the suspended materials in a water column that cause light to be scattered and absorbed rather than directly transmitted. This is a pollutant parameter, measured by nephelometry (the units are nephelometric turbidity units, or NTU). The measured light is what is scattered at a 90° angle from the source lamp. This is a commonly measured water quality parameter; for example, in North Carolina the freshwater turbidity standard is 50 NTU for streams and 25 NTU for lakes and reservoirs; the brackish and saltwater standard is 25 NTU; the trout water standard is 10 NTU; and the drinking water standard is 1 NTU.

Some of the deleterious impacts of elevated TSS and turbidity in streams, rivers, and estuaries are listed below (compiled from Gardner (1981); Berg and Northcote (1985); Rothschild et al. (1994); Waters (1995); Wood and Armitage (1997); Baudardt et al. 2000; Donohue and Garcia Molinos (2009); Mallin et al. (2009)):

- *Interferes with shellfish filter feeding*: shellfish are filter feeders and obtain their nutrition by filtering large amounts of water to obtain foods such as microscopic algae. Suspended materials, particularly inorganic "dirt," can clog gills, cause them to reject large amounts of particulates (called pseudofeces), and spend way too much energy in feeding to survive.
- *Interferes with sight-feeding finfish*: they cannot see their prey at appreciable distances, their capture success decreases, and their ingestion rate decreases; normal social structure also breaks down for salmonids.
- *Can directly injure fish*: by clogging gill rakers and gill filaments and reducing resistance to disease.
- *Can adversely impact benthic invertebrates*: through altering living habitat, reducing effective filter feeding, changing bottom water dissolved oxygen (DO), and reducing available food items.
- *Reduces rooted aquatic plant photosynthesis:* submersed aquatic vegetation beds are fish and crustacean habitats and lengthy increases in turbidity will cause losses in this resource.
- *Changes bottom habitat for fish and invertebrates:* eroded soils from upland land disturbance tend to be fine textured and will cover natural sediments; for instance salmonids require aerated gravel for spawning sites (called redds) and fine suspended sediments will smother these sites. Also, upland-sourced sedimentation can bury shellfish beds in estuarine areas.
- *Suspended sediment particles accumulate pollutants:* including fecal bacteria, phosphate, ammonium, and other pollutants, and transport them downstream. Concentrations of TSS and turbidity have been strongly correlated with many pollutants, especially fecal bacteria, phosphorus, and biochemical oxygen-demanding (BOD) materials.

A direct human health issue is that fecal bacteria (including pathogens) will attach to small particles (particularly clays). Bonding of substances to sediment particles occurs by both physical and chemical sorption (Olsen et al. 1982). The "reactivity" of sediment particles increases along with decreasing grain size due to the increased specific surface area of the particle, relative to volume. Clays are most reactive, sands least reactive. This has implications for river transport of pollutants; i.e. clay soils are often more abundant far inland and can thus be carried far downstream to coastal areas in a river subject to at least periodically high discharge. Particle composition is important—sediments that are rich in Fe and Al oxides can sorb more phosphate than "purer" clays.

Sediments can contain 10–1000× the fecal bacteria of the overlying water column. Bottom sediments provide organic substrate and nutrients (C, N, P) and protect the bacteria from ultraviolet (UV) radiation. Fecal bacteria can survive for weeks to months in sediments, and possibly reproduce (Box 1.1).

---

**Box 1.1 Examples of microbial pollutants associated with suspended and deposited sediments**

After a 25-million-gallon swine waste lagoon burst and polluted the New River in North Carolina, high fecal bacteria counts were found 20 miles downstream in the water column for about two weeks. However, settling of fecal bacteria to the river sediments caused excessively high sediment fecal bacteria counts to persist for over two months in the river (Burkholder et al. 1997).

Following a raw sewage spill of 3 million gallons into Hewletts Creek in North Carolina in July 2005, water column fecal coliform bacteria counts declined over a week from 270,000/100 ml to about 500/100 ml. However, sediment bacteria (background levels about 500/cm$^2$) increased after the spill to > 5,000/cm$^2$ and remained > 1,800/cm$^2$ for six weeks after the spill, forming a reservoir for resuspension of polluting bacteria (Mallin et al. 2007)

A study on the Tech River in France showed that *Salmonella* and fecal coliform bacteria were mainly associated with clays < 2 μm in size. During storms, the increased current would erode these particles from the bottom and carry them downstream where they would pollute the Mediterranean Sea with pathogenic microbes (Baudart et al. 2000).

---

## 1.9 Physico-Chemical Factors Influencing the Stream

### 1.9.1 Water Temperature

Temperature is a major physical force acting upon a stream and its biota. Locally, water temperature is lower in shaded steams and higher in open stream reaches. Also, temperature of groundwater inputs can either increase or decrease stream water temperature, depending on season (groundwater temperatures will be more consistent throughout the year than surface temperatures). As noted earlier, temperature controls the rates of evaporation and transpiration, critical to stream discharge. This is also critical regarding flooding. When a storm that drives floodwaters occurs in summer, the waters will recede much more rapidly than during a winter storm, when the water table is high and evaporation and transpiration rates are too low to significantly help in floodwater removal and system reset. In northern and mountain areas, air temperature controls the rate of snowmelt. Thus, spring high water periods for streams receiving significant snowmelt can be delayed or moved up in time depending upon temperature (Naiman et al. 2005). Increasing temperatures associated with climate change are already affecting snowmelt and stream discharge in mountain areas (see Chapter 17). Temperature is also an important driver of decomposition. Metabolic rates of microorganisms increase along with temperature, so that in summer, decomposition of organic matter, including polluting organic materials, is much more rapid than in colder months.

Of course, water itself will freeze when temperatures are low enough, affecting numerous biotic interactions. An important concept related to freezing is the density of water. As water temperature decreases, density increases until its maximum density is reached at 4.0 °C; and denser water sinks to the bottom. This is important since less-dense water freezes more readily, so water at the surface exposed to the air will freeze downward, while the densest water near the bottom will freeze last. A river with a frozen surface will still flow and perform its ecological functions below, rather than becoming a solid block of ice. Temperature also affects the behavior of cold-blooded stream biota, such as fish, reptiles, and amphibians. Since their internal temperatures change in accordance with external temperatures, they become more metabolically active in warmer weather and less so in colder water (Gordon et al. 2004).

Temperature impacts gas solubility, especially that of oxygen, necessary for stream animal life to survive. In general, oxygen solubility increases along with decreasing water temperature. This has major biogeographical impacts. For instance, trout are restricted to mountain areas where stream water temperatures remain relatively low even in summer, since these fish have little tolerance for low DO. A lowland stream in the same watershed may be unpolluted, yet due to increased water temperature alone the DO concentration within the water may be too low to support trout. In practical terms for the stream sampler, proper readings of DO need to be reconciled with its solubility based on water temperature (most instruments will do this automatically).

Water temperature differences can cause **stratification** in quiescent waters such as reservoirs, or even slow-moving rivers. When the surface layer heats up relative to lower layers, the warmer waters stay at the surface due to lower density. This has the net effect of reducing vertical mixing of the water column. As discussed in Chapter 11 regarding reservoirs, the upper water layer may be well oxygenated while the lower layers, especially near the bottom, can become depleted in DO concentrations.

## 1.10 Salinity

Salinity increases where the lower river becomes an estuarine system, and tidal creek tributaries to lower rivers are often brackish as well. In addition to seawater-induced salinity, streams that receive discharges from certain industries can have measurable salinity levels, thus locally affecting stream reaches well away from the coast. Salinity affects organism distribution due to osmotic constraints; this is discussed in Chapter 9 regarding tidal creeks. Additionally, salinity impacts mixing as water masses of greater salinity are denser than freshwater and can stratify to form saline layers

below freshwater. This impedes mixing of plankton and DO between layers. In some estuaries, these salinity-stratified regions can persist for many river miles. Salinity also impacts water column concentrations of nutrients and other chemical constituents that are chemically bound to clay particles. Critically, regarding nutrient distribution, as river water encounters measurable salinity, the more abundant anions present in brackish waters (including $OH^-$, $F^-$, $SO_4^{2-}$) displace bound phosphate ($PO_4^{3-}$) from suspended particles (Froelich 1988).

## 1.11 Dissolved Oxygen

DO is impacted by many physical factors associated with streams. As discussed above, water temperature controls its solubility and thus availability to organisms in space and time. The amount of DO in a stream is also controlled by diffusion into the stream water from the air above it. A slow-moving lowland stream will have far less diffusion into it than a turbulent, roiling mountain stream plunging over rapids, cascades, and falls. Many bottom-dwelling organisms are relatively sedentary and have to depend upon a flowing stream current to bring them DO across their gills.

One of the biological processes that it is critical to introduce at this point is **photosynthesis**, which is the sunlight-catalyzed conversion of carbon dioxide and water to organic matter that is essential to microscopic and macroscopic plants. Photosynthesis and the organisms that conduct it in streams are detailed in Chapters 3 and 4. However, this process is another means of introducing DO into streams and other water bodies. On the opposite side of the coin is **respiration**, in which DO is utilized (consumed) during the processing of organic matter by microorganisms (bacteria and fungi) as well as macroscopic organisms including invertebrates and fish. This process removes DO from stream waters. Since photosynthesis is light-dependent, in a productive stream or river DO concentration will peak during daylight hours and decline, sometimes precipitously, after dark when respiration continues to occur. A term that will be used frequently later in this book is biochemical oxygen demand (BOD) which is the amount of oxygen used by the microbiotic flora in the water while they break down (consume) a load of organic material, such as sewage for instance. The BOD of a pollutant load is directly related to the amount of DO that a water body loses when the load enters it.

When a water column (in a reservoir, estuary, or slow-moving river) stratifies due to temperature or salinity differences, this can impact DO distribution. Phytoplankton receive the greatest amount of light in the upper water column, photosynthesize, and produce DO to the water column. In the lower water column, little photosynthesis may occur due to low light conditions caused by turbidity or water color. Additionally, organic debris and dead organisms settle to the bottom where they decompose and exert a BOD. In well-mixed waters this may not be a problem, but in slow-moving stratified waters the lack of mixing will lead to low DO (hypoxia) or even no DO (anoxia) in the lower water column.

As noted, subsequent chapters discuss primary producers (photosynthetic organisms of various size and habit). At this point, it is useful to discuss how physical processes influence stream primary productivity. Since plants depend upon sunlight in the photosynthetic process, some basics on the light field are useful here and then will be related to our suspended sediment discussion above.

## 1.12 Light and the Aquatic Environment

Solar energy is critical for the study of streams and rivers. It supplies the energy to perform all the temperature-related functions listed above. Solar radiation consists of a continual stream of photons. A photon is a single quantum of radiation. Photons have wavelength and frequency. Wavelength is expressed in nanometers (nm): $1 \text{ nm} = 10^{-9} \text{ m}$

As wavelength increases, the frequency decreases (Lobban et al. 1985). Energy decreases with increasing wavelength per this relationship:

$$\gamma = c/v,$$

where c = speed of light (constant), $v$ = frequency, and $\gamma$ = wavelength.

Short-wave radiation contains larger quanta of energy. UV radiation is short wave, with wavelengths ranging from 100 to 400 nm; these wavelengths are very energetic, causing skin cancer to humans and various types of damage to aquatic organisms that are exposed for too long. On the good side, UV radiation kills or deactivates fecal microbes in water bodies, and UV treatments are used in wastewater disinfection (see Chapter 13). The most damaging wavelengths are UV-B, which are from 280 to 320 nm; somewhat less energetic are UV-A wavelengths, from 320 to 400 nm. Although very damaging, recent research (reviewed by Williamson and Rose 2010) indicates that these wavelengths perform beneficial functions as well, with a broad variety of organisms able to use them in orientation, communication, navigation, foraging, and mate selection. Infrared radiation is long wave (> 700 nm), with lower energy. This has implications with the greenhouse effect; while shorter wavelengths penetrate the Earth's atmosphere, upon reflection from the Earth the longer wavelengths lack the energy to penetrate greenhouse gases and water vapor in the atmosphere, trapping heat. Photosynthetically active radiation (PAR) is what plants utilize—wavelengths from about 400 to 700 nm (Lobban et al. 1985). PAR itself thus consists of different wavelengths, and plants have evolved a variety of different pigments to take advantage of these wavelengths when acquiring sufficient light for photosynthesis. The most common and widespread photosynthetic pigment is chlorophyll *a*, which has peak light absorbance at 640 and 405 nm. Other pigments are discussed in Chapter 3 regarding phytoplankton communities.

## 1.13 Measurement of Irradiance

Solar radiation entering the water is termed irradiance. Aquatic biologists studying the interactions between irradiance and photosynthesis measure photon flux density (PPFD). The unit of measurement is the Einstein, which refers to a mole ($6.02 \times 10^{23}$) of photons. PPFD is measured by quantum sensors, which measure the number of quanta, usually as $\mu E/m^2/s$. For many aquatic science purposes, such as primary production studies, PAR (400–700 nm) is what is measured. For measuring PAR at the sediment surface a $2\pi$ collector is used (Fig. 1.14), while for measuring PAR that suspended phytoplankton cells experience a $4\pi$ collector is used (Fig. 1.14).

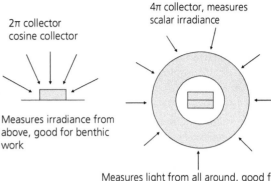

**Fig. 1.14** Stylized depictions of PDF collectors used to measure PAR in water bodies; the collectors are wired to an on-deck integrator. (Figure by the author.)

## 1.14 Absorbance and Transmittance

Light (irradiance) entering the water has three fates (Fig. 1.15). First, it can be absorbed by the water itself, the photopigments of phytoplankton suspended within the water, nonpigmented particles, or chromophoric (colored) dissolved organic matter (CDOM) (Durako et al. 2010). CDOM is important in absorbing UV radiation in water columns (Williamson and Rose 2010). Irradiance can also be scattered by reflection from suspended solids or transmitted through the water. The most energetic wavelengths (short waves) are transmitted deepest, while less-energetic long-wave irradiance is attenuated earlier (closer to the surface).

Ecologists are usually most interested in measuring PAR, although various light meters can measure whichever wavelengths are most useful to the researcher. A cheap and easy way to get an idea of water clarity (considering all visible wavelengths) is to use a Secchi disk, which was developed by P.A. Secchi, an Italian scientist in 1865. This is a 20-cm-diameter disk painted with alternating white and black quarters that is lowered down through the water with a calibrated line (off the shady side of the boat) until the operator can no longer see it (a larger-diameter disk is used for oceanographic work). This is very common and used throughout the world. As one can imagine, it is rather subjective, depending upon the operator's eyes, and difficult to use in rough water and strong currents. We can objectively measure how deep PAR is transmitted through the water column by computing the **light attenuation**

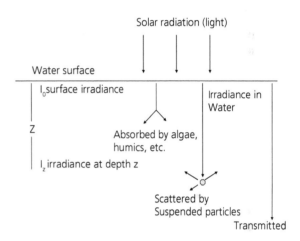

**Fig. 1.15** The fate of solar radiation entering surface water. Note that some of the scattered light is also absorbed by algae or other substances. (Figure by the author.)

**coefficient** ($k$), also called the **light extinction coefficient** (Strickland 1958). The greater the value of $k$, the more turbid or highly colored is the water.

$$k = (\ln I_0 - \ln I_z) / z : \text{this is the same } as$$
$$k = 2.3 (\log I_0 - \log I_z) / z \qquad (1.1)$$
$$(\ln \text{ is the natural log; which is } = 2.3 \ (\log)).$$

where $z$ = water depth of interest, $I_0$ = surface irradiance, measured just below the water surface, and $I_z$ = irradiance at depth $z$. $k$ itself does not have units; it is expressed on a per meter basis as $k/m$.

Let's determine $k$ for a well-mixed estuary, the Neuse River in North Carolina, using data collected by the author. At the location in question, the depth was approximately 3.5 m, and we were interested in collecting data from the 3-m depth; thus $z = 3$. Surface irradiance ($I_0$) was 1200 $\mu E/m^2/s$ and measured irradiance at our 3-m depth ($I_z$) was 50 $\mu E/m^2/s$. Computing the natural log (ln) of $1200 = 7.09$ and ln of $50 = 3.91$. Inserting the results into eqn 1.1 gives $k = 7.09 - 3.91/3 = 1.06\ m^{-1}$.

This $k$ value is quite low, indicating that the waters are relatively clear for estuarine systems, and since the depth of the system in that region is only 3.5 m, there is abundant PAR for phytoplankton production. As this is a highly productive estuary (Mallin and Paerl 1992), the phytoplankton obviously respond to the available light.

Since phytoplankton are normally not static in a riverine or estuarine water column, but move with the water motion, it is also useful to get an idea of the average irradiance a phytoplankton cell might experience in a well-mixed water body like the Neuse River of our example. We can rearrange the light attenuation eqn 1.1 above to determine $I_x$, mean water column irradiance. $I_x$ is important in that it can tell you what a phytoplankton cell in a well-mixed water column is receiving in terms of irradiance. It can be calculated as:

$$I_x = I_0 \left(1 - e^{-kz}\right) / kz \tag{1.2}$$

Inserting our measured data from above gives

$$I_x = [1,200\,(1 - 0.0415)] / [(1.06)\,(3)]$$
$$= 1150/3.18 = 361.6\ \mu E/m^2/s.$$

In a well-mixed situation, this then is the approximate average solar irradiance a phytoplankton cell might experience near mid-day; i.e. abundant PAR for photosynthesis.

Suppose the river or estuary or reservoir becomes stratified by temperature or a salt wedge. The $k$ value can differ in different water layers if the water is stratified because there is turbidity or phytoplankton layers beneath the surface. Thus, if there is a layer of phytoplankton at a given depth, say $z = 2.5$ m, we might be interested in determining how much irradiance that phytoplankton layer experiences. We can rearrange our equation above

to determine irradiance at any depth $z$ ($I_z$) as:

$$I_z = I_0 \times e^{-kz}\quad I_z = 1200 \times e^{(-1.06)(2.5)} = 84.8\ \mu E/m^2/s \tag{1.3}$$

Light attenuation coefficients can be quite low in very clear water situations such as Lake Superior or the open ocean. Generally, the more impact the water body receives from land-based suspended solids or water color or nutrient-stimulated phytoplankton blooms, the more that $k$ will increase. Examples of computed light attenuation coefficients in the southeastern USA that the author has been associated with are as follows:

- Open ocean is about 0.20/m (very clear water)
- Coastal ocean is about 0.70/m
- Pamlico Sound, North Carolina is about 0.80/m
- Neuse Estuary, North Carolina is about 1.1/m

  Cape Fear River, North Carolina is about 3.0–4.0/m (i.e. very turbid water from upstream suspended solids loading; also influence from organic color from this river's blackwater tributaries)

Within a water column, primary productivity and plant distribution are affected by the light field and the current. The light field (solar irradiance) is affected by turbidity, which is controlled primarily by runoff, erosion, and suspended sediments (although phytoplankton itself as well as bacteria contribute to measured turbidity). Increases in turbidity lead to decreased solar irradiance, which leads to reduced photosynthesis. Thus, in many rivers draining catchments of disturbed soils (especially clay), turbidity is often high and phytoplankton productivity low.

$\uparrow$ Turbidity $\Rightarrow \downarrow$ Irradiance

$\Rightarrow \downarrow$ Phytoplankton production

Water color (CDOM) results from leaching of organic materials. It primarily results from humic substances, principally humic and fulvic acids, and is dissolved material that passes through a 0.45-$\mu$m membrane filter (Walling and Webb 1992). This is frequently associated with acidic conditions such as coastal plain blackwater river systems (the Ogeechee in Georgia and the Black and Northeast Cape Fear Rivers in North Carolina are well-studied examples). As water color increases, solar irradiance decreases and photosynthesis decreases as well. In US coastal plain regions, water color intensifies during periods of high flow, when riparian swamp forests drain large quantities of leached dissolved organic material into main river channels (Meyer 1992). In peat moorland areas in Great Britain, the highest color concentrations occur in autumn when dissolved organic material is leached from wetted peat (Walling and Webb 1992).

$$\uparrow \text{Water color} \Rightarrow \downarrow \text{Irradiance}$$

$$\Rightarrow \downarrow \text{Phytoplankton production}$$

Current speed is another physical force that exerts control on stream phytoplankton production. Elevated current can reduce stream primary production by reducing the available irradiance to the water column in several ways: 1) higher current erodes the streambank and bed and carries more material in solution downstream, increasing turbidity and TSS; 2) higher current may also lead to higher water color from increased (highly colored) swamp water inputs (Meyer 1992); and 3) current can control the type of plant species or functional groups. The latter is explained because with elevated current there are fewer floating plants and less phytoplankton in the water to contribute to stream primary production. Additionally, to withstand higher currents, macroscopic plants need to be rooted or have holdfasts or stick to substrata by mucilage production. However, it is also notable that established rooted and submersed aquatic macrophytes can baffle and reduce current velocity and help reduce turbidity (Madsen and Warncke. 1983; Petticrew and Kalff 1992; Leonard and Luther 1995). Chapters 3 and 4 go into more detail on how currents and other physical forces influence stream primary productivity.

## 1.15 Summary

- The headwaters of rivers consist of springs, ephemeral streams, and drainage ditches. Streams are classified by the Strahler stream order system, with first order the smallest temporary stream; a second-order stream is formed by the joining of two first-order streams; and so on. The world's largest rivers are orders 10–12.

- Streamflow and bedrock geology together structure streams, forming falls and cascades in mountain areas; riffles, pools, and runs in mountains, foothills, and Piedmont areas; whereas streams in low-lying coastal regions have little slope and well-sorted sediments. Lower rivers may be braided, will form meanders, and oxbow lakes are created when a meander is cut off from the main river flow.

- Sediments are classified by size, with boulders and cobbles the largest and clays the smallest. Larger particles are mainly moved during strong storms; at some point along the route to the sea, the largest particles settle out and the sediments of lower rivers are largely well-sorted sands and silts. Clays are lightest and hard to settle out and are physically and chemically associated with pollutants such as excess metals, ammonium, phosphate, and fecal bacteria.

- When rain falls to Earth as rain, it has three fates: infiltration into soils to the water table, evaporation and plant transpiration back to the atmosphere, and running off the landscape downslope to the nearest body of water. Rainfall brings considerable erosion both to the landscape and within the stream, continually reshaping streambeds and impacting water quality by increasing suspended solids loads. Groundwater within the surficial aquifer will also move laterally to the nearest stream.

- Water in lower rivers, estuaries, and reservoirs can stratify due to differences in water temperature and salinity. Stratified waters impede mixing and can lead to stressed conditions such as bottom-water hypoxia, dangerous for finfish and shellfish.

- Erosion-driven suspended solids will contribute to turbidity, which affects stream biota by carrying pollutants, shading plant growth, covering fish nests and shellfish beds, and interfering with fish feeding and behavior. Turbid rivers will attenuate solar irradiance; this attenuation can be computed by the light extinction coefficient $k$, a useful tool to assess living conditions for phytoplankton, periphyton, and rooted aquatic macrophytes.

# References

Allan, J.D. 2004. Landscapes and riverscapes: The influence of land use on stream ecosystems. *Annual Review of Ecology Evolution and Systematics* 35:257–284.

Alley, W.M., T.E. Reilly and O.L. Franke. 1999. Sustainability of ground-water resources. US Geological Survey Circular 1186, Denver, CO.

Baudardt, J., J. Grabulos, J.-P. Barussean and P. Lebaron. 2000. *Salmonella* spp. and fecal coliform loads in coastal waters from a point vs. non-point source of pollutants. *Journal of Environmental Quality* 29:241–250.

Berg, L. and T.G. Northcote. 1985. Changes in territorial, gill-flaring, and feeding behavior in juvenile Coho salmon (*Onchorhynchus kisutch*) following short-term pulses of suspended sediments. *Canadian Journal of Fisheries and Aquatic Sciences* 42:1410–1417.

Blöschl, G., J. Hall, J. Parajka, et al. 2017. Changing climate shifts timing of European floods. *Science* 357: 588–590.

Burkholder, J.M., M.A. Mallin, H.B. Glasgow Jr, et al. 1997. Impacts to a coastal river and estuary from rupture of a swine waste holding lagoon. *Journal of Environmental Quality* 26:1451–1466.

Carlough, L.A. 1994. Origins, structure, and trophic significance of amorphous seston in a blackwater river. *Freshwater Biology* 31:227–237.

Cole, G.A. 1994. *Textbook of Limnology*, 4e. Waveland Press, Prospect Heights, IL.

Dodds, W.K. 2002. *Freshwater Ecology: Concepts and Environmental Applications*. Academic Press, San Diego.

Donohue, I. and J. Garcia Molinos. 2009. Impacts of increased sediment loads on the ecology of lakes. *Biological Reviews* 84:517–531.

Durako, M.D., P. Kowalczuk, M.A. Mallin, W.J. Cooper, J.J. Souza and D.H. Wells. 2010. Spatial and temporal variation in photosynthetically-significant optical properties and water quality in a coastal blackwater river plume. *Estuaries and Coasts* 33:1430–1441.

Enserink, M. 2011. Despite sensitivities, scientists seek to solve Haiti's cholera riddle. *Science* 331:388–389.

Froelich, P.N. 1988. Kinetic control of dissolved phosphate in natural rivers and estuaries: a primer on the phosphate buffer mechanism. *Limnology and Oceanography* 33:649–668.

Gardner, M.B. 1981. Effects of turbidity on feeding rates and selectivity of bluegills. *Transactions of the American Fisheries Society* 110:446–450.

Gordon, N.D., T.A. McMahon, B.L. Finlayson, C.J. Gippel and R.J. Nathan. 2004. Stream Hydrology, 2e. Wiley, London.

Gregory, S.V., F.J. Swanson, W.A. McKee and K.W. Cummins. 1991. An ecosystem perspective of riparian zones. *BioScience* 41:540–551.

Hjulström, F. 1939. Transportation of detritus by moving water. Part 1. Transportation. In: Trask, P.D. (ed.) *Recent Marine Sediments: A Symposium. Special Publication 10*, pp. 7–31. AAPG Technical Publications.

Hupp, C.R. and W.R. Osterkamp. 1996. Riparian vegetation and fluvial geomorphic processes. *Geomorphology* 14:277–295.

Keeney, D. 1986. Sources of nitrate to ground water. *CRC Critical Reviews in Environmental Control* 16:257–304.

Lacombe, G., A. Pierret, C.T. Hoanh, O. Sengtaheuanghoung and A.D. Noble. 2010. Conflict, migration and land-cover changes in Indochina: a hydrological assessment. *Ecohydrology* 3:382–391.

Leonard, L.A. and M.E. Luther. 1995. Flow hydrodynamics in tidal marsh canopies. *Limnology and Oceanography* 40:1474–1484.

Leopold, L.B. and M.G. Wolman. 1957. River channel patterns: braided, meandering and straight. Geological Survey Professional Paper 282-B. US Department of the Interior, Washington, DC.

Lobban, C.S., P.J. Harrison and M.J. Duncan. 1985. *The Physiological Ecology of Seaweeds*. Cambridge University Press, New York.

Madsen, T.V. and E. Warncke. 1983. Velocities of currents around and within submerged aquatic vegetation. *Archiv für Hydrobiologie* 97:389–394.

Mallin, M.A. and H.W. Paerl. 1992. Effects of variable irradiance on phytoplankton productivity in shallow estuaries. *Limnology and Oceanography* 37:54–62.

Mallin, M.A., L.B. Cahoon, M.R. McIver, D.C. Parsons and G.C. Shank. 1999. Alternation of factors limiting phytoplankton production in the Cape Fear Estuary. *Estuaries* 22:985–996.

Mallin, M.A., L.B. Cahoon, B.R. Toothman, et al. 2007. Impacts of a raw sewage spill on water and sediment quality in an urbanized estuary. *Marine Pollution Bulletin* 54:81–88.

Mallin, M.A., V.L. Johnson and S.H. Ensign. 2009. Comparative impacts of stormwater runoff on water quality of an urban, a suburban, and a rural stream. *Environmental Monitoring and Assessment* 159: 475–491.

Meyer, J.L. 1992. Seasonal patterns of water quality in blackwater rivers of the coastal plain, southeastern United States. In: Becker, C.D. and D.A. Neitzel (eds) *Water Quality in North American River Systems*, pp. 249–276. Batelle Press, Columbus.

Morens, D.M., G.K. Folkers and A.S. Fauci. 2004. The challenge of emerging and re-emerging infectious diseases. *Nature* 430:242–249.

Mueller, J.E. 1968. An introduction to the hydraulic and topographic sinuosity indexes. *Annals of the Association of American Geographers* 58: 371–385.

Naiman, R.J., H. Decamps and M.E. McClain. 2005. *Riparia: Ecology, Conservation and Management of Streamside Organisms*. Elsevier, Amsterdam.

Olsen, C.R., N.H. Cutshall and I.L. Larsen. 1982. Pollutant-particle associations and dynamics in coastal marine environments: a review. *Marine Chemistry* 11: 501–533.

Patz, J.A., P. Daszak, G.M. Tabor, et al. 2004. Unhealthy landscapes: policy recommendations on land use change and infectious disease emergence. *Environmental Health Perspectives* 112: 1092–1098.

Pearce, F. 2021. When the levees break. *Science* 372: 676–679.

Petticrew, E.L. and J. Kalff. 1992. Water flow and clay retention in submerged macrophyte beds. *Canadian Journal of Fisheries and Aquatic Sciences* 49:2483–2489.

Robinson, P.J. 2005. *North Carolina Weather and Climate*. University of North Carolina Press, Chapel Hill.

Rothschild, B.J., J.S. Ault, P. Goulletquer and M. Heral. 1994. Decline of the Chesapeake Bay oyster population: a century of habitat destruction and overfishing. *Marine Ecology Progress Series* 111:29–39.

Savidge, T.W. and L.B. Cahoon. 2002. Correlations between river flow in the southeastern United States and El Niño/Southern Oscillation events. *Journal of the North Carolina Academy of Science* 118:70–78.

Steward, A.L., D. von Schiller, K. Tockner, J.C. Marshall and S.E. Bunn. 2012. When the river runs dry: human and ecological values of dry riverbeds. *Frontiers in Ecology and the Environment* 10:202–209.

Strahler, A.N. 1952. Dynamic basis of geomorphology. *Geological Society of America Bulletin* 63:923–938.

Strickland, J.D.H. 1958. Solar radiation penetrating the ocean: a review of requirements, data and methods of measurement, with particular reference to photosynthetic productivity. *Journal of the Fisheries Research Board of Canada* 15:453–493.

Walling, D.E. and B.W. Webb. 1992. Water quality: physical characteristics. In: Calow. P. and G.E. Petts (eds) *The Rivers Handbook: Hydrological and Ecological Principals*, Vol I, pp. 48–72. Blackwell Scientific, Oxford.

Waters, T.F. 1995. Sediment in Streams: Sources, Biological Effects, and Control. American Fisheries Society monograph 7. American Fisheries Society, Bethesda, MD.

Wentworth, C.K. 1922. A scale of grade and class terms for clastic sediment. *Journal of Geology* 30:377–392.

Williamson, C.E. and K.C. Rose. 2010. When UV meets fresh water. *Science* 329:637–639.

Wood, P.J. and P.D. Armitage. 1997. Biological effects of fine sediment in the lotic environment. *Environmental Management* 21:203–217.

# CHAPTER 2

# Nutrients and River Ecosystems

Numerous chemical elements are carried along in streams and rivers. Amounts vary depending on groundwater inputs, weathering of the soils and rocky substrata the stream passes through and erodes, and of course the elements humans contribute to the system, called anthropogenic loading. Where river life and ecosystems are concerned, two of the most important elements influencing a system are nitrogen (N) and phosphorus (P). These are sometimes referred to as macronutrients, since relative to most other elements their energetic and structural importance is essential to aquatic life. Inorganic forms of N and P are taken up and used (assimilated) by **photosynthetic** organisms (**phototrophs** or **primary producers**), which are plant life including macrophytes (aquatic vascular plants visible to the human eye), macroalgae (visible to the human eye), benthic microalgae, phytoplankton, and streamside (or riparian) vegetation, including plant life well up into the floodplain. N, P, and other macronutrients are also required and assimilated by **mixotrophs**—algae that simultaneously use both photosynthesis and animal-like **heterotrophy** to obtain their food (Burkholder et al. 2008 and references therein)—and by heterotrophic microbes—the fungi and both benign and pathogenic bacteria that decompose dead biota for their nutrients. Macronutrients are thus critical for both supporting the growth of plant life and enhancing the decomposition of dead organic matter.

Aquatic ecosystems characterized by low nutrient supplies are **oligotrophic**, such as rivers draining granite bedrock soils in mountainous areas. Some river systems drain watersheds with rich soils and are naturally **eutrophic**, or nutrient enriched. Too many nutrients from pollution (sometimes referred to as **cultural eutrophication**) can cause macrophyte, macroalgal, or phytoplankton nuisance blooms in streams, rivers, reservoirs, lakes, and estuaries. Silica (Si) is a structural requirement of diatoms, one of the most abundant groups of algae across aquatic ecosystems. Hydrated Si is a key nutrient in maintaining a healthy taxonomic distribution and diversity among riverine and estuarine algal communities. N and P are generally recycled rapidly through biota, whereas Si is recycled more slowly. The relative remineralization rates of major elements in aquatic ecosystems are P, remineralized faster than N, remineralized faster than C, with Si remineralized slowest. Zooplankton readily consumes diatoms; whereas N and P are assimilated more fully, zooplankton fecal pellets contain mostly carbon (C) and Si. Carbon, a major structural and energetic macronutrient, is discussed in Chapters 6, 8, and 13.

## 2.1 Nitrogen

Nitrogen exists in numerous forms in water, soils, and the atmosphere, so its cycling is relatively complex. In riverine and estuarine systems, within both the water column and the sediments, there is continuous cycling of N from one form to another. Much of the N cycle involves oxidation–reduction reactions (see Burgin et al. 2011 for details). In these reactions, a substance that receives electrons becomes reduced, and the electron donor is a reducing agent. A substance donating electrons becomes oxidized, and the electron recipient is an oxidizing agent. The term redox potential, or redox condition, describes whether a given location in the water column or sediments is a reducing or oxidizing environment,

*River Ecology*. Michael A. Mallin, Oxford University Press. © Michael A. Mallin (2023). DOI: 10.1093/oso/9780199549511.003.0003

important in predicting reaction directions. Reducing conditions dominate when DO is low or absent, such as in river sediments, the hyporheic zone, or riparian wetlands. The dominant inorganic N form in reducing conditions is ammonia (Glibert et al. 2016). Oxidizing conditions dominate when DO is available through mixing or photosynthesis. Redox reactions can occur abiotically but are enhanced by microbial enzymes (Burgin et al. 2011).

Atmospheric N exists as dinitrogen ($N_2$) and is the most abundant gas, comprising approximately 78% of the atmosphere. In terms of impacting primary producers, it is largely inert in this form (due to its strong triple bond) and must be "fixed" or chemically transformed to ammonia via microbial mediation. Legumes include common crop plants in association with certain bacteria that perform this function. In aquatic ecosystems, cyanobacteria (blue-green algae) are the best-known N fixers, although some eubacterial N fixers also occur. In the atmosphere, $N_2$ can be converted to chemically and biologically reactive nitrogen (Nr) compounds (nitrate, for example) through the energy released by lightning. Reactive N also occurs as a combustion byproduct of auto exhaust and fossil-fuel-fired power plants and can become airborne and deposited far downwind (hundreds of kilometers) from the source. The principal immediate compounds formed by this combustion are nitric oxide (NO) and nitrogen dioxide ($NO_2$), jointly referred to as NOx compounds. These are components of smog and acid precipitation (discussed in Chapter 12).

Various N compounds can be used by plants and bacteria, and these occur in both inorganic and organic forms. N is both directly critical to human nutrition and indirectly critical because it is a key component of fertilizers used in food production. As the global human population increases, fertilizer use has escalated (Glibert 2020). Primarily due to its use in fertilizer, global Nr creation increased from 15 Tg (teragrams; 1 Tg = $10^{12}$ g) N in 1860, to 156 Tg N in 1995, to 187 Tg N/year in 2005 (Galloway et al. 2008). Unfortunately, much of the N in fertilizers escapes farms, lawns, and gardens and enters streams, rivers, reservoirs, lakes, and coastal oceans, resulting in excessive nutrient loading that greatly accelerates the eutrophication process.

### 2.1.1 Inorganic Nitrogen

In aquatic ecology, the term DIN refers to "dissolved inorganic nitrogen," the two major forms of which are nitrate ($NO_3^-$) and ammonium ($NH_4^+$). DIN is assimilated by primary producers and decomposers (bacteria and fungi) to form organic N compounds such as amino acids, proteins, and nucleic acids. Both inorganic N forms are bioavailable and used by phototrophs, but ammonia is much less energetically costly for phototrophs to take up. Ammonia tends to adsorb to sediments and other particles, whereas nitrate is highly soluble and can travel long distances (hundreds of kilometers) if not taken up by phototrophs (Mallin et al. 1993; Houser et al. 2010; Houser and Richardson 2010).

Nitrate is an oxidized form of N and it is very mobile in soils. Nitrate has a low tendency to complex with metals or to be sorbed onto surfaces (Keeney 1986). It is negatively charged, and soil solids are negatively charged, enhancing mobility. This soil mobility has important implications for the loading of N to waterways. Nitrate co-occurs with nitrite ($NO_2^-$); normally nitrate is much more abundant than nitrite and the combination is often referred to as simply nitrate. Nitrite may increase in abundance under low DO (i.e. reducing) conditions. Common anthropogenic sources to waterways are fertilizers (agricultural, garden, lawn. and ornamental plant), wastewater effluents, and septic tank drainfields. A common term regarding the N cycle is "new" nitrogen, which refers to nitrate or $N_2$ gas, which is brought into surface waters by currents, runoff from land, or the atmosphere.

The other major inorganic N form, ammonia (grams) or ammonium (ionized), is a reduced form of N. It is a product of the decomposition of organic materials (called mineralization) as well as excretion by humans and other animals. It chemically binds to clays and other soils and can be carried downstream with suspended sediments. Above pH 10 it exists primarily as un-ionized $NH_3$ (ammonia) and is readily aerosolized to become airborne and deposited downwind from a source. A major N pollution issue is ammonia offgassing from swine- and poultry-concentrated (confined) animal feeding operations (CAFOs) (discussed in Chapter 13) and subsequently deposited on land or in a water body down-airshed (Walker et al. 2000). Ammonium in low concentration is often the preferred form of DIN for plant uptake (Glibert et al. 2016) because it is energetically less costly to take up than nitrate. Nitrate uptake requires a high-energy enzyme system and then must be reduced to ammonium for use (Canfield et al. 2010). Important to wastewater treatment is that ammonium is a cause of nitrogenous biochemical oxygen demand (N-BOD): when ammonium enters the water, bacteria use oxygen to transform it to nitrite and then nitrate. This process, called **nitrification**, causes DO in the water to be used up more quickly and can be the principal mechanism increasing hypoxia in riverine areas receiving ammonium from sewage treatment plants (Lehman et al. 2004).

Ammonium recycling is a key process maintaining primary production in oligotrophic waters characterized by low nutrient supplies. Zooplankton and other aquatic herbivores consume algae, and in some cases they consume bacteria as well. Ammonium is excreted from zooplankton, sessile shellfish, and various other fauna as waste and can be rapidly assimilated by the primary producers. A good example of this closely coupled situation involves the tidal creeks in the North Inlet, South Carolina (Lewitus et al. 2004). These creeks are remote from developed areas and sources of waterborne or airborne new nutrients, and the phytoplankton assemblage is primarily supported by recycled ammonium.

## 2.1.2 Organic Nitrogen

Various dissolved organic N (DON) compounds can be used by aquatic biota. Some of these compounds are labile (**bioavailable**), which means they can be easily broken down by microbes or easily assimilated by algae. Labile components of the DON pool include urea, dissolved free amino acids, nucleic acids, and methylamines. Urea, for example, is an animal waste product that is often used in crop fertilizers. Use of urea has been increasing, as has its transport into waterways, and it can be a significant contributor to cultural eutrophication (Glibert et al. 2006). Urea is known to be used by phytoplankton in freshwater (Antia et al. 1991; Mallin et al. 2004), estuaries, and marine waters (Wawrick et al. 2009).

Other labile DON compounds have been studied to determine their bioavailability to phytoplankton, but much of this research has been conducted in estuarine and marine waters (Wawrick et al. 2009; Bradley et al. 2010).

Beyond aquatic ecosystems, urea can be used to cause the denitrification of NOx compounds in diesel truck exhaust. A 32.5%-by-weight urea solution called diesel exhaust fluid (DEF) is injected into the hot exhaust gases of a vehicle, where the urea breaks down into ammonia. In the presence of a catalyst, the ammonia then reacts with NOx and oxygen to form $N_2$ and water ($H_2O$) (CEN 2011). This process can translate into 4–8% improvement in fuel efficiency and is leading to large commercial increases in DEF production.

Semilabile DON includes proteins, dissolved combined amino acids (AA), amino polysaccharides (i.e. chitins and peptidoglycans), and refractory compounds including humic acids, fulvic acids, porphyrins, and amides (Bronk et al. 2007). Humic and fulvic acids are prominent components of "blackwater"—the darkly colored waters found in Coastal Plain rivers like the Black and Northeast Cape Fear systems in North Carolina and the Ogeechee River in Georgia (Meyer 1992). Other organic N forms can be large, complex molecules, often referred to as **refractory** compounds, are difficult to break down by heterotrophic microbes and cannot be readily used by primary producers.

## 2.1.3 Component Processes of the Nitrogen Cycle

The N cycle involves several key processes (Box 2.1), many of which may be occurring simultaneously in proximity (Fig. 2.1). Much of the cycling involves oxidation–reduction reactions and specific microbes with certain enzymes as catalysts. As mentioned earlier, new N can enter a system through N fixation. N fixation is the microbially driven "capture" of atmospheric N and conversion into ammonium, catalyzed by the enzyme nitrogenase (Box 2.1). This process is inhibited by oxygen. It is accomplished by certain bacteria associated with legume plants including some aquatic/wetland macrophytes (Rivas et al. 2002 and references therein) and by certain cyanobacteria. In

---

**Box 2.1  At a glance: processes involved in aquatic nitrogen cycling**

**Nitrogen fixation:** $N_2 \rightarrow NH_4^+$

Performed by eubacteria and cyanobacteria; requires an energy source and near-anoxic conditions; catalyzed by enzyme nitrogenase. A key reason why some cyanobacteria can maintain blooms.

**Ammonification:** organic $N_2 \rightarrow NH_4^+$

A decomposition process, it is the remineralization of ammonium from dead organisms or waste products.

**Nitrification:** $NH_4^+ \rightarrow NO_2^- \rightarrow NO_3^-$

An oxidation process, bacterially mediated by *Nitrosomonas* and *Nitrobacter*.

**Denitrification:** $NO_3^- \rightarrow NO_2^- \rightarrow NO \rightarrow N_2O \rightarrow N_2$

A reduction process; requires nitrate, low DO (< 0.2 mg/L), and an organic matter substrate. Critical enzymes include nitrate reductase, nitrite reductase, and reductase.

**Anammox:** $NO_2^- \rightarrow NO \ NO + NH_4^+ \rightarrow N_2H_2 \ N_2H_2 \rightarrow N_2$

An enzyme-driven process causing the removal of Nr from the system to dinitrogen.

**Assimilation:** $NO_3^-, NH_4^+ \rightarrow$ organic nitrogen compounds

The uptake of DIN by plants, bacteria, and fungi.

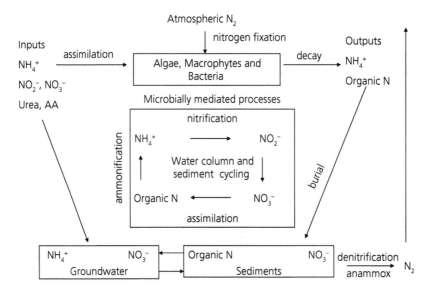

**Fig. 2.1** Generalized schematic of nitrogen cycling in aquatic environments. (Figure by the author.)

filamentous cyanobacteria it is often associated with the presence of a **heterocyte** (formerly known as a heterocyst), a highly specialized cell in which low DO allows for this process (see Chapter 3).

The recycling (remineralization) of DIN maintains system productivity even when new N is unavailable. Ammonification is the mineralization of organic N compounds during the decomposition of dead animal and plant material. This microbially driven process can occasionally be detected in

forests by hunters, birders, and wildlife biologists who encounter a strong ammonia odor in a remote area. A search will usually uncover the nearby carcass of a dead deer, dog, or wild hog undergoing decomposition. Ammonification also occurs during the breakdown of animal manure. In surface waters, most ammonia gas is ionized to ammonium except at very high pH and warm temperatures.

**Nitrification** (Box 2.1) is the oxidation of ammonium to nitrite and then to nitrate. This microbially driven process requires aerobic conditions and is carried out by the bacterial genera *Nitrosomonas* and *Nitrobacter*.

How is Nr removed from an aquatic system? **Denitrification** is the enzyme-driven process that converts nitrate to nitrite to nitric oxide to nitrous oxide and ultimately to dinitrogen gas under anoxic or near-anoxic conditions (Box 2.1). Denitrification requires the availability of nitrate, sufficient organic matter to serve as a reducing agent (electron donor), and low DO conditions (less than about 0.2 mg $O_2$/L). In this microbial oxidation of organic matter, nitrate or nitrite is the terminal electron acceptor. Denitrification is measureable at temperatures down to zero but is much more rapid in the 20 to 25°C range (Keeney 1986). A wide array of aquatic/wetland biomes have appropriate conditions for aquatic denitrification to occur, including reservoirs and lakes, rivers, estuaries, marshes, and oceanic sediments (Seitzinger et al. 2006).

Strong oxygen gradients occur near the surface of aquatic sediments. Nitrate produced from aerobic sediments or moving from overlying water into the sediments will diffuse into suboxic zones where denitrification occurs. These areas include hyporheic zones beneath and adjacent to stream channels. In rivers, denitrification rates generally are highest within the top 5 cm of the benthic sediments (Inwood et al. 2007). Low rates of denitrification occur in stream sediments consisting of large particles and low organic matter content, whereas elevated rates occur in finer-grained sediments with more organic content and areas of anoxia (Inwood et al. 2007). Groundwater nitrate can be denitrified in riparian wetland sediments with floodplain inundations or in estuarine wetlands. Stream floodplains and wetlands often sustain varying water levels due to river discharge or tidal changes, forming periodically suboxic sediments that favor denitrification. In tidal freshwater river stretches, Ensign et al. (2008) found that tidal inundation of riparian wetlands and floodplains led to much higher denitrification than within the river channel itself. Rivers, reservoirs, and estuaries that become stratified can also form suboxic bottom-water zones where sedimentation of organic matter occurs to fuel this process. Regarding current speed, an analysis by Howarth et al. (2006) suggested that less denitrification occurs in watersheds with higher precipitation and river discharge due to faster flushing through low- order streams and riparian wetlands.

Because removal of N from streams and rivers is a key means to combat cultural eutrophication, natural wetlands perform a valuable service in N removal (among their many other values). There is growing use of constructed wetlands in urban and suburban areas to treat stormwater runoff for multiple pollutants (see Chapter 16). To address the problem of excess N in runoff, such wetlands can be designed to enhance denitrification (Fig. 2.2). Note that in constructed wetlands, designers can manipulate water flow, choice of macrophyte species planted, and sediment type (if the area is limited). Soil particle size influences anoxia, but nitrate concentration depends on landscape factors and the amount of stormwater runoff.

Total denitrification in rivers on a global scale was estimated at 35 Tg N/year, accounting for the removal of about 13% of all land-based N sources (Seitzinger et al. 2006). Estuarine denitrification was estimated to remove about 8 Tg/year, or 3% of land-based sources annually. The majority of denitrification worldwide occurs on continental shelves, in terrestrial soils, and in oceanic oxygen minimum zones. It should be noted that, while denitrification is generally considered to be a

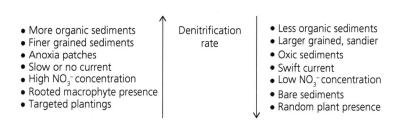

- More organic sediments
- Finer grained sediments
- Anoxia patches
- Slow or no current
- High $NO_3^-$ concentration
- Rooted macrophyte presence
- Targeted plantings

Denitrification rate

- Less organic sediments
- Larger grained, sandier
- Oxic sediments
- Swift current
- Low $NO_3^-$ concentration
- Bare sediments
- Random plant presence

**Fig. 2.2** Factors influencing rate of denitrification in aquatic ecosystems.

beneficial process in combatting eutrophication of water bodies, some byproducts can be considered pollutants. NO contributes to the formation of harmful tropospheric ozone. $N_2O$, produced in variable amounts, is a powerful greenhouse gas that has 300 times the warming potential of $CO_2$ (Canfield et al. 2010). Another reduction process is dissimilatory nitrate reduction to ammonium (**DNRA**). DNRA maintains the system N, whereas denitrification is a net loss (Korum 1992). DNRA produces ammonium, which is a more biologically available form of DIN than nitrate (see Burgin and Hamilton 2007 for review).

A more recently recognized process with growing relevance to N removal is anaerobic ammonium oxidation (**anammox**) which combines ammonium and nitrite to convert to $N_2$ gas (Box 2.1). This is a microbial/enzyme-mediated process. The microbes use nitrite reductase enzyme to convert $NO_2$ to NO. Hydrazine hydrolase combines NO and $NH_4^+$ to form hydrazine ($N_2H_4$), and then a hydrazine oxidoreductase enzyme converts $N_2H_4$ to $N_2$ (Burgin and Hamilton 2007; Jetten et al. 2009). While most work to date has occurred in marine systems, this process clearly occurs in river systems as well (Dale et al. 2009). Anammox is being investigated as a way to remove N during wastewater treatment and plays a role in N removal from stormwater runoff entering constructed wetlands (Song et al. 2014). Finally, it is important to note that once a stream or river becomes excessively loaded with N, it will overwhelm the collective nitrate removal processes

addressed above and cause water quality problems in the lower river and receiving estuary (Mulholland et al. 2008).

## 2.2 Phosphorus

Phosphorus is integral to the metabolism of all life and is important in numerous compounds involved in metabolic processes (Wetzel 2001). As examples, it is a key component of photosynthesis and respiration (in adenosine triphosphate [ATP] and nicotinamide adenine dinucleotide phosphate/nicotinamide adenine dinucleotide phosphate hydrogen [NADP/NADPH]) and cell membranes. Unlike N, there are no changes in oxidation state with P; it exists as either inorganic or organic P (Fig. 2.3). Bacteria have a greater need for P as opposed to N, structurally as well as for ATP (Kirchman 1994) and are considered better competitors for P than microalgae (Cotner et al. 2000; Wetzel 2001). P is slowly sourced naturally from rock via weathering, but in the past two centuries it has been heavily mined as well, especially for fertilizer ("fossil P;" Forsberg et al. 2003). Natural weathering of P amounts to about 3 Tg/year (Falkowski et al. 2000), while estimates of human mining of P is about 25 Tg/year (Schlesinger and Bernhardt 2013); i.e. mining has increased P flux by approximately 800% over natural weathering. P shortages for various human-related activities are predicted for the near future (Alewell et al. 2020).

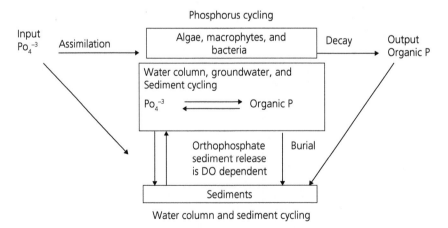

**Fig. 2.3** Generalized phosphorus cycling in aquatic environments. (Figure by the author.)

## 2.2.1 Inorganic Phosphorus

The main inorganic P form, highly bioavailable and preferred by primary producers, is orthophosphate ($PO_4^{-3}$), often measured as soluble reactive P (SRP). Important sources are animal manure, human wastewater effluent, fertilizers, and, as noted, weathering from natural lithogenic sources. Orthophosphate chemically binds to soil particles on land and can be carried well downstream from its origin by suspended sediments. Phosphate complexes with metals associated with suspended sediments (i.e. iron [Fe], aluminum [Al], calcium [Ca]), especially in clays. There are strong statistical correlations between turbidity and P concentrations in rivers and estuaries, such as the Delaware River (Lebo and Sharp 1993) and the Cape Fear River Estuary (Mallin et al. 1999), and strong correlations between TSS, turbidity, and P in urban streams (Mallin and Cahoon 2020). Under aerobic conditions, stream sediments are a sink for P, but under anaerobic conditions, $PO_4^{-3}$ can mobilize and leach out into the water column (Mulholland 1992). Thus, especially during summer in some rivers and estuaries, DO is very low or negligible near the bottom and P will subsequently increase in the water column (Rudek et al. 1991; Mallin et al. 1999). This phenomenon also commonly occurs in stratified reservoirs and lakes (Wetzel 2001). Indeed, hypoxia can impact some shallow, turbid eutrophic reservoirs throughout the water column, sometimes for days to weeks (Jones et al. 2011; Burkholder et al. 2022). Intense riverine algal blooms can also cause sharp temporal decreases in water-column P concentrations (Lebo and Sharp 1993; Knowlton and Jones 2000).

## 2.2.2 Organic Phosphorus

Both labile and refractory organic P compounds are common. Organic P compounds are substrates for aquatic bacteria and fungi. The lability or bioavailability of P compounds to bacteria and phytoplankton is poorly understood. Particulate P in rivers and estuaries can be associated with organic matter, aluminum oxides, iron oxides, and apatite (Lebo 1990). Measurement of the P forms within the total

P (fractions, operationally defined) generally follows various sequential fractionation methods that are used to quantify loosely adsorbed P, redox-extractable P, base-extractable P, acid-extractable P, and residual P. Dissolved P can also exist in organic forms (as dissolved inorganic phosphorus [DOP]) with highly variable bioavailability. A survey of 27 Midwestern aquatic systems found the bioavailability of DOP varying from 0 to 100%, with a median value of 78% (Thompson and Cotner 2018).

Agriculture is a major source of non-point-source P to US rivers and estuaries. Fields receiving fertilizers or animal manures will have both dissolved and particulate-bound P available for potential runoff. Caution must be taken when assuming that dissolved P is the only fraction to worry about in terms of bioavailability to algae and bacteria in receiving waters, because part of the particulate-bound P later becomes bioavailable, depending on physical and chemical factors (Daniel et al. 1998). Usually, minimal P moves (is leached) down into groundwater due to sorption of P by soil particles, and as mentioned, soil sorption of P is enhanced by Al and Fe oxides bound within clay particles. In some situations, however, rainfall-induced runoff from agricultural fields will lead to both surface and subsurface P movement into water bodies.

P is most susceptible to leaching down through the soil profile to groundwater when soils are sandy, or highly organic, or saturated, or when fertilization or surface applications of organic wastes have overloaded the soil P carrying capacity (Sims et al. 1998). Of these, the most common agricultural situation leading to overloading and surface and subsurface runoff is spraying or spreading animal wastes onto fields and exceeding their carrying capacity (Sims et al. 1998; note that this can also occur from municipal sewage wastes). Subsurface transport of P off-site is enhanced by artificial drainage structures in or beneath cultivated fields (Sims et al. 1998). During rainfall and subsequent runoff, fine-sized fractions of source soils are preferentially eroded; thus, reactivity and P content of eroded soils generally is higher than the source soils (Sharpley et al. 1993). Both dissolved and particulate P is entrained in rainfall-driven runoff from cultivated areas, with the bioavailable portion of particulate

P varying from 10 to 90% (Daniel et al. 1998). As with suspended sediments, most P transported in runoff downstream occurs in one or two intense storms during a given year (Sharpley et al. 1993). Much of the particulate-bound P thus enters river or estuarine sediments, and upon appropriate physical/chemical conditions later becomes released and bioavailable. P concentrations in US waterways have been increasing over time (Stoddard et al. 2016), and such pollution is widespread (Paulsen et al. 2008).

P concentrations along lower rivers and estuaries vary depending on the source, salinity, DO, and pH. For instance, in the Delaware River and Estuary, Lebo (1990) found that sewage effluent discharges at Philadelphia caused an increase in dissolved and particulate P concentrations in the river. Downstream where salinity is encountered (and pH increases), P associated with particulate Al, Fe, and Ca is released. This desorption occurs because with increased salinity, there is more competition for sites on clay particles, with the more abundant anions such as hydroxide [$OH^-$] and sulfate [$SO_4^{-2}$], which are orders-of-magnitude more abundant in salty water than freshwater (Froelich 1988).

## 2.3 Silica

Silica is important to certain important groups of aquatic planktonic and benthic primary producers. It occurs as silicon dioxide ($SiO_2$, the same as glass) but is more simply referred to as Si (symbol for silicon). Hydrated silica, $Si(OH)_4$, is the form taken up by diatoms (freshwater to marine, class Bacillariophyceae), certain other algae, macrophytes, and siliceous sponges for sponge spicule formation (Schoelynck et al. 2010; Graham et al. 2016). Si is a macronutrient for diatoms; the amount contained within diatom cell walls ranges from 2% (rare) to 85% (more common) as dry weight. Most diatoms are unable to survive without their siliceous cell walls (Graham et al. 2016). Si that is "locked up" in the remains of diatoms and other organisms is considered "biogenic" silica. Si is lithogenic, i.e. it comes from terrestrial weathering and is supplied by riverine inputs to estuaries. Human activities generally do not contribute Si to waterways, so it

is not a pollutant, although humans have caused major Si limitation (see below). Turner et al. (2003) found no statistical relationship between watershed human populations and the amount of Si delivered from watersheds to a large selection of rivers. The Si cycle, until recently, was considered to be "geologically controlled" because Si slowly dissolves from granitic soils, diatom cell walls, etc. and slowly becomes bioavailable (Wetzel 2001). Particulate forms of Si consist of that bound up in organic materials such as diatoms, while some portion of Si will be adsorbed to inorganic materials (Wetzel 2001). The dissolution (recycling) of biogenic Si back into the water column from sediments increases along with increasing water temperature and increasing salinity (D'Elia et al. 1983).

Seasonally, Si in the water column often is rapidly removed by diatoms in late winter/early spring, when concentrations are minimal in rivers and estuaries that develop strong diatom blooms. This phenomenon occurs across a wide array of aquatic ecosystems, from small streams (e.g. Sheath and Burkholder 1988) to large estuaries such as Chesapeake Bay (D'Elia et al. 1983). Lack of dissolved Si availability can then limit further diatom growth, so that other species become dominant. Both riverine inputs of dissolved Si and dissolution of biogenic Si are important in supplying Si to aquatic ecosystems to support diatom growth (D'Elia et al. 1983; Wetzel 2001).

The molar ratio of macronutrients in diatoms (N:Si:P) is approximately 16:16:1. Experiments have shown that changing the ratio of N:Si or N+P:Si or P:Si, depending on the aquatic system, can alter the phytoplankton community. High ratios increase the abundance of flagellates or cyanobacteria to the detriment of diatoms, whereas low ratios favor diatoms as a beneficial food web effect (Doering et al. 1989; Schelske et al. 2006). In a variety of situations around the world, including freshwater Lake Michigan (Schelske and Stoermer 1971) and estuarine and marine locations (Chesapeake Bay, Gulf of Mexico, Adriatic Sea, Kattegatt, Baltic Sea, and others—reviewed by Smayda 1989; Conley et al. 1993), increasing inputs of N and P from cultural eutrophication have stimulated diatom blooms that have eventually exhausted the dissolved Si in the water. An analysis of many rivers and estuaries has demonstrated that N:Si, or P:Si, or N+P:Si ratios have substantially increased since the 1960s, as have eutrophication symptoms (Smayda 1989; Justić et al. 1995; Turner et al. 2003; Schelske et al. 2006). Diatoms have been replaced by flagellates (e.g. cryptomonads and golden-brown flagellates in freshwaters, and dinoflagellates and other species, including various harmful taxa, in estuarine/coastal marine waters) and cyanobacteria in fresh or low-salinity waters. Thus, chronic N and P enrichment can slowly, fundamentally alter the food quality at the base of the food web. Besides immediate, obvious, negative effects such as fish kills, over time the food web shifts to algae that are less desirable food species for higher trophic levels.

In addition to eutrophication effects, significant decreases in the Si load delivered to coastal waters (e.g. from alumino-silicate clays) have occurred through dam construction (see Chapter 10) through the mechanisms of TSS settling and uptake by/subsequent sedimentation of diatom blooms— for example, in major rivers such as the Danube (Humborg et al. 2000). That retention of Si, coupled with increases in N and P, has been shown to cause blooms of noxious algae such as toxic flagellates (Humborg et al. 2000). Besides major dams, the countless small reservoirs, farm ponds, and beaver ponds will also retention of Si by dams, coupled with increases in N and P, Si from downstream movement into coastal waters (Turner et al. 2003).

## 2.4 Micronutrients

Micronutrients are elements or compounds required in small amounts by biota for various processes. Examples include iron, manganese, molybdenum, and vitamins. While needed at very low levels, many of them become toxic as their concentrations increase. Micronutrients have been very poorly studied in comparison to macronutrients. Depending on the situation, some micronutrients can limit phytoplankton production. In lotic ecosystems, micronutrients are thought to limit algal growth rarely. In urbanizing coastal creeks in South Carolina, Fe can occasionally be limiting (Kawaguchi et al. 1997).

## 2.5 Nutrient Spiraling

Under normal conditions, nutrients entering river headwaters go through many transformations as they are transported downstream (Fig. 2.4). Nutrient molecules pass through different forms, organic and inorganic, particulate and dissolved. For instance, a nitrate molecule (DIN) is taken up by a primary producer (benthic microalga) and becomes algal material, particulate organic N that is later eaten by a grazer (a snail, for example). Some of the N is excreted as organic waste (such as urea) or inorganic waste (such as ammonia). The grazer may in turn be eaten by a predator (such as a fish) that will also excrete N. Excretion is sometimes termed **egestion**. Organic waste N compounds can be remineralized by biochemical means to ammonium and then oxidized to nitrate. These DIN compounds are available for uptake by primary producers. The nutrients in dead biota materials are also remineralized over time. This process has been termed nutrient spiraling (Webster 1975); see detailed reviews by Newbold 1992 and Ensign and Doyle 2006).

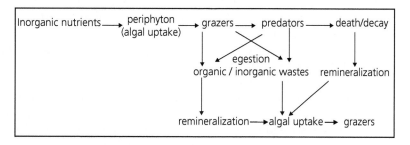

**Fig. 2.4** Simplified schematic of some nutrient spiraling components. (Figure by the author.)

**Spiraling length** is the linear distance required for a dissolved nutrient molecule to complete one cycle from its dissolved inorganic form in the water column through a particulate phase and finally through a consumer phase to be returned to the water column in a dissolved inorganic form (Fig. 2.3). Spiraling length can be very long (e.g. for nitrate, many kilometers) under periods of high discharge (Peterson et al. 2001). With fast-moving water, increased erosion, and subsequent limitation of photosynthesis by light and current in suspended-sediment-laden waters, nutrient spiraling will be limited (i.e. spirals are greatly lengthened) and much DIN and less so P, may reach and even pass through a river system into an estuary unassimilated or otherwise transformed. Conversely, in slowly moving waters (with long hydraulic residence time), much biological activity and numerous transformations may occur across a short river distance, which implies a short spiraling length.

Peterson et al. (2001) conducted an N-processing study using stable isotope techniques on 12 streams located in various biomes across the USA. The researchers found that headwaters streams are intense sites of the biological activity that is critical to spiraling (i.e. assimilation, storage, regeneration, export). Ammonium was removed rapidly, within ten to hundreds of meters, while nitrate traveled five- to tenfold longer distances than ammonium. Principal removal of ammonium (70–80%) occurred on stream bottoms through uptake by algae, bryophytes (mosses and liverworts), and decomposers (bacteria and fungi), and 20–30% of the removal was due to nitrification. Nitrate was removed from the stream water by biological assimilation and denitrification processes. Headwaters streams retained and transformed much of the inorganic N that entered the reach, generally > 50% of inputs. During stream passage, multiple cycles of uptake, storage, and regeneration occurred, and eventually denitrification. Peterson et al. (2001) concluded that headwaters streams are very important in processing N because of large surface-to-volume ratios. Unfortunately, these streams are also most subject to anthropogenic disturbance and transformation in watershed development.

Considerable spiraling also occurs in larger rivers. In an analysis of 52 papers about nutrient spiraling, Ensign and Doyle (2006) concluded that P recycling was relatively constant longitudinally within a stream continuum. Ammonium was recycled more rapidly in mid-order streams, whereas nitrate was recycled most rapidly in higher-order (larger) streams where mass flux (nitrate concentration × discharge) was greatest. The authors described difficulty in identifying and quantifying the mechanisms that influence transient nutrient storage (areas where localized stream conditions lead to temporary storage of dissolved nutrients) within a given stream. In an experiment in Coastal Plain streams, Ensign and Doyle (2005) found that coarse woody debris and aquatic vegetation increased the amount of transient storage area and increased contact between the water column and sediments, which, in turn, considerably enhanced sediment uptake of ammonium and phosphate.

Regarding spiraling compartments, most studies focus on nutrients taken up by algae and vascular plants along the river–estuary continuum. Yet, bacteria can take up considerable amounts of N and P, especially orthophosphate (Lebo 1990) and organic P. In streams, the decomposition of organic detritus by microbes has been found to be enhanced by nutrient additions (Elwood et al. 1981). The bacteria, in turn, can be directly eaten by some consumers (e.g. blackfly larvae in freshwater streams or filter-feeding bivalves in rivers and estuaries) or by various organisms (harpacticoid copepods, amphipods, isopods, shrimp, etc.) when they (and fungi) are colonizers of detrital particles. Experimental additions of P to Coastal Plain streamwaters indicated that P directly stimulated bacterial abundance, and subsequent BOD, relative to bacterial abundance and BOD in unenriched control (Mallin et al. 2004; Mallin and Cahoon 2020; see Chapter 8).

## 2.6 Impact of Animals on Nutrient Dynamics

Aquatic animals (fauna—zooplankton, protozoans, and other microfauna; macroinvertebrates from insect larvae to shellfish; finfish, waterbirds, muskrats, beavers, otters, and other mammals;

and megafauna; see Chapter 7) recycle nutrients back into the water column as excreta. These products include ammonia, phosphate, and organic compounds such as urea. As noted above, the importance of recycling to maintain the resident algal and macrophyte populations increases in situations where anthropogenic nutrient inputs are low (oligotrophic or minimally impacted situations).

Another important function of animals in streams, rivers, reservoirs, and estuaries is **translocation** of nutrients, which occurs when an animal feeds in one location and goes elsewhere to excrete or die (Vanni 2002). An example is fish grazing on or among bottom sediments and excreting nutrients in the upper region of a stratified water column, where the phytoplankton are located. On a diel basis, many large zooplankton vertically migrate to the lower water column (hypolimnion) in the daytime to avoid predators and migrate into the upper water column (epilimnion) at night to feed. Thus, consumers may feed in one stratum and excrete in another. Some piscivorous waterfowl feed in a river during daytime but return inland to a forest or reservoir area to roost at night, where they excrete. A critical example of translocation occurs in streams and rivers of the US Pacific Northwest, where anadromous fish like salmon spend years feeding and growing at sea, then return inland to streams to spawn and die. Their carcasses supply large amounts of nutrients to feed these low-order streams (Schindler et al. 2003), as well as the bears.

## 2.7 Estimated Boundaries for Trophic Status

What nutrient concentrations might be expected in unimpacted streams and rivers? Surface waters minimally impacted by nutrient pollution are increasingly difficult to find in general. Pristine waters, with negligible nutrient inputs from human activities, are now extremely difficult to find and do not exist in some regions (e.g. the US Environmental Protection Agency [EPA]'s "Corn Belt" nutrient ecoregion; see below). Some pristine waters still can be found in remote areas well away from population or agricultural centers, or in preserved

parklands. These generalizations regarding undeveloped systems are supported by Clark et al. (2000), who analyzed data from 85 stream and river sites and determined median concentrations of nutrients as follows: ammonium-N (0.020 mg/L), nitrate-N (0.087 mg/L), total nitrogen (TN) (0.260 mg/L), orthophosphate-P (0.010 mg/L), and total phosphorus (TP) (0.022 mg/L). The authors noted that in some regions, true unimpacted N levels were essentially unavailable due to airborne deposition of N, even in remote areas.

Based on approximately ten years of data, the US EPA (2000a, b) offered recommendations for use by states to develop protective numeric nutrient water quality criteria for TN and TP concentrations in minimally impacted waters, along with associated response variables chlorophyll *a* and turbidity. Two sets of recommendations were published—one for lakes and reservoirs, the other for rivers and streams—in 14 different ecoregions (geographic areas). The recommended EPA criteria were designed "to reduce water quality problems associated with excess nutrients" (see US EPA 2014). The realistic approach was based on present-day land uses.

Table 2.1 gives examples of how these recommended numeric nutrient criteria can vary widely by ecoregion within the USA. Mountainous ecoregions (II—Western Forested Mountains, XI—Central and Eastern Forested Uplands), with granitic bedrock soils and relatively limited anthropogenic N and P loads, generally have the lowest recommended criteria. Similarly, the Xeric American West (III) has limited agriculture and industry. In marked contrast is the heavily fertilized Midwestern "Corn Belt" nutrient ecoregion (VI—Corn Belt and Northern Great Plains) with much higher criteria, reflecting present-day land use, whereas the Eastern Coastal Plain (XIV) has more moderate criteria.

## 2.8 Nutrients as Pollutants

Nutrients naturally enter a stream as seasonal loads during fall as leaf leachate, as breakdown products of upland or in-stream decomposition of organic matter, or (in the case of P) leached or eroded from

**Table 2.1** Numeric nutrient criteria (TN, TP, rounded to the nearest integer) recommended by the US EPA (2000a, b) for rivers and streams in some nutrient ecoregions based on waters "minimally impacted" by nutrient pollution (not pristine). Corresponding levels of response variable chlorophyll *a* (chl *a*) are also included, along with associated turbidity.

| Ecoregion | TP (µg/L) | TN (µg/L) | Chl *a* (µg/L) | Turbidity (NTU) |
|---|---|---|---|---|
| XI Central and Eastern Forested Uplands | 10 | 310 | 1.61 | 2.30 |
| II Western Forested Mountains | 10 | 120 | 1.08 | 1.30 |
| III Xeric American West | 22 | 380 | 1.78 | 2.34 |
| VI Corn Belt and Northern Great Plains | 76 | 2100 | 2.70 | 6.36 |
| XIV Eastern Coastal Plain | 31 | 710 | 3.75 | 3.04 |

lithogenic sources, and through a constant interaction between the water column and the riparian zone (Mulholland 1992). As mentioned, however, numerous anthropogenic sources of nutrients all too frequently lead to extreme, excessive nutrient loading in comparison to natural sources.

Anthropogenic nutrients enter streams, rivers, reservoirs, and estuaries from **point** and **non-point** sources. Point sources are municipal and industrial wastewater treatment plants. The nutrients exit the facility as part of the waste stream through a pipe and are easily measured where they enter the receiving stream. Anthropogenic nutrient loading to surface waters is not a recent phenomenon. It has been estimated that considerable loads of nutrients entered rivers and coastal areas from factory discharges in the late nineteenth century during the Industrial Revolution (Billen et al. 1999). Excessive nutrient loading causes cultural eutrophication, exemplified by toxic and otherwise noxious algal blooms, low DO (hypoxia and anoxia), fish kills, and changes beginning with primary producers at the base of the food web and affecting all trophic levels (Glibert et al. 2011; Burkholder and Glibert 2023). For instance, a large-scale study showed that the biotic integrity of fish communities in low- to middle-order streams is strongly negatively correlated with stream nutrient concentrations (Miltner and Rankin 1998).

Non-point sources are much more difficult to measure and control than point (pipe, single-location) sources. Various non-point sources supply nutrients to receiving surface waters. The largest source is agricultural fertilizer runoff during/after storms. Throughout much of human history, animal waste, and in some cases human waste, has been applied to crops as fertilizer. Guano (bird droppings

in large quantities) was shipped into industrialized nations from South America or various islands to serve as fertilizer for large-scale agriculture. In 1905, however, Fritz Haber and Carl Bosch developed an industrial process that reduces $N_2$ to ammonium (Haber was awarded the 1918 Nobel Prize in Chemistry for this work). The Haber–Bosch process vastly reduced the need for guano and was rapidly embraced by the world community as the primary means of obtaining fertilizer. This process has been responsible for a major increase in fixed N to the biosphere. Research by Howarth et al. (2002) demonstrated that use of inorganic fertilizer is the largest source of Nr in the USA. From 1961 to 1999, N fertilizer use increased from 3.1 to 11.2 Tg N/year. On a per capita basis, the rate of US inorganic N fertilizer use (1300 kg N/km$^2$/year) has been approximately threefold higher than the global average. As of 1999, the largest N input to US watersheds was from inorganic fertilizer use (47%), while N fixation in agricultural systems (legumes) contributed 25% and NOx emissions comprised ~28% of the total. About 5.0 Tg N/year is exported by rivers to the coastal oceans (Howarth et al. 2002). N (and less so P) fertilizer use in many regions of the world is rapidly increasing (Glibert et al. 2014). According to large-scale studies, high concentrations of N, especially nitrate, are common in agriculturally influenced and urban streams (Mulholland et al. 2008); Stanley and Maxted 2008). In livestock-producing areas, runoff of nutrients from animal manure (mainly swine, poultry, and cattle) can be extreme (see Chapter 13).

In urbanized watersheds, stormwater runoff-driven inputs of nutrients to streams and rivers can be substantial as well. These nutrients are sourced from lawn and garden fertilizers, domestic and wild animal wastes, and N from vehicle exhausts. In an analysis of 19 coastal creeks from North Carolina, South Carolina, and Georgia, Sanger et al. (2015) found significant positive relationships between the degree of watershed development and in-stream concentrations of both nitrate and ammonium.

In rural areas as well as some urbanized regions, septic systems can be major contributors to groundwater pollution (see Chapter 13). A faulty septic system alone can be a defined source of pollution, but when many such systems are located in an area with a high water table and porous soils, the sum of the untreated septic leachate (especially the ammonium/nitrate and P) is essentially nonpoint-source pollution. This nutrient load enters the upper groundwater table and moves downslope; thus, groundwater can be a major source of DIN to streams, rivers, reservoirs, and estuaries, especially when in close proximity to the receiving surface waters. Either nitrate or ammonium can be the form, depending upon redox conditions and sources of the N. In a diverse array of locations such as the karst areas of Kentucky, the Outer Banks of North Carolina, and the west coast of Florida, high levels of ammonium enter groundwater from poorly functioning septic systems, and the groundwater then moves into surface waters, promoting algal blooms. Other important sources of ammonium to groundwater include agriculture as mentioned, leaking sewer systems, and landfill leakage (Wakida and Lerner 2005). Large amounts of N can then be advected upward from river sediment porewater into the water column to feed the eutrophication process (Null et al. 2011). Nutrients entrained in groundwater tend to remain in concentrated plumes rather than dispersing (Keeney 1986). Other nutrient sources include runoff from clearcuts for forestry or land development (housing, commercial areas—high organic and inorganic nutrients), airborne nutrients from smokestacks, vehicle exhausts, and CAFOs.

Turner et al. (2003) analyzed large sets of worldwide river nutrient data for patterns driving concentrations and nutrient ratios. They assessed N yield for watersheds (Y; kg N/ha/year) as a function of human population density (X; persons/km$^2$) and found strong statistical relationships:

Rivers worldwide: Nitrate-N yield = $24.1 + 2.12X$

$$\left(n = 49;\ R^2 = 0.39;\ p = 0.0001\right);$$

North American/European rivers:

Nitrate-N yield = $-55.4 + 4.3X$

$$\left(n = 24;\ R^2 = 0.82;\ p = 0.001\right).$$

Increases in nitrate, rather than other N species, were thought to be responsible for the increased N yield in urbanizing watersheds.

During the latter twentieth century and early twenty-first century, N discharge in the USA to coastal waters had increased by sixfold over the previous few decades (due to human activities); in the North Sea and Yellow Sea, it had increased 10- to 15-fold; and in remote, poorly populated areas, N flow to coastal waters had minimally changed (Howarth and Marino 2006).

Turner et al. (2003) also found significant statistical relationships between human population density and dissolved P yield from watersheds (here, Y = kg P/ha/year), although the relationships were weaker than those for nitrate:

Rivers worldwide: Phosphate-P yield = $14.7 + 0.22X$

$$\left(n = 40;\ R^2 = 0.15;\ p = 0.007\right);$$

North American and European rivers:

Phosphate-P yield = $10.3 + 0.44X$

$$\left(n = 22;\ R^2 = 0.37;\ p = 0.002\right).$$

Thus, N and P availability to phytoplankton and bacteria in rivers and estuaries can be somewhat predictable spatially, with elevated concentrations near point sources and turbidity maximum zones, and seasonally during summer periods of bottom water hypoxia.

What nutrient concentrations can be considered representative of different trophic states (eutrophic, mesotrophic, oligotrophic) in streams and rivers? Dodds et al. (1998) used large datasets from North America and New Zealand to create cumulative distributions of the values. They then divided the data into thirds to suggest boundaries for trophic states; see Table 2.2.

**Table 2.2** Suggested boundaries for stream trophic classification based on seasonal algal biomass as mean and maximum chlorophyll *a* (chl *a*), mean TN, and mean TP (Dodds et al. 1998; also see Dodds and Smith 2016).

| Variable | Oligotrophic-mesotrophic | Mesotrophic-eutrophic | n |
|---|---|---|---|
| Mean benthic chl *a* (mg/m$^2$) | 20 | 70 | 286 |
| Maximum benthic chl *a* (mg/m$^2$) | 60 | 200 | 176 |
| Mean suspended chl *a* ($\mu$g/L) | 10 | 30 | 292 |
| Mean TN ($\mu$g/L) | 700 | 1500 | 1070 |
| Mean TP ($\mu$g/L) | 25 | 75 | 1366 |

In the USA, some major point-source dischargers ($\geq$ 4.55 million L/day or $\geq$ 1 million gallons/day) have effluent nutrient limits (usually only ammonium) they are supposed to follow in order to control pollution levels in receiving waters. Point sources are supposed to be regulated by a permitting process under the National Pollutant Discharge Elimination System (NPDES) through the EPA, which usually cedes regulatory power to the individual states. Such permits can be used to regulate the amount of nutrients or other pollutants that can be discharged daily into ambient waters, an often-contentious process that can involve politicians, industry lobbyists, and concerned citizens groups. The states are also required by the EPA, under the Clean Water Act, to assess individual water bodies for impairment, that is, to see if they are in violation of state water quality standards based on available data. If impaired, the state must develop a management plan called a total maximum daily load (TMDL) for the pollutant in violation and reduce loading of that pollutant so that the water body meets water quality standards. Entire rivers, streams, lakes, reservoirs, and estuaries can thus be legally targeted for specific quantitative reductions in N or P.

If an individual state fails to do its job properly the EPA can rescind that state's regulatory authority and take over permitting from the state in question. State regulatory agencies are ultimately under the authority (and funding) of state legislatures, who can and will underfund the agency and appoint antiregulatory agency heads because of pressure from industrial dischargers, agribusiness lobbyists, or simply politicians who blindly support what they perceive as business growth (at the expense of water and air protections). Likewise, if

the EPA is too lenient in its permitting and a specific water body suffers from eutrophication, concerned citizens or citizen's groups can sue the agency to seek more stringent controls. There is currently an effort underway in a number of states to come up with nutrient criteria, which would then be subject to approval by the federal EPA. These efforts are complicated by the vast variety of geomorphic, meteorological, and hydrological variables that can influence natural and anthropogenic nutrient loading, as well as local and national politics.

## 2.9 Toxic Effects of Nitrogen Pollution

### 2.9.1 Nitrate Toxicity

The USA nitrate standard for drinking water is 10 mg $NO_3$-N/L to protect the public from a disease called methemoglobinemia, also called blue-baby syndrome. This is a potentially fatal condition (mainly to infants) that is caused by ingestion of elevated nitrate concentrations in drinking water or food. The nitrate is reduced to nitrite by gut microflora, which then serves as an oxidizing agent and oxidizes the iron in hemoglobin (which carries oxygen in the blood). The oxidized product, methemoglobin, cannot transport oxygen and can lead to infant death (Keeney 1986; Johnson and Koss 1990). In the USA all documented cases of methemoglobinemia have been from consumption of water with nitrate concentrations in excess of this standard (Fan and Steinberg 1996).

Nitrate concentrations much lower than 10 mg/L have caused toxic effects on aquatic life; as for most toxic substances, young life history stages are most sensitive. Nitrate toxicity to fish and invertebrates is mainly due to conversion of oxygen-carrying pigments (hemoglobin and hemocyanin) to forms incapable of carrying oxygen (Camargo et al. 2005). Mortality and developmental changes in macroinvertebrates have been documented from exposure to concentrations of nitrate as low as 1.4–2.4 mg/L (early instar caddisfly larvae; Camargo and Ward 1995). For adult freshwater invertebrates, the lowest reported chronic toxicity levels promoting adverse impacts are 2.8–4.4 mg/L for two species of amphipods (Camargo et al. 2005).

Regarding fish, eggs and fry of some salmonids (rainbow trout, *Oncorhynchus mykiss*; cutthroat trout, *Salmo clarki*; chinook salmon, *Oncorhynchus tshawytschia*) sustained liver damage and/or mortality at nitrate concentrations less than 5.0 mg/L (Camargo et al. 2005). In controlled experiments, nitrate has interfered with fish steroid hormone synthesis, adversely affected fish fecundity and sperm motility/viability, and been toxic to fish embryos (mosquitofish, *Gambusia holbrooki*; Edwards et al. 2004; 0.2–5.1 mg/L). On a more speculative note, results of studies on alligators suggest that exposure to nitrate/nitrite can act directly on steroidogenesis, particularly by impacting cell mitochondria, as well as in the liver (Guillette and Edwards 2005). Some work suggests that synergism with other pollutants, such as pesticides, enhances toxicity (Guillette and Edwards 2005). Sublethal developmental effects on amphibians (three common frog species within the genus *Rana*) have been detected at nitrate concentrations as low as 2.3–2.5 mg/L, depending on the species (Hecnar 1995; Orton et al. 2006). Such concentrations are commonly seen in streams draining agricultural fields, CAFO-influenced drainages, and streams impacted by faulty sewage treatment, so these are ecologically relevant nitrate concentrations (Stanley and Maxted 2008; Mallin et al. 2015). Thus, from a review of published data, Rouse et al. (1999) concluded that nitrate concentrations in about 20% of Great Lakes watersheds were high enough to cause death and developmental anomalies among various amphibians and other aquatic fauna.

Some species of algae and aquatic plants can develop nitrate toxicity due to luxury uptake (that is, the characteristic in many algae and plants of taking up much more of a given nutrient than is needed; Salisbury and Ross 1992; Wetzel 2001).

Nitrate uptake in plants requires high cellular energy and carbon skeletons for amino acid synthesis (Turpin 1991; Larsson 1994). The sustained high carbon demand from extended nitrate uptake appears to drive the plants into severe internal carbon limitation (Osborne et al. 2017 and references therein). Nitrate inhibition of certain algal and macrophyte species has been documented at water-column concentrations ranging from 0.05 mg/L (for sensitive species that thrive in low-nutrient habitats, such as brackish areas affected by riverine inputs) to 0.50 mg/L (for species found in more nutrient-enriched habitats).

## 2.9.2 Ammonia Toxicity

Ammonia/ammonium is generally about 100- to 1000-fold more toxic to aquatic fauna, especially finfish, than nitrate (Camargo and Alonso 2006; US EPA 2013). Toxicity is mostly generated by the un-ionized form (ammonia, $NH_3$), the abundance of which is pH- and temperature-dependent, with the highest percentages of this form found in warmer waters (>25°C) at pH > 8 (Emerson et al. 1975). The ionized form ($NH_4^+$) is much less toxic. Mortality has occurred with un-ionized $NH_3$ at ~0.30–3.0 mg/L (Boyd 2000). Some streams are subject to such concentrations as a result of sewage works (aging infrastructure, malfunctions, bypasses during moderate rain events, etc.; see Glibert et al. 2016 and references therein; also see Burkholder et al. 2022). Such un-ionized $NH_3$ levels are also common in streams near industrialized animal operations, not only following waste spills (e.g. Burkholder et al. 1997) but also as a result of normal CAFO maintenance procedures (e.g. Mallin et al. 2015).

Toxic levels of ammonia can cause major physiological stresses to fish (Camargo and Alonso 2006), including damage to gill epithelium causing asphyxiation; stimulation of glycolysis and suppression of the Krebs cycle, leading to acidosis and reduction in oxygen-carrying capacity by the blood; inhibition of ATP production and ATP depletion in the basilar region of the brain; disruption of blood vessels and osmoregulatory activity in the liver and kidneys; and immune system suppression; and many of these ailments can result in higher susceptibility to bacterial and parasitic diseases. Not surprisingly, these major impacts also can suppress feeding activity, fecundity, and survivorship, with population-level implications.

Although algae tend to prefer ammonia over nitrate as an N source at low ammonia concentrations, higher levels can be inhibitory (Glibert et al. 2016). As for fauna, aquatic macrophytes generally are more sensitive to ammonia than to nitrate. Ammonia toxicity is known for many vascular plants, with die-off occurring at ~0.05–3.0 mg/L (Britto and Kronzucker 2002). Un-ionized ammonia can decrease chlorophyll content, inhibit respiration, and adversely affect the plant electron transport system. Similarly, as for nitrate, high ammonia concentrations can influence carbon and N metabolism in submersed macrophytes, exacerbated by low light availability and by reducing carbohydrates internally due to high demand for amino acid synthesis to prevent ammonia toxicity (Tan et al. 2019 and references therein). Un-ionized ammonia can also be toxic to *Nitrosomonas* and *Nitrobacter* bacteria, inhibiting nitrification, which can sometimes lead to increased accumulation of total $NH_3$ (un-ionized + ionized forms) (Russo 1985; Burkholder et al. 2022).

## 2.10 Precipitation and Hydrology Impacts

Precipitation, the driver of river discharge, is the major factor controlling riverine N loads that subsequently make their way to estuaries and coastal waters. The US Geological Survey (USGS) completed a large-scale modeling effort involving 63 minimally impacted reference basins to determine what could be considered "reference" levels of TN and TP or what should be expected under conditions minimally influenced by humans (Smith et al. 2003). Precipitation-driven runoff was the strongest explanatory variable in predicting TN and TP concentrations in reference sites. With increasing human development in a watershed comes increasing N and P loading to rivers. For example, during the above USGS study, the authors compared data from an unimpacted site with a large-scale dataset that represented all levels of human impact (Dodds et al. 1998). Smith et al. (2003) found that, for TN, the actual median concentration (0.89 mg/L) in the large dataset exceeded median background (unimpacted rivers concentration, 0.14 mg/L) by a factor of 6.4. For TP, the large-scale median concentration (0.045 mg/L) exceeded the unimpacted rivers concentration (0.023 mg/L) by a factor of about 2.

A study of 16 Northeastern US watersheds indicated that riverine N export during 1988–1993 ranged from 310 to 1760 kg $N/km^2$/year, while remote rivers with undeveloped watersheds in that north temperate region were estimated to export approximately 100 kg $N/km^2$/year (Howarth et al. 2006) (Fig. 2.5). N delivery was strongly correlated with both precipitation and river discharge. Net anthropogenic N inputs to the watersheds for the 16 rivers were highly significantly related to average

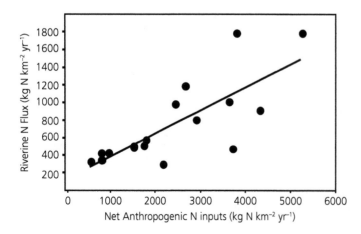

**Fig. 2.5** Relationship between watershed N inputs from the landscape and riverine N fluxes for 16 watersheds (redrawn from Howarth et al. (2006).

annual riverine N fluxes ($Y = 0.26X + 107$; $R^2 = 0.62$; $p = 0.003$; Fig. 2.5). Even so, the analysis also indicated that on average only ~26% of human N inputs to the landscape were exported to the lower river, while the rest was retained in the watershed or lost through denitrification. As indicated above, Peterson et al. (2001) assessed the intense processing of N that occurs in lower-order streams with a large surface to volume ratio for sediment contact.

Another example of riverine N load driving estuarine eutrophication is the eutrophic Neuse River Estuary in North Carolina, which has been plagued by hypoxia, toxic algal blooms, and fish kills (Burkholder et al. 2006). In this system rainfall in the headwaters region (Piedmont region) was positively correlated with river discharge 145 km downstream in the middle reaches (Upper Coastal Plain), which, in turn, was positively correlated with nitrate concentrations 120 km downstream in the estuary (Mallin et al. 1993). Experiments indicated that nitrate inputs directly stimulate algal blooms in the estuarine waters (Rudek et al. 1991). Furthermore, river discharge was positively correlated with measured primary productivity pulses in the Neuse Estuary, given a two- to four-week lag period between rainfall in the Piedmont area and algal blooms in the lower estuary (Mallin et al. 1993). Runoff from upper watersheds moving downstream into the estuary is considered "new" N, as mentioned. In temperate river systems pulses of nitrate often occur in late winter/early spring, associated with increased river discharge (termed the "freshet"), especially when coupled with organic nutrient remineralization and farm field fertilization (followed by subsequent rainfall runoff); these freshets initiate spring algal blooms in estuaries. Thus, river reaches do not operate as independent entities; what occurs in headwaters areas can strongly influence the estuary even hundreds of kilometers downstream.

On truly large scales it has been well documented that precipitation effects in river headwaters can be measured as nutrient increases or biological stimuli hundreds of kilometers downstream in the estuary. A well-known example is the Chesapeake Bay ecosystem, where the Susquehanna River discharge controls N loading to the Bay. The amount of N loading from that river controls the annual magnitude of algal blooms in the Bay, which in turn controls the extent of hypoxia (via bloom decomposition and subsequent BOD) in the Bay bottom waters (Hagy et al. 2004). Similarly, the Gulf of Mexico develops a benthic hypoxia ("dead") zone every year, which expands in areal coverage during periods of high river flow and shrinks in years of low flow (Mitsch et al. 2001), depending upon the magnitude of the N load (concentration × flow) reaching the Gulf phytoplankton. It is notable that during the previous century the nitrate concentration in the lower Mississippi River tripled (Goolsby and Battaglin 2001).

Precipitation can be an important direct source of new N in some situations where N is very limiting. Bioassay experiments adding rainwater to Neuse Estuary water and Bogue Sound (Paerl et al. 1990) as well as offshore water (Willey and Cahoon 1991) found that phytoplankton growth could be significantly stimulated by such additions. The degree of stimulation depended upon the DIN concentration in the rainwater, which was largely dependent upon pH (i.e. continental rain has lower pH [and higher DIN] than rain from offshore—acid precipitation forms as a "byproduct" of continental vehicle exhausts).

Thus, there are numerous sources of nutrients to lotic systems. Nutrients can leave the river/estuary ecosystem if:

- taken up by riparian vegetation (sequestered at least for a while; woody vegetation can lock up nutrients for very long periods);
- turned into $N_2$ gas by denitrification or anammox of DIN by bacteria in wetlands;
- removed as organic N in a prey item (fish, clam, crab, etc.) by a fisherman, raptor, raccoon, etc.; and/or
- buried for long periods under sediment loads.

The remainder, largely as organic N and P compounds, ends up in the ocean.

## 2.11 Nutrient Limitation and Excess Phytoplankton Growth

Resource managers often want to know "which nutrient, N or P" is limiting so that they can direct limited funding to control one or the other, rather than both macronutrients. Some of the treatments used for control are very different, and expensive. For instance, P is often controlled in wastewater treatment plants by alum treatment (precipitation and sludge removal), whereas N removal is a complex biological process, initially more costly, involving both nitrification and denitrification (US EPA 2007a, b).

The nutrient status of an aquatic ecosystem is generally considered using the productivity or abundance of primary producers, usually the phytoplankton. The typical approach is too narrowly focused because several other major functional groups of primary producers (benthic microalgae, floating and benthic macroalgae and macrophytes) also respond to nutrient inputs. Phytoplankton biomass, indicated by the concentration of the "universal plant pigment" chlorophyll *a*, is often the only target of sampling by resource managers, even in cases where the water body is sustaining massive blooms of noxious macrophytes or benthic microalgae rather than phytoplankton. This section is included to explain historic generalizations and present-day realities about nutrient limitation versus excess. Nutrient status also can directly affect the efficacy of stream microbial decomposers (bacteria, fungi) (Elwood et al. 1981; Padgett et al. 2000; Mallin and Cahoon 2020).

The abundance of a given nutrient (N, P, Si, or other element) can limit (control) algal growth in a given river, reservoir, or estuary. Liebig's Law of the Minimum was original developed for crops. It essentially states that the plant nutrient present in lowest concentration with respect to the plant's requirements will limit growth. This concept is frequently applied to aquatic ecosystems— unfortunately, often wrongly, in situations where the nutrients considered are in surplus, not in limiting supply, for the naturally occurring phytoplankton assemblages. For phytoplankton the limiting nutrient is usually N or P, sometimes Si, and on occasion trace elements such as Fe (in offshore marine waters mostly).

The limiting nutrient is often inferred by first estimating the molar ratio of N:P in the water column and comparing the result to the "Redfield ratio" for healthy conditions, or by performing controlled experiments to assess when/where a given nutrient (supposedly; see below) limits phytoplankton growth. The Redfield ratio (Redfield 1958), developed from the approximate ratio of certain major nutrients in healthy, rapidly growing marine phytoplankton, has been applied across aquatic ecosystems. For C:N:P it is about 106:16:1 on an atomic (molar) basis or 42:7:1 on a mass basis (Hecky and Kilham 1988). The Redfield ratio has become a commonly used bellwether to estimate the limiting nutrient for a given stream, reservoir, estuary, or coastal marine area from chemical data alone. It is assumed that N:P molar ratios above 16 indicate P limitation of phytoplankton and N:P ratios less than 16 indicate N limitation.

But what if *both* nutrients are actually in excess relative to natural background? That is the common present-day situation for many surface waters. Unfortunately, resource managers often conduct nutrient addition bioassays in such situations to assess which nutrient is limiting. In a heavily eutrophied system, determining which nutrient the system will run out of first becomes irrelevant, as both supplies are extreme. In the same way, N:P ratios and nutrient addition bioassays logically should only be used to infer "nutrient limitation" under limiting conditions by one nutrient or another. For example, the extreme N and P contamination near a sewage outfall often stimulates growth of noxious algae such as the cyanobacterium phytoplankter *Microcystis* (see Chapter 3). The natural assemblage that was once N or P limited is long gone. Nutrient addition bioassays applied under such conditions only effectively inquire of the remaining noxious, highly pollution-tolerant taxa, "Would more N or P make life even better?" Resource managers commonly misuse such bioassays to conclude, mistakenly, that "the algae did not respond to more N but grew more with additional P, so that means N control is not needed"—when both N and P should be decreased. The noxious algae were saturated with N relative to P, but in a very unhealthy aquatic system where both N and P supplies were already extreme and the balance of N:P was highly skewed relative to healthy conditions.

Under limiting conditions, testing for the limiting nutrient by use of nutrient addition bioassays provides a more accurate indication than inferences from N:P ratios, especially when combined with field chemical data. And clearly this can be useful to prevent nutrient overload to saturated conditions, provided action is taken. There are various types and levels of such assays, ranging from single-culture algal species conducted in arrays of small flasks, through cubitainer bioassays of different sizes, and mesocosms small and large, up to whole-lake studies such as in the Experimental Lakes Area of Canada (Schindler and Fee 1973), with larger testing systems providing better information (Hecky and Kilham 1988).

Where nutrient limitation occurs, patterns historically have been documented to change along a river–estuary gradient moving seaward from freshwater. The following generalizations are suggested based on physical and chemical differences (Ryther and Dunstan 1971; Hecky and Kilham 1988; Howarth and Marino 2006). These generalizations still apply to waters minimally impacted by nutrient pollution (sensu US EPA 2000a, b).

## 2.11.1 Freshwater Systems

Phytoplankton in most minimally impacted freshwater systems tend to be P limited due to the abundance of N compared with P inputs:

- Terrestrial N fixation supplies copious amounts of N to freshwater watersheds. Agricultural N fixation associated with crop legumes provides new N. Following plant decomposition, the organic N is remineralized and added by runoff to waterways.
- Aquatic N fixation could make up the deficit when P concentrations are high—there are many species of N-fixing cyanobacteria and other N-fixing bacteria in freshwaters.
- P is lost readily to the sediments by being sorbed to clay and/or complexed with Al, Ca, and Fe and precipitated to the sediments.
- Crop fertilizers have elevated N:P ratios, and they are a major source of runoff-driven nutrient loading to streams and other water bodies (Daniel et al. 1998).
- Atmospheric deposition of N has been determined to more-than-double the N:P ratios in mountain lakes that are remote from land-based

sources, driving them from N limitation to P limitation (Elser et al. 2009).

Among the major exceptions to the freshwater P limitation concept are minimally impacted southeastern blackwater systems (see Chapter 8). These systems are low in DIN compared to other freshwaters. Cyanobacteria are generally much less common in these systems, so there is less N fixation. In summer, southeastern blackwaters often have low DO concentrations due to elevated water temperatures combined with high bacterial respiration and drainage from organic riparian wetlands. The hypoxic conditions promote release of sediment-associated P. Many of these systems are located in areas with high industrialized swine and poultry production. Animal wastes have low N:P ratios (about 1:1) in comparison to fertilizers, which have higher N concentrations (Daniel et al. 1998). An example of this occurs in the Black and Northeast Cape Fear Rivers and tributaries, where experiments demonstrated that N stimulates algal production rather than P (Mallin et al. 2004).

## 2.11.2 Estuaries

Minimally impacted riverine estuaries are mostly N limited, but P limitation or N+P co-limitation occurs when N is elevated, such as during spring rains (arrival of the spring freshet):

- In estuaries there is less N fixation than in freshwaters because there are fewer N-fixing cyanobacterial species (due to salinity constraints). Thus, overall N-fixation rates are lower in estuaries than farther upstream.
- In estuaries with significant water exchange, the water is moved out by tides or river discharge, so N fixed by microbes is not retained.
- Considerable denitrification occurs in freshwater tidal and estuarine marshes, so there is relatively high loss of N relative to P. The amount of N denitrified increases with increasing residence time in the wetlands (Seitzinger et al. 2006).
- Phosphate sorbed onto sediment particles is released upon entry into estuaries (minutes to days). There is more competition for sites on clay particles with the more abundant anions

such as OH$^-$ and SO$_4^{-2}$, which are orders-of-magnitude more abundant in seawater than freshwater (Froelich 1988).

- There is lower P storage in estuarine sediments, possibly a result of higher sulfate concentrations and subsequent higher sulfate reduction rates with more sequestration of iron by sulfides. Iron sulfides adsorb less P than other iron minerals (Howarth and Marino 2006).

There can be a strong temporal component to determining nutrient limitation in estuaries, especially riverine estuaries where N can be limiting to phytoplankton growth in summer, but P can limit or co-limit in spring. This results from strong spring freshets bringing high concentrations of N from uplands to the estuary, driving spring phytoplankton to P limitation (or N+P co-limitation). With lower river discharge in summer, N then becomes limiting due to the reasons listed above. This temporal change occurs in Chesapeake Bay tributaries (D'Elia et al. 1986), the Neuse River Estuary (Rudek et al. 1991), and the Cape Fear River Estuary (Mallin et al. 1999). The latter system is also light limited in winter due to high turbidity from storm events and increased discharge.

### 2.11.3 Coastal Waters

N is almost always limiting in minimally impacted coastal marine systems for several reasons:

- There are relatively few N fixers in marine waters. N-fixing cyanobacteria are sparse in the water column due to high salinity, although some N fixation occurs by benthic cyanobacteria.
- Decomposition of terrestrial-sourced organic matter can release P more readily than N (Ryther and Dunstan 1971).
- DIN from river and coastal creek inputs are assimilated by algae and marshes within the lower river, estuary, or river plume, i.e. marine phytoplankton assemblages can be relatively far from point and non-point N sources (exception, continental-sourced precipitation). Much of the N that does enter these coastal waters is in the form of refractory organic compounds (Dafner et al. 2007) and, thus, not readily assimilated by phytoplankton.

- Substantial denitrification occurs on the continental shelf (Seitzinger et al. 2006); thus, offshore waters that enter estuaries are depauperate in N relative to P (Howarth and Marino 2006).

N and P can limit different parts of the microbial community at the same time. In South Carolina marshes, for example, experiments have shown that while N limits *Spartina* growth, the associated bacteria are limited by P (Sundareshwar et al. 2003). In North Carolina blackwater systems, N limits phytoplankton growth while P limits growth of aquatic bacteria (Mallin and Cahoon 2020). Research has also verified that P (organic or inorganic) can have a stimulatory effect on survival and production of fecal bacteria entering water bodies from sewage or runoff (Chudoba et al. 2013; Mallin and Cahoon 2020).

## 2.12 Importance of Nutrient Ratios and Management of Both N and P

An emerging line of research, ecological stoichiometry (Sterner and Elser 2002; Glibert et al. 2011; Burkholder and Glibert 2023), deals with nutrient ratios in the water, sediments, and biota, including how nutrient ratios can directly affect algal physiology and assemblage composition, with impacts on the entire aquatic food web. Grazers are impacted and changed by their algal/plant food quality—that is, by the ratios as well as the levels of nutrients within their prey. Whole ecosystems can be gradually shifted in response to stoichiometric imbalances in available nutrient supplies (Glibert et al. 2011 and references therein).

Strong arguments can be made for co-management, that is, control of *both* N and P along a river continuum (Fisher et al. 1992; Burkholder et al. 2006; Conley et al. 2009). Controlling P alone in the upper watershed can lead to eutrophication of the estuary downstream, where N is often the limiting nutrient. Controlling N alone in freshwater areas can lead to blooms of N-fixing cyanobacteria in fresh and oligohaline waters, since there is abundant P to support them. Additionally, bacteria are generally P limited rather than N limited, and controlling N alone ignores P stimulation of bacteria, which can lead to increased BOD and increased survival and production of fecal bacteria (see above). Fortunately, various available techniques can help control both nutrients simultaneously, including use of streamside buffer zones, stormwater treatment wetlands, improved livestock manure control, and reduced use of fertilizers (Conley et al. 2009).

## 2.13 Summary

- Major plant nutrients include macronutrients, most notably N and P, while Si is critical for diatoms and some other taxa groups. The N cycle consists of several processes that are microbially and enzyme driven and controlled by redox conditions. The P cycle is much simpler, but there are P compounds for which little is known about bioavailability and use.
- DIN consists mostly of ammonia/ammonium and nitrate, which are assimilated by algae and commonly promote nuisance and harmful blooms (eutrophication) when in excess. DIN is added to surface waters by wastewater treatment plant effluents, septic systems, atmospheric NOx compounds, CAFOs, fertilizers, and general stormwater runoff. Additionally, urea is a labile form of DON that is used in large quantities of fertilizers.
- Atmospheric $N_2$ can be "fixed" into ammonia for use by cyanobacteria in water and legumes on land. Nitrate can be converted to atmospheric $N_2$ through the process of denitrification by bacteria in sediments, marshes, and wetlands; this is an important means of removing excess N for eutrophic situations. Anammox is an enzyme-driven process that also converts reactive N to atmospheric $N_2$.
- High concentrations of nitrate (e.g. 10 mg/L, US drinking water standard) can be toxic to human infants, but much lower concentrations can be toxic to sensitive macroinvertebrate and fish species (1.5–2.5 mg/L). Among the most sensitive aquatic organisms to nitrate pollution are certain macrophyte species, which have been physiologically stressed or killed at chronic water-column concentrations commonly found in polluted rural and urban environments.

- Ammonia generally is much more toxic to aquatic life than nitrate, although at low concentrations it can be preferred over nitrate by various primary producers. Un-ionized ammonia is highly toxic especially to freshwater mussels and fish, exacerbated by low DO and warm temperatures.
- Phosphate binds readily to sediments, especially clays, and is added to surface waters by stormwater runoff in both dissolved and particulate forms. Excessive P readily fuels algal blooms in rivers, reservoirs, and lakes. Stimulatory effects often have occurred or increased under both P and N enrichment. P can directly stimulate bacterial growth, including harmful fecal bacteria, and in some situations can lead to increased BOD.
- In most freshwater streams phytoplankton production is limited by P, estuaries by P or N, and the coastal ocean by N. However, some streams, such as Coastal Plain streams, are N limited for phytoplankton and P limited for bacterial production. It is important to realize that for heavily eutrophic systems it is no longer a case of there being a limiting nutrient; they are likely both in excess and the primary producer community has been greatly altered to noxious, pollutant-tolerant species. Thus, in such situations reductions in both N and P are required.

## References

Alewell, C., B. Ringeval, C. Ballabio, D.A. Robinson, P. Panagos and P. Borrelli. 2020. Global phosphorus shortage will be aggravated by soil erosion. *Nature Communications* 11:4546.

Antia, N.J., P.J. Harrison and L. Oliveira. 1991. The role of dissolved organic nitrogen in phytoplankton nutrition, cell biology and ecology. *Phycologia* 30:1–89.

Billen, G., J. Garnier, C. Deligne and C. Billen. 1999. Estimates of early-industrial inputs of nutrients to river systems: implications for coastal eutrophication. *Science of the Total Environment* 243:43–52.

Boyd, C.E. 2000. *Water Quality: An Introduction*. Kluwer Academic Publishers, Boston.

Bradley, P.B., M.W. Lomax and D.A. Bronk. 2010. Inorganic and organic nitrogen use by phytoplankton along Chesapeake Bay, measured using a flow cytometric sorting approach. *Estuaries and Coasts* 33:971–984.

Britto, D.T. and H.J. Kronzucker. 2002. NH4+ toxicity in higher plants: a critical review. *Journal of Plant Physiology* 159:567–584.

Bronk, D.A., J.H. See, P. Bradly and L. Killberg. 2007. DON as a source of bioavailable nitrogen for phytoplankton. *Biogeosciences* 4:283–296.

Burgin, A.J. and S.K. Hamilton. 2007. Have we overemphasized the role of denitrification in aquatic ecosystems? A review of nitrate removal pathways. *Frontiers in Ecology and the Environment* 5:89–96.

Burgin, A.J., W.H. Yang, S.K. Hamilton and W.L. Silver. 2011. Beyond carbon and nitrogen: how the microbial energy economy couples elemental cycles in diverse ecosystems. *Frontiers in Ecology and the Environment* 9:44–52.

Burkholder, J.M. and P.M. Glibert. 2023. Eutrophication and oligotrophication. In Levin, S. (ed.), *Encyclopedia of Biodiversity*, Vol. 2, 3rd edition. Academic Press, New York. doi:10.1016/B978-0-12-822562-2.00052-9.

Burkholder, J.M., M.A. Mallin, H.B. Glasgow, et al. 1997. Impacts to a coastal river and estuary from rupture of a large swine waste holding lagoon. *Journal of Environmental Quality* 26:1451–1466.

Burkholder, J.M., D.A. Dickey, C. Kinder, et al. 2006. Comprehensive trend analysis of nutrients and related variables in a large eutrophic estuary: A decadal study of anthropogenic and climatic influences. *Limnology and Oceanography* 51:463–487.

Burkholder, J.M., P.M. Glibert and H.M. Skelton. 2008. Mixotrophy, a major mode of nutrition for harmful algal species in eutrophic waters. *Harmful Algae* 8: 77–93.

Burkholder, J.M., C.A. Kinder, D.A. Dickey, et al. 2022. Classic indicators and diel dissolved oxygen vs. trend analysis in assessing eutrophication of potable-water reservoirs. *Ecological Applications* 32:e2541.

Camargo, J.A. and A. Alonso. 2006. Ecological and toxicological effects of inorganic nitrogen pollution in aquatic ecosystems: a global assessment. *Environment International* 32:831–849.

Camargo, J.A. and J.V. Ward. 1995. Nitrate ($NO_3^-$-N) toxicity to aquatic life: a proposal of safe concentrations for two species of Nearctic freshwater invertebrates. *Chemosphere* 31:3211–3216.

Camargo, J.A., A. Alonso and A. Salamanca. 2005. Nitrate toxicity to aquatic animals: a review with new data for freshwater invertebrates. *Chemosphere* 58:1255–1267.

Canfield, D.E., A.N. Glazer and P.G. Falkowski. 2010. The evolution and future of Earth's nitrogen cycle. *Science* 330:192–196.

CEN. 2011. Urea's unlikely role. Chemical & Engineering News June 2011:32.

Chudoba, E.A., M.A. Mallin, L.B. Cahoon and S.A. Skrabal. 2013. Stimulation of fecal bacteria in ambient waters by experimental inputs of organic and inorganic phosphorus. *Water Research* 47:3455–3466.

Clark, G.M., D.K. Mueller and M.A. Mast. 2000. Nutrient concentrations and yields in undeveloped stream basins of the United States. *Journal of the American Water Resources Association* 36:849–860.

Conley, D.J., C.L. Schelske and E.F. Stoermer. 1993. Modification of the biogeochemical cycle of silica with eutrophication. *Marine Ecology Progress Series* 101: 179–192.

Conley, D.J., H.W. Paerl, R.W. Howarth, et al. 2009. Controlling eutrophication: nitrogen and phosphorus. *Science* 323:1014–1015.

Cotner, J.B., R.H. Sada, H. Bootsma, T. Johengen, J.F. Cavaletto and W.S. Gardner. 2000. Nutrient limitation of bacteria in Florida Bay. *Estuaries* 23:611–620.

Dafner, E.V., M.A. Mallin, J.J. Souza, H.A. Wells and D.C. Parsons. 2007. Nitrogen and phosphorus species in the coastal and shelf waters of southeastern North Carolina, Mid-Atlantic U.S. coast. *Marine Chemistry* 103: 289–303.

Dale, O.R., C. Tobias and B. Song. 2009. Biogeographical distribution of diverse anaerobic ammonium oxidizing (anammox) bacteria in the Cape Fear River Estuary. *Environmental Microbiology* 11:1194–1207.

Daniel, T.C., A.N. Sharpley and J.L. Lemunyon. 1998. Agricultural phosphorus and eutrophication: a symposium overview. *Journal of Environmental Quality* 27: 251–257.

D'Elia, C.F., D.M. Nelson and W.R. Boynton. 1983. Chesapeake Bay nutrient and plankton dynamics: III. The annual cycle of dissolved silicon. *Geochimica et Cosmochimica Acta* 47:1945–1955.

D'Elia, C.F., J.G. Sanders and W.R. Boynton. 1986. Nutrient enrichment studies in a Coastal Plain estuary: phytoplankton growth in large-scale, continuous cultures. *Canadian Journal of Fisheries and Aquatic Sciences* 43: 397–406.

Dodds, W.K. and V.H. Smith. 2016. Nitrogen, phosphorus, and eutrophication in streams. *Inland Waters* 6:155–164.

Dodds, W.K., J.R. Jones and E.B. Welch. 1998. Suggested classification of stream trophic state: distributions of temperate stream types by chlorophyll, total nitrogen, and phosphorus. *Water Research* 32:1455–1462.

Doering, P.H., C.A. Oviatt, L.L. Beatty, et al. 1989. Structure and function in a model coastal ecosystem: silicon, the benthos and eutrophication. *Marine Ecology Progress Series* 52:287–299

Edwards, T.M., L.J. Guillette Jr., K. McCoy and T. Barbeau. 2004. Effects of Nitrate/Nitrite on Two Sentinel Species Associated with Florida's Springs. Final Report. Florida Department of Environmental Protection, Tallahassee.

Elser, J.J., T. Anderson, J.S. Baron, et al. 2009. Shifts in lake N:P stoichiometry and nutrient limitation driven by atmospheric nitrogen deposition. *Science* 326:835–837.

Elwood, J.W., J.D. Newbold, A.F. Trimble and R.W. Stark. 1981. The limiting role of phosphorus in a woodland stream ecosystem: effects of P enrichment on leaf decomposition and primary producers. *Ecology* 62: 146–158.

Emerson, K., R.C. Russo, R.E. Lund and R.V. Thurston. 1975. Aqueous ammonia equilibrium calculations: effects of pH and temperature. *Journal of the Fisheries Research Board of Canada* 32:2379–2383.

Ensign, S.H. and M.W. Doyle. 2005. In-channel transient storage and associated nutrient retention: evidence from experimental manipulations. *Limnology and Oceanography* 50:1740–1751.

Ensign, S.H. and M.W. Doyle. 2006. Nutrient spiraling in streams and river networks. *Journal of Geophysical Research* 111:G04009.

Ensign, S.H., M.F. Piehler and M.W. Doyle. 2008. Riparian zone denitrification affects nitrogen flux through a tidal freshwater river. *Biogeochemistry* 91:133–150.

Falkowski, P., R.J. Scholes, E. Boyle, et al. 2000. The global carbon cycle: a test of our knowledge of Earth as a system. *Science* 290:291–296.

Fan, A.M. and V.E. Steinberg. 1996. Health implications of nitrate and nitrite in drinking water: an update on methemoglobinemia occurrence and reproductive and developmental toxicity. *Regulatory Toxicology and Pharmacology* 23:35–43.

Fisher, T.R., E.R. Peele, J.W. Ammerman and L.W. Harding Jr. 1992. Nutrient limitation of phytoplankton in Chesapeake Bay. *Marine Ecology Progress Series* 82:51–63.

Forsberg, C., O. Savchuk and L. Rydén. 2003. A new regime for nutrient turnover—eutrophication. Pages 256–293 in Ryden, L., P. Migula, M. Andersson and M. Lehman (eds.), *Environmental Science: Understanding, Protecting, and Managing the Environment in the Baltic Sea Region.* Baltic University Programme, Uppsala.

Froelich, P.N. 1988. Kinetic control of dissolved phosphate in natural rivers and estuaries: a primer on the phosphate buffer mechanism. *Limnology and Oceanography* 33:649–668.

Galloway, J.N., A.R. Townsend, J.W. Erisman, et al. 2008. Transformation of the nitrogen cycle: recent trends, questions and potential solutions. *Science* 320:889–892.

Glibert, P.M. 2020. From hogs to HABs: impacts of industrial farming in the US on nitrogen and phosphorus and greenhouse gas production. *Biogeochemistry* 150: 139–180.

Glibert, P.M. J. Harrison, C. Heil and S. Seitzinger. 2006. Escalating worldwide use of urea—a global change contributing to coastal eutrophication. *Biogeochemistry* 77:441–463

Glibert, P.M., D, Fullerton, J.M. Burkholder, J.C. Cornwell and T.M. Kana. 2011. Ecological stoichiometry, biogeochemical cycling, and aquatic food webs: San Francisco Estuary and comparative systems. *Reviews in Fisheries Science* 19:358–417.

Glibert, P.M., R. Manager, D.J. Sobota and L. Bouwman. 2014. The Haber–Bosch-harmful algal bloom (HB-HAB) link. *Environmental Research Letters* 9:105001.

Glibert, P.M., F.P. Wilkerson, R.C. Dugdale, et al. 2016. Pluses and minuses of ammonium and nitrate uptake and assimilation by phytoplankton and implications for productivity and community composition, with emphasis on nitrogen-enriched conditions. *Limnology and Oceanography* 61:165–197.

Goolsby, D.A. and W.A. Battaglin. 2001. Long-term changes in concentrations and flux of nitrogen in the Mississippi River Basin, USA. *Hydrological Processes* 15:1209–1226.

Graham, L.E., J.M. Graham, L.W. Wilcox and M.E. Cook. 2016. *Alga*e, 3rd edition. LJLM Press, Madison, WI.

Guillette, L.J. Jr. and T.M. Edwards. 2005. Is nitrate an ecologically relevant endocrine disruptor in vertebrates? *Integrative Comparative Biology* 45:19–27.

Hagy, J.D., W.R. Boynton, C.W. Keefe and K.V. Wood. 2004. Hypoxia in Chesapeake Bay, 1950–2001: long-term changes in relation to nutrient loading and river flow. *Estuaries* 27:634–658.

Hecky, R.E. and P. Kilham. 1988. Nutrient limitation of phytoplankton in freshwater and marine environments: a review of recent evidence on the effects of enrichment. *Limnology and Oceanography* 33:796–822.

Hecnar, S.J. 1995. Acute and chronic toxicity of ammonium nitrate fertilizer to amphibians from southern Ontario. *Environmental Toxicology and Chemistry* 14:2131–2137.

Houser, J.N. and W.B. Richardson (2010) Nitrogen and phosphorus in the Upper Mississippi River: transport, processing, and effects on the river ecosystem. *Hydrobiologia* 640:71–88.

Houser, J.N., D.W. Bierman, R.M. Burdis and L.A. Soeken-Gittinger. 2010. Longitudinal trends and discontinuities in nutrients, chlorophyll, and suspended solids in the Upper Mississippi River: implications for transport, processing, and export by large rivers. *Hydrobiologia* 651:127–144.

Howarth, R.W. and R. Marino. 2006. Nitrogen is the limiting nutrient for eutrophication in coastal marine ecosystems: evolving views over three decades. *Limnology and Oceanography* 51:364–376.

Howarth, R.W., E.W. Boyer, W.J. Pabich and J.N. Galloway. 2002. Nitrogen use in the United States from 1961–2000 and potential future trends. *AMBIO* 31:88–96.

Howarth, R.W., D.P. Swaney, E.W. Boyer, R. Marino, N. Jaworski and C. Goodale. 2006. The influence of climate on average nitrogen export from large watersheds in the Northeastern United States. *Biogeochemistry* 79:163–186.

Humborg, C., D.J. Conley, L. Rahm, F. Wulff, A. Cociasu and V. Ittekkot. 2000. Silicon retention in river basins: far-reaching effects on biogeochemistry and aquatic food webs in coastal marine environments. *AMBIO* 29:45–50.

Inwood, S.E., J.L. Tank and M.J. Bernot. 2007. Factors controlling sediment denitrification in Midwestern streams of varying land use. *Microbial Ecology* 53:247–258.

Jetten, M.S.M., L. van Niftrik, M. Strous, B. Kartal, J.T. Keltjens and H.J.M. Op den Camp. 2009. Biochemistry and molecular biology of anammox bacteria. *Critical Reviews in Biochemistry & Molecular Biology* 44:65–84.

Johnson, C.J. and B.C. Kross. 1990. Continuing importance of nitrate contamination of groundwater and wells in rural areas. *American Journal of Industrial Medicine* 18:449–456.

Jones, J.R., M.F. Knowlton, D.V. Obrecht and J.L. Graham. 2011. Temperature and oxygen in Missouri reservoirs. *Lake and Reservoir Management* 27:173–182.

Justić, D., N.N. Rabalais and R.E. Turner. 1995. Stoichiometric nutrient balance and origin of coastal eutrophication. *Marine Pollution Bulletin* 30:41–46.

Kawaguchi, T., A.J. Lewitus, C.M. Aelion and H.N. McKellar. 1997. Can urbanization limit iron availability to estuarine algae? *Journal of Experimental Marine Biology and Ecology* 213:53–69.

Keeney, D. 1986. Sources of nitrate to ground water. *CRC Critical Reviews in Environmental Control* 16:257–304.

Kirchman, D.L. 1994. The uptake of inorganic nutrients by heterotrophic bacteria. *Microbial Ecology* 28:255–271.

Knowlton, M.F. and J.R. Jones. 2000. Seston, light, nutrients and chlorophyll in the lower Missouri River, 1994–1998. *Journal of Freshwater Ecology* 15:283–297.

Korum, S.F. 1992. Natural denitrification in the saturated zone; a review. *Water Resources Research* 28:1657–1668.

Larsson, C.M. 1994. Responses of the nitrate uptake system to external nitrate availability: a whole-plant perspective. Pages 31–45 in Roy, J. and E. Garnier (eds.), *A Whole Plant Perspective on Carbon–Nitrogen Interactions*. SPB Academic Publications, The Hague.

Lebo, M. 1990. Phosphate uptake along a Coastal Plain estuary. *Limnology and Oceanography* 35:1279–1289.

Lebo, M.E. and J.H. Sharpe. 1993. Distribution of phosphorus along the Delaware, and urbanized Coastal Plain estuary. *Estuaries* 16:290–301.

Lehman, P.W., J. Sevier, J. Giulianoti and M. Johnson. 2004. Sources of oxygen demand in the lower San Joaquin River, California. *Estuaries* 27:405–418.

Lewitus, A.J., T. Kawaguchi, G.R. DiTullio and J.D.M. Keesee. 2004. Iron limitation of phytoplankton in an urbanized vs. forested U.S. salt marsh estuary. *Journal of Experimental Marine Biology and Ecology* 298: 233–254.

Mallin, M.A. and L.B. Cahoon. 2020. The hidden impacts of phosphorus pollution to streams and rivers. *BioScience* 70:315–329.

Mallin, M.A., H.W. Paerl, J. Rudek and P.W. Bates. 1993. Regulation of estuarine primary production by rainfall and river flow. *Marine Ecology-Progress Series* 93: 199–203.

Mallin, M.A., L.B. Cahoon, M.R. McIver, D.C. Parsons and G.C. Shank. 1999. Alternation of factors limiting phytoplankton production in the Cape Fear Estuary. *Estuaries* 22:985–996.

Mallin, M.A., M.R. McIver, S.H. Ensign and L.B. Cahoon. 2004. Photosynthetic and heterotrophic impacts of nutrient loading to blackwater streams. *Ecological Applications* 14:823–838.

Mallin, M.A., M.R. McIver, A.R. Robuck and A.K. Dickens. 2015. Industrial swine and poultry production causes chronic nutrient and fecal microbial stream pollution. *Water, Air and Soil Pollution* 226:1–13.

Meyer, J.L. 1992. Seasonal patterns of water quality in blackwater rivers of the Coastal Plain, southeastern United States. Pages 249–276 in Becker, C.D. and D.A. Neitzel (eds.), *Water Quality in North American River Systems*. Battelle Press, Columbus, OH.

Miltner, R.J. and E.T. Rankin. 1998. Primary nutrients and the biotic integrity of rivers and streams. *Freshwater Biology* 40:145–158.

Mitsch, W.J., J.W. Day Jr., J.W. Gilliam, et al. 2001. Reducing nitrogen loading to the Gulf of Mexico from the Mississippi River Basin: strategies to counter a persistent ecological problem. *BioScience* 51:373–388.

Mulholland, P.J. 1992. Regulation of nutrient concentrations in a temperate forest stream: roles of upland, riparian and instream processes. *Limnology and Oceanography* 37:1512–1526.

Mulholland, P.J. A.M. Helton, G.C. Poole, et al., 2008. Stream denitrification across biomes and its response to anthropogenic nitrate loading. *Nature* 452:202–205.

Newbold, J.D. 1992. Cycles and spirals of nutrients. Chapter 18 in Calow, P. and G.E. Petts (eds.), *The Rivers Handbook: Hydrological and Ecological Principles*. Blackwell Scientific, Oxford.

Null, K.A., D.R. Corbett, D.J. DeMaster, J.M. Burkholder, C.J. Thomas and R.E. Reed. 2011. Porewater advection of ammonium into the Neuse River Estuary. *Estuarine, Coastal and Shelf Science* 95:314–325.

Orton, F., J.A. Carr and R.D. Handy. 2006. Effects of nitrate and atrazine on larval development and sexual differentiation in the northern leopard frog *Rana pipiens*. *Environmental Toxicology and Chemistry* 25:65–71.

Osborne, T.Z., R.A. Mattson and M.F. Coveney. 2017. Potential for direct nitrate–nitrite inhibition of submersed aquatic vegetation (SAV) in Florida springs:

a review and synthesis of current literature. *Water* 8: 30–46.

Padgett, D.E., M.A. Mallin and L.B. Cahoon. 2000. Evaluating the use of ergosterol as a bioindicator for assessing water quality. *Environmental Monitoring and Assessment* 65:547–558.

Paerl, H.W., J. Rudek and M.A. Mallin. 1990. Stimulation of phytoplankton productivity in coastal waters by natural rainfall inputs; nutritional and trophic implications. *Marine Biology* 107:247–254.

Paulsen, S.G., A. Mayio, D.V. Peck, et al. 2008. Condition of stream ecosystems in the US: an overview of the first national assessment. *Journal of the North American Benthological Society* 27:812–821.

Peterson, B.J., W.M. Wollheim, P.J. Mulholland, et al. 2001. Control of nitrogen export by headwater streams. *Science* 292:86–90.

Redfield, A.C. 1958. The biological control of chemical factors in the environment. *American Scientist* 46:205–222.

Rivas, R., E. Velázquez, A. Willems, et al. 2002. A new species of *Devosia* that forms a unique nitrogen-fixing root-nodule symbiosis with the aquatic legume Neptunia natans (L.f.) druce. *Applied and Environmental Microbiology* 68:5217–5222.

Rouse, J.D., C.A. Bishop and J. Struger. 1999. Nitrogen pollution: an assessment of its threat to amphibian survival. *Environmental Health Perspectives* 107:799–803.

Rudek, J., H.W. Paerl, M.A. Mallin and P.W. Bates. 1991. Seasonal and hydrological control of phytoplankton nutrient limitation in the Neuse River Estuary, North Carolina. *Marine Ecology-Progress Series* 75: 133–142.

Russo, R.C. 1985. Ammonia, nitrite, and nitrate. Pages 455–471 in Rand, G.M. and S.R. Petrocelli (eds.), *Fundamentals of Aquatic Toxicology: Methods and Applications*. Hemisphere Publishing Group, Washington, DC.

Ryther, J.H. and W.M. Dunstan. 1971. Nitrogen, phosphorus, and eutrophication in the coastal marine environment. *Science* 171:1008–1013.

Salisbury, F.B. and C.W. Ross. 1992. *Plant Physiology*, 4th edition. Wadsworth Publishing, Belmont, CA.

Sanger, D., A. Blair, G. DiDonato, et al. 2015. Impacts of coastal development on the ecology of tidal creek ecosystems of the U.S. southeast including consequences to humans. *Estuaries and Coasts* 38(Suppl. 1): 49–66.

Schelske, C.L. and E.F. Stoermer. 1971. Eutrophication, silica depletion, and predicted changes in algal quality in Lake Michigan. *Science* 173:423–424.

Schelske, C.L., E.F. Stoermer and W.F. Kenney. 2006. Historic low-level phosphorus enrichment in the Great Lakes inferred from biogenic silica accumulation in sediments. *Limnology and Oceanography* 51:728–748.

Schindler, D.E., M.D. Scheuerell, J.W. Moore, S.M. Gende, T.B. Francis and W.J. Palen. 2003. Pacific salmon and the ecology of coastal ecosystems. *Frontiers in Ecology and the Environment* 1:31–37.

Schindler, D.W. and E.J. Fee. 1973. Experimental Lakes Area: whole-lake experiments in eutrophication. *Journal of the Fisheries Research Board of Canada* 31: 937–953.

Schlesinger, W.H. and E.S. Bernhardt. 2013. *Biogeochemistry: An Analysis of Global Change*. Academic Press, New York.

Schoelynck, J., K. Bal, H. Backx, T. Okruszko, P. Meire and E. Struyf. 2010. Silica uptake in aquatic and wetland macrophytes: a strategic choice between silica, lignin and cellulose? *New Phytologist* 186:385–391.

Seitzinger, S., J.A. Harrison, J.K. Bohlke, et al. 2006. Denitrification across landscapes and waterscapes: a synthesis. *Ecological Applications* 16:2064–2090.

Sharpley, A.N., T.C. Daniel and D.R. Edwards. 1993. Phosphorus movement in the landscape. *Journal of Production Agriculture* 6:453–500.

Sheath, R.G. and J.M. Burkholder. 1988. Stream macroalgae. Pages 53–59 in Sheath, R.G. and M.M. Harlin (eds.), *Freshwater and Marine Plants of Rhode Island*. Kendall Hunt, Dubuque, IA.

Sims, J.T., R.R. Simard and B.C. Joern. 1998. Phosphorus losses in agricultural drainage: historical perspective and current research. *Journal of Environmental Quality* 27:277–293

Smayda, T.J. 1989. Primary production and the global epidemic of phytoplankton blooms in the sea: a linkage? Pages 449–484 in Cosper, E.M., V.M. Bricelj and E.J. Carpenter (eds.), *Novel Phytoplankton Blooms*. Coastal and Estuarine Studies No. 35. Springer, New York.

Smith, R.A., R.B. Alexander and G.E. Schwarz. 2003. Natural background concentrations of nutrients in streams and rivers of the conterminous United States. *Environmental Science and Technology* 37:3039–3047.

Song, B., M.A. Mallin, A. Long and M.R. McIver. 2014. *Factors Controlling Microbial Nitrogen Removal Efficacy in Constructed Stormwater Wetlands*. Report No. 443. UNC Water Resources Research Institute, Raleigh, NC.

Stanley, E.H. and J.T. Maxted. 2008. Changes in dissolved nitrogen pool across land cover gradients in Wisconsin streams. *Ecological Applications* 18:1579–1590.

Sterner, R.W. and J.J. Elser. 2002. *Ecological Stoichiometry: The Biology of Elements from Molecules to the Biosphere*. Princeton University Press, Princeton, NJ.

Stoddard, J.L., J. Van Sickle, A.T. Herlihy, et al. 2016. Continental-scale increase in lake and stream phosphorus: are oligotrophic systems disappearing in the United States? *Environmental Science and Technology* 50: 3409–3415.

Sundareshwar, P.V., J.T. Morris, E.K. Koepfler and B. Forwalt. 2003. Phosphorus limitation of coastal ecosystem processes. *Science* 299:563–565.

Tan, X., G. Yuan, H. Fu, et al. 2019. Effects of ammonium pulse on the growth of three submerged macrophytes. *PLoS ONE* 14(7):e0219161.

Thompson, S.K. and J.B. Cotner. 2018. Bioavailability of dissolved organic phosphorus in temperate lakes. *Frontiers in Environmental Science* 6:62.

Turner, R.E., N.N. Rabalais, D. Justić and Q. Dortch. 2003. Global patterns of dissolved N, P and Si in large rivers. *Biogeochemistry* 64:297–317.

Turpin, D.H. (1991) Effects of inorganic N availability on algal photosynthesis and carbon metabolism. *Journal of Phycology* 27:14–20.

US EPA. 2000a. Nutrient Criteria Technical Guidance Manual: Lakes and Reservoirs. Report EPA-822-B00-001. Office of Water and Office of Science and Technology, US EPA, Washington, DC.

US EPA. 2000b. Nutrient Criteria Technical Guidance Manual: Rivers and Streams. Report EPA-822-B00-002. Office of Water and Office of Science and Technology, US EPA, Washington, DC.

US EPA. 2007a. Biological Nutrient Removal Processes and Costs. Fact Sheet EPA-823-R-07-002. Office of Water, US EPA, Washington, DC.

US EPA. 2007b. Advanced Wastewater Treatment to Achieve Low Concentration of Phosphorus. Report EPA 910-R-07-002. Office of Water & Watersheds, EPA Region 10, Seattle, WA.

US EPA. 2013. Aquatic Life Ambient Water Quality Criteria for Ammonia—Freshwater 2013. Report EPA 822-R-18-002. Office of Water, US EPA, Washington, DC.

US EPA. 2014. Office of Water. Website, http://www2.epa.gov/sites/production/files/2014-08/documents/criteria-nutrient-ecoregions-sumtable.pdf

Vanni, M.J. 2002. Nutrient cycling by animals in freshwater ecosystems. *Annual Review of Ecology and Systematics* 33:341–370.

Wakida, F.T. and D.N. Lerner. 2005. Non-agricultural sources of groundwater nitrate: a review and case study. *Water Research* 39:3–16.

Walker, J.T., V.P. Aneja and D.A. Dickey. 2000. Atmospheric transport and wet deposition of ammonium in North Carolina. *Atmospheric Environment* 34:3407–3418.

Wawrik, B., A.V. Callaghan and D.A. Bronk. 2009. Use of inorganic and organic nitrogen by *Synechococcus* spp. and diatoms on the West Florida Shelf as measured using stable isotope probing. *Applied and Environmental Microbiology* 75:6662–6670.

Webster, J. 1975. Analysis of potassium and calcium dynamics in stream ecosystems on three Southern Appalachian watersheds of contrasting vegetation. PhD dissertation, University of Georgia, Athens.

Wetzel, R.G. 2001. *Limnology: Lake and River Ecosystems*, 3rd edition. Academic Press, San Diego.

Willey, J.D. and L.B. Cahoon. 1991. Enhancement of chlorophyll *a* production in Gulf Stream surface seawater by rainwater nitrate. *Marine Chemistry* 34:63–75.

# CHAPTER 3

# Lotic Primary Producers

Phytoplankton and Periphyton

This book recognizes four major functional groups of primary producers in streams and rivers: phytoplankton (suspended microalgae living in the water column), periphyton (including benthic microalgae), macroalgae, and macrophytes. For the purposes of this book the phytoplankton community consists of both potamoplankton (microalgae living and reproducing in the water column) and tychoplankton (benthic microalgae swept into the current and entrained in the water column for a time). Phytoplankton and periphyton are discussed in this chapter, while macroalgae and macrophytes are discussed in Chapter 4.

Filamentous forms of algae are difficult to categorize because they can be part of the phytoplankton and suspended in the water column or they can be associated with the benthos; they can be both microscopic and macroscopic. Thus we will follow Lapointe et al.'s (2018) approach, wherein macroalgae were considered as algal forms visible to the human eye, with the additional caveat that they are greater in length than what is generally considered as maximal periphyton biofilm thickness (2 cm; e.g. Burkholder 1996 and references therein).

## 3.1 Phytoplankton

Energy to support the organisms living within or using a stream or river can come from two sources. Energy produced within the stream is derived from photosynthesis performed by primary producers, including algae, bryophytes (mosses and lichens), and vascular plants. This is sometimes termed autochthonous production or autotrophy. The second energy source is from materials that enter the stream from the surrounding terrestrial system, referred to as allochthonous production. That material is considered in Chapters 5 and 6. This chapter and Chapter 4 focus on the primary producers as defined above.

As noted, tychoplankton are microalgae that have been moved into the water column from benthic habitats by disturbance (Hynes 1970; Whitton 1975). This dislodging and suspension can occur through disturbance from water motion or fauna, or when benthic microalgal photosynthesis creates oxygen bubbles that lift the algal cells into the overlying water (Whitton 1975). Potamoplankton and tychoplankton are part of seston, or suspended material. Often in rapidly flowing smaller rivers and streams, the suspended microalgae usually do not have time to reproduce before they are moved downstream. By contrast, larger, slowly moving rivers generally have true phytoplankton, able to reproduce before they are transported from the system by water motion, together with the tychoplankton displaced from benthic habitats.

Phytoplankton drifting or floating in the water are unable to control their position when confronting wind and current (Wetzel 2001). As such, the physical conditions of the stream or river determine whether or not the phytoplankton are abundant versus only a minor portion of the primary producers. Many of these algae have benthic life history stages such as cysts or resting cells that are used to survive harsh seasonal conditions. They can be solitary cells, filaments, or colonial forms (see Section 3.3), and they sometimes are subdivided into size groups including picoplankton (major dimension < 2.0 μm), nanoplankton (2–20 μm),

*River Ecology.* Michael A. Mallin, Oxford University Press. © Michael A. Mallin (2023). DOI: 10.1093/oso/9780199549511.003.0004

microplankton (20–200 μm), and macroplankton (> 200 μm) (Bellinger and Sigee 2010).

Riverine phytoplankton often have high species diversity, partly because of their various source waters, which may include tributary streams that drain farm ponds and wet detention ponds, drainage ditches serving urban or agricultural areas, overland runoff carrying algal cells during storms, oxbow lakes that are periodically filled by flooding rivers, or algal cells carried into rivers by birds or mammals using lentic water sources elsewhere (Reynolds and Descy 1996). The river or stream offers various habitats such as turbulent and quiescent waters, shallow and deep areas, the main channel, or side channels and backwater areas.

### 3.1.1 Productivity and Biomass

#### 3.1.1.1 Productivity

Primary production is defined as the synthesis of organic matter by organisms with plant biochemistry, usually over long-term (e.g. annual) periods (Wetzel 2001). Productivity is defined as the short-term (e.g. hours) *rate* of primary production via photosynthesis. Photosynthesis is the metabolic process by which algae and plants use light energy, electrons from water, and the "universal plant pigment" chlorophyll *a* to convert carbon dioxide into carbohydrates, with oxygen also produced as a byproduct:

$$6CO_2 + 12H_2O \xrightarrow{\text{Light energy}} C_6H_{12}O_6 + 6H_2O + 6O_2$$

Since photosynthesis requires energy from sunlight, it is a daytime phenomenon except that the "dark reactions" of this process can occur for a short period after sunset. Respiration (R) in aerobic (oxygen-containing) waters is metabolically the opposite of primary production in that it consumes oxygen while it breaks down organic matter, leaving water, $CO_2$, and energy as products. Respiration occurs both day and night; the general equation at its simplest is:

$$C_6H_{12}O_6 + 6O_2 = 6CO_2 = 6H_2O + ATP$$

Algal productivity in a steam or river can be estimated from oxygen production or carbon consumption over time, on either a volumetric basis (e.g. mg

$C/m^3/h$) or a surface-area basis (mg $C/m^2/h$) if considering algae in benthic habitats. It is not measured for individual algal cells but, rather, for a known volume of water which includes bacteria, fungi, and some zooplankton (which consume oxygen through respiration), as well as algal cells. Algae produce oxygen but also consume some oxygen through respiration. The combined amount of oxygen produced and respired is referred to as gross primary production (GPP). Net primary production (NPP), that is, what is left for the algae to use, is defined as the GPP minus respiration.

Respiration occurs during both day and night, and when short-term dissolved oxygen "bottle experiments" are performed researchers usually simplistically assume similar respiration rates during day and night. Because stream water includes many other microorganisms in addition to microalgae, the oxygen measurements include respiration from the total microbial community; that is, they are not sufficiently fine-tuned to focus only on the microalgae. Thus, the ratio of GPP/R (or more commonly production/respiration [P/R]) is used to describe the overall metabolism of the microbial community in the water column. When P/R ratio is 1, photosynthesis equals respiration. A P/R ratio greater than 1 has been interpreted to indicate a net autotrophic system, while a P/R ratio less than 1 has been interpreted as a net heterotrophic system (Odum 1956; Vannote et al. 1980)—but note that these interpretations can be in error because they represent only an instantaneous snapshot of the system and, in particular, do not account for high productivity of phytoplankton and periphyton (see below) that is rapidly removed by grazing (e.g. regarding phytoplankton see Sellers and Bukaveckas 2003 and references therein).

The most common experimental techniques for measuring phytoplankton productivity and respiration over short periods (hours) include measuring carbon-14 ($^{14}C$) uptake (as labeled carbon dioxide or, more commonly, bicarbonate), or dissolved oxygen changes in light and dark bottles over time. Another technique involves measurements of dissolved oxygen by in-situ instruments to assess water-column stream metabolism. Recommended texts for detail on these and various other methods include Shearer et al. (1985), Wetzel and Likens

(1991), Hauer and Lamberti (1996), and Boyd (2000). Phytoplankton primary production and productivity are both descriptive and comparative measures, and there are large databases of these parameters worldwide. In rivers, phytoplankton productivity can vary considerably according to stream order (see Chapter 6) and is influenced by shading, depth, currents, turbidity, water color, and latitude, among many other factors (e.g. Vannote et al. 1980).

### 3.1.1.1.1 *The Light Regime*

Light is a major factor influencing algal and plant productivity in rivers and streams (Fig. 3.1). In relatively quiet waters such as slowly flowing rivers, UV radiation can be strong near the water surface and can lead to surface inhibition of photosynthesis (exceptions to this will be discussed subsequently). Recall from Chapter 1 how irradiance is attenuated through the water column, either absorbed or reflected. Depending upon the phytoplankton group, at some point in the water column an optimal depth for maximum photosynthesis ($P_{max}$) occurs,

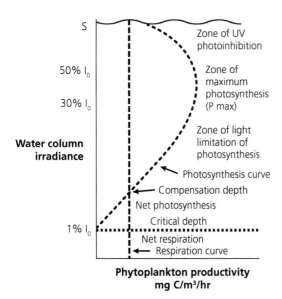

**Fig. 3.1** Idealized water-column photosynthesis vs. respiration depending on the penetrating solar irradiance in quiescent water, showing surface inhibition by UV radiation, the approximate depth coinciding with $P_{max}$, and light limitation at depth. (Figure by the author.)

often at a depth with 30–50% of surface radiation ($I_0$). Light continues to attenuate as the water depth increases, and it becomes a limiting factor for phytoplankton growth. The critical depth (about 1% of $I_0$) is defined as the depth where water-column photosynthesis equals respiration. This is often considered to be the depth limit for photosynthesis, although many algae such as diatoms and cyanobacteria can survive in even more light-attenuated situations (Tilzer 1987; Cahoon 1999). Simplistically, above the critical depth there is net water-column photosynthesis; below that depth is net water-column respiration. In reality, however, the situation becomes much more complex depending on the phytoplankton species characteristics (pigments, motility, light optima, etc.), depth, currents, mixing, temperature, turbidity, and water color (see Sections 3.1.3). The compensation depth is where the rate of photosynthesis equals the rate of respiration.

### 3.1.1.2 Biomass

Biomass of the suspended microalgal assemblage (sometimes referred to as *sestonic* algae) has been determined as dry weight of filtered material, but that is a very rough measurement because the sample will also include zooplankton and abiotic small particulates, and detritus (dead remains of algae and other microbes). Instead, microalgal biomass is often estimated using the chlorophyll *a* content as an indicator, since this essential pigment for photosynthesis is found in all photosynthetic algae. Corrected chlorophyll, or chlorophyll *a* measurements corrected to subtract pheophytin (decomposing chlorophyll from dead cells that can erroneously overestimate the chlorophyll within living cells), can be measured by fluorometry, spectrometry, or high-performance liquid chromatography. There is a vast amount of comparative chlorophyll *a* data available from rivers and streams all over the world, and many governments and regulatory agencies use chlorophyll *a* as a standard water quality parameter to assess the health of water bodies. Additionally, chlorophyll *a* has been used in models to predict potential elevated levels of algal toxin production (Yuan et al. 2014). Chlorophyll *a* as a metric is problematic, however, because cell content varies considerably among species, and even according to time of day within the same species; and because methods are not well standardized from laboratory to laboratory.

## 3.1.2 Major Taxonomic Groups of Stream and River Phytoplankton

The most abundant suspended microalgae in temperate river systems are the cyanobacteria, diatoms, and green algae; other common taxa include cryptomonads, dinoflagellates, chrysophyceans, and euglenoids (Lizotte and Simmons 1985; Reynolds and Descy 1996; Wehr and Thorp 1997; Wehr and Descy 1998). Much of the information below is taken from Graham et al. (2016) and references therein. The phytoplankton assemblages of rivers and reservoirs have been analyzed by numerous techniques, depending upon the goal. Species identification used to be accomplished by light microscopy with emphasis on cell morphology, scanning electron microscopy was used to identify some taxa, and transmission electron microscopy was needed to confirm identifications of many small flagellates. The advent of molecular science in the 1990s dramatically changed algal taxonomy, and molecular techniques increasingly are used for identifications (such methods require pure cultures for accuracy to avoid contamination by other taxa). Note that for practical purposes such as rapid identification of algal blooms needed by water treatment plants or beach managers, molecular techniques are usually too time consuming or impractical due to shipping to a university laboratory, and taxonomic expertise is required.

Because phytoplankton cells come in numerous shapes and sizes—sometimes within the same population, such as diatoms—biovolume is another measure used; i.e. literature biovolumes by taxa or body shape are multiplied by cell counts (see Wetzel 2001, for examples). The following key ecological information concerns microalgal groups that are common in streams and rivers, along with examples. As taxonomic schemes vary according to texts and naming is in constant flux, these groups are considered in groupings at the phylum (algae) or division (cyanobacteria) level. More detailed recent taxonomic information can be found in phycological journals (Bellinger and Sigee 2010; Wehr et al. 2015).

### 3.1.2.1 Cyanobacteria

The name cyanobacteria is under the Bacteriological Code of Nomenclature, and these organisms, formerly known as blue-green algae (phylum Cyanophyta under the Botanical Code of Nomenclature), are now generally included there because they are prokaryotes, although it has been argued that in their ecology they function almost entirely as primary producers (Lewin 1976; Golubic 1979). This widespread and highly adaptable group is claimed by phycologists and bacteriologists, the latter because they are prokaryotic and have many traits in common with bacteria, *with one major, extremely important exception*: cyanobacteria have higher plant photosynthetic biochemistry and they generally function in aquatic ecosystems as primary producers, as do the phototrophic and mixotrophic eukaryotic algae. These organisms have a very simple cell structure without organelles such as a nucleus or chloroplasts, yet they have the "universal plant pigment" chlorophyll *a* and the same basic photosynthetic biochemistry as higher plants.

Other important pigments that help harness light for chlorophyll *a* in photosynthesis are blue and red pigments called phycobilins (phycobiliproteins including blue phycocyanins and red phycoerythrins; Graham et al. 2016 and see below). Phycobilins are called accessory pigments because they can capture green light wavelengths and fluoresce the light at a much lower energy, the energy of red light, to chlorophyll *a* molecules for use in photosynthesis. Chlorophyll *a* is green; thus it reflects green light wavelengths and cannot use green light for photosynthesis. Instead, it absorbs red and blue light wavelengths for use in this important process. Greenish and greenish-yellow light is often referred to as the "green window" in aquatic systems because chlorophyll *a* cannot use it in photosynthesis. As depth increases in rivers—often, just a few centimeters below the water surface in turbid systems—the only light wavelengths available after light passes through the upper water column are greenish and greenish-yellow. Thus, phycobilin pigments give cyanobacteria a major advantage in turbid, low-light environments characteristic of many streams, rivers, and run-of-river impoundments.

The morphology of cyanobacteria is also simplistic: they occur as colonial forms in mucilage—the most notorious toxic bloom formers in freshwaters are *Microcystis* (Fig. 3.2a), as filaments including *Aphanizomenon*, *Fischerella*, and *Anabaena* (Fig. 3.2b; also called *Dolichospermum*), and as coccoid solitary or colonial forms (e.g. *Chroococcus*; *Gloeothece* [Burkholder 2002; Gobler et al. 2016 and references therein]). In temperate regions cyanobacteria are often especially abundant in nutrient overenriched waters during summer. Cyanobacteria have many representatives in the phytoplankton assemblage as well as among the benthic microalgae (see Chapter 4). This group is known for many adaptations that enhance survival and periodic dominance (see Whitton and Potts 2000 for a detailed review of cyanobacteria physiology and ecology).

Although cyanobacteria lack flagella, in calm waters of impoundments and larger rivers with dependable external pressure, some species can vertically migrate to optimal depths in the water column for light and nutrients. This is accomplished by the use of gas vacuoles, comprised of groups of balloon-like structures called gas vesicles that are impermeable to water but permeable to gases such as nitrogen gas. During the day, many cyanobacterial species in the phytoplankton photosynthesize near the water surface. The increasing carbohydrates from photosynthesis increase the internal cell turgor pressure, collapsing gas vesicles, so that the cells begin to sink as the light wanes. In the evening and night, they can be found at deeper depths where nutrient supplies are greater from decomposition processes. The cells take in nitrogen gas, which expands the gas vesicles, and the cells begin to rise again in the water column so that they are near the surface again by dawn (Reynolds and Walsby 1975).

When blooms of cyanobacteria occur at or near the water surface (Fig. 3.2c) they can competitively exclude other algae through shading. This adaptation also allows cyanobacteria to survive turbidity pulses that severely reduce available light. Other means of cyanobacterial survival under turbid conditions include colony fragmentation and temporary adsorption to particles (e.g. Burkholder 1992). Cyanobacteria are adept at thriving in

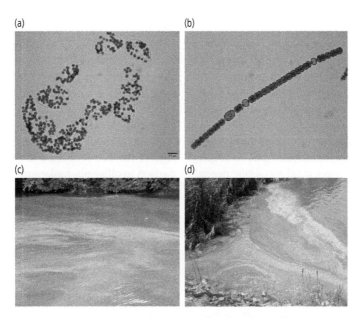

**Fig. 3.2**  Cyanobacteria. Fig. 3.2a (top left) *Microcycstis* colony within a gelatinous matrix. Fig. 3.2b (top right) *Anabaena* (also called *Dolichospermum*) filament showing round heterocytes (sites of N fixation) and oblong akinetes (resting cells). Fig. 3.2c (bottom left) Surface toxic *Microcystis* bloom in turbid Cape Fear River, NC; maximal total microcystins concentration was ~390 μg/L. Fig. 3.2d (bottom right) Decaying cyanobacterial bloom in Sutton Reservoir (which receives makeup water from Cape Fear River) showing pigmented bands common to decaying blooms. (*Microcystis* and *Anabaena* photos courtesy of Amy E. Grogan, University of North Carolina [UNC] Wilmington; Cape Fear River bloom photo courtesy of Matthew McIver, UNC Wilmington; decaying cyanobacterial bloom photo by the author.)

low-light environments because of their phycobilin pigments, which capture green light wavelengths and fluoresce the light at a much lower energy, the energy of red light, to chlorophyll *a* molecules, just as fucoxanthin can do in diatoms (explained below). In fact, phycobilins—especially the red pigment, phycoerythrin—are excellent at absorbing light in the "green window." Because of these accessory phycobilin pigments, cyanobacteria can do well in very low light regimes, unlike many other algae, giving them a competitive edge.

During the decay of a surface cyanobacterial bloom, bands of several colors are often seen in the surface biomass (Fig. 3.2d). An important adaptation used by cyanobacteria (and by other algae such as euglenoids—see below) to survive in harsh, high-UV light environments at the water surface is to produce more carotenoids for photoprotection. The carotenoids absorb UV light, and also protect from destructive photooxidation by consuming oxygen. In that process, carotenoids combine with oxygen and form xanthophylls. Experiments have also shown that the carotenoids can increase chlorophyll *a*-specific photosynthetic $O_2$ production (Paerl 1984).

Another key physiological trait that helps cyanobacteria to proliferate is the ability of various species to "fix" inert nitrogen gas into a form of inorganic nitrogen, ammonium, which can be readily used in cell metabolism. That is, such cyanobacteria are diazotrophs (see N cycle in Chapter 2). No other algal group is capable of making inorganic nitrogen from nitrogen gas. Nitrogen gas cannot be used to meet algal metabolic needs for nitrogen; only inorganic nitrogen (ammonia or its ionized form, ammonium, and nitrate) and certain forms of organic nitrogen such as urea can be used. Thus, in situations where useable forms of nitrogen supplies are limiting but phosphorus is abundant, these cyanobacteria basically can make more useable nitrogen on their own. In certain filamentous blue-green algae (*Anabaena*, *Aphanizomenon*, *Nostoc*), specialized cells called **heterocytes** (formerly heterocysts) are the locations where nitrogen fixation occurs (Graham et al. 2016) (Fig. 3.2b). Heterocytes have thick walls to protect the main enzyme involved in nitrogen fixation, called nitrogenase, which cannot function in the

presence of oxygen. Thus, heterocytes provide the low-oxygen environment needed for nitrogen fixation to occur. Some taxa (e.g. *Aphanizomenon*—Carlton and Paerl 1989) have thick clumps of filaments, so that the interior areas of the clumps have little or no oxygen and nitrogen fixation can occur there. Another thick-walled cell found on various filaments of some taxa is called an **akinete**, which is a resting structure that allows the cell to survive unfavorable conditions for months to years (Fig. 3.2b).

While some cyanobacteria can be safely consumed by selected zooplankton taxa, they are generally considered a poor food source for zooplankton, other invertebrates, and herbivorous fish, for several reasons: they have gelatinous sheaths that are hard to digest (Fig. 3.2a); some make chemical poisons or toxins that can hurt or kill the grazers; and many have generally poor food quality (Arnold 1971; de Bernardi and Guissani 1990). Regarding toxins, many taxa are toxigenic (potentially toxic), including various species of common genera such as *Anabaena*, *Aphanizomenon*, *Cylindrospermopsis*, *Microcystis*, *Nostoc*, *Oscillatoria*, and *Planktothrix*. Thus, not all of the populations or strains within each of these species are toxic; rather, it means that *some* populations are capable of making cyanobacteria toxins (cyanotoxins). Within the same species, even the same bloom, some cells can be highly toxic, while others have minimal or negligible cyanotoxins.

Cyanobacterial abundance in many geographic regions has been reported to be increasing in comparison to other phytoplankton and are expected to continue to increase (Huisman et al. 2018) due to favorable factors such as warming trends in climate change and escalating nutrient pollution. Various toxigenic, high-biomass, bloom-forming cyanobacteria historically have been widely regarded as the most noxious, undesirable algae of freshwater ecosystems. Their toxic outbreaks have been described by observers at least since the late 1800s (Gorham and Carmichael 1980; Carmichael 1994; Burkholder et al. 2018 and references therein). Since then, many more incidents have been recorded of wild and domesticated animals such as cattle, horses, dogs, and birds dying after being exposed to cyanobacterial blooms, especially in warmer months. There are several major groups of cyan-otoxins, considering their target organ(s). Neuro-toxins such as anatoxin-A damage the nervous system in mammals and birds (Ibelings and Havens 2008); there are an array of potent dermatotoxins and cytotoxins; hepatotoxins such as microcystins (MCs) damage the liver and gastrointestinal tract in mammals, birds, fish, etc.; MCs can also occur as irritants to the skin, mucous membranes, and respiratory system (Falconer 1989; Carmichael 1994; Ibelings and Havens 2008; van der Merwe 2012). More than a hundred known forms (congeners) of MCs have been characterized (Meriluoto and Spoof 2008); at the time of writing, at least 246 MC variants are known, and the number is expected to continue to rise in this fertile research area for chemists (Diez-Quijada et al. 2019). There are no known antidotes to cyanotoxins. Various hypotheses have been suggested as to why cyanobacteria produce toxins; they may be protective metabolites against grazing, for example. Yet, cyanobacterial bloom toxins have caused sublethal illness and death to "nontarget" organisms such as cows and humans (Carmichael et al. 2001; Burkholder et al. 2010 and references therein).

### 3.1.2.2 Diatoms

Diatoms (phylum Heterokontophyta, class Bacillariophyceae) are credited with contributing a remarkable 25% of all of the primary productivity in the world on an annual basis (Graham et al. 2016). They are the major primary producers seasonally in running waters and reservoirs. Photosynthetic pigments include chlorophylls $a$ and $c_2$ (usually also with chlorophyll $c_1$ or $c_3$), and fucoxanthin which predominates, usually making them golden-brown in color. Fucoxanthin, an accessory pigment, is adept at capturing light in the "green window" and fluoresces the energy as red light to chlorophyll $a$ for direct use in photosynthesis. Similarly as for cyanobacteria with their phycobilins, because of fucoxanthin, diatoms can do well in very low light regimes unlike many other algae, giving them a competitive edge.

Diatoms (Figs 3.3 and 3.4) are unusual in that their cell walls, called frustules, are made of beautifully ornamented silica (silicon dioxide, literally glass), and diatoms have a very high silica requirement. The siliceous structures and ornamentation are obscured in living cells by an overlying membrane (Fig. 3.3a, c). The frustule in side view (Fig. 3.4c) resembles the side view of a petri dish; it consists of an overlapping top half called the epivalve and a lower half called the hypovalve. A diatom cell looks very different depending on whether the view is from the top (valve view) or the side (girdle view); the girdle view (Fig. 3.4c) makes identification particularly hard. The two halves of the frustule are held together with organic bands of materials called girdle bands. The pores in the silica walls (punctae) that allow chemical communication with

(a)  (b)  (c)

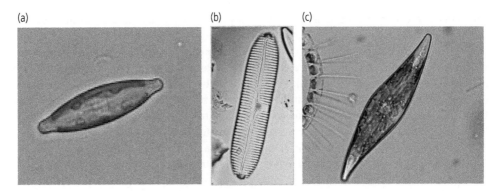

**Fig. 3.3** Pennate diatoms. Fig. 3.3a (left) Common pennate diatom *Navicula*. Fig. 3.3b (center) *Pinnularia*, showing striae and raphe. Fig. 3.3c (right) Pennate diatom *Pleurosigma*. (Photos: *Navicula* by the author; *Pleurosigma* by Amy E. Grogan, UNC Wilmington; *Pinnularia* by JoAnn M. Burkholder, North Carolina State University.)

(a)    (b)

(c)    (d)

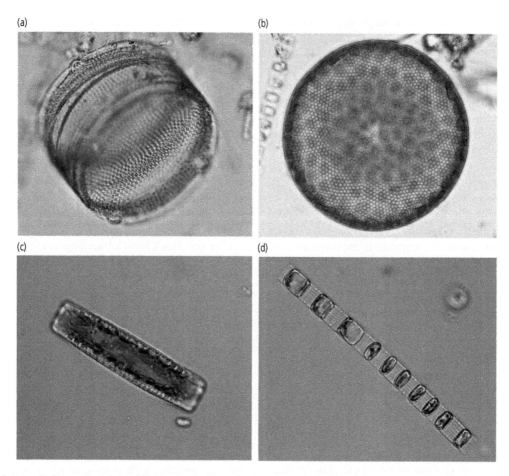

**Fig. 3.4** Centric diatoms. Fig. 3.4a (top left) Empty centric diatom showing punctae. Fig. 3.4b (top right) Valve (top) view of centric diatom *Coscinodiscus*. Fig. 3.4c (bottom left) Girdle (side) view of centric diatom. Fig. 3.4d (bottom right) Chain-forming centric diatom *Skeletonema*, extremely common and abundant in riverine estuaries. (Photos courtesy of Amy E. Grogan, UNC Wilmington.)

the external environment are arranged in patterns (striae) used for species identifications; or there can be regularly arranged silica deposits.

The most common diatom shapes in valve view are pennate (Fig. 3.3) and centric (Fig. 3.4). Pennate diatoms are generally elongated and bilaterally symmetrical; many of them appear boat-shaped in valve (top) view. Examples include solitary cells such as *Navicula* (Fig. 3.3a) and *Pinnularia* (Fig. 3.3b) and colonial forms such as some species of *Tabellaria* and *Fragilaria*. They are common among the phytoplankton, but generally more characteristic of periphyton (see Section 3.2). As mentioned, benthic diatoms are readily swept up into overlying waters

by currents or other disturbance and become part of the suspended assemblage (Wehr and Thorp 1997; Beaver et al. 2015). A crack or fissure in the siliceous frustule, called a raphe (Fig. 3.3b), occurs in some pennate diatoms. Secretion of mucus through the raphe by the cell enables the cells to move, often fairly rapidly, over surfaces, despite the fact that pennate diatoms have no motile organelles such as flagella.

Centric diatoms (Fig. 3.4) are generally circular in valve view (Fig. 3.4a, b), cylindrical in girdle view (Fig. 3.4c), and radially symmetrical like a petri dish, as mentioned. They have no motile organelles, except that their sperm cells each have one flagellum. Centric taxa may be solitary cells such as *Cyclotella* and *Stephanodiscus*, connected in chains like *Skeletonema* (Fig. 3.4d), or colonial forms such as *Aulacoseira*. They tend to be more common in the phytoplankton than in the periphyton. In fact, changes in the centric diatoms to pennate diatoms ratio in the sediment record of slowly moving rivers, lakes, and estuaries have been used to track eutrophication over time in those water bodies. Centric diatoms are less abundant in nutrient-poor waters, but as nutrient overenrichment occurs, they become much more abundant and reduce the available light for benthic pennate diatoms. Thus, dramatic increases in the centrics to pennates ratio over time can indicate the period when eutrophication became more pronounced, e.g. with European settlement of the Upper Midwestern USA (Engstrom et al. 2009).

Diatoms tend to be cold-optimal and predominate during late winter–early spring in temperate waters when nutrients are abundant in stormwater or snowmelt runoff from urban and agricultural sources. Diatoms are excellent competitors for nutrients such as nitrogen and phosphorus when silica is abundant (Sommer 1988). They generally are good food for zooplankton and other grazers (Porter 1977; Willen 1991). There are no known toxic species in freshwaters. Note that most diatom species in estuarine/marine waters are also benign, but some strains (populations) of certain species within the genus *Pseudo-nitzschia* and close relatives can produce a potent neurotoxin, domoic acid, which is harmful to waterbirds such as pelicans and mammals including humans. Toxigenic species of *Pseudo-nitzschia* have been increasing in some polluted areas, such as at the mouth of the Mississippi River. In that area, sediment records have strongly correlated their increased abundance with increasing nitrate pollution and decreasing silica to nitrate ratios in Mississippi River water (Parsons et al. 2002).

### 3.1.2.3 Greens

The phyla Chlorophyta and Streptophyta, the largest, most diverse algal group, are commonly referred to as **green algae** and comprise two separate phyla, as indicated above. Their photosynthetic pigments include chlorophylls *a* and *b*, and most apparently lack accessory pigments other than chlorophyll *b* that can transfer light to chlorophyll *a* for photosynthesis. Planktonic green microalgae can occur as unicells (e.g. *Cosmarium*, *Staurastrum*, *Pediastrum* [Fig. 3.5a, b, d]) or as colonial forms (e.g. *Volvox* [Fig. 3.5d]). They may or may not have flagella. Many green algal species are filamentous and commonly form noxious blooms in freshwater rivers. Their biomass can increase to such an extent, often in response to nutrient pollution such as sewage, that they are sometimes considered as macroalgae rather than microalgae (Lapointe et al. 2018 and references therein), so they are discussed in Chapter 4.

Green algae that do not have thick, tough gelatinous sheaths can be good food for zooplankton and various other fauna. However, those encased in such outer mucilaginous coverings are poorly grazed by zooplankton or, if consumed, some cells can survive passage through the consumer due to the sheath and even absorb nutrients within the grazer digestive track, stimulating growth after being excreted (Porter 1977). Freshwater green microalgae do not produce toxins but can form high-biomass, noxious blooms.

### 3.1.2.4 Cryptophytes

Cryptophytes (cryptomonads; phylum Cryptophyta) are unicells bearing two uneven flagella, with a gullet visible in some species (Fig. 3.6). As indicated by their name ("crypto," meaning cryptic or secret, hidden), cryptophytes are often overlooked but can be abundant in the water column, often near the sediments of streams and rivers. Pigments in phototrophic taxa (cryptophytes) include chlorophyll *a* and $c_2$, α and β carotene, alloxanthin, diadinoxanthin, and several phycobiliproteins. Some species are colorless and heterotrophic; some are mixotrophic and consume bacteria (phagotrophy), or dissolved organics (osmotrophy). This group can form cysts to survive poor environmental conditions. Cryptomonads are generally considered excellent food for aquatic microfauna such as zooplankton (Knisely and Geller 1986; Kerfoot et al. 1988) and for some mixotrophic and heterotrophic dinoflagellates (Burkholder et al. 2008; Carty and Parrow 2015).

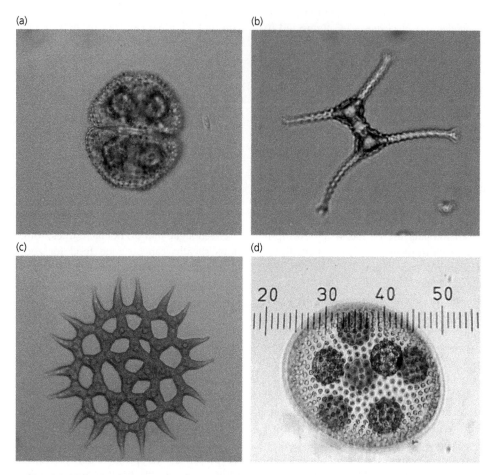

**Fig. 3.5** Green algae. Fig. 3.5a (top left) *Cosmarium*, Fig. 3.5b (top right) *Staurastrum*, Fig. 3.5c (bottom left) *Pediastrum*, Fig. 3.5d (bottom right) *Volvox* colony from a bloom on the Northeast Cape Fear River, North Carolina, July 2019. (Photos courtesy of Amy E. Grogan, UNC Wilmington.)

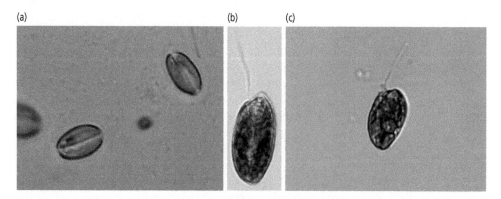

**Fig. 3.6** Cryptomonads. Fig. 3.6a (left) Cryptomonads from a Southeastern USA impoundment showing gullet. Fig. 3.6b (middle) *Cryptomonas* sp. showing gullet and flagella. Fig. 3.6c (right) *Chroomonas* with flagella. (Photo Fig. 3.6a by the author; Fig. 3.6b, c courtesy of Mark Vander Borgh, NC Department of Environmental Quality.)

### 3.1.2.5 Euglenoids

Euglenoids (phylum Euglenophyta) generally are motile unicells (Fig. 3.7) characterized by a single emergent flagellum and a red eyespot (Graham et al. 2016). Their cell covering, just beneath the cell membrane, is called a pellicle, which consists of flat strips of protein wound helically around the cell from anterior to posterior. The pellicle can be flexible and allows the cell to greatly change shape. Some euglenoids such as *Trachelomonas* also form a rigid mucilaginous shell called a lorica, often containing ferric hydroxide or manganese compounds that impart an orange, brown, or black color. Euglenoids can be green phototrophs or mixotrophs (phagotrophic on bacterial prey), or colorless, obligate heterotrophs. Photosynthetic pigments include chlorophyll *a* and *b*; others include β-carotene, neoxanthin, and diadinoxanthin. The three major euglenoid genera are *Euglena*, *Phacus*, and *Trachelomonas* (Fig. 3.7).

Euglenoids are widespread in freshwaters and can also be found on beaches and within saturated sediments moving among the sand grains. They can vertically migrate through sediments, apparently for nutrient acquisition and also in response to circadian rhythms in light availability. They especially

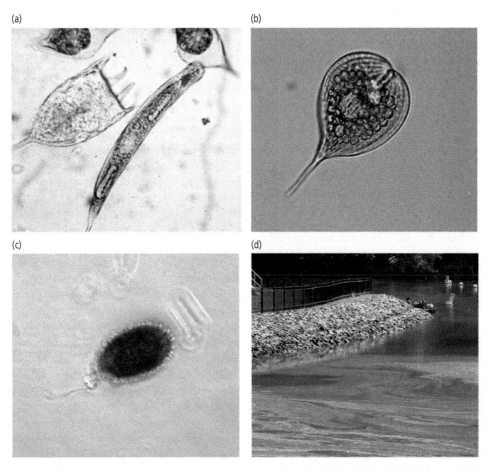

(a)  (b)  (c)  (d)

**Fig. 3.7** Euglenoids. Fig. 3.7a (top left) *Euglena*, with red eyespot and other euglenophytes above, next to *Keratella*, a rotifer (see Chapters 5 and 11) to the left. Fig. 3.7b (top right) *Phacus*, another major euglenoid genus. Fig. 3.7c (bottom left) *Trachelomonas*, the third major euglenoid genus, often showing an orange or brown coloration. Fig. 3.7d (bottom right) Red-colored euglenoid bloom mixing with a cyanobacterial bloom in a North Carolina reservoir. (*Euglena* and *Trachelomonas* photos by the author; *Phacus* by Amy E. Grogan, UNC Wilmington; euglenoid bloom photo courtesy of Elle Allen, North Carolina State University.)

thrive in waters with organic enrichment and are generally considered to be pollution-tolerant (Box 3.1). Strains of various common euglenoid species (*Euglena sanguinea*, *E. socialis*, and *E. stellata*; *Euglenaria anabaena* and *E. clavata*; *Lepocinclus acus*; *Strombomonas borysteniensis*; and *Trachelomonas ellipsodalis*) recently have been found to produce a potent ichthyotoxin, euglenophycin, that is lethal to multiple tested fish species (Zimba et al. 2010, 2017).

---

### Box 3.1 The versatility and persistence of euglenoids

Euglenoids thrive in areas with dissolved organic enrichment and they can be highly tolerant to pollution and other stressors. This author's first exposure to these organisms was during an undergraduate phycology field trip in southeastern Ohio, where we observed a stream with a pH of 2 that flowed from a defunct coal mine. I commented on the appearance of a green "lawn" on the stream bottom to the professor, Dr. Herb Graffius of Ohio University. He said "Mallin, collect some of that, I'll bet it's *Euglena*." I did, and it was. Many years later my laboratory was investigating a 3.8-million-L (1-million-gallon) swine waste lagoon spill into a series of North Carolina blackwater streams. We sampled a site for phytoplankton near the spill location, where snails had crawled out of the polluted waters to die along the shoreline. The chlorophyll *a* concentration at that site was 100 µg/L and *Euglena* was abundant (Mallin et al. 1997). Note that many euglenoid blooms have been missed because when stressed, the populations often throw off their flagella and form palmelloid (gelatinous) masses and sink down to the sediments (Zimba et al. 2017).

---

### 3.1.2.6 Dinoflagellates

Dinoflagellates (phylum Dinophyta) are especially challenging with respect to general classification because about half of the known species are photosynthetic (phototrophic) (Fig. 3.8), whereas the other half are animal-like heterotrophs. Thus, they are claimed by both phycologists and protozoologists. Moreover, most of the photosynthetic species have well-developed heterotrophic capabilities in nutrient acquisition. The use of both phototrophy and heterotrophy by a cell is referred to as mixotrophy, and many if not most photosynthetic

dinoflagellates are mixotrophs (Burkholder et al. 2008; Carty and Parrow 2015).

Although they exist in a wide array of forms, most are considered to be biflagellate unicells (Carty and Parrow 2015). Structurally, they are usually characterized by an epitheca and a hypotheca (upper and lower areas of the cell, respectively), separated by a horizontal groove around the cell called a cingulum (Fig. 3.8a, b). A second groove from the cingulum down into the hypotheca on the cell ventral surface is called the sulcus (Fig. 3.8b). A helically constructed transverse flagellum lies within the cingulum, which causes the cell to spin. A longitudinal flagellum (Fig. 3.8b) occurs in the sulcus and serves as a rudder when the cell swims.

Photosynthetic dinoflagellates exhibit diel vertical migration wherein they swim down to deeper depths in the water column at night where nutrient concentrations are higher during the night, and move to shallower depths during the day where light is optimal for photosynthesis (Pollingher 1988). Dinoflagellates have an outer covering called an amphiesma, which consists of several layers of membranes that may or may not include deposits of cellulose called plates. If plates are present in vesicles beneath the outer membrane, the dinoflagellate is referred to as thecate and "armored;" if not, it is referred to as "naked." Some species, such as *Ceratium*, have elaborate armored projections called "horns," also used in taxonomy (Fig. 3.8a).

Dinoflagellates can form short-term temporary cysts when stressed (Burkholder 1992) or longer-term, highly resistant cysts (Carty and Parrow 2015). Phototrophic cells contain the green photosynthetic pigments chlorophyll *a* and chlorophyll *c*, but the cells are often golden-brown in color because of a predominance of xanthophylls such as peridinin.

Dinoflagellates obtain nutrition in a remarkable number of ways and can be photosynthetic, mixotrophic (usually via phagotrophy), or heterotrophic; occasionally all three modes of nutrition can occur within strains of the same species. An organelle called the peduncle is present in some species and is emitted from a longitudinal, elastic canal or groove (the sulcus). The peduncle is involved in acquiring nutrition by suctioning the contents from various other organisms as prey. Some species can form a protoplasmic feeding net

(a)                                                         (b)

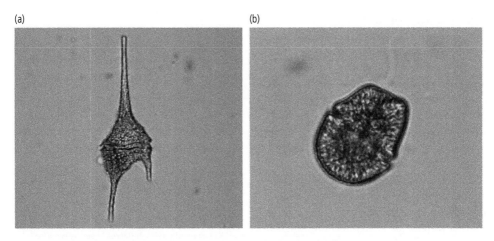

**Fig. 3.8** Dinoflagellates. Fig. 3.8a (left) *Peridinium*, a common dinoflagellate; this genus has horns that show a great variety of forms, depending upon species. Fig. 3.8b (right) Gymnodinoid dinoflagellate showing epitheca, hypotheca, cingulum, sulcus, and longitudinal flagella. (Photos courtesy of Amy E. Grogan, UNC Wilmington.)

or veil called a pallium to collect bacteria and other minute organic particles, or to surround much larger prey and secrete enzymes which dissolve the prey contents as food. These phenomena are much better studied from estuarine and marine (e.g. Jacobsen and Anderson 1986) than freshwater species (e.g. Burkholder 1992; Carty and Parrow 2015).

Another trait that is well developed in about 60 estuarine/marine dinoflagellate species, but has not yet been verified in freshwater species, is the production of potent toxins that can cause major food web impacts as well as human disease and death (Burkholder 1998; Burkholder and Marshall 2012; Shumway et al. 2018). Note that toxins vary greatly in potency, making their distinction from other allelochemicals a "gray area," not well defined (Burkholder et al. 2018). Allelochemicals are defined as substances produced by one species that elicit physiological or behavioral responses in other species (Dicke and Sabelis 1988; Ianora et al. 2011). With these caveats in mind, it can be said that so far in the published science literature, one dinoflagellate species, *Peridinium aciculiferum*, has been described to be capable of producing "a toxic algicidal substance(s)" which allows it to outcompete other phytoplankton (World Health Organization [WHO] 2003).

### 3.1.2.7 Haptophytes

Haptophytes (phylum Haptophyta) in freshwaters are mostly known from softwater lakes (Nicholls 1995), and their ecology in streams and rivers is poorly understood.

Haptophytes are morphologically diverse, but, absent information about most morphs, they usually are considered as small biflagellate photosynthetic unicells. Although they are often difficult to discern or identify under light microscopy, it is thought that the most common species in freshwaters, including streams and rivers, are very small (sometimes only ~3–5 μm in diameter), oval biflagellate phototrophs, *Chrysochromulina* spp., with a signature organelle, the haptonema, which protrudes between the two flagella. The haptonema can be used for bacterial prey acquisition in mixotrophy, as well as for attachment (Nicholls 2003). Photosynthetic pigments include chlorophylls *a* and $c_2$, and the accessory pigment fucoxanthin and derivatives. Haptophytes are known for their coverings of one to several layers of scales, often ornamented, including organic scales; more rarely, they have outer siliceous scales and/or calcareous scales. The scale characteristics are helpful diagnostics in identifications.

Freshwater haptophytes include phototrophs, mixotrophs, and heterotrophs, although their nutrition has mostly been studied in estuarine and marine waters rather than in freshwaters. Given their small size, rapid growth, digestibility, and nutritional content, most haptophytes are regarded as high-quality food for zooplankton (Graham et al. 2009). There are no known toxic freshwater species. Beneficial effects of estuarine haptophytes as food for shellfish (*Isochrysis galbana*) in aquaculture operations and for production of omega-3 fatty acids (e.g. *Pavlova lutheri*) have not been explored for freshwater taxa, although *Pavlova* spp. are known to occur in freshwaters (Simon et al. 2012). One species in particular, *Prymnesium parvum* (nicknamed the "golden alga"), first known from brackish and marine waters, has come into prominence in some eutrophic inland brackish or high-specific-conductance rivers and reservoirs in the Western USA (Van Landeghem et al. 2013)—but not in a good way (Box 3.2). Their toxic blooms turn the water a golden color (Fig. 3.9) and they have caused massive kills of gill-breathing organisms.

---

## Box 3.2 *Prymnesium parvum*—a recent deadly toxic bloom-former to the USA

The haptophyte *Prymnesium parvum* entered the collective consciousness of phycologists, fishery scientists, and managers in 1985, when it began causing huge fish kills in saline Texas reservoirs and rivers. In this region natural conditions and brines associated with oil production produce saline water bodies, and blooms mainly occur in salinities between 1 and 12 ppt. *P. parvum* can sustain itself through both photosynthesis and heterotrophy, and releases exotoxins to the waters it impacts that deter or kill bacteria, some zooplankton, and other algae to gain a competitive advantage. Toxic blooms generally occur above densities > 50,000 cells/ml (but can be far more dense) and color the water with a yellow, copper, or golden hue, hence its common name "golden algae." At a minimum, such blooms in Texas have caused deaths of over 34 million fish, with economic losses in the tens of millions of dollars. Massive blooms and kills of this species are becoming more common in western states such as in the Colorado River basin.

Whereas some researchers believe that *P. parvum* may have existed in the Southern USA for a long time and only recently found proper conditions to bloom, other researchers believe it is an introduced species; indeed there is genetic evidence of similarities of various strains

to populations in several European nations, indicating it may have arrived in multiple invasions. Regardless, its rapid movement through the southwest was exacerbated by drought and increased salinities in inland waters in the late twentieth century, as well as interbasin transfer of waters (see Chapter 10). As noted, blooms of this organism have been especially prominent in southern and southwestern states, although it has been reported as far north as Washington State. A truly unusual tribute to this fish-killer's adaptability was a massive 2009 bloom and subsequent fish, salamander, and mussel kill in Dunkard Creek, in the mountains of West Virginia and Pennsylvania. In this location mining waste has changed the normally fresh conditions to brackish levels that could support the bloom. How *P. parvum* itself was carried there (by persons, mining equipment, or other means) is still debatable. Some progress has occurred in controlling *P. parvum* blooms in ponds and aquaculture situations, but control of such blooms in reservoirs and rivers is truly challenging.

(Compiled from Reynolds 2009; Weber and Janik 2010; Roelke et al. 2011; 2016; Flood and Burkholder 2018.)

### 3.1.2.8 Chrysophyceans and Synurophyceans

Chrysophyceans and synurophyceans (phylum Heterokontophyta, classes Chrysophyceae and Synurophyceae) are sometimes referred to as golden-brown algae. They are a morphologically diverse group, although they are considered mostly as solitary, biflagellate unicells (Fig. 3.10a) or colonies (Fig. 3.10b, c) of biflagellate cells (Graham et al. 2016). Their photosynthetic pigments include chlorophylls *a* and *c* and the accessory pigment fucoxanthin. Chrysophyceans include both phototrophs and heterotrophs, with mixotrophy also well represented. Many taxa have an elaborate covering of organic scales, and siliceous resting cysts (stomatocysts) are also common in both subgroups. Chrysophyceans and synurophyceans can be either unicells (e.g. *Chrysamoeba, Mallomonas*) (Fig. 3.10a) or colonial forms (e.g. *Dinobryon, Synura*) (Fig. 3.10b, c). These algae are nearly all freshwater inhabitants, mostly in mildly acidic softwaters low in nutrients, although some species occur in eutrophic alkaline waters. Little is known about their ecology in rivers (Nicholls and Wujek 2015; Siver 2015). Chrysophyceans and synurophyceans are often most common in temperate streams and rivers during late winter/early spring, and some

**Fig. 3.9** Example of a fish kill caused by *Prymnesium parvum* (Wichita, TX, February 2009). Note the golden color of the water. Inset: Toxic *Prymnesium parvum* cell, showing the golden color and characteristic haptonema sticking straight up between the two flagella (scale bar = 10 μm). (Fish kill photo courtesy of Paul Zimba, Texas A&M, Corpus Christi, TX; inset photo by Stacie L. Flood in Flood (2017), with permission.)

(a)                              (b)                              (c)

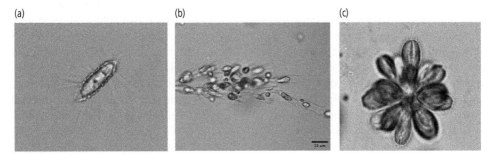

**Fig. 3.10** Chrysophytes. Fig. 3.10a (left) *Mallomonas*. Fig. 3.10b (center) *Dinobryon* colony. Fig. 3.10c (right) *Synura* colony. (Photos courtesy of Amy E. Grogan, UNC Wilmington.)

species impart a "fish" odor and cause taste-and-odor problems for potable water treatment plants during periods of high abundance (Watson 2004).

## 3.1.3 Controls on Phytoplankton in Running Waters

The primary controls on riverine abundance and species composition of phytoplankton assemblages are water temperature, current, light, and nutrient concentrations and ratios. These factors are all interlinked, however, and several themes arise

regarding the productivity and diversity of riverine phytoplankton.

### 3.1.3.1 Water Temperature

One of the most basic physical parameters controlling the growth of phytoplankton in streams and rivers is water temperature. As examples, this parameter has been positively correlated with phytoplankton productivity in larger rivers such as the Potomac River (Jones 1998), with chlorophyll *a* in the Missouri River (Knowlton and Jones 2000), and with phytoplankton cell density in the Ohio

River (Wehr and Thorp 1997). Water temperature also shapes phytoplankton assemblage structure. Researchers studying a variety of large temperate rivers have described a phytoplankton assemblage dominated by diatoms in spring and by cyanobacteria and chlorophytes in summer (Lizotte and Simmons 1985; Wehr and Thorp 1997). Another major factor controlling the abundance and diversity of suspended microalgae is solar irradiance, which generally increases seasonally along with water temperature. As shown below, water-column solar irradiance is, in many cases, also dependent upon water current and depth. Current velocity, a component of river discharge, tends to be inversely related to temperature because during winter there is little evapotranspiration (evaporation plus plant transpiration; see Chapter 1) and watershed runoff is maximal. In summer, watershed plant biomass is maximal, so evapotranspiration is highest, reducing watershed runoff.

### 3.1.3.2 Currents and Mixing

Current speed alone can shape the assemblage structure of phytoplankton. The short water residence time selects for fast-growing cells able to cope with varying light conditions (Wehr and Descy 1998). River conditions tend to select for diatoms that are capable of rapid growth, a high surface-to-volume ratio, and tolerance of variable light conditions and accessory pigments to aid in that purpose. Slower flows and enhanced irradiance allows for other taxa such as chlorophyceans, cryptomonads, and euglenoids to survive and even thrive.

Elevated river discharge increases suspended sediment loads and turbidity (see Chapter 1). With the increase in turbidity comes stronger light attenuation within the water column. The increased light attenuation may only leave a very shallow zone near the water surface (the euphotic zone) that provides enough light for microalgal production. While productivity decreases in high discharge periods, water-column (sestonic) microalgal biomass still can be relatively high. That is because concurrently with the increase in light attenuation, the increased discharge has the effect of entraining more benthic microalgae into the upper water column (Wehr and Thorp 1997; Beaver et al. 2013, 2015). In addition, it has been suggested that rainfall/flooding periods wash a considerable amount of algae from drainage ditches and agricultural fields into the river channels (Knowlton and Jones 2000). In a selection of 31 Canadian rivers, Basu and Pick (1996) found no significant relationship between water residence time and chlorophyll $a$ and suggested that the suspended microalgal flora were dominated by small picoplankton and nanoplankton which have rapid doubling times that are unaffected by high flow. The impact of elevated river discharge on the light field and abundance and productivity of suspended microalgae can be described as follows:

$$\uparrow \text{Current} \Rightarrow \uparrow \text{Turbidity} \Rightarrow \downarrow \text{Irradiance}$$
$$\Rightarrow \downarrow \text{Potamoplankton} \Rightarrow \uparrow \text{Tychoplankton}$$

Current effects on turbidity are thus described above as an important constraint on irradiance. However, in lowland or Coastal Plain situations the rivers are less turbid, but irradiance can be severely constrained by water color produced by dissolved organic material (DOM) (see Chapter 8). River discharge impacts water color too, however. This is because in lowland river systems DOM is derived from riparian swamp forests and drains into the river channels during elevated rainfall/runoff periods. In summer there is a general reduction in light attenuation in lowland rivers due to reduced inputs from riparian areas (Meyer 1990) as well as breakdown of DOM by UV radiation (Wetzel 2001).

The slowing of current (reduction in river discharge) increases water residence time in a given river segment. Increases in water residence time occur seasonally as a general summer decrease in discharge, as described above, or by man-made structures impounding and slowing sections of river, thereby creating a more lake-like effect. The slower currents allow for increased suspended particle sedimentation (see Chapter 1), so that more light becomes available. Koch et al. (2004) conducted a comparative study among the Ohio (regulated by low dams) and the Tennessee and Cumberland Rivers (regulated by high dams). In the heavily regulated Tennessee and Cumberland rivers, the high dams reduce discharge and subsequent turbidity and allow sufficient light penetration to cause significant growth of phytoplankton. Thus, with increased light, phytoplankton production in the

Cumberland River was often P limited and in the Tennessee River it was co-limited by P and N. By contrast, in the Ohio River, the low dams do not significantly reduce current velocity, so phytoplankton growth is usually light limited by turbidity particles entrained by the currents. Current speed also regulated microalgal entrainment into the upper water column; in a taxonomic study of the phytoplankton of the Ohio River, Wehr and Thorp (1997) found that resuspended benthic diatom taxa were least common near the dams where water currents were slowed. Hence, the relationship between current reduction, lower light attenuation, and increased phytoplankton production can be generalized as follows:

$\downarrow$ Current $\Rightarrow \downarrow$ Turbidity $\Rightarrow \uparrow$ Irradiance

$\Rightarrow \uparrow$ Potamoplankton $\Rightarrow \downarrow$ Tychoplankton

Researchers can gain an idea of the propensity of a river or stream toward net autotrophy or net heterotrophy by considering information on the light field and determining light attenuation $k$ on various dates, also knowing the depth of a reach and having some idea of current (which drives vertical mixing). Exercise 3.1 walks the reader through some useful computations with data from two river systems separated by about 150 km.

**Exercise 3.1 Light, mixing, photosynthesis, and respiration: two contrasting situations.** *An illustration of how current-induced mixing, water depth, and light attenuation interact to determine the degree of phytoplankton productivity in riverine environments.*

*The lower Neuse River/upper Neuse Estuary of Eastern North Carolina is an environment that is well mixed by current and tides (Fig. 3.11). Its bottom depth (z) = 3.4 m and the light attenuation coefficient k = 1.1/m in this reach (Mallin and Paerl 1992). For the purposes of this exercise let us assume that surface irradiance (I$_0$) = 1500 µE/m$^2$/s. From Equation 1.3, irradiance at a given depth z (I$_z$) = I$_0$ × e$^{-kz}$ . Thus,*

$$I_z = 1{,}500\,\mu E/m^2/s \times e^{-\,(1.1/m)(3.4m)} = 35.6\,\mu E/m^2/s$$

*This is the average irradiance reaching the Neuse River bottom, at 3.4 m depth.*

*The percent of incident solar irradiance reaching the bottom depth of 3.4 m is thus computed as (I$_z$/I$_0$) or (35.6 µE/m$^2$/s/1500 µE/m$^2$/s) = 2.4% I$_0$. 2.4% is greater than our theoretical photosynthesis limit of 1% I$_0$; thus the entire water column in this stretch of the Neuse River is within the euphotic zone, and P/R > 1.0, which provides a conducive environment for a net photosynthetic or net autotrophic situation.*

*Now consider the Northeast Cape Fear River (Fig. 3.11), a fifth-order lowland blackwater river in eastern North Carolina that is also well mixed by current and tides, but much deeper (z = 9 m). The abundant DOM attenuates surface irradiance rapidly, so k = 3.6/m. With the same surface irradiance (I$_0$) = 1500 µE/m$^2$/s as above, let's examine the fate of light and phytoplankton productivity in the water column. First compute light*

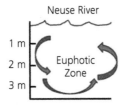
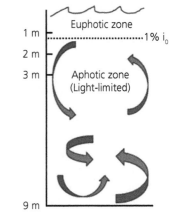

**Fig. 3.11** Current-driven turbulent mixing in the clearer Piedmont-derived Neuse River keeps phytoplankton cells suspended in the euphotic zone where high phytoplankton productivity is maintained (net photosynthesis). In the deeper lowland Northeast Cape Fear River light is strongly attenuated by DOM, and suspended microalgal cells are mixed into the light-limited aphotic zone most of the time, a net heterotrophic situation. (Figure by the author.)

*availability at the bottom (9 m):*

$$I_z = (1,500\mu E/m^2/s)\left(e^{-\,(3.6/m)(9.0m)}\right)$$

$$= 1.3 \times 10^{-11}\mu E/m^2/s$$

(i.e., no light for photosynthesis).

*How much light would be available at 1 m depth?*

$$I_z = (1,500\mu E/m^2/s)\left(e^{-\,(3.6/m)(1.0m)}\right) = 41.0\mu E/m^2/s$$

So, there is essentially the same amount of available light at only 1 m depth in the blackwater Northeast Cape Fear River as at the bottom of the much clearer Neuse River.

To find the limit of the defined euphotic zone of 1% $I_0$ for the Northeast Cape Fear River we compute 1% of 1500 $\mu E/m^2/s = 15\ \mu E/m^2/s$. The natural log ln of surface irradiance 1500 = 7.31 and the natural log ln of 15 = 2.71. Recall the light attenuation coefficient equation, Equation 1.1 ($k = [ln\ I_0 - ln\ I_z]/z$). Into Equation 1.1 substitute:

$$3.6/m = (7.31-2.71)/z, \text{rearranged as: (z)}$$

$$3.6/m = 4.6, \text{thus } z = 1.28m.$$

With 1.28 m representing the approximate 1% $I_0$ (bottom of the euphotic zone), this means that about 86% of the water column is in the aphotic zone. Under current-driven well-mixed conditions in the Northeast Cape Fear River, phytoplankton cells will spend far more time in severely light-limited conditions than in favorable conditions. Thus, phytoplankton productivity is likely to be very low and P/R < 1.0, a net heterotrophic situation. In this example the strong light attenuation is provided by DOM. Turbidity would serve a similar purpose in a non-lowland river system (Knowlton and Jones 2000; Koch et al. 2004).

### 3.1.3.3 Nutrients

It has been suggested that phytoplankton nutrient limitation is unlikely in rivers because nutrients are often in excess of algal requirements (Wehr and Descy 1998). In many situations, however, sometimes regardless of light attenuation, strong statistical relationships have been discerned between river nutrient and chlorophyll a concentrations (Van Nieuwenhuyse and Jones 1996; Heiskary et al. 2013). For example, Heiskary and Markus (2001)

sampled five large Minnesota rivers and found a significant, strong statistical relationship between sestonic chlorophyll a and TP as well as a strong relationship between sestonic chlorophyll a and total Kjeldahl nitrogen (TKN). In the Embarrass River in Illinois, Figueroa-Nieves et al. (2006) found a significant, although weak correlation between sestonic chlorophyll a and both TP and nitrate. Basu and Pick (1996) analyzed data from 31 rivers, fifth order or larger, in southern Ontario and western Quebec. They reported strong statistical relationships between chlorophyll a and both TP and TN. Saul et al. (2019) used several statistical models to determine that chlorophyll a in the lower Cape Fear River, North Carolina was strongly associated with nitrate concentrations within the river 85–105 km upstream, where a series of large point-source dischargers are located. Since chlorophyll a in the lower Cape Fear River is strongly correlated with BOD (Mallin et al. 2006), these relationships between nutrients and chlorophyll, and chlorophyll and BOD provide both justification and guidance for nutrient reduction schemes for running waters (see Section 3.1.5.1.2).

Nutrients, as well as physical and temporal factors, also control the species composition of phytoplankton in rivers and downstream estuaries. Rothenberger et al. (2009) used multivariate ordination techniques to assess influences on phytoplankton taxa composition from nutrients and other factors in the upper Neuse River Estuary. In addition to seasonal factors, elevated TP concentrations were related to abundances of certain toxigenic dinoflagellate species; other dinoflagellates were related to high TN and TKN; the filamentous cyanobacteria *Anabaena* and *Oscillatoria* were related to high orthophosphate; and certain diatoms were related to elevated DIN. Heterokontophytes such as the toxigenic flagellate *Heterosigma akashiwo*, haptophytes, chlorophytes, and the bloom-forming dinoflagellate *Heterocapsa rotundata* increased concomitant with increasing ammonium concentrations.

### 3.1.4 Phytoplankton Assemblages in Lower-Order Streams

Phytoplankton assemblages can develop in slow-moving rivers since they are not current-limited, but nonetheless require sufficient light and nutrients to become a major part of the producer community. Thus, during periods when a stream is shaded by riparian (i.e. streamside) forests, phytoplankton will be sparse. In contrast, phytoplankton can become abundant in slowly moving, unshaded streams—for example, following a clearcut, a stream in the Goshen Swamp of North Carolina was affected by cyanobacteria blooms where none had been known to occur previously (Ensign and Mallin 2001). Phytoplankton can also bloom in streams following a major nutrient pollution event—for example, following a 3.8-million-L (1-million-gallon) spill of swine wastes into Harris Creek, eastern North Carolina (Box 3.1). As noted earlier, large-scale evidence for nutrients impacting the phytoplankton community of streams comes from Van Nieuwenhuyse and Jones (1996). These researchers used data from 116 temperate streams to determine that, in summer, a strong, statistically significant curvilinear relationship existed between TP and sestonic chlorophyll *a* concentrations. Lower-order streams, however, are subject to "flashiness" or rapid changes in discharge due to rainstorms (see Chapter 1), which can prevent nutrient-driven bloom formation. For instance, in Illinois small stream studies Figueroa-Nieves et al. (2006) noted that frequent rainstorms had the effect of scouring phytoplankton from the streams, obscuring predictive relationships.

## 3.1.5 Problematic Phytoplankton Blooms in Running Waters

Nutrient loading to streams and rivers comes from sewage treatment facility discharges, septic system effluent leachate, leached or spilled animal wastes, cropland runoff, urban runoff, and airborne sources (see Chapter 2). Too much will cause eutrophic conditions, which are often most visible with phytoplankton. The term "harmful algal blooms" (HABs) refers to blooms that are toxic to fish, terrestrial animals, and humans, and/or to nontoxic "noxious" algal blooms that cause ecological or economic damage through other means such as altered food web dynamics. Both toxic and nontoxic blooms are generally high biomass in freshwaters.

### 3.1.5.1 Noxious Phytoplankton Blooms in Streams and Rivers

#### 3.1.5.1.1 Socioeconomic Costs

Economic assessment of impacts from noxious algal blooms historically has been difficult but has been attempted (Dodds et al. 2009). From a human point of view, river algal blooms, especially cyanobacterial blooms, can cause severe economic damage, but much more work is needed to estimate realistically the overall economic impacts of blooms worldwide, including US waters. Blooms commonly impart foul tastes and odors to drinking water that is sourced from bloom-impacted locations (Burkholder et al. 2010). Besides cyanobacteria, other taxonomic groups of microalgae that can produce taste-and-odor problems include certain diatoms, chrysophyceans, synurophyceans, cryptomonads, dinoflagellates, and green algae (Watson 2004; Burkholder et al. 2010).

In the Ohio River system, for example, phytoplankton blooms during low-flow periods have been known to cause severe impacts for potable water treatment plants since at least the 1930s (WHO 2003; Wines 2015). Potable treatment facilities that must draw water from bloom-affected areas incur the additional costs (passed on to the customers or taxpayers) of enhanced water treatment such as carbon filtration or ozonation as well as enhanced monitoring of the source waters for future noxious blooms (Chorus and Bartram 1999). Other noxious and economically costly effects of excessive algal biomass include physical blockage by thick cyanobacteria blooms, interference with recreational uses of water such as swimming, boating, and fishing, and loss of waterfront property values (Chorus and Bartram 1999). The shading of rooted aquatic plants by surface blooms impairs fishery habitat as well as reducing phytoplankton biodiversity. Dodds et al. (2009) assessed data concerning the economic impact of HABs in freshwaters. They estimated that the cost of bottled water usage by consumers was about US$813 million (based on surveyed taste and odor issues). Dodds et al. (2009) determined that the overall freshwater impact of HABs was conservatively estimated as $2.2 billion, including loss of waterfront property value, recreational use loss, issues with endangered species, and higher costs of drinking water treatment.

### 3.1.5.1.2 Ecological Impacts

One of the most common ecological results of noxious algal blooms is effects on the dissolved oxygen supply. BOD is a measure of the organic matter available for consumption by bacteria in a body of water during respiration. Algal blooms vary greatly in duration, but the bloom-forming algae eventually die, leaving high biomass of labile (easily digested) organic matter for microbes such as bacteria and fungi to decompose. During this decomposition process, these microbes proliferate and can rapidly consume all of the available dissolved oxygen in the water through their respiration. They also release chemicals such as ammonia which is then converted to nitrate by oxygen still available in the surrounding water. All of this oxygen consumption is referred to as BOD. It is a direct result of stimulation of algal blooms by excessive nutrient loading.

The biomass of suspended microalgae (as the indicator, chlorophyll a) has been strongly, positively correlated with BOD concentrations in various water bodies (e.g. Heiskary and Markus 2001; Mallin et al. 2006). As discussed in Chapter 2, increases in BOD drive down dissolved oxygen, sometimes to hypoxic or even anoxic levels. The BOD concentration in milligrams per liter, usually measured over five days (BOD5), is used as measure of the health of waterways. For example, Hynes (1960) noted that during 1898 in England the Royal Commission on Sewage Disposal was set up, and one of their later reports had the following classification for BOD5 in rivers: Very Clean ($\leq 1.0$ mg/L), Clean ($\leq 2$ mg/L), Fairly Clean ($\leq 3$ mg/L), Doubtful ($\geq 5$ mg/L), and Bad ($\geq 10$ mg/L). Based on sampling of multiple lotic systems in the Southeastern USA Mallin et al. (2006) considered BOD5 concentrations $\leq 2.0$ mg/L to represent normal background conditions in various urban and rural streams.

Downstream from rivers and reservoirs—sometimes hundreds of kilometers distant—this riverine nutrient loading causes estuarine noxious algal blooms. A prime example is the Susquehanna River, which drains massive cropland areas into the Chesapeake Bay, the largest estuary on the US mainland. Springtime nutrient loading supports persistent estuarine phytoplankton blooms that die, become labile organic matter for bacterial/fungal consumption, increase BOD, and cause bottom waters to become hypoxic or even anoxic (Officer et al. 1984; Boesch et al. 2001). Hypoxia in deeper waters of this massive system was first reported in the 1930s, then greatly intensified along with phytoplankton blooms between the mid-1950s and mid-1980s. The blooms have caused seagrass habitat loss by decreasing light availability, while the coinciding hypoxia depresses the abundance, diversity, and productivity of beneficial benthic animals. For decades there has been an ongoing effort to reduce N and P loading to the Bay by 30%, now raised to 40%. There has been good success with reducing nutrient inputs from point sources, but much poorer progress with non-point sources. This is because the reduction efforts are countered by a continually increasing population leading to more watershed urbanization; a massive poultry production industry contributing pollution from poultry wastes in runoff (see Chapter 13); and airborne nutrient inputs as well (e.g. N inputs from automobiles).

A second well-known example of impacted estuaries far downstream of nutrient sources is the Gulf of Mexico. The Mississippi River drains 40% of the continental USA, with N loading to the Gulf having tripled from the 1970s through the 1990s (Goolsby et al. 2001). The increasing N inputs have led to major increases in phytoplankton production, followed by sedimentation to the bottom and high BOD from decomposition. The Gulf waters stratify in summer, and hypoxic benthic waters form an ecologically stressful area popularly called a Dead Zone, which has low densities of fish and benthic organisms. On average, this hypoxic area spans $17,000 \text{ km}^2$ on the inner continental shelf. An interagency task force has undertaken efforts to guide resource managers in an attempt to reduce this zone to an average of $5000 \text{ km}^2$. A losing battle? In July 2017 the dead zone reached a new record of 22,007 $\text{km}^2$ (Van Meter et al. 2018).

What is the good side of drought? In 2012 the fourth smallest dead zone on record ($7480 \text{ km}^2$) for the Gulf occurred; this was in response to a massive

Midwestern drought that reduced nitrate runoff to the Gulf as well as stratification-forming conditions (Wickham 2012). Of course one of the many bad sides of drought is that low water severely endangers shipping on the Mississippi, which can be a huge economic punch! Remedial activities are ongoing in the watershed. However, note that much of the current N reaching the Gulf is "legacy" N, that which has been in soils and sediments for as long as 30 years or more (Van Meter et al. 2018). Thus, ongoing remedial actions in the watershed need to factor such legacy N in before significant improvements occur.

### 3.1.5.1.3 Riverine Toxic Algal Blooms

Nutrient loading can directly stimulate or indirectly stimulate (by creating favorable habitat) certain species of toxic algae, as mentioned earlier. Considering estuaries as well as freshwaters, these toxic species 1) can injure humans directly (consumption, aerosols, skin contact); or 2) can injure humans through consumption of contaminated shellfish or drinking water; and 3) the toxins from some of these algae cause serious disease and/or death in fish, birds, and/or mammals, including humans (Carmichael 1994; Burkholder et al. 2018). As mentioned, in freshwater and inland brackish ecosystems, known toxin-producing species have been verified among the cyanobacteria, haptophytes, and euglenoids; and many more species, mostly of dinoflagellates and also including some diatoms, have been verified in estuarine waters receiving land-based nutrient pollution (Heisler et al. 2008).

Cyanotoxins from suspended blooms are generally less toxic to aquatic fauna than to terrestrial fauna that ingest or otherwise encounter toxic blooms. Most reported animal deaths have involved dogs and livestock, but as noted earlier some human deaths have occurred (Chorus and Bartram 1999; Carmichael et al. 2001). Human illnesses attributed to cyanobacterial toxins have occurred in many countries, including Australia, Brazil, Canada, China, England, and the USA (compiled in WHO 2003). Various cyanobacteria have strains capable of producing MC cyanotoxins. The concentration of one form of MC, MC-LR, has been used by the WHO to develop a recommended protective guideline of $\leq 1$ µg/L for human consumption in

drinking water over a lifetime. The US EPA (2019a) proposed a total MC concentration of 8.0 µg/L as a recommended guideline for recreational waters. The WHO (2003) suggested that there may be a low probability for adverse human health effects in recreational waters if a cyanobacteria-dominated water body has a chlorophyll *a* concentration below 10 µg/L, but a moderate probability of adverse health impacts if a cyanobacterial-dominated water body has chlorophyll *a* above 50 µg/L. The state of Minnesota recommended that such water bodies should have less than 30 µg/L of chlorophyll *a* to avoid adverse impacts (Lindon and Heiskary 2008). Unfortunately, it is not that easy because chlorophyll *a* concentration has been shown to be unreliable in indicating potential danger to humans from cyanotoxins. There is no consistent relationship between cyanobacteria biomass (as chlorophyll *a* or as other measures) and toxin production; in fact, small, developing blooms can be much more toxic than large, aging blooms (Boyer 2007; Gobler et al. 2016 and references therein). The toxins themselves must be measured, in every location of interest, repeatedly during a bloom to assess the danger from cyanotoxin exposure.

Euglenoid blooms are common in nutrient-polluted waters. Because a general dogma operative for decades was that euglenoids are benign, their blooms have received little attention, so this is a pioneer area of research. As noted earlier, ichthyotoxic euglenoids are now known to be common, based on recent studies wherein toxic strains have been isolated from various freshwaters (Zimba et al. 2017). Toxic euglenoid blooms have also been documented in aquaculture ponds, causing kills of catfish, tilapia, and striped bass (Zimba et al. 2010). Such incidents represent a potentially significant economic burden for aquaculturists.

### 3.1.5.1.4 Stream Trophic Status: Chlorophyll a and Nutrients—How Much Is Too Much?

Assessment of trophic status for stream ecosystems and determination of boundaries or breakpoints for oligotrophic, mesotrophic, and eutrophic/hypereutrophic conditions have been much more challenging in streams, rivers, and reservoirs than in the natural lakes that are the foundation of limnology because phytoplankton growth is often controlled primarily by current velocity/residence time and/or by light availability, with nutrients of secondary importance. Under favorable conditions such as low flow, nutrients can strongly control phytoplankton growth. Dodds et al. (1998) suggested 10 μg/L of sestonic chlorophyll *a* as a breakpoint between oligotrophic and mesotrophic streams and 30 μg/L as a breakpoint between mesotrophic and eutrophic streams (Table 2.2; see also Wetzel 2001). In the USA, chlorophyll *a* water quality standards are not set by the federal government but, rather, left to the individual states, some of which have adopted eutrophication standards based on planktonic chlorophyll *a*. As examples, North Carolina has a water quality standard for chlorophyll *a* of 40 μg/L (or 40 ppb); this standard is also used for lakes in South Carolina. Minnesota recently adopted "river eutrophication criteria" ranging from 7 to 40 μg of chlorophyll *a*/L depending on the region of the state; the northern, more pristine area was assigned the lowest standard and the "corn belt" of southern Minnesota was assigned the highest (Heiskary et al. 2013).

As mentioned in Chapter 2, the US EPA (2000) has nutrient ecoregion-based guidelines that are recommended as targets for states to use in developing numeric TN and TP criteria that protect the designated uses of streams and rivers (Table 2.2). The ecoregions were based on soil/geological features and present-day land uses. The basic approach is to use the 75th percentile of reference or minimally disturbed streams and rivers in a given nutrient ecoregion or, if such streams were no longer available, then the EPA advised selecting numeric nutrient criteria from the 25th percentile of all-streams data in the nutrient ecoregion. If a sub-ecoregion is known to be strongly impacted by agriculture or urbanization, the EPA recommended using the 5th to 15th percentile of all streams data, depending on the area affected. The EPA also left room for states to develop numeric nutrient criteria based on stressor–response relationships (e.g. Stevenson et al. 2008) and other approaches if sufficient data were available to support more sophisticated statistical models. Unfortunately, the EPA's recommendations thus far have been considered too protective by most states, and few have yet developed

protective TN and TP numeric criteria, statewide, for their streams and rivers (US EPA 2019b).

Many states still use narrative standards for nutrient pollution based on vague aesthetics (e.g. "unsightly" or "unnatural" algal and plant growth), which are subjective and, therefore, difficult to enforce. Furthermore, the main response variable recommended by the EPA in developing protective numeric nutrient criteria is water-column chlorophyll *a*, which includes only phytoplankton (sestonic) biomass and excludes or is unclear about floating (or benthic) filamentous macroalgae, surface scums from planktonic cyanobacteria such as *Microcystis*, and macrophyte overgrowths.

## 3.2 Periphyton

A ubiquitous additional assemblage of primary producers, especially in streams and shallow rivers, is the periphyton. Periphyton are defined as the microbial consortium, including bacteria, fungi, algae, and microfauna, growing on various types of substrata, from sand, rocks, and macrophytes to zooplankton and even turtle backs. They are sometimes referred to as *Aufwuchs*, a German term which has no direct English translation; rather, it is a term that refers to all the attached microalgae and attached and free-living organisms within a benthic mat (Fig. 3.12), but not penetrating into the substrata (Ruttner 1958). Periphyton biofilms are commonly considered to be up to about 2.5 cm (1 inch) thick (e.g. Burkholder 1996 and references therein). Filamentous algae exceeding that length are considered macroalgae (see Chapter 4). Within the matrix are complex nutrient interactions as well as trophic (grazing) activity (Fig. 3.12a).

Periphyton can contain solitary, colonial, and filamentous algae. These algae are usually considered benthic (or edaphic), but the substrata they use as habitat actually may occur in the water column (e.g. attached to macrophytes or aquatic animals) or along the bottom of the water body. They may be epizoic (attached to animals), but mostly live on unvegetated sediments or sediments beneath macrophyte beds (see Chapter 4), or on rocks and other nonliving substrata at the bottom of streams, rivers, and reservoirs (Fig. 3.13). A set of terms describes these microalgae: they may be

(a)                                    (b)

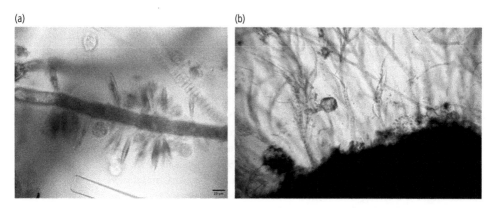

**Fig. 3.12** Periphyton matrices. Fig. 3.12a (left) Periphyton matrix on a green filament. Note loosely attached diatoms; additionally there are stalked ciliated protozoans *Vorticella* sp. within the matrix feeding on bacteria and picoplankton. Fig. 3.12b (right) Periphyton matrix on a surface creating a forest-like appearance. (Fig. 3.12a by the author; Fig. 3.12b by the late L.A. Whitford.)

(a)                                    (b)

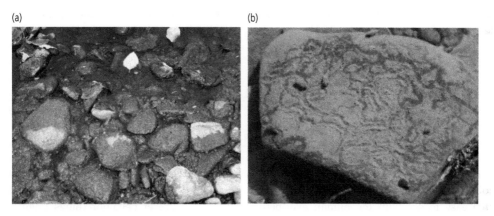

**Fig. 3.13** Fig. 3.13a (left) Epilithic periphyton in a North Carolina mountain stream. Fig. 3.13b (right) Tracks left by snails grazing periphyton in the Cumberland River, Tennessee. (Photos by the author.)

*epipelic* (living on or among sediments), *epidendric* (on living or dead wood or roots), *epipsammic* (on or among sand), or *epilithic* (on rock substrata—Fig. 3.13). Another category, *epiphytic* benthic microalgae, lives on other algae or plants and is discussed separately below; and still another, *epizoic* microalgae, live attached to fauna.

### 3.2.1 Benthic Microalgae

Benthic microalgae may be *adnate*, that is, immediately adjacent to the substratum along their major cell axis, or *loosely attached* among the biofilm matrix. Some loosely attached microalgae secrete long stalks of glycocalyx material to elevate the cells to the top of the biofilm for easier access to water-column nutrients and light (Burkholder 1996; Burkholder et al. 1990 and references therein). Benthic microalgae become entrained into the water column when dislodged by current or other disturbances. In the water column, they are considered part of the tychoplankton (see Chapter 11 for more information on water motion and suspended microalgae).

While the basic methods used to assess production and productivity for suspended microalgae can be applied to assess benthic microalgae, a fundamental problem involved in sampling them must first be addressed, which is their extreme habitat and assemblage heterogeneity (Burkholder 1996).

Artificial substrata such as glass slides or clay tiles are often used in an attempt to address that major challenge, but they fall far short of simulating the many microhabitats on natural substrata that are vitally important in controlling benthic microalgal colonization (e.g. Cattaneo and Amireault 1992). Compounding the problem, as a biofilm continues to develop, much of the microalgae included within it are dead, which is why a dry weight or even an ash-free dry weight measure of periphyton as "microalgal biomass" can greatly overestimate the biomass of the living microalgae within it (Burkholder et al. 1990). It is a fundamental challenge to sample the living benthic microalgae, usually among high quantities of debris, enough to achieve reasonable error bars in statistical analyses (e.g. ±20%; Burkholder and Wetzel 1989). Regardless, novel techniques have helped to accomplish more insights about benthic microalgae over the past few decades (see Biggs and Kilroy 2000; Consalvey et al. 2004),

It is well known that benthic microalgae can be rich in species diversity; they are often the dominant algal biomass and the major primary producers in open streams, upper river segments, and shallow reservoirs; and they are hugely important to food webs in such systems (Whitton 1975; Burkholder 1996). Although periphyton biofilms are often easy to see with the human eye, benthic microalgae can usually only be identified using light microscopy and/or scanning electron microscopy. These microalgae consist mostly of diatoms (more heavily silicified, in general, than planktonic diatoms), cyanobacteria, and green algae, along with locally abundant cryptomonads, euglenoids, dinoflagellates, chrysophyceans, and synurophyceans. Due in part to their accessory pigments that capture light in the "green window," diatoms and cyanobacteria can thrive at the extreme edge of the euphotic zone, down to 0.5% $I_o$ (Cahoon 1999).

Some benthic microalgae can vertically migrate through the sediments, which is thought to help protect from abrasive wave action and access nutrients from the deeper, richer sediments (Palmer and Round 1965; Kingston 1999; Consalvey et al. 2004). Thus, the surface sediments in shallows and exposed mud flats can appear greenish (e.g.

euglenoids) or golden-brown (diatoms). This phenomenon can occur in mud flats and shallows from fresh to marine waters. In freshwaters, vertical migration of the benthic microalgae (sometimes called microphytobenthos) follows a circadian rhythm, with most cells photosynthesizing at the sediment surface during daylight hours, then going down into the sediments apparently to more easily obtain nutrients at night (Palmer and Round 1965). In tidal creek areas and brackish to marine mud flats, the major cues are both tidal cycles and light periods: the microalgal populations migrate up to the sediment surface during low tide in daylight and back down before the first wave from the incoming tides moves over the exposed sediment or near darkness (Kingston 1999 and references therein). This vertical migration, sometimes over a depth of several centimeters or more, is thought to help protect microalgal populations from being swept away by tidal action and also facilitates nutrient (at depth) and light (at the surface) acquisition (Kingston 1999; Consalvey et al. 2004).

Benthic microalgae can maintain relatively constant, low biomass under frequent disturbance or undergo cycles of accumulation and loss under less frequent disturbance. Overlying that, their abundance and species composition is subject to seasonal variability resulting from light availability, temperature, and grazer behavior (Biggs 1996). Grazing effects on the seasonal cycles (succession) of benthic microalgae apparently are highly habitat-specific (Steinman 1996). Nutrient limitation and herbivory both become seasonally important as major influences on benthic microalgal communities in some systems. Growth of benthic microalgae depends on the season, altitude, and geographic location. Small streams in temperate areas, for example, tend to be heavily shaded in late spring through early fall, limiting light availability, whereas in winter the lack of shading from terrestrial vegetation alleviates light limitation. Unshaded streams in prairies or mountain areas can have high benthic microalgal biomass, whereas heavily shaded forest streams will have much lower biomass. Substrata is important too; in streams with shifting, sandy bottoms there will be less biomass compared with rocky substrata where rich epilithic communities may develop.

Benthic microalgae are an important energy source for herbivores (grazers) in stream food webs, such as various nematodes, ostracods, cladocerans, rotifers, harpacticoid copepods, gastropods, isopods, amphipods, mollusks, shrimp, finfish, and insect larvae. These herbivores, in turn, are eaten by fauna in higher trophic levels such as amphibians, reptiles, fish, and waterfowl.

The productivity and respiration rates of benthic microalgae are affected by the architecture of the stream bed itself. Experiments have shown that increasing heterogeneity (i.e. vertical complexity supplied by boulders and cobbles within riffles) promotes increases in both productivity and respiration (Cardinale et al. 2002). The enhanced metabolic rates are thought to result from increases in microcurrent velocity and turbulence in and around the complex habitat structures, which have the net effect of reducing boundary layer thickness (Box 3.3).

---

### Box 3.3 Benthic diatom immigration

Consider a nonliving substrata in the benthos, such as rock of various size, muddy or sandy sediments, or organic debris. In Chapter 1 we discussed how current velocity increases farther away from the sediments, and close to the substrata there is a laminar benthic boundary layer that impedes movement of particles and nutrients into the substrata. As current and turbulence increases, this boundary layer decreases, allowing easier impingement of materials, including diatom colonizers, to the substrata. Thus, the greater the current, there should be greater impingement and colonization. Of course too high a current induces a shear stress that entrains the diatom back into the water column. However, microzones of low shear stress can be created by attached diatoms themselves, which adhere to the substrata with mucilage. Diatoms already present, through their architecture, create eddies downstream of them that allow for more immigration into the matrix of new diatoms and other living components of the matrix.

(From Stevenson 1983.)

---

Benthic microalgae in headwaters (lower-order) streams generally often have low standing biomass, but moderate to high primary productivity per unit area *and* high removal by grazers (Graham and Wilcox 2000). In Exercise 3.1 we discussed P/R ratios for phytoplankton in different riverine situations. Regarding shallow streams, it has long been considered that headwaters stream ecosystems are predominantly heterotrophic rather than autotrophic, having more respiration than NPP (e.g. Vannote et al. 1980; Schade et al. 2011). These streams were thought to be allochthonous—that is, acquiring most of the energy needed to drive the food web from external sources, mainly the abundant leaf fall from terrestrial plants. In some systems that may not necessarily be the case. A classic study by Minshall (1978), in a so-called net heterotrophic stream with low apparent benthic microalgal production, showed that while primary *production* at a given time was low, primary *productivity* was very high, and high herbivore grazing activity maintained low benthic microalgal biomass. In Minshall's work, when productivity (rate of production) was accounted for, the net heterotrophic stream actually was found to be autotrophic or autochthonous, gaining the major proportion of its energy from in-stream primary producers.

Periphyton biofilm architecture (physiognomy) is temporary. Where current velocities are low (but nutrients sufficient), the matrix architecture can grow outward and upward (complex formation) and achieve high biomass (Luttenton and Rada 1986). The current brings a constant resupply of nutrients to the attached microalgae, and stronger currents reduce the thickness of the boundary layer (Box 3.3; see Chapter 1), leading to an increase in the delivery rate of nutrients and $CO_2$ and increased microalgal productivity (Biggs 1996; Cardinale et al. 2002). When a spate (rapid current increase from rain and stormwater runoff) occurs, however, loosely attached microalgae are often scoured away, leaving the adnate, low-profile microalgae. Thus, in general, benthic microalgal biomass is negatively related to current velocity (Biggs 1996). Following a disturbance, cyanobacteria and diatoms are often among the first colonizers to repopulate an area (in days or weeks), whereas vascular plants may require months to recolonize (Biggs 1996). How often a stream experiences spates is critical; average periphyton biomass (as chlorophyll *a*) is also negatively related to the frequency of flood (spates) disturbance (Biggs 1996).

Benthic microalgae occur on and among unconsolidated sediments as well as on solid surfaces. Such sediments easily shift with changes in current or tidal movement, potentially burying the microalgae or impeding exposure to light and nutrients. Epipelic and epipsammic microalgae are motile, which helps them to survive shifting sediments and evade burial. Pennate diatoms and cyanobacteria slide along on mucilaginous materials they secrete, whereas other types of benthic microalgae such as *Euglena* have flagella they use for propulsion. Thus, many taxa can regulate their position on and within uncertain substrata using motility (see vertical migration in sediments, above). Also, finer sediments have smaller interstitial spaces between particles than do larger, sandier sediments. Smaller interstitial spaces allow less light penetration into the sediments, which limits photosynthesis. Cahoon et al. (1996) reported that in brackish streams, benthic microalgal biomass declines as the percentage of finer sediments increases relative to larger grain sizes. An important implication of this finding is that erosion of fine sediments into streams, rivers, and estuaries (such as from logging, housing, or highway construction) can lead to lower benthic microalgal biomass and less food for beneficial consumers in the aquatic food web.

Overall, then, benthic microalgae can be the major source of fixed carbon in lower-order streams and in reservoir shallows. They sequester nutrients and help to control nutrient fluxes from sediments (Carlton and Wetzel 1988), and can thereby indirectly reduce phytoplankton biomass. The physiognomy and species complexity of benthic microalgal assemblages vary greatly depending on the degree of disturbance (including suspended sediments), current velocity, substratum, temperature, light, nutrients, and grazing pressure. Benthic microalgal communities are *extremely* patchy (heterogeneous) in colonization. In currents slower than 10 cm/s, growth of benthic microalgae generally is limited by boundary layer effects that restrict nutrient acquisition (Stevenson 1996). Thus, benthic microalgal density tends to be higher in intermediate current velocities.

## 3.2.2 Epiphytic Microalgae

Epiphytes are a special case among periphyton. They often have extremely high productivity due to their small size and relatively rapid growth rates and also because of the enormous substratum area available for colonization (i.e. macrophyte leaves and stems), higher light because host plants or macroalgae lift them toward the water surface, and nutrient supplies from both the overlying water and the plant substrata below (Moeller et al. 1988; Burkholder 1996). Their seasonal population dynamics are much more complex because the surface area available for colonization changes continually as the host macrophyte or macroalga develops, grows, senesces, and dies (Burkholder 1996). The leaves of many macrophytes are also continually replaced throughout the plants' growing season. In addition, epiphyte assemblage composition can differ greatly depending on the underlying plant or macroalga and the general water quality conditions of the habitat (Eminson and Moss 1980). Studies of the seasonal dynamics of the living epiphytic algal assemblage on natural plant substrata are rare, even at the phylum level; most information on stream-inhabiting epiphytes is based instead on artificial substrata and parameters such as chlorophyll *a* or ash-free dry weight.

Host macrophyte stems and leaves first become colonized by bacteria (Burkholder 1996 and references therein). Microalgae are secondary colonizers; diatoms and other algae secrete mucilage that provides anchoring materials for other organisms to settle, such as fungi and protozoans, and for minute organic and inorganic particles to stick. These organisms and nonliving particles form a microbial consortium among copious hydrated glycocalyx materials (e.g. Fig. 3.12); in fact, much of epiphyte biofilms is water.

In nutrient-poor waters, nitrogen-fixing cyanobacterial epiphytes can likely supply some inorganic N to host plants. In nutrient overenriched waters, however, overgrowth of loosely attached epiphytes can stress the host plant by blocking light needed for photosynthesis. Light stress due to epiphytic overgrowth in nutrient-polluted waters is a common mechanism for the loss of submersed plant meadows (Wetzel 2001), leading in turn to loss of beneficial fauna that depended on the macrophytes for critical habitat.

### 3.2.3 Nutrients and Noxious Periphyton Blooms

The importance of nutrients in controlling the growth of benthic microalgae is often secondary to disturbance, light availability, and herbivory (Borchardt 1996). Thus, benthic microalgal biomass is often more weakly related to water-column nutrient supplies than that of phytoplankton. Epipelic and epiphytic benthic microalgae additionally have two major sources of nutrient supplies, namely, the overlying water and the substrata. As the biofilm thickness increases, adnate taxa become relatively isolated from the water column and appear to rely mostly on nutrients provided by the underlying substratum and nutrients recycled within the biofilm layer, and are relatively isolated from the overlying water except where fissures in the layer allow water-column nutrients to penetrate (Burkholder 1996; Wetzel 2001).

Many authors consider any benthic algal growth as "periphyton," meaning that the term is misused to include not only periphyton but also benthic filamentous macroalgae that can be 2 m or more long(!) (e.g. *Cladophora*; see Chapter 4). Noxious levels of periphyton biomass have been suggested at chlorophyll *a* concentrations exceeding 100 mg/m$^2$, as general guidance for streams across geographic regions (Dodds 2006). To maintain relatively healthy water quality, periphyton biomass levels have been proposed as $\leq$ 150 mg/m$^2$ for Minnesota rivers (Heiskary et al. 2013 and references therein), or < 107 mg/m$^2$ to protect relatively pristine Ohio streams from degradation and < 182 mg/m$^2$ to manage impacted Ohio streams to prevent further degradation (Miltner 2010). Remember, however, that the suggested maximal accepted values for benthic algal biomass above mesotrophic conditions may include filamentous macroalgae *as well as* periphyton (e.g. see figures cited in Heiskary et al. 2013).

Assessment of relationships between nutrients and benthic microalgal biomass in streams and rivers is challenging, however, even more than for lakes and reservoirs, because of the "biofilm factor"—that is, the internal nutrient regime that is of primary importance to periphyton, so that water-column influences are weakened and often indirect. Other reasons include variability in elevation, geology, meteorology, discharge regimes, shading, temperature, slope, and other factors along a stream course. On a broad scale (286 streams), Dodds et al. (1998) suggested trophic classifications based on frequency distributions, with mean and maximum benthic chlorophyll *a* for the oligotrophic–mesotrophic boundary of 20 and 60 mg/m$^2$, respectively, and the mesotrophic–eutrophic boundary of 70 and 200 mg/m$^2$, respectively, with accompanying average stream nutrient concentrations (see Table 2.2). Another statistical approach by Stevenson et al. (2008) involved 149 streams in the Mid-Atlantic Highlands region and focused on TP concentrations. Their models determined that minimally impacted streams had $\leq$ ~10 μg TP/L. Based on responses of benthic biomass and taxonomic changes, Stevenson et al. (2008) suggested TP thresholds of 10–12 μg P/L as a protective approach for stream ecosystems.

Storm-induced spates wash out significant amounts of accumulated benthic biomass, even in eutrophic streams. Thus the intervals of reasonably calm water between spates (i.e. accrual time) has been modeled as a key factor in benthic biomass accumulation (Biggs 2000). Based on data from 25 New Zealand streams of generally similar physical character and considering the trophic state boundaries of Dodds et al. (1998), Fig. 3.14 depicts how accrual time interacts with

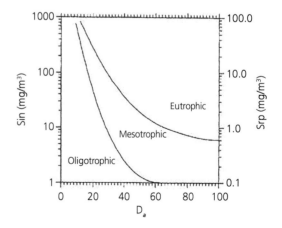

**Fig. 3.14** Depiction of interaction among average dissolved P, dissolved N, and time between spates (accrual interval $d_a$) in determining trophic state of a stream. (Adapted from Biggs (2000), Fig. 7.)

(a)                                    (b)

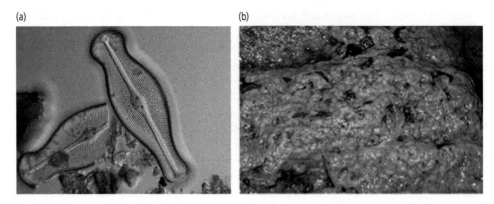

**Fig. 3.15** *Didymosphenia geminata*. Fig. 3.15a (left) Microscopic view of *D. geminata* from Rio Espolon, Chile. Fig. 3.15b (right) *D. geminata* bloom in River Yuso, León, Spain. (Microscopic view photo by Sarah Spaulding, USGS, public domain photo; photo of bloom by David Perez, Creative Commons Attribution 3.0.)

nutrient concentrations to control eutrophication symptoms (from Biggs 2000). Unfortunately, use of this compelling approach is confounded because relatively relatively few streams in any nation have adequate hydrological data, which would require considerable field effort and expense. Control of stream flow to minimize adverse effects of nutrient pollution is not usually possible unless the stream or river is dammed, which is politically difficult due to many competing interests for water (see Chapter 10).

This chapter, and much of this book in general, discuss the ecosystem and public health dangers of stream, river, and reservoir eutrophication. However, there is a seemingly contradictory story of a nuisance blooming benthic diatom, *Didymosphenia*

*geminata* (also known as Didymo, as well as less complimentary names) (Fig. 3.15), which forms extreme overgrowths (Fig. 3.15b) on rock substrata in oligotrophic streamwaters that can completely carpet large bottom areas. These blooms are apparently caused by low abundance of phosphorus, a process sometimes alluded to as *oligotrophication* as explained in Box 3.4.

### 3.2.3 Benthic Microalgae as Indices of Stream Conditions

Because benthic microalgae are attached (relatively sessile) and thought to integrate water quality at least over periods between spates, they historically

---

**Box 3.4 The Didymo story—the oligotrophic bloomer**

Since the 1990s there have been reports of strange benthic pennate diatom blooms (Fig. 3.15) from a broad span of areas including Vancouver Island in Canada, the US Mountain West, the US Northeast, New Zealand, Patagonia, Europe, and Asia. The perpetrator, the stalk-producing benthic diatom *Didymosphenia geminata*, has formed thick (up to 20-mm) mats covering areas as wide as the stream and up to 20 km long, mostly on rocky substrata. Most blooms have occurred in headwaters streams in forested areas away from sewage or mass livestock production,

generally where P concentrations are 2 ppb or 2.0 µg/L or less. Some blooms have occurred in dam tailwaters as well. This species, reported in North America in the 1890s, is nicknamed *Didymo* (and called "rock snot" by annoyed fishermen and stream ecologists) and is native to North America but is spread as a local invasive, apparently largely by adhering to fishermen's felt-bottomed stream boots and transferred from stream to stream in that fashion. However, invasions may have spread it to other continents.

---

**Box 3.4** *Continued*

Because Didymo mats (Fig. 3.15b) can completely cover a benthic area, it would appear that the natural ecosystem is disrupted. And indeed it has been demonstrated by researchers that benthic invertebrate larval taxa generally considered to represent good quality water, such as members of the Ephemeroptera (mayflies), Plecoptera (stoneflies), and Trichoptera (caddisflies; see Chapter 5), are greatly reduced under the mats in favor of chironomids (midge larva) and oligochaetes (aquatic worms). However, the story regarding fish is more nuanced. While some studies have found reductions in salmonids, other researchers have found no difference in growth rate or abundances of salmonids in Didymo-impacted reaches versus control areas, and no dietary concerns. What is critically understudied is the impact of Didymo mats on the spawning behavior of salmonids and other fishes and early life history survival.

A recent question is "Does Didymo have any value or uses?" The stalks of Didymo are made of proteins, sulfated polysaccharides with uronic acid, with the diatom cells containing biosilicate. Researchers have proposed that the stalks and the extracellular biopolymers they produce have the ability to absorb harmful metal ions from ambient waters, especially lead (Pb), cadmium (Cd), and nickel (Ni). How this would be accomplished for wastewater treatment for instance has not been determined, i.e. mass production in laboratory cultures, or stream biomass removal? The fact that the mats also contain N-fixing bacteria has been noted. Which brings us to conditions enhancing the blooms

and potential removal. As noted, blooms and enhanced stalk production are hypothesized to occur as a response to low P near the substrata, and stalk growth upwards may be a response to absorb more P from the flowing streamwaters away from the benthic boundary layer. Thus, the blooms occur under low-P conditions, normally considered a good thing in freshwater eutrophication protection. In fact, Didymo colonies and blooms dissipate under P-enriched conditions. It is also possible that Didymo blooms may be stimulated by the N:P imbalance that occurs when low P conditions are impacted by N inputs such as atmospheric deposition, silviculture, agriculture, or fewer salmon returning upstream to spawn and die (see Chapter 5). The increased N throws the system off balance, promoting the mat creation.

What about bloom removal? Sure, herbicides have been tested to control the blooms but invariably have impacts on nontarget organisms. Dam creation reduced P loading downstream via trapping of sediments (see Chapter 10), enhancing Didymo production. However, dam releases of high-discharge water will reduce the bloom formation through either P additions or physical flushing. Ultimately, better control of stream N pollution would be the ultimate means of restoring balance in these headwaters systems and reducing Didymo blooms.

(Compiled from Sundareshwar et al. 2011; Bothwell et al. 2014; Coyle 2015; Flood and Burkholder 2018; New York Sea Grant 2019; Clancy et al. 2021; Ejaz et al. 2021.)

---

have been used as biomarkers. Diatoms, as individual species or groups, especially have been used in this way, commonly after colonization on artificial substrata such as glass slides for ease of replication and in collection. The slides are placed into a specially designed rack called a diatometer, and then secured in locations of interest such as upstream and downstream of a polluted discharge, or in nutrient-rich versus nutrient-poor streams. After a set period (days to weeks), the slides are collected and the diatoms are removed and quantified. Researchers have compiled lists of species that are characteristic of high-nutrient versus low-nutrient situations, or pollution tolerant versus more "natural" flora (see Patrick and Palavage 1994 for examples of the latter). Applying diatom species (or taxa groups)

that have known responses to nutrients, organic pollution, or other stressors and combining their presence and abundance in various statistical formulas produces indices that integrate and elucidate environmental effects and track changes over time. There are numerous such indices, some more specific to global geographical areas than others. Some of the more well-known indices include the Trophic Diatom Index (TDI), the diatom Biological Index (IBD), and the specific Pollution Sensitivity Index (IPS). For details and comparisons see Descy and Coste (1991), Kelly and Whitton 1995), Coste et al. (2009), and Tan et al. (2017). Note that, owing to biogeography and combined pollutant effects, no one method is best for all needs.

## 3.7 Summary

- The most abundant planktonic microalgal taxonomic groups in streams and rivers generally are diatoms, cyanobacteria, and green algae. In addition, cryptomonads, euglenophytes, dinoflagellates, haptophytes, chrysophyceans, and synurophyceans are periodically important on a seasonal basis.

- Phytoplankton abundance and production increases under low discharge conditions when settling of suspended solids improves water column light conditions. Under high discharge turbidity increases, light penetration decreases, and microalgal production declines, but the increased currents entrain more benthic microalgae into the water column.

- In many lotic situations, nutrient inputs are strongly related to phytoplankton biomass accumulation. Under low flow and high nutrient conditions, nuisance microalgal blooms can cause increased BOD and low DO, taste-and-odor problems to water intakes, and adverse impacts on waterfront property values.

- Toxic cyanobacteria blooms occur in freshwater rivers (and in estuaries). Their outbreaks, apparently increasing, are known to kill livestock and pets and to cause human illness. A growing number of riverine and reservoir fish kills are also being caused by the haptophyte *Prymnesium parvum* (the golden alga).

- The periphyton community consists of tightly (adnate) and loosely attached microalgae, mainly diatoms, cyanobacteria, and chlorophytes, living on plants as epiphytes or nonliving substrata in a hydrated biofilm matrix (up to ~2.5 cm or 1 inch thick) with other microbiota.

- These benthic microalgae are hugely important to stream and river food webs and are the dominant primary producers in headwaters streams and reservoir shallows. Their taxonomic structure and biomass are controlled by the light regime, nutrient concentrations, season, currents, and storm-driven spates of flow.

- Control of benthic microalgal production in response to nutrient regimes during low flow and other favorable conditions has mostly emphasized attempts to minimize nutrient overenrichment; but the story of Didymo, the recent naturally occurring diatom-turned invader, reveals a more general need to restore balance in nutrient supplies for aquatic communities so that they are neither too high nor too low.

## References

Arnold, D. 1971. Ingestion, assimilation, survival, and reproduction by *Daphnia pulex* fed seven species of cyanobacteria. *Limnology and Oceanography* 6:907–920.

Basu, B.K. and F.R. Pick. 1996. Factors regulating phytoplankton and zooplankton biomass in temperate rivers. *Limnology and Oceanography* 41:1572–1577.

Beaver, J.R., D.E. Jensen, D.A. Casamatta, et al. 2013. Response of phytoplankton and zooplankton communities in six reservoirs of the middle Missouri River (USA) to drought conditions and a major flood event. *Hydrobiologia* 705:173–189.

Beaver, J.R., K.C. Scotese, E.E. Manis, S.T.J. Juul, J. Carroll and T.R. Renicker. 2015. Variation in water residence time is the primary determinant of phytoplankton and zooplankton composition in a Pacific Northwest reservoir ecosystem (Lower Snake River, USA). *River Systems* 21:261–275.

Bellinger, E.G. and D.C. Sigee. 2010. *Freshwater Algae: Identification and Use as Bioindicators*. Wiley-Blackwell, Hoboken, NJ.

Biggs, B.J.F. 1996. Hydraulic habitat of plants in streams. *Regulated Rivers: Research and Management* 12:131–144.

Biggs, B.J.F. 2000. Eutrophication of streams and rivers: dissolved nutrient–chlorophyll relationships to benthic algae. *Journal of the North American Benthological Society* 19:17–31.

Biggs, B.J.F. and C. Kilroy. 2000. *Stream Periphyton Monitoring Manual*. National Institute of Water and Atmospheric Research, Christchurch, New Zealand.

Boesch, D.F., R.B. Brinsfield and R.E. Magnien. 2001. Chesapeake Bay eutrophication: scientific understanding, ecosystem restoration, and challenges for agriculture. *Journal of Environmental Quality* 30:303–320.

Borchardt, M.A. 1996. Nutrients. Pages 184–227 in Stevenson, R.J., M.L. Bothwell and R.L. Lowe (eds.), *Algal Ecology: Freshwater Benthic Ecosystems*. Academic Press, San Diego.

Bothwell, M.L., B.W. Taylor and C. Kilroy. 2014. The Didymo story: the role of low dissolved phosphorus in the formation of *Didymosphenia geminata* blooms. *Diatom Research* 29:229–236.

Boyd, C.E. 2000. *Water Quality: An Introduction*. Kluwer Academic Publishers, Boston.

Boyer, G.L. 2007. The occurrence of cyanobacterial toxins in New York lakes: lessons from the MERHAB-Lower Great Lakes Program. *Lake and Reservoir Management* 23:153–160.

Burkholder, J.M. 1992. Phytoplankton and episodic suspended sediment loading: phosphate partitioning and mechanisms for survival. *Limnology and Oceanography* 37:974–988.

Burkholder, J.M. 1996. Interactions of benthic algae with their substrata. Pages 253–297 in Stevenson, R.J., M. Bothwell and R.L. Lowe (eds.), *Benthic Algae in Freshwater Ecosystems*. Academic Press, New York.

Burkholder, J.M. 1998. Implications of harmful microalgal and heterotrophic dinoflagellates in management of sustainable marine fisheries. *Ecological Applications* 8:S37–S62.

Burkholder, J.M. 2002. Cyanobacteria. Pages 952–982 in Bitton, G. (ed.), *Encyclopedia of Environmental Microbiology*. Wiley, New York.

Burkholder, J.M. and H.G. Marshall. 2012. Toxigenic *Pfiesteria* species—updates on biology, ecology, toxins and impacts. *Harmful Algae* 14:196–230.

Burkholder, J.M. and R.G. Wetzel. 1989. Epiphytic microalgae on natural substrata in a hardwater lake: seasonal dynamics of community structure, biomass, and ATP content. *Archiv für Hydrobiologie Supplement* 83:1–56.

Burkholder, J.M., R.W. Wetzel and K.L. Klomparens. 1990. Direct comparison of phosphate uptake by adnate and loosely attached microalgae within an intact biofilm matrix. *Applied and Environmental Microbiology* 56:2882–2890.

Burkholder, J.M., P.M. Glibert and H.M. Skelton. 2008. Mixotrophy, a major mode of nutrition for harmful algal species in eutrophic waters. *Harmful Algae* 8:77–93.

Burkholder, J.M., W. Frazier and M.B. Rothenberger. 2010. Source water assessment and treatment strategies for harmful and noxious algae. Pages 299–328 in AWWA (ed.), *Algae: Source to Treatment. Manual 57*. American Water Works Association, Denver.

Burkholder, J.M., S.E. Shumway and P.M. Glibert. 2018. Food web and ecosystem impacts of harmful algae. Pages 243–336 in Shumway, S.E., J.M. Burkholder and S.L. Morton (eds.), *Harmful Algal Blooms and Their Management: A Compendium Desk Reference*. John Wiley & Sons, Hoboken, NJ.

Cahoon, L.B. 1999. The role of benthic microalgae in neritic ecosystems. *Oceanography and Marine Biology: An Annual Review* 37:47–86.

Cahoon, L.B., J.E. Nearhoof and C.L. Tilton. 1996. Sediment grain size effect on benthic microalgal biomass in shallow aquatic ecosystems. *Estuaries* 22:735–741.

Cardinale, B.J., M.A. Palmer, C.M. Swan, S. Brooks and N.L. Poff. 2002. The influence of substrate heterogeneity on biofilm metabolism in a stream ecosystem. *Ecology* 83:412–422.

Carlton, R.G. and H.W. Paerl. 1989. Oxygen-induced changes in morphology of aggregates of *Aphanizomenon flos-aquae* (Cyanophyceae): implications for nitrogen fixation potentials. *Journal of Phycology* 25: 326–333.

Carlton, R.G. and R.G. Wetzel. 1988. Phosphorus flux from lake sediments: effect of epipelic algal oxygen production. *Limnology and Oceanography* 33:562–571.

Carmichael, W.W. 1994. The toxins of cyanobacteria. *Scientific American* January:2–9.

Carmichael, W.W., S.M.F.O. Azevedo, J.S. An, et al. 2001. Human fatalities from cyanobacteria: chemical and biological evidence for cyanotoxins. *Environmental Health Perspectives* 109:663–668.

Carty, S. and M.W. Parrow. 2015. Dinoflagellates. Pages 773–807 in Wehr, J., R.G. Sheath and J.P. Kociolek (eds.), *Freshwater Algae of North America*. Academic Press, New York.

Cattaneo, A. and M.C. Amireault. 1992. How artificial are artificial substrata for periphyton? *Journal of the North American Benthological Society* 11:244–256.

Chorus, I. and J. Bartram (eds.) 1999. *Toxic Cyanobacteria in Water—A Guide to Their Public Health Consequences, Monitoring and Management*. E & FN Spon for the World Health Organization, New York.

Clancy, N.G., J. Brahney, J. Dunnigan and P. Budy. 2021. Effects of diatom ecosystem engineer (*Didymosphenia geminata*) on stream food webs: implications for native fishes. *Canadian Journal of Fisheries and Aquatic Sciences* 78:154–164.

Consalvey, M., D.M. Paterson and G.J.C. Underwood. 2004. The ups and downs of life in a benthic biofilm: migration of benthic diatoms. *Diatom Research* 19: 181–202.

Coste, M., S. Boutry, J. Tison-Rosebury and F. Delmas. 2009. Improvements of the Biological Diatom Index (BDI): description and efficiency of the new version (BDI-2000). *Ecological Indicators* 9:621–650.

Coyle, M. 2015. Your ugly algae neighbor: rock snot, a native nuisance. *Lakeline* Summer:42–45.

de Bernardi, R. and G. Guissani. 1990. Are blue-green algae a suitable food for zooplankton? An overview. *Hydrobiologia* 200/201:29–41.

Descy, J.P. and M.A. Coste. 1991. A test of methods for assessing water quality based on diatoms. *International Association of Theoretical and Applied Limnology* 24: 2112–2116.

Dicke, M. and W.M.W. Sabelis. 1988. Infochemical terminology: should it be based on cost-benefit analysis rather than origin of compounds? *Functional Ecology* 2:131–139.

Díez-Quijada, L., A.I. Prieto, R. Guzmán-Guillén, A. Jos and A.M. Ceán. 2019. Occurrence and toxicity of microcystin congeners other than MC-LR and MC-RR: a review. *Food and Chemical Toxicology* 125:106–132.

Dodds, W.K. 2006. Eutrophication and trophic state in rivers and streams. *Limnology and Oceanography* 51: 671–680.

Dodds, W.K., J.R. Jones and E.B. Welch. 1998. Suggested classification of stream trophic state: distributions of temperate stream types by chlorophyll, total nitrogen, and phosphorus. *Water Research* 32:1455–1462.

Dodds, W.K., W.W. Bouska, J.L. Eitzmann, et al. 2009. Eutrophication of U.S. freshwaters: analysis of potential economic damages. *Environmental Science and Technology* 43:12–19.

Ejaz, H., E. Somander, U. Dave, H. Ehrlich and M.A. Rahman. 2021. Didymo and its polysaccharide stalks: beneficial to the environment or not? *Polysaccharides* 2:69–79.

Eminson, D.F. and B. Moss. 1980. The composition and ecology of periphyton communities in freshwaters. I. The influence of host type and external environment on community composition. *British Phycological Journal* 15:429–446.

Engstrom, D.R., J.E. Almendinger and J.A. Wolin. 2009. Historical changes in sediment and phosphorus loading to the upper Mississippi River: mass-balance reconstructions from the sediments of Lake Pepin. *Journal of Paleolimnology* 41:563–588.

Ensign, S.E. and M.A. Mallin. 2001. Stream water quality following timber harvest in a Coastal Plain swamp forest. *Water Research* 35:3381–3390.

Falconer, I.R. 1989. Effects on human health of some toxic cyanobacteria (blue-green algae) in reservoirs, lakes and rivers. *Toxicity Assessment: An International Journal* 4:175–184.

Figueroa-Nieves, D., T.V. Royer and M.B. David. 2006. Controls on chlorophyll *a* in nutrient-rich agricultural streams in Illinois, USA. *Hydrobiologia* 568:287–298.

Flood, S.L. 2017. Ecotoxicology of estuarine phytoplankton growth and toxicity in response to atrazine exposures. PhD Thesis, Department of Plant and Microbial Biology, North Carolina state University, Raleigh.

Flood, S.L. and J.M. Burkholder. 2018. Imbalanced nutrient regimes increase *Prymnesium parvum* resilience to herbicide exposure. *Harmful Algae* 75:57–74.

Gobler, C.J., J.M. Burkholder, T.W. Davis, et al. 2016. The dual role of nitrogen supply in controlling the growth *and* toxicity of cyanobacterial blooms. *Harmful Algae* 54:87–97.

Golubic, S. 1979. Cyanobacteria (blue-green algae) under the bacteriological code? An ecological objection. *Taxon* 28:387–389.

Goolsby, D.A., W.A. Battaglin, B.T. Aulenbach and R.P. Hooper. 2001. Nitrogen input to the Gulf of Mexico. *Journal of Environmental Quality* 30:329–336.

Gorham, P.R. and W.W. Carmichael. 1980. Toxic substances from freshwater algae. *Progress in Water Technology* 12:189–198.

Graham, L.E. and L.W. Wilcox. 2000. *Algae*. Prentice Hall, Upper Saddle River, NJ.

Graham, L.E., J.M. Graham and L.W. Wilcox. 2009. *Algae*, 2nd edition. Pearson Benjamin Cummings, San Francisco.

Graham, L.E., J.M. Graham, L.W. Wilcox and M.E. Cook. 2016. *Algae*, 3rd edition. LJLM Press, Madison.

Hauer, F.R. and G.A. Lamberti. 1996. *Methods in Stream Ecology*. Academic Press, San Diego.

Heiskary, S. and H. Markus. 2001. Establishing relationships among nutrient concentrations, phytoplankton abundance, and biochemical oxygen demand in Minnesota, USA, rivers. *Journal of Lake and Reservoir Management* 17:251–267.

Heiskary, S., R.W. Bouchard Jr. and H. Markus. 2013. Minnesota Nutrient Criteria Development for Rivers— Update of November 2010 Report. Minnesota Pollution Control Agency, St. Paul.

Heisler, J., P. Glibert, J. Burkholder, et al. 2008. Eutrophication and harmful algal blooms: a scientific consensus. *Harmful Algae* 8:3–13.

Huisman, J., G.A. Codd, H.W. Paerl, B.W. Ibelings, J.M.H. Verspagen and P.M. Visser. 2018. Cyanobacterial blooms. *Nature Reviews Microbiology* 16:471–483.

Hynes, H.B.N. 1960. *The Biology of Polluted Waters*. University of Toronto Press, Toronto.

Hynes, H.B.N. 1970. *The Ecology of Running Waters*. Liverpool University Press, Liverpool.

Ianora, A., M.G. Bentley, G.S. Caldwell, et al. 2011. The relevance of marine chemical ecology to plankton and ecosystem function: an emerging field. *Marine Drugs* 9:1625–1648.

Ibelings, B.W. and K.E. Havens. 2008. Cyanobacterial toxins: a qualitative meta-analysis of concentrations, dosage and effects in freshwater, estuarine and marine biota. Pages 675–732 in Hudnell, K. (ed.), *Cyanobacterial Harmful Algal Blooms: State of the Science and Research Needs*. Springer, New York.

Jacobsen, D.M. and D.M. Anderson. 1986. Thecate heterotrophic dinoflagellates—feeding behavior and mechanisms. *Journal of Phycology* 22:249–258.

Jones, R.C. 1998. Seasonal and spatial patterns in phytoplankton photosynthetic parameters in a tidal freshwater river. *Hydrobiologia* 364:199–208.

Kelly, M.G. and B.A. Whitton. 1995. The trophic diatom index: a new index for monitoring eutrophication in rivers. *Journal of Applied Phycology* 7:433–444.

Kerfoot, W.C., C. Levitan and W.R. DeMott. 1988. *Daphnia*–phytoplankton interactions: density-dependent shifts in resource quality. *Ecology* 69:1806–1825.

Kingston, M.B. 1999. Wave effects on the vertical migration of two benthic microalgae: *Hantzschia virgata* var. *intermedia* and *Euglena proxima*. *Estuaries* 222: 81–91.

Knisely, C.A. and W. Geller. 1986. Selective feeding of four zooplankton species on natural lake phytoplankton. *Oecologia* 69:86–94.

Knowlton, M.F. and J.R. Jones. 2000. Seston, light, nutrients and chlorophyll in the lower Missouri River, 1994–1998. *Journal of Freshwater Ecology* 15:283–297.

Koch, R.W., D.L. Guelda and P.A. Bukaveckas. 2004. Phytoplankton growth in the Ohio, Cumberland and Tennessee Rivers, USA: inter-site differences in light and nutrient limitation. *Aquatic Ecology* 38:17–26.

Lapointe, B.E., J.M. Burkholder and K. Van Alstyne. 2018. Harmful macroalgal blooms in a changing world: causes, impacts, and management. Pages 515–560 in Shumway, S.E., J.M. Burkholder and S.L. Morton (eds.), *Harmful Algal Blooms and their Management: A Compendium Desk Reference*. Wiley, New York.

Lewin, R.A. 1976. Naming the blue-greens. *Nature* 259:360.

Lindon, M.J. and S.A. Heiskary. 2008. Blue-Green Algal Toxin (Microcystin) Levels in Minnesota Lakes. Environmental Bulletin No. 11. Minnesota Pollution Control Agency, St. Paul.

Lizotte, M.P. and G.M. Simmons Jr. 1985. Phytoplankton populations and seasonal succession in the Kanawha River, West Virginia. *Castanea* 50:7–14.

Luttenton, M.R. and R.G. Rada. 1986. Effects of disturbance on epiphytic community architecture. *Journal of Phycology* 22:320–326.

Mallin, M.A. and H.W. Paerl. 1992. Effects of variable irradiance on phytoplankton productivity in shallow estuaries. *Limnology and Oceanography* 37:54–62.

Mallin, M.A., J.M. Burkholder, M.R. McIver, et al. 1997. Comparative effects of poultry and swine waste lagoon spills on the quality of receiving streamwaters. *Journal of Environmental Quality* 26:1622–1631.

Mallin, M.A., V.L. Johnson, S.H. Ensign and T.A. MacPherson. 2006. Factors contributing to hypoxia in rivers, lakes and streams. *Limnology and Oceanography* 51: 690–701.

Meriluoto, J.A. and L.E. Spoof. 2008. Cyanotoxins: sampling, sample processing and toxin uptake. *Advances in Experimental Medicine and Biology* 619:483–499.

Meyer, J.L. 1990. A blackwater perspective on riverine ecosystems. *BioScience* 40:643–651.

Miltner, R.J. 2010. A method and rationale for deriving nutrient criteria for small rivers and streams in Ohio. *Environmental Management* 45:842–855.

Minshall, G.W. 1978. Autotrophy in stream ecosystems. *BioScience* 28:767–771.

Moeller, R.E., J.M. Burkholder and R.G. Wetzel. 1988. Significance of sedimentary phosphorus to a rooted submersed macrophyte (*Najas flexilis*) and its algal epiphytes. *Aquatic Botany* 32:261–281.

New York Sea Grant. 2019. New York Invasive Species Information. http://nyis.info/invasive_species/rock-snot-didymo/.

Nicholls, K.H. 1995. Chrysophyte blooms in the plankton and neuston of marine and freshwater systems. Pages 181–213 in Sandgren, C.D., J.P. Smol and J. Kristiansen (eds.), *Chrysophyte Algae: Ecology, Phylogeny and Development*. Cambridge University Press, Cambridge.

Nicholls, K.H. 2003. Haptophyte algae. Pages 511–521 in Wehr, J. and R.G. Sheath (eds.), *Freshwater Algae of North America—Ecology and Classification*. Elsevier, New York.

Nicholls, K.H. and D.E. Wujek. 2015. Chrysophyceae and Phaeothamniophyceae. Pages 537–586 in Wehr, J., R.G. Sheath and J.P. Kociolek (eds.), *Freshwater Algae of North America—Ecology and Classification*, 2nd edition. Elsevier, New York.

Odum, H.T. 1956. Primary production in flowing waters. *Limnology and Oceanography* 1:102–117.

Officer, C.B., R.B. Biggs, J.L. Taft, L.E. Cronin, M.A. Tyler and W.R. Boynton. 1984. Chesapeake Bay anoxia: origin, development, and significance. *Science* 223:22–27.

Paerl, H.W. 1984. Cyanobacterial carotenoids: their roles in maintaining optimal photosynthetic production among aquatic bloom forming genera. *Oecologia* 61:143–149.

Palmer, J.D. and F.E Round. 1965. Persistent, vertical-migration rhythms in benthic microflora. I. The effect of light and temperature on the rhythmic behavior of *Euglena obtusa*. *Journal of the Marine Biological Association of the United Kingdom* 45:567–582.

Parsons, M.L., Q. Dortch and E. Turner. 2002. Sedimentological evidence of an increase in *Pseudo-nitzschia* (Bacillariophyceae) abundance in response to coastal eutrophication. *Limnology and Oceanography* 47:551–558.

Patrick, R. and D.M. Palavage. 1994. The value of species as water quality indicators. *Proceedings of the Philadelphia Academy of Natural Sciences of Philadelphia* 145: 55–92.

Pollingher, U. 1988. Freshwater armored dinoflagellates: growth, reproduction strategies and population dynamics. Pages 134–174 in Sandgren, C.D. (ed.), *Growth and Reproductive Strategies of Freshwater Phytoplankton*. Cambridge University Press, New York.

Porter, K.G. 1977. The plant–animal interface in freshwater ecosystems. *American Naturalist* 65:159–170.

Reynolds, C.S. and J.-P. Descy. 1996. The production, biomass and structure of phytoplankton in large rivers. *Archiv für Hydrobiologie Supplement* 113:161–187.

Reynolds, C.S. and A.E. Walsby. 1975. Water blooms. *Biological Reviews* 50:437–481.

Reynolds, L. 2009. *Update on Dunkard Creek*. Environmental Analysis and Innovation Division, Office of Monitoring and Assessment, US EPA Region 3, Philadelphia.

Roelke, D.L., J.P.G Rover, B.W. Brooks, et al. 2011. A decade of fish-killing *Prymnesium parvum* blooms in Texas: roles of inflow and salinity. *Journal of Plankton Research* 33:243–253.

Roelke, D.L., A. Barkoh, B.W. Brooks, et al. 2016. A chronicle of a killer alga in the west: ecology, assessment, and management of *Prymnesium parvum* blooms. *Hydrobiologia* 764:29–50.

Rothenberger, M.B., J.M. Burkholder and T.R. Wentworth. 2009. Use of long-term data and multivariate ordination techniques to identify environmental factors governing estuarine phytoplankton species dynamics. *Limnology and Oceanography* 54:2107–2127.

Ruttner, F. 1958. Fundamentals of Limnology. University of Toronto Press, Toronto.

Saul, B., M.G. Hudgens and M.A. Mallin. 2019. Upstream causes of downstream effects. *Journal of the American Statistical Association* 114:1493–1504.

Schade, J.D., K. MacNeill, S.A. Thomas, et al. 2011. The stoichiometry of nitrogen and phosphorus spiraling in heterotrophic and autotrophic streams. *Freshwater Biology* 56:424–436.

Sellers, T. and P.A. Bukaveckas. 2003. Phytoplankton production in a large, regulated river: a modeling and mass balance assessment. *Limnology and Oceanography* 48:1476–1487.

Sherarer, J.A., E.R. DeBruyn, D.R. deClerck, D.W. Schindler and E.J. Fee. 1985. *Manual of Phytoplankton Primary Production Methodology*. Canadian Technical Report of Fisheries and Aquatic Sciences No. 1341. Department of Fisheries and Oceans, Winnipeg, Manitoba.

Shumway, S.E., J.M. Burkholder and S.L. Morton (eds.). 2018. *Harmful Algal Blooms and their Management: A Compendium Desk Reference*. John Wiley & Sons, Hoboken, NJ.

Simon, M., P. López-García, D. Moreira and L. Jardillier. 2012. New haptophyte lineages and multiple independent colonizations of freshwater ecosystems. *Environmental Microbiology Reports* 5:322–332.

Siver, P.A. 2015. Synurophyte algae. Pages 607–652 in Wehr, J., R.G. Sheath and J.P. Kociolek (eds.), *Freshwater Algae of North America—Ecology and Classification*. Elsevier, New York.

Sommer, U. 1988. Growth and survival strategies of planktonic diatoms. Pages 227–260 in Sandgren, C.D. (ed.), *Growth and Reproductive Strategies of Freshwater Phytoplankton*. Cambridge University Press, New York.

Steinman, A.D. 1996. Effects of grazers on freshwater benthic algae. Pages 341–374 in Stevenson, R.J., M. Bothwell and R.L. Lowe (eds.), *Benthic Algae in Freshwater Ecosystems*. Academic Press, New York.

Stevenson, R.J. 1983. Effects of current and conditions simulating autogenically changing microhabitats on benthic diatom immigration. *Ecology* 64:1514–1524.

Stevenson, R.J. 1996. The stimulation and drag of current. Pages 321–340 in Stevenson, R.J., M. Bothwell and R.L. Lowe (eds.), *Benthic Algae in Freshwater Ecosystems*. Academic Press, New York.

Stevenson, R.J., B.H. Hill, A.T. Herlihy, L.L. Yuan and S.B. Norton. 2008. Algae–P relationships, thresholds, and frequency distributions guide nutrient criterion development. *Journal of the North American Benthological Society* 27:783–799.

Sundareshwar, P.V., S. Upadhayay, M. Abessa, et al. 2011. *Didymosphenia geminata*: algal blooms in oligotrophic stream and rivers. *Geophysical Research Letters* 38:LI0405.

Tan, X., Q. Zhang, M.A. Burford, F. Sheldon and S.E. Bunn. 2017. Benthic diatom based indices for water quality assessment in to subtropical streams. *Frontiers in Microbiology* 7:601.

Tilzer, M.M. 1987. Light-dependence of photosynthesis and growth in cyanobacteria: implications for their dominance in eutrophic lakes. *New Zealand Journal of Marine and Freshwater Research* 21:401–412.

US EPA. 2019a. Recommended Human Health Recreational Ambient Water Quality Criteria or Swimming Advisories for Microcystins and Cylindrospermopsin. Fact Sheet EPA 822-F-19-001. US EPA, Washington, DC.

US EPA. 2019b. State Progress Toward Developing Numeric Nutrient Water Quality Criteria for Nitrogen and Phosphorus. https://www.epa.gov/nutrient-policy-data/state-progress-toward-developing-numeric-nutrient-water-quality-criteria (accessed September 2019).

US EPA. 2000. Nutrient Criteria Technical Guidance Manual: Rivers and Streams. Report EPA-822-B00-002. Office of Water and Office of Science and Technology, US EPA, Washington, DC.

Van der Merwe, D. 2012. HAB impacts on domestic animals. *Lakeline* Fall:23–25.

Van Landeghem, M.M., M. Farooqi, B. Farquhar and R. Patiño. 2013. Impacts of golden alga *Prymnesium parvum* on fish populations in reservoirs of the upper Colorado River and Brazos River basins, Texas. *Transactions of the American Fisheries Society* 142:581–595.

Van Meter, K.J., P. Van Cappellen and N.B. Basu. 2018. Legacy nitrogen may prevent achievement of water quality goals in the Gulf of Mexico. *Science* 360:427–430.

Van Nieuwenhuyse, E.E. and J.R. Jones. 1996. Phosphorus–chlorophyll relationship in temperate streams and its variation with stream catchment area. *Canadian Journal of Fisheries and Aquatic SciencesFisheries and Aquatic Sciences* 53:99–105.

Vannote, R.L., G.W. Minshall, K.W. Cummins, J.R. Sedell and C.E. Cushing. 1980. The River Continuum concept. *Canadian Journal of Fisheries and Aquatic Sciences* 37: 130–137.

Watson, S.B. 2004. Aquatic taste and odor: a primary signal of drinking-water integrity. *Journal of Toxicology and Environmental Health* 67:1779–1795.

Weber, S. and J. Janik. 2010. *P. parvum* bloom in a Nevada reservoir. *Lakeline* Summer:27–31.

Wehr, J.D. and J.-P. Descy. 1998. Use of phytoplankton in large river management. *Journal of Phycology* 34: 741–749.

Wehr, J.D., R.G. Sheath and J.P. Kociolek (eds.). 2015. *Freshwater Algae of North America—Ecology and Classification*, 2nd edition. Academic Press, San Diego, CA.

Wehr, J.D. and J.H. Thorp. 1997. Effects of navigation dams, tributaries, and littoral zones on phytoplankton communities in the Ohio River. *Canadian Journal of Fisheries and Aquatic Sciences* 54:378–395.

Wetzel, R.G. 2001. *Limnology*, 3rd edition. Academic Press, New York.

Wetzel, R.G. and G.E. Likens. 1991. *Limnological Analysis*, 2nd edition. Springer, New York.

Whitton, B.A. (ed.). 1975. *River Ecology*. Blackwell Scientific Publications, Oxford.

Whitton, B.A. and M. Potts (eds.). 2000. *The Ecology of Cyanobacteria—Their Diversity in Time and Space*. Kluwer Academic Publishers, Boston.

Wickham, J. 2012. 2012 Gulf of Mexico "Dead Zone" size. *Coastal and Estuarine Research Federation Newsletter* October:26.

Willen, E. 1991. Planktonic diatoms—an ecological review. *Algological Studies* 62:69–106.

Wines, M. 2015. Toxic algae outbreak overwhelms a polluted Ohio River. *New York Times* September 30. https://www.nytimes.com/2015/10/01/us/toxic-algae-outbreak-overwhelms-a-polluted-ohio-river.html.

WHO. 2003. *Guidelines for Safe Recreational Water Environments. Vol. 1: Coastal and Fresh Waters*. WHO, Geneva.

Yuan, L.L., A.I. Pollard, S. Pather, J.L. Oliver and L. D'Anglada. 2014 Managing microcystin: identifying national-scale thresholds for total nitrogen and chlorophyll *a*. *Freshwater Biology* 58:1979–1981.

Zimba, P.V., P.D. Moeller, K. Beauchesne, H.E. Lane and R.E. Triemer. 2010. Identification of euglenophycin—a toxin found in certain euglenoids. *Toxicon* 55: 100–104.

Zimba, P.V, I.S. Huang, D. Gutierrez, W. Shin, M.S. Bennett and R.E. Triemer. 2017. Euglenophycin is produced in at least six species of euglenoid algae and six of seven strains of *Euglena sanguinea*. *Harmful Algae* 63: 79–84.

# CHAPTER 4

# Lotic Primary Producers

## Macroalgae and Macrophytes

This chapter describes the other two major groups of primary producers in streams, rivers, and reservoirs: macroalgae and macrophytes. Both are technically "macrophytes," primary producers that are visible to the human eye, including aquatic mosses and ferns, and angiosperms. Nevertheless, the term is usually restricted to refer to vascular plants, with macroalgae considered separately (Fox 1992; Wetzel 2001).

## 4.1 Macroalgae

Macroalgae are loosely defined as algae that are visible to the human eye. They are generally considered to have a simple plant-like body, called a thallus, that is macroscopic (visible to the human eye) and ranges from simple to complex in structure (Graham et al. 2016). Freshwater macroalgae, unlike macrophytes (aquatic vascular plants), have no vascular tissues. Beyond thalloid taxa, the term "macroalgae" has been broadened to include filamentous, colonial, tuft-forming, crustose, tissue-like, and coenocytic algae or cyanobacteria that can be visibly discerned (Lapointe et al. 2018 and references therein). The distinction is somewhat artificial because many filamentous macroalgae begin growth in a benthic habit but proliferate in the water column, and because many of these organisms would, in low abundance, be considered as microalgae until they become clearly visible to the human eye.

A major constraint on stream macroalgal biomass is current velocity (Sheath and Hambrook 1988). Most attached macroalgae do well in slow or moderate currents, but storm-induced spates can tear them from the substrata, and current-entrained inorganic suspended solids can abrade and shred macroalgae (Biggs 1996). Notable exceptions to this generalization are freshwater reds, which commonly form thick carpets that provide excellent habitat for chironomids and other macroinvertebrates even in whitewater-fast currents (Everitt and Burkholder 1991). Slowly moving streams and rivers can support floating macroalgal mats, but during strong rain/flooding events they are swept downstream. Herbivory can control the growth of some taxa such as the net-forming green alga *Hydrodictyon*, but cyanobacteria and grazer-resistant filamentous green algal taxa can be little affected by grazers (Burkholder 2009).

Depending on the species, macroalgae can thrive floating in the water column and/or attached to various substrata in a benthic habit. In streams, rivers, and reservoirs, macroalgae are mostly beneficial, providing food and valuable habitat for many other organisms. Some species, however—especially some green algae (e.g. *Cladophora*; see Chapter 3)—proliferate rapidly in rivers, streams, and reservoirs when stimulated by nutrient pollution (see below). The night-time respiration of the resulting excessive biomass commonly exerts a BOD and robs the water of oxygen, causing fish kills, degrading habitats, and reducing species diversity. Some benthic cyanobacteria are capable of producing potent toxins as well. Excessive macroalgal biomass can cause an array of adverse socio-economic impacts for people, such as decreased aesthetics and recreational uses of waterways, taste-and-odor problems in drinking water supplies, clogging of water intakes, increased costs

*River Ecology*. Michael A. Mallin, Oxford University Press. © Michael A. Mallin (2023). DOI: 10.1093/oso/9780199549511.003.0005

of managing aquatic resources, and loss of desirable finfish and shellfish.

In freshwaters, macroalgae mostly consist of widespread green algae (Chlorophyta, Streptophyta) and floating and benthic mat-forming cyanobacteria (Figs. 4.1; 4.2). Thalloid red algae (Rhodophyta) and a few golden-brown and brown forms (Heterokontophyta) can be abundant in more restricted habitats. These phylum-level groupings are based on molecular science, life histories, the fine structure of their cells under transmission electron microscopy, and other general features such as pigments and food storage products. Selected stream-inhabiting macroalgae are described below, beginning with cyanobacteria as the most primitive group.

### 4.1.1 Benthic Mat-Forming Cyanobacteria

Blooms of benthic cyanobacterial macroalgae generally are much more cryptic than the green, obvious appearance of many planktonic cyanobacterial microalgal blooms, and much less is known about them. There is a critical need to fill major knowledge gaps about their ecology and impacts, because recent information indicates that many populations of cyanobacterial macroalgae in benthic habitats are highly toxic.

At present, among the best-known noxious benthic mat formers in freshwaters is *Lyngbya wollei*, also called *Microsiera wollei* (Fig. 4.1), found in temperate to tropical, alkaline to mildly acidic rivers, reservoirs, and wetlands in many regions of the world (Speziale and Dyck 1992; Hudon et al. 2014). This native species recently has begun to act as an opportunistic invader or initial colonizer following natural or human-related disturbances (Lapointe et al. 2018 and references therein). It can fix atmospheric nitrogen into ammonia, giving it an advantage over many potential competitors; and it can produce many cyanotoxins. Initially, *L. wollei* grows attached to various benthic substrata. As mats develop, oxygen bubbles from photosynthesis and other disturbances commonly dislodge them so that they become free-floating (Fig. 4.1a, b). Biomass apparently is maximal in summer–early fall, but *L. wollei* can overwinter with high biomass, as well, in warm temperate climates (Hudon et al. 2014

and references therein). This cyanobacterium is well adapted to low light and can also withstand high light. Mats of *L. wollei* can be up to 10 m long, 1 m wide, and 0.5 m thick (Fig. 4.1a, b). Reported harmful ecological impacts have included declines in beneficial fauna, lower species richness, and toxicity to would-be grazers. Among various harmful socio-economic effects are clogged water intakes, offensive odors, and highly compromised recreational and potable water use (Hudon et al. 2014 and references therein). It is believed that *L. wollei* blooms will continue to increase because it thrives at high temperatures and can be favored over competitors by nutrient pollution (e.g. Vis et al. 2008).

Much less ecological information is available about other benthic mat-forming cyanobacteria (such as *Phormidium*; Fig. 4.2a, b), but at least 34 taxa found in various rivers worldwide have been confirmed to produce numerous toxins with multiple modes of action and target organs (Quiblier et al. 2013; McAllister et al. 2016; Shalygin et al. 2019). Most of the taxa are impossible to identify without molecular techniques (Fig. 4.2b). They thrive in habitats ranging from oligotrophic to eutrophic, including streams, rivers, quiet areas of reservoirs, and downstream estuarine/coastal areas (Lapointe et al. 2018). The mats can be found on soft sediments or rock substrata (Fig. 4.2a) and are usually about 1 cm in thickness, but can be up to 70 cm thick. Water-column nutrient concentrations (both N and P) during initial colonization can strongly influence initial mat formation, but, over time, the microenvironment within the mat biofilm becomes separated and distinct from that of the overlying water (Wood et al. 2015). Filamentous cyanobacteria are motile, and when in benthic mats, they can exude mucus to use gliding motility (see Chapter 3) to move to different levels within the mat to regulate their exposure to light or nutrients.

The prevalence of some taxa is increasing in rivers worldwide (e.g. Wood and Puddick 2017) and "reports of their involvement in animal poisonings have also increased markedly during the last decade" (Quiblier et al. 2013). Dog deaths have occurred following ingestion of benthic *Phormidium* mats containing anatoxin-producing strains and by ingesting benthic *Oscillatoria* mats (Lapointe et al. 2018 and references therein).

(a)    (b)    (c)

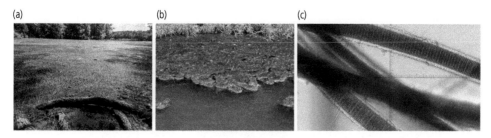

**Fig. 4.1** *Lynbya wollei* (also called *Microsiera wollei*). Fig. 4.1a (left) Bloom over a wide stretch of a run-of-the-river reservoir. Fig. 4.1b (middle) Close-up view of a bloom. Fig. 4.1c (right) Microscopic view of *L. wollei* showing individual filaments. (Photos courtesy of D. Wiltsie, North Carolina Department of Environmental Quality.)

(a)    (b)

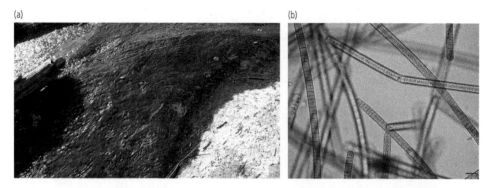

**Fig. 4.2** Filamentous cyanobacterium *Phormidium*. Fig. 4.2a (left) Dense mat on rocky substrata in a North Carolina Piedmont stream. Fig. 4.2b (right) Microscopic view of filaments. (Mat photo by the author; microscopic view photo by Amy E. Grogan, UNC Wilmington.)

## 4.1.2 Stream-Inhabiting Red Algae

The contribution of red algae (Fig. 4.3) to production in small to mid-sized, clear, fast-flowing, relatively pristine softwater streams is often substantial in mountainous areas, New England, and even parts of the Piedmont in the Southeastern USA. The productivity of two *Batrachospermum* (Fig. 4.3a) species (20–50 mg $O_2$/g dry wt/h) is among the highest measured for benthic algae (Westlake 1980). Most freshwater reds are found in running waters, with an apparent threshold requirement of about 30 cm/s (Sheath 1984). They thrive in nutrient-poor conditions and provide excellent habitat for beneficial fauna (e.g. aquatic insects such as caddisfly larvae).

Freshwater reds are a misnomer with respect to color, since most of them are not red at all but, rather, may appear olive to bluish-green (Fig. 4.3a, b). In their upstream habitats there is high potential for washout to harsh downstream areas where they could not survive. Their fascinating life histories meet this challenge in two basic ways: first, they have a perennial crustose stage ("chantransia") that steadfastly holds onto suitable space on rocks and stays there whether inundated by floodwaters or baking in hot sun when exposed during droughts; and second, under seasonally favorable conditions, the other two major stages in the life history attach directly on top of the crustose stage, ensuring that the entire life history can be completed in the favorable upstream habitat.

## 4.1.3 Green Algae

Stream-inhabiting green macroalgae are mostly filamentous, although some thalloid and colonial forms also occur. While most are beneficial and important in providing desirable habitat for invertebrates and fish, this diverse group also includes some notorious responders to nutrient pollution

(a)

(b)

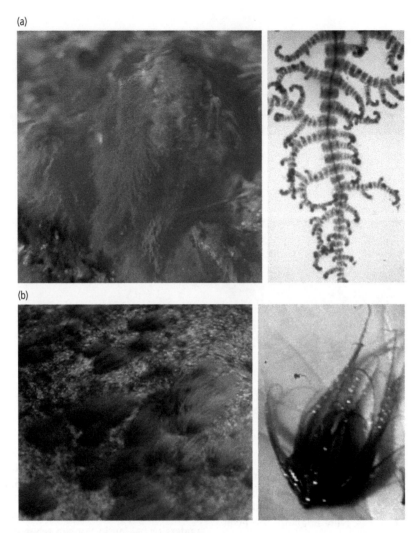

**Fig. 4.3** Stream views and close-up views of stream-inhabiting thalloid red macroalgae. Fig. 4.3a (top) *Batrachospermum*. Fig. 4.3b (bottom) *Paralemanea*. (Photos courtesy of JoAnn M. Burkholder of North Carolina State University except for *Batrachospermum* close-up by the late Robert Sheath.)

(Burkholder 2009). They commonly overgrow stream and river habitats, causing adverse changes in food quality and habitat for other biota, such as elimination of more desirable algal food resources, and major diel DO swings from supersaturation of oxygen during the day to DO sags at night.

### 4.1.3.1 Thalloid Greens

The thalloid green algae in streams and the shallows of some reservoirs are mostly beneficial charaleans (Streptophyta, order Charales) that live in habitats that span the pH gradient from hardwaters (*Chara*

[Fig. 4.4a] and other taxa) to mildly acidic softwaters (*Nitella* spp.). Their rhizoids, structures that anchor them in sediments, are root-like in that they can take up nutrients. Their eggs and sperm develop in protective jackets of vegetative cells, an advanced feature that sets them apart from most algae. On the other hand, like many algae, they can reproduce rapidly by fragmentation or vegetatively (asexually) (Graham et al. 2016 and references therein). One species, starry stonewort (*Nitellopsis obtusa*), is an exotic invasive that has proliferated via asexual reproduction in rivers of the Midwestern and

(a)                                          (b)

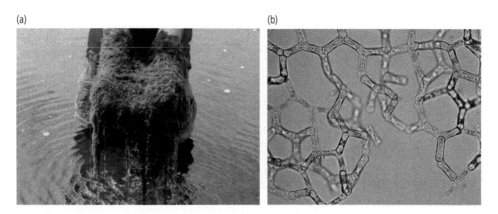

**Fig. 4.4**  Fig. 4.4a (left) Green alga *Chara* from a North Carolina impoundment. Fig. 4.4b (right) Net-like *Hydrodictyon* in microscopic view; in streams, large net-like masses can form. (Fig. 4.4a courtesy of US Fish & Wildlife Service; microscopic view photo by Amy E. Grogan, UNC Wilmington.)

Eastern USA, and its range is expanding (Larkin et al. 2018). Charaleans can sometimes form high biomass (Fig. 4.4a) with growth reaching the water surface, undesirable for fishermen and other recreationists, in shallow waters (Lembi 2003).

Other stream-inhabiting green macroalgae have varied morphologies, such as *Tetraspora*, which is found in second- to third-order softwater streams in winter/early spring. This taxon resembles greenish amorphous, thick "filaments" when seen in stream habitat. Under light microscopy it resembles greenish "Swiss cheese," consisting of groups of cells among mucus that has obvious holes in it. Another example, water net (*Hydrodictyon reticulatum*), produces colonies called coenobia that, to the human eye, look just like green nets (Fig. 4.4b). Although often cryptic, these colonies can grow to 1 m or more in length (Burkholder 2009 and references therein).

### 4.1.3.2 Filamentous Greens

This group thrives in lower- to mid-order, slowly flowing, and quieter areas of streams with moderate to high available light. They often show strong seasonality; many taxa occur in late spring/summer, but some are obligate winter/early-spring forms. They are highly diverse and mostly beneficial, providing habitat and food for higher trophic levels.

Unfortunately, however, filamentous green macroalgae in streams, rivers, and reservoirs are well known mainly because some taxa are highly stimulated by nutrient pollution from sewage, agricultural runoff, and other sources. They commonly form major noxious overgrowths and when these "blooms" decompose, the result can be elevated BOD and hypoxia/anoxia. They commonly cause substantial ecological and economic impacts (Lapointe et al. 2018 and references therein).

Harmful high-biomass blooms of filamentous green macroalgae can be free-floating near/at the water surface, and/or near the sediments (metaphytic), and/or attached to benthic substrata. As examples, streptophyte greens (e.g. *Spirogyra* and *Mougeotia* spp.; Fig. 4.5a) commonly form seasonal blooms in alkaline and mildly to moderately acidic, nutrient-enriched waters in many regions of the world. Their rapid overgrowth sometimes occurs in impoundments and slowly flowing streams and ditches affected by nutrient pollution (Burkholder 2009). For example, *Spirogyra* spp. form both benthic and floating mats, and some taxa respond rapidly to N and P enrichment especially in high light.

Chlorophyte greens such as *Oedogonium*, *Pithophora*, and *Cladophora* also commonly respond to nutrient pollution by producing high-biomass, floating blooms. Of these, *Cladophora* can be floating or attached to substrata (Fig. 4.5b) and historically has been thought to be the most widely distributed macroalga throughout the world's

(a)                               (b)

**Fig. 4.5** Fig 4.5a (left) *Mougeotia* + *Spirogyra* bloom covering a eutrophic North Carolina impoundment. Fig 4.5b (right) *Cladophora* bloom in nutrient-enriched Stocking Head Creek, draining CAFOs in eastern North Carolina. (Photos by the author.)

freshwater ecosystems (extending into estuaries and marine coastal areas as well) and the most abundant macroalga in alkaline streams worldwide (Dodds and Gudder 1992; Lapointe et al. 2018). Various *Cladophora* spp. have been shown to be strong responders to P pollution and accumulate in large masses in eutrophic streams (Fig. 4.5b) and rivers (Burkholder 2009 and references therein). *Cladophora* spp. may be grazed when small, but they generally are a poor, nonpreferred food source (Zulkifly et al. 2013). They have a relatively high light optimum for photosynthesis and can rapidly acclimate to low or high light.

Filamentous green algal blooms can be operationally problematic for industrial and municipal water intakes, as well as ecologically harmful, with impacts similar to those caused by harmful blooms of suspended microalgae. The blooms cause noxious odors and taste-and-odor problems in potable water supplies, shading and loss of beneficial macrophytes and phytoplankton taxa beneath surface macroalgal accumulations, reduction and loss of beneficial habitat for fish, clogging of water intake systems for municipalities and industries, crowding out of more beneficial plant species, interference with boating, swimming, and other recreation, and decrease of waterfront property values (Lapointe et al. 2018 and references therein). The high BOD from decaying blooms leads to hypoxia or anoxia and, in turn, physiological stress and death of fish and other aquatic life.

### 4.1.4 Diatoms

The filamentous diatom *Eunotia pectinalis* (Fig. 4.6) is abundant in mildly acidic softwater streams worldwide (Patrick and Reimer 1966; Sheath and Burkholder 1985 and references therein). For example, in the spring season, its period of maximal growth, *E. pectinalis* strongly dominates New England streams, and historically dominated the stream flora of Coastal Plain streams in the Southeastern USA (Whitford and Schumacher 1963) and some river systems of central Finland (e.g. Eloranta and Kunnas 1979). This macroalga has the appearance of golden-brown hair (Fig. 4.6), but when collectors pick it up and rub the clump of filaments between their fingers, the filaments "disappear" as the cells separate. It is believed to be a beneficial species in stream food webs, but, despite its seasonal predominance in many softwater streams throughout the world, very little is known about its autecology or its roles in aquatic food webs.

### 4.1.5 Ecological Roles of Lotic Macroalgae

Macroalgae perform various beneficial roles, especially in low to moderate nutrient regimes. They can provide valuable habitat and nursery areas for small fish and invertebrates to hide from predators, and some of them provide substrata for benthic microalgae and their invertebrate grazers. Some macroalgae serve as food for fish or invertebrates, although

(a)                                                    (b)

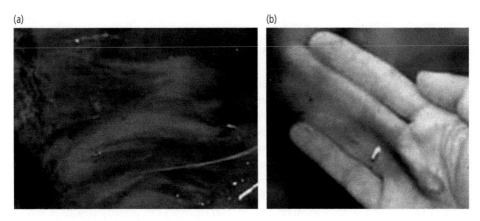

**Fig. 4.6** Two perspectives of the filamentous diatom *Eunotia pectinalis* in its benthic habitat within a softwater Rhode Island stream. (Photos courtesy of JoAnn M. Burkholder, North Carolina State University.)

others such as *Cladophora* (and many cyanobacterial taxa) are avoided by consumers. Macroalgae can slow local currents to allow for biotic deposition of food materials or living habitat. They also are useful to scientists and the regulatory community in that they can be used as bioindicators of stream water quality (reviewed by Stancheva and Sheath 2016), preferably in conjunction with diatoms.

The genus *Cladophora* is probably the best studied of the freshwater macroalgal taxa (reviewed by Dodds and Gudder 1992 and Zulkifly et al. 2013 among others). Zulkifly et al. (2013) described *Cladophora* as follows: "By virtue of its robust construction and environmental persistence, *Cladophora* functions as a forest-like scaffold that provides numerous ecological niches." It provides substrata for epiphytic diatoms and its clumps are also associated with several filamentous green algal and cyanobacterial taxa. *Cladophora* is generally resistant to grazing except by larger fish (Power 1990), and can reproduce sexually to form zygotes or asexually to form flagellated spores and can also form akinetes (resting cells). Because *Cladophora* growths can provide diverse habitat to hard substrata, modify currents, provide habitat for microalgal periphyton and numerous associated minute animals, and modify gas exchange, this genus has been referred to as ecological engineers (Zulkifly et al. 2013). *Cladophora* is being used in the biofuel industry as a carbon feedstock for genetically engineered bacteria and is being assessed for other economically valuable uses as well.

### 4.1.6 Influences on Macroalgal Growth in Lotic Habitats

Many of the factors that control the growth of suspended microalgae in running waters also control macroalgal growth. Along with the major influence of flow, most riverine macroalgae are strongly controlled by temperature and photoperiod, so that they have clear seasonality. In temperate regions, macroalgal abundance and diversity are most abundant during winter/early spring when tree canopy is minimal (e.g. Sheath and Burkholder 1985, 1988; Everitt and Burkholder 1991). For macroalgae to persist or thrive in a given stream or river habitat they first require adequate light. In heavily shaded streams or in turbid or tannin-colored waters, growth typically is light-limited. Attachment sites will also limit the abundance of benthic macroalgae, which can grow densely in rocky streams and on human-made surfaces, but poorly in sandy-bottomed streams. Macroalgae can attain moderate and even lush growth in streams with light and dependable substrata (e.g. Everitt and Burkholder 1991). Lotic environments are inherently more nutrient-rich than lentic habitats because, even if the nutrient concentrations are very low, the current constantly carries renewed supplies near the algal cells and reduces the thickness of the boundary layer, facilitating nutrient uptake (Whitford and Schumacher 1964; Whitton 1975). Under elevated nutrient loading, noxious growth of macroalgae can occur, as previously mentioned.

Human management attempts to minimize noxious macroalgal blooms in flowing waters can be tricky (see also Chapter 14). Applications of herbicides will kill bloom-forming species, but a rapid collapse of a bloom will impart a BOD load that may lead to anoxia and fish kills, as discussed in Chapter 3 regarding phytoplankton blooms. In a regulated river, increasing dam release to increase flow to bloom-disruptive levels will reduce or eliminate downstream noxious macroalgal blooms, but reducing water levels behind the dam may interfere with other water uses. Reducing nutrient inputs to a given stream or river would be the logical choice, but such reductions are fought by point-source dischargers upstream as well as sources of non-point nutrient pollution (see Chapter 13).

## 4.2 Macrophytes

Aquatic macrophytes spend at least a portion of the year in water, partially or wholly. Various species occur across the salinity gradient from freshwaters to brackish and marine waters, mostly on or near shorelines, growing attached in sediments or unattached below the water surface, on the water surface, or on various hard surfaces within the water or on the water surface. Hydrophilous submersed plants have highly reduced flowers and complete pollination underwater; others elevate their flowers at or above the water surface and are wind pollinated. Moderate macrophyte abundance provides many benefits to aquatic communities in general (see below), but their overabundance leads to many of the same water quality issues previously described for microalgae and macroalgae.

### 4.2.1 Importance

Aquatic macrophytes are hugely important to stream and river water quality and ecosystem function. They can stabilize shorelines with their root masses, and their physical presence protects streambanks from direct, abrasive effects of rainfall and enhanced currents. They help to minimize both onshore and within-stream erosion; they also reduce turbidity and protect in-stream habitats from burial. Macrophytes on or near shore intercept overland runoff before it enters the main stream, river, or

reservoir, thus acting as filters of sediment and other pollutants. As runoff encounters macrophyte beds, the velocity slows and suspended sediments with associated pollutants settle out or are physically filtered by the plant material. This is also true within the stream itself, as macrophytes buffer currents and cause particle sedimentation (see Chapter 1). Macrophytes reduce water-column nutrient loads and concentrations. As described earlier, runoff from urban or agricultural areas commonly carries excessive nutrient loads into receiving waters that can cause algal blooms and degrade water quality. Macrophytes can directly absorb N and P from the sediments and water column. The bacterial consortium in the rhizosphere (root–rhizome complex) of some species greatly enhance denitrification (Song et al. 2014), which causes direct N loss from aquatic ecosystems to the overlying air as $N_2$ gas.

Aquatic communities greatly benefit directly from macrophyte growth as well. Some species are used as food by various organisms (see Section 4.2.7) depending upon their nutrient content and palatability. In addition, macrophytes provide enormous surface area as substrata for periphyton (attached algae and the associated microbial consortium) which are a major food resource consumed by many organisms in higher trophic levels (see Chapters 3, 5, and 6). Macrophytes create critical habitat for beneficial aquatic life such as many species of littoral zooplankton, macroinvertebrates, crustaceans, and fish. Many of these organisms consume the periphyton attached to macrophyte leaves; in addition, macrophyte beds serve as nursery areas for young stages of many organisms to hide from larger predators (i.e. to avoid becoming food for fish, wading birds, etc.). Smaller adult fish also use macrophyte beds to hide from larger fish, who work the outskirts of macrophyte beds to hunt prey.

### 4.2.2 Primitive Aquatic Vascular Plants: Mosses, Liverworts, and Ferns

Primitive vascular plants—the aquatic mosses and liverworts (phylum Bryophyta) and ferns (phylum Pteridophyta)—are often overlooked but sometimes abundant (Fig. 4.7) in stream ecosystems of many geographic regions, such as the reported

(a)

(b)

**Fig. 4.7** Fig 4.7a (left) Giant salvinia *Salvinia molesta* bloom in North Carolina wetland. Fig. 4.7b (right) Close-up of giant salvinia. (Photos courtesy of Rob Emens of North Carolina Department of Environmental Quality.)

dominance of the common water moss *Fontinalis* sp. during warm seasons in some Northeastern US streams (Everitt and Burkholder 1991). Some bryophytes (phylum Bryophyta, mosses, and liverworts) are especially well adapted to lotic habitats because they can thrive in low light, are eurythermal, have rapid nutrient uptake, and are resistant to scouring and spates. The largest plants of this group are some of the aquatic mosses, which can reach 100 cm or more in length. Although they are not nearly as well studied as advanced (flowering) aquatic macrophytes, the production of primitive vascular plants can exceed that of microalgal periphyton, macroalgae, and other submersed macrophytes (Glime 2017).

Aquatic bryophytes can strongly influence the community structure of stream fauna in general. These plants decrease stream velocity at the rock surface; trap substantial detritus which serves as food for shredder invertebrates; provide refuge habitats from predators for some aquatic fauna and better background coloration for camouflage; provide major surface area for colonization of algae that are food resources for various fauna; and retain water longer than other stream substrata during exposure to the air and desiccation (Glime 2017 and references therein).

In some waters the most abundant members of this group are ferns, including the small water fern (or mosquito fern, *Azolla*) (see Fig. 4.13) which can be found in often-nutrient-poor softwater habitats;

these ferns have cyanobacterial endosymbionts that help provide N through their N fixation (Peters and Meeks 1989). In marked contrast, the exotic invasive fern giant salvinia (*Salvinia molesta*) (Fig. 4.7) grows to massive nuisance coverage and thrives in nutrient-polluted waters (Kay 2001; see Chapter 14).

### 4.2.3 Advanced Aquatic Vascular Plants: Angiosperms

Angiosperm communities in streams, rivers, reservoirs, and adjacent wetlands often are characterized by strong dominance of one taxon (Sculthorpe 1967). The roles of flowering macrophytes in streams and rivers can best be examined through a functional group perspective (Fox 1992). These macrophytes include four major functional groups: emergent, floating-leaved, submersed, and free-floating (Fig. 4.8). Conditions such as stream flow velocity, light availability, and substrata determine where in the river ecosystem members of these functional groups are likely to be found. Emergent vegetation mostly occurs along the shores of lower rivers, in river backwater areas, and in fresh, brackish, and salt marshes. Floating-leaved rooted vegetation is confined to quiet freshwater areas such as stream pools, river backwaters and oxbows, and shallow reservoir arms and coves. Submersed vegetation is common in shallow midriver main channels and along the banks where there is sufficient light, in reservoir arms and coves,

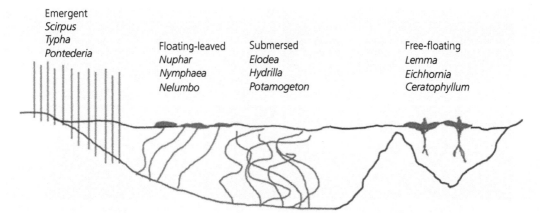

Emergent
*Scirpus*
*Typha*
*Pontederia*

Floating-leaved
*Nuphar*
*Nymphaea*
*Nelumbo*

Submersed
*Elodea*
*Hydrilla*
*Potamogeton*

Free-floating
*Lemma*
*Eichhornia*
*Ceratophyllum*

**Fig. 4.8** Stylized view of habitat preferences of four major functional groups of aquatic macrophytes, along with common examples of each. (Figure by the author.)

and in upper stream reaches where adjacent tree canopy opens and creates light-sufficient conditions. A special case, riverweed (*Podostemum*), the world's most highly modified aquatic angiosperm, anchors its holdfast-like roots to rocks in upper stream reaches. Free-floating vegetation is common in oxbows, nutrient-enriched reservoir protected areas, low-velocity Coastal Plain streams, and other quiet waters. In general, the amount of macrophyte vegetation in a stream, particularly during maximal biomass periods, is negatively related to average stream flow (Biggs 1996 and references within).

### 4.2.3.1 Emergent Macrophytes

Most emergent macrophytes ("emergents") are structurally robust angiosperms that are anchored by their roots in periodically wet, exposed soils or submersed aquatic sediments (Fig. 4.8). They have well-developed vascular tissue that enables them to remain upright and withstand winds, rain, and other abrasive conditions. If submersed, some species can survive down to a depth of ~1.5 m. Many taxa are adept at thriving in habitats ranging from nearly dry to submersed (Crawford 1987). The leaves of these versatile plants may range from broad in an aerial habit to narrow and strap-shaped below the water surface, and even the same plant can show this range in leaf shape (heterophylly).

Emergents have aerial reproductive organs—that is, they are elevated above the water and are pollinated by either wind or insects (anemophilous

or entomophilous, respectively). Because of their location (Fig. 4.8) they can use resources characteristic of both terrestrial and aquatic habitats, which is why the most productive vascular plants in the world are emergent macrophytes (Wetzel 2001). Atmospheric $CO_2$, the dominant source of carbon for emergent macrophytes, generally is taken up by aquatic plants via the slow process of diffusion, rather than by active uptake enzyme systems. It is much easier to access by emergent plants than by plants submersed below the water surface, because $CO_2$ diffuses 10,000 times more slowly through water than through air. As another benefit of the emergent habit, light is plentiful, especially because the lance-like, straight leaf shape of many emergents minimizes self-shading as often-dense growth develops (Sculthorpe 1967).

The substrata where emergent macrophytes are anchored are usually anaerobic, which makes nutrients much more soluble and available for uptake. Emergent plants have a ventilation system of air tubes, sometimes collectively called a lacunar system, that are oriented horizontally and vertically from the leaves down through the stem to the roots. These plants use the lacunar system to shunt oxygen from leaf photosynthesis down to oxygenate the root zone. This trait gives them another, major advantage because it enables the roots to use aerobic respiration to take up nutrients from the rich but anaerobic soil. To defend against excessive water entry from wounding (e.g. by grazers) the lacunar

**Fig. 4.9** Common emergent macrophytes. Fig. 4.9a (top left) Close-up of cattail *Typha* sp. Fig. 4.9b (top right) Cattail monoculture from phalanx growth. Fig. 4.9c (middle left) Alligatorweed *Alternanthera philoxeroides*. Fig. 4.9d (middle right) Creeping water primrose *Ludwigia hexapetala*. Fig. 4.9e (bottom left) Pickerelweed *Pontederia* sp. Fig. 4.9f (bottom right) Pennywort *Hydrocotyle* sp. (Photos by the author.)

system is equipped with diaphragms, sheets of multiperforated cells that stretch across the air tubes at some locations.

Many species of emergent macrophytes occur along riverbanks, the shores of reservoirs with dependable water levels, and in adjacent wetlands. Cattails (*Typha* spp.) are widely distributed, anemophilous, and easily recognized by their dense flowering heads (Fig. 4.9a). They can be beneficial or noxious, locally invasive species that can rapidly crowd out and replace native flora via their

phalanx growth (Fig. 4.9b). Their recent takeover of large portions of the Florida Everglades, for example, was linked to phosphorus pollution from human activities. Nutrient pollution afforded them an advantage over the native sawgrass (the sedge *Cladium jamaicense*) which thrives in nutrient-poor habitats (Belanger et al. 1989). Some species of emergents such as alligatorweed (*Alternanthera philoxeroides*) (Fig. 4.9c) and creeping water primrose (*Ludwigia hexapetala*) (Fig. 4.9d) can form dominant monocultures in shallow water (see Chapter 14).

Pickerelweed (*Pontederia cordata*) has broad leaves and attractive purple flowers (Fig. 4.9e); this species has been valuable when used in constructed wetlands adjacent to streams and rivers because the bacteria associated with its rhizomes are highly effective in denitrification (Song et al. 2014). Pennywort (*Hydrocotlye*I is widespread and abundant in shallow water and in moist soils (Fig. 4.9f).

Grasses, sedges, and rushes are among the most common emergent taxa worldwide (Fig. 4.10) and often form dense growths, facilitated by minimal self-shading of their narrow strap-shaped leaves. Many species of these three groups closely resemble one another, so that their reproductive structures often are needed for identifications. These plants are mostly beneficial as habitat for wildlife, but a few are notorious exotic invasives, such as common reed (*Phragmites australis*) (Fig. 4.10d) which also has native benign populations. Exotic invasive common reed has overgrown areas along riverbanks, ditches, and other habitats in many geographic regions

because it can thrive in disturbed areas and outcompetes more sensitive native species (see Chapter 14). As the salinity gradient along a river changes from fresh to oligohaline and then to higher salinity, various species of salt-tolerant sedges, rushes, and grasses replace obligate freshwater emergents.

Moderate abundance of emergent macrophytes is essential to achieve the water quality improvement function of natural and constructed wetlands along streams and rivers (see Chapter 16). Emergent macrophytes slow currents (Madsen and Warncke 1983; Petticrew and Kalff 1992; Leonard and Luther 1995; see Chapter 1), so that algae and other suspended materials settle out and become food for benthic fauna while water quality improves. Through this mechanism, emergent macrophytes can remove excess nutrients from the overlying water that might otherwise promote noxious algal blooms. Their rhizosphere takes up nutrients directly from the interstitial water in the sediments. In addition, the consortium of bacteria in the

**Fig. 4.10** Common emergent grasses and sedges. Fig. 4.10a (top left) Giant cutgrass *Zizaniopsis* sp. Fig. 4.10b (top right) Rush *Juncus* sp. Fig. 4.10c (bottom left) Bur-reed *Sparganium*. Fig. 4.10d (bottom right) Invasive common reed *Phragmites australis*. (Photos by the author.)

rhizosphere remove N from the system as $N_2$ gas through denitrification (see Section 4.2.6), which is a very important reason to protect riparian wetlands.

### 4.2.3.2 Floating-Leaved Macrophytes

Floating-leaved macrophytes (Fig. 4.11), many of which are commonly called water lilies or lily pads, are among the most attractive aquatic plants because various species produce spectacular flowers (Fig. 4.11a, c). These plants thrive in quiet, relatively shallow waters (depths 1–3 m) such as shoreside pools and reservoir coves characterized by minimal abrasive flow and protection from wind. In rivers they are especially important in tropical habitats. The largest known species is the giant water lily (*Victoria amazonica*), which can be 8 m long. Floating-leaved macrophytes can be highly abundant (Fig. 4.11b) and, if crowded, the floating

leaves can become aerial in some species. They are firmly rooted with extensive rhizome systems, and their long, flexible petioles can grow rapidly to elevate the specially designed floating leaves up to the water surface (Sculthorpe 1967). Nutrients are moved efficiently from the root system in the sediments up to the leaves (Wallace and O'Hop 1985). The flowers are floating or aerial and usually entomophilous (Sculthorpe 1967). Floating-leaved macrophytes have a lacunar-based efficient ventilation system to transport $O_2$ from the atmosphere to the roots and $CO_2$ and methane ($CH_4$) from the roots to the atmosphere (Dacey 1981; Bubier and Moore 1994).

Lotus (*Nelumbo*), sometimes called sacred lotus (Shen-Miller 2002), has long factored in eastern religious symbolism, in part because of the inspirational beauty of its flowers. Lotus is also cultivated

**Fig. 4.11** Floating-leaved macrophytes. Fig. 4.11a (top left) Spatterdock *Nuphar* with yellow flower. Fig. 4.11b (top right) Profusion of *Nuphar* in freshwater tidal area of a coastal South Carolina river. Fig. 4.11c (bottom left) Water lily *Nymphaea* with white flower. Fig. 4.11d (bottom right) Narrow-leaved *Nuphar* from fifth-order Black River showing variation in form. (Photos by the author.)

for human consumption, especially in China (e.g. Guo 2008). Other common examples of floating-leaved macrophytes include spatterdock or yellow pond lily (*Nuphar*) (Fig. 4.11a, b, d), water lily (*Nymphaea*) (Fig. 4.11c), and watershield (*Brasenia*). These plants are generally rooted rosettes, but with highly developed floating leaves. Among very different species even from different taxonomic orders, there is remarkable consistency in the structure and shape of floating leaves—probably parallel evolution in a restricted and sharply defined habitat, the air–water interface.

Various physiological and physical factors enable this functional group to be successful in calm, protected aquatic habitats. The floating habit makes plenty of light available for photosynthesis. Their floating leaves are usually positioned horizontally on the surface, which exposes maximum leaf area to incident light so that water turbidity is not a problem for light acquisition. In addition, atmospheric $CO_2$ is abundant and easily accessible. Their nearshore location provides ready access to nutrients from overland runoff, and in shallow habitats they contend with little stress from wave action or wind.

They have strong, leathery, broad leaves, oval or circular in shape, with a simple margin (Fig. 4.11). The leaves have water-repellent upper surfaces and hydrophilic lower surfaces, with a long flexible petiole that can respond (by altering length) to water depth (Sculthorpe 1967). Still, the near-surface stresses are so challenging that these plants mostly occur in sheltered habitats with minimal turbulence.

### 4.2.3.3 Submersed Macrophytes

Submersed macrophytes (sometimes mistakenly called submerged) live primarily below the water surface and mostly have strap-shaped or highly dissected leaves (Fig. 4.12). The commonly used

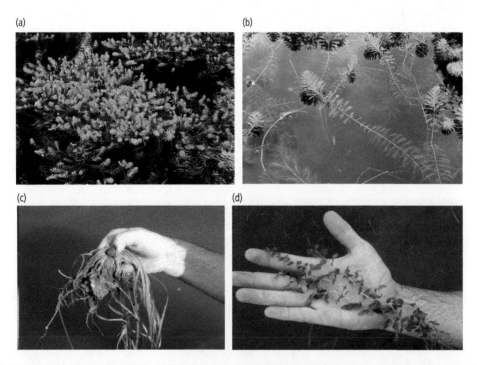

(a)  (b)  (c)  (d)

**Fig. 4.12** Submersed macrophytes. Fig. 4.12a (top left) Parrot feather *Myriophyllum* with pink flowers, growing in a dense patch. Fig. 4.12b (top right) *Myriophyllum* showing heterophylly; thick leaves above surface, thinner and less dense subsurface, and very fine leaves to gather scant light farther below surface. Fig. 4.12c (middle left) Wild celery or tapegrass *Vallisenaria americana*. Fig. 4.12d (middle right) Pondweed *Potamogeton* (this genus has numerous species). (*Myriophyllum* photos by the author; *Vallisenaria* and *Potamogeton* photos courtesy of US Fish & Wildlife Service.)

acronym SAV refers to submersed aquatic vegetation (submersed aquatic macrophytes). As noted earlier, primitive vascular plants, the mosses and liverworts, occur in streams and rivers and are sometimes well represented. For example, aquatic mosses (*Fontinalis* spp.) are dominant macrophytes seasonally in some softwater New England streams (Sheath and Burkholder 1985, 1988). Angiosperms, the most abundant and widespread submersed plants inhabiting streams and rivers, are emphasized here. Most are benign and provide critical habitat for beneficial higher-trophic-level representatives such as finfish and shellfish. For instance, tapegrass (*Vallisenaria*) (Fig. 4.12c) is a major fish and invertebrate habitat in shallow lower riverine areas; various seagrasses perform the same functions in estuarine and marine waters. Occasionally, notorious invasive aquatic weed species such as hydrilla and its lookalike Canadian water weed elodea (*Elodea canadensis*) can cause problematic overgrowths (see Chapter 14).

The most limiting resources for submersed plants are usually carbon and light, essential for photosynthesis (Wetzel 2001) (Box 4.1). Their leaves are often only two to three cells thick, with most chloroplasts in the epidermis, adaptations thought to maximize carbon acquisition and light capture (Sculthorpe 1967). To increase the efficiency of light gathering, gas exchange, and nutrient absorption (see below) leaves can increase their surface area by elongating and becoming more finely dissected with depth from the surface, called heterophylly (Fig. 4.12b). The thin, pliable leaf structures also provide a mechanical advantage in moving water. The presence of submersed macrophytes can reduce current velocity as well (Madison and Warncke 1983; also see Box 4.1 and Chapters 1 and 16), which allows sedimentation of inorganic and organic particles, reduces turbidity, and provides food for benthic organisms.

---

### Box 4.1 Habitat modifiers

SAV beds can be resilient and even improve the nearby water conditions for further expansion. Beds of SAV have long been known to stabilize the physical environment and improve ecological health through the production of cover, the production of algal and macrophyte food, as areas of accumulation of organic matter for grazers and decomposers, and as preferred habitat for numerous vertebrate and invertebrate species.

Using several time-series datasets, researchers in the tidal freshwater upper Chesapeake Bay demonstrated that beds of SAV can preserve their own integrity and even improve physical habitat of adjoining waters (Gurbisz et al. 2016). A large bed of wild celery (*Vallisneria americana*) (Fig. 4.12c), water stargrass (*Heteranthera dubia*), Eurasian watermilfoil *Myriophyllum spicatum*, and hydrilla *Hydrilla verticillata* was studied before, during, and after major 2011 storm/flooding events. High flow from the storms caused SAV loss at the edges of the bed, but the interior of the bed maintained its integrity. Prior to the storms, chlorophyll *a* and turbidity within the bed were lower than areas outside the bed. However, during and after the flooding the concentrations of these parameters within the bed rose due to wind-driven resuspension (with turbidity exceeding 600 NTU), thus causing light-limiting conditions to the macrophyte bed. Later, concentrations of chlorophyll *a* and turbidity again decreased within the SAV bed and the increased light availability led to increased plant growth in summer. Evidently, once the size of the SAV bed reached a critical biomass, the drag caused by the canopy dissipated the wind-driven wave energy, reducing resuspension of sediments and reducing turbidity. This "positive feedback" mechanism continued to improve light conditions to support more photosynthesis. Furthermore, as the tide moves out, clear water from within the SAV bed spills out of the bed into the adjoining waters, creating another positive feedback effect that makes the adjoining area more amenable to additional plant growth. Thus, large, contiguous SAV beds have greater storm resilience through protection of the inner core of the bed. However, such SAV beds require protection from anthropogenic impacts such as destructive boating or fishing techniques.

---

Sexual reproduction is well known in submersed aquatic plants, but vegetative reproduction of new shoots and whole-plant reproduction by fragmentation are much more common (Sculthorpe 1967). Small fragments of many submersed aquatic plants can produce entire new plants, generally making mechanical harvesting methods for control of nuisance overgrowth highly ineffective in flowing environments. Fragmentation also enables high dispersal, which has resulted in noxious growths of some species outside of their native habitats (see Chapter 14). Some taxa can also form turions (winter buds from leaf material) and tubers (specialized sections of roots or stems) as means of vegetative propagation.

Like other wetland/aquatic plants, submersed macrophytes typically have a well-developed lacunar system to oxygenate their roots; in fact, they consist mostly (up to 98%) of air and water. The many air passageways in their leaves, stems, and roots restrict the depth where they can thrive to about 10 m or less; in deeper waters, the pressure is too great for survival. As pressure increases moving down toward 10 m, leaves shorten, stems become thinner, and the lacunar system diminishes.

Submersed macrophytes are very efficient at nutrient uptake, especially through their roots, less so through their foliage (Barko et al. 1991; Wetzel 2001). They commonly take up N and P from the sediments in excess of their needs and translocate nutrients to their aboveground tissues (Barko and Smart 1981). Healthy growing plants lose minimal nutrients whereas senescent plants readily release nutrients, thus effectively moving sediment-based nutrients into the water column (Barko et al. 1991).

#### 4.2.3.4 Free-Floating Macrophytes

Free-floating macrophytes (Fig. 4.13) are a highly variable functional group living unattached on the water surface, including angiosperms, ferns, and liverworts. Examples range from the very large and showy to near-microscopic in size. They can thrive in slow-moving, nutrient-enriched water bodies, covering the surface and shading out phytoplankton and vascular plants below in the water column or at the sediments. A well-known example is duckweed (*Lemna*), a small green plant that can bloom in calm, nutrient-enriched waters (Fig. 4.13a, b). This group also includes tiny floating

(a)

(b)

(c)

(d)

**Fig. 4.13** Free-floating macrophytes. Fig. 4.13a (top left) Duckweed *Lemna* bloom on nutrient-polluted Stocking Head Creek draining multiple swine CAFOs in eastern North Carolina. Fig. 4.13b (top right) Close-up of *Lemna*. Fig. 4.13c (bottom left) *Azolla*, a small free-floating fern. Fig. 4.13d (bottom right) Water hyacinth *Eichornia crassipes*, an invasive that can be problematic in profusion—see also Chapter 14. (Photos by the author.)

plants with no or reduced roots; examples are *Azolla*, a reddish-colored fern that can cover large areas of isolated calm water bodies (Fig. 4.13c), and watermeal (*Wolffia*), which is the smallest known angiosperm (0.5–1.5 mm across). A large, showy, free-floating macrophyte is the invasive water hyacinth (*Eichornia*), which features large flowers and a well-developed root system (Figs 4.13d and also see 14.5a in Chapter 14 on Invasive Species). Filamentous free-floating macrophytes will often have considerable amounts of the plant mass submerged and finely dissected leaves. Such taxa include coontail (*Ceratophyllum*) and bladderwort (*Utrichularia*), which is a carnivorous plant that has specialized organs that trap zooplankton and digest them.

Free-floating macrophytes that live on the surface may experience high water stress; thus they prefer quiet or slow-moving waters (Biggs 1996). There is abundant light, so turbidity is not an issue for their success. As noted, they become abundant (and even problematic) in nutrient-rich waters. A prime example of this is water hyacinth (detailed in Chapter 14 regarding invasive species). Water hyacinth is a major nuisance in Florida, such as the interconnected nutrient-rich Oklawaha chain of lakes north of Orlando. This species multiplies rapidly by using fragmentation or a lateral stolon for vegetative propagation.

### 4.2.4 Macrophyte Communities: Role of Currents, Substrata, and Light Availability

In flowing waters, current velocity is considered the most important influence on the types and abundances of macrophytes in a given system because of the abrasive effects of flow and washout. In general, as velocity increases, fewer taxa are able to withstand the increased current stress. Free-floating and floating-leaved taxa can thrive in gentle current or sheltered pools, but these taxa as well as many submersed taxa cannot survive in high-flow areas. Some taxa, particularly bryophytes and hornleaf riverweed, are able to anchor to rocky substrata so that they can thrive in rapid currents (Sculthorpe 1967; Biggs 1996). Storm-driven surges of flow, or spates, can be strong enough to completely disrupt an established macrophyte community. Following a spate, rocks and other hard substrata, as well

as bottom sediments, are relatively rapidly colonized by periphyton and benthic microalgae (see Chapters 3 and 17). However, most macrophytes may take months to recolonize a given reach (Biggs 1996).

The types of substrata also dictate the types of macrophyte flora present. Sandy sediments provide little stability and are often bare of macrophytes (Fig. 4.14a); pebbly sediments likewise provide little stability. In areas with currents, rooted plants are best suited to stable soils with some organic content. In rocky habitats or in areas where large, usually submersed areas of granite outcrops are available (Fig. 4.14b–d), populations of rooted plants can be maintained in depositional areas between boulders, in protected pools, or on the highly dependable granite flatrock that currents cannot move (Everitt and Burkholder 1991). Adequate light for photosynthesis is a third critical factor controlling macrophyte distribution. Light often limits macrophyte growth in streams that are deeply shaded by forest growth or narrow canyon walls. High water-column turbidity (Fig. 4.14b) and highly colored waters often limit submersed macrophyte growth to immediate nearshore environments, as well.

### 4.2.5 Macrophyte Net Primary Production and Productivity

The primary production contributed by macrophytes in rivers, streams, and adjacent wetlands is highly variable and depends on the mix of species, season, latitude, light, temperature, current velocity, and other factors. Methods and terminology for macrophyte production (longer-term net biomass, that is, excluding death, loss to grazing, etc., usually per year) and productivity (short-term rate, usually per hour or day) are reviewed in Dennis and Isom (1984), Lipkin et al. (1986), and Wetzel and Likens (1991). These references describe comparisons among groups or ecosystems as problematic because of substantial differences in sampling techniques and frequencies. The measurements often focus only on shoots (stems and leaves), yet many species—especially emergents and floating-leaved taxa—produce substantial biomass (half of the total or more) below ground as roots and rhizomes, which should be included in

(a)　　　　　　　　　　　　　　(b)

(c)　　　　　　　　　　　　　　(d)

**Fig. 4.14** Substrata in streams and rivers can control macrophyte growth. Fig. 4.14a (top left) Coastal Plain sandy-bottom stream provides an unstable substratum that provides unfavorable habitat for rooted macrophytes. Fig. 4.14b (top right) Turbid Haw River after a spate of elevated discharge over organic substrata with occasional boulders—this habitat limits the area where free-floating and floating-leaved rooted species can survive scouring; turbidity can cause light limitation to benthic algae as well. Fig. 4.14c (lower left) Rapid currents and rocky substrata in this mountain river (Ocoee River) provide some substrata for macrophyte growth, but its location downstream from a power plant can lead to rapidly changing currents due to human water use. Fig. 4.14d (bottom right) Wide areas of granite outcrop or flatrock effectively separate the tree canopy, widening this third-order stream and providing an expansive area of dependable substratum for macrophytes such as hornleaf riverweed *Podostemum ceratophyllum*—darkened areas, which anchor to the rock. (Figs 4.14a–c by the author; Fig. 4.14d by JoAnn M. Burkholder of North Carolina State University.)

production/productivity estimates (Wetzel 2001). Besides functions of anchoring, bank stabilization, gas exchange, and water and nutrient uptake, the root–rhizome area is host to bacteria responsible for denitrification and anammox (see Chapter 2).

In flowing waters, macrophyte production is usually highest in mid-order streams characterized by low to moderate flow, where the tree canopy has separated and turbidity is low so that light is adequate for photosynthesis (Allan and Castillo 2007; see Chapter 6 regarding the River Continuum Concept). Measurements are often reported as either dry weight (wt) or grams of carbon per unit area per year. Maximal net production of rooted submersed macrophytes in streams and rivers draining nutrient-rich agricultural areas can approach ~ 2000 g dry wt/m$^2$ during the growing season (e.g. Mebane et al. 2013). More generally in temperate rivers, production was reported to range from 320 to 3700 g/m$^2$/year for emergent plants and 8 to 400 g/m$^2$/year for submersed plants, whereas in a tropical river, production of a floating meadow of macrophytes ranged from 2430 to 4050 g/m$^2$/year (Wetzel 2001 and references therein). As a short-term rate, some submersed macrophytes in streams and rivers were estimated to have maximal net primary productivity of about 3 g $C/m^2/day$, versus about 10–20 g $C/m^2/day$ for emergents (Westlake 1975).

Freshwater wetlands and their macrophyte communities are among the most highly productive ecosystems in the world (Wetzel 2001). They have been much better studied for macrophyte production than streams and rivers. For example, estimates from freshwater marsh macrophyte communities ranged from 1000 to 3500 g $C/m^2/year$ (Odum et al. 1984; also see Wetzel 2001 and references therein). In another study, whole-plant production (including both above- and belowground biomass) of a cattail stand exceeded 4300 g $C/m^2/year$ (Roberts and Ganf 1986). Marsh and swamp communities were reported to range from 1500 to 7000 $g/m^2/year$ in aboveground production and from 150 to 3000 $g/m^2/year$ in belowground production (Wetzel 2001 and references therein). Species having a higher proportion of tough structural material (cattail, wild rice, common reed, etc.) produce considerably more biomass than less lignified taxa such as spatterdock, pickerelweed, etc. (Simpson et al. 1983; Odum 1984).

## 4.2.6 Macrophytes and Nutrients

In streams and rivers, macrophytes rooted in the sediments or attached with holdfasts have access to abundant nutrients brought to them by the current; even if the concentrations are low, supplies continually move past (Whitton 1975). Nonrooted macrophytes positioned in backwaters, oxbows, or wetlands draining into river systems are more dependent upon rainfall and stormwater-driven nutrients. Both rooted and nonrooted submersed macrophytes can proliferate in nuisance masses, which is especially problematic if the macrophyte in question is a non-native species with no local natural grazers (see Chapter 14). As noted earlier, senescing macrophytes recycle nutrients into the water column. Thus, a rapid "crash" or die-off of noxious macrophyte growth can release nutrients to the overlying water that can stimulate growth of phytoplankton, including harmful species.

On the other hand, as noted earlier, rooted macrophytes can reduce nutrient pollution in eutrophying waters. Their root–rhizome complex hosts bacteria

that are integral to the N-removal processes of denitrification and anammox (see Chapter 2). The rates of denitrification and anammox vary depending on the macrophyte species and sediment environmental conditions. Rates of N loss have been experimentally compared from the root–rhizome complex of various macrophyte species in a wetland constructed to treat stormwater runoff (Table 4.1).

Rates of denitrification and anammox varied considerably among species, with highest efficiency in both processes shown by the bacterial consortium associated with the roots/rhizomes of pickerelweed (Song et al. 2014; Table 4.1). Furthermore, denitrification removed much more N from the wetland than anammox. Such data have been used to guide species selection for planting to achieve maximum N removal from polluted runoff in constructed wetlands. It is noteworthy that some macrophyte species that can produce noxious growths (in this study, alligatorweed and cattail) functioned beneficially in removing excess N.

## 4.2.7 Herbivory on Macrophytes

Macrophytes are a major source of energy to river ecosystems, both directly and indirectly. Historically it was thought that macrophytes contributed to river food webs and other aquatic ecosystems mostly as detritus (see Wetzel 2001 and references therein). It is now recognized that a diverse array of herbivores directly consume living macrophytes, including some species of waterfowl, mammals, fish, reptiles, crustaceans, gastropods, and insects (Table 4.2). Many macrophyte species are as N-rich as algae and tree leaves, and some species are specifically targeted by grazers (Lodge 1991). Waterfowl are a highly visible group of macrophyte grazers, particularly Canada geese, wood ducks, widgeon, green-winged teal, mallards, black ducks, and pintails (summarized in Odum et al. 1984). Plants favored (especially seeds and rhizomes) are various sedges, rushes, and grasses, and broad-leaved macrophytes including arrow-arum, pickerelweed, and yellow water lily.

African cichlids (e.g. some *Tilapia* spp.) have been introduced to help control macrophyte overgrowths in reservoirs with suitably high temperatures in the Southeastern USA (Crutchfield et al. 1992). The

**Table 4.1** Comparison of N loss rates among some macrophyte species. Note that pickerelweed, bur-reed, and giant cutgrass were planted in the wetland following construction, whereas the other species were opportunistic invaders. All species listed are emergent, except parrot feather which is submersed. (From Song et al. 2014.)

| Species | N loss rates (nmol N/g sediment wet wt/h) |
|---|---|
| **Denitrification** | |
| Pickerelweed *Pontederia cordata* | 27.39±2.68 |
| Alligatorweed *Alternanthera philoxeroides* | 16.90±3.64 |
| Bur-reed *Sparganium americanum* | 13.65±9.33 |
| Giant cutgrass *Zizaniopsis miliacea* | 12.37±2.40 |
| Cattail *Typha angustifolia* | 11.92±3.90 |
| Soft rush *Juncus effuses* | 11.66±5.49 |
| Parrot feather *Myriophyllum aquaticum* | 4.59±0.14 |

Pickerelweed > alligatorweed, bur-reed, giant cutgrass, cattail, soft rush > parrot feather ($p = 0.0037$)

| Species | N loss rates (nmol N/g sediment wet wt/h) |
|---|---|
| **Anammox** | |
| Pickerelweed *Pontederia cordata* | 3.72±2.34 |
| Cattail *Typha angustifolia* | 2.91±1.49 |
| Giant cutgrass *Zizaniopsis miliacea* | 2.90±1.55 |
| Bur-reed *Sparganium americanum* | 2.22±1.33 |
| Alligatorweed *Alternanthera philoxeroides* | 1.50±2.13 |
| Parrot feather *Myriophyllum aquaticum* | 0.61±0.13 |
| Soft rush *Juncus effuses* | 0.33±0.26 |

Pickerelweed, cattail, giant cutgrass, bur-reed > alligatorweed, soft rush ($p = 0.048$)

**Table 4.2** Direct grazers of macrophytes. (From Odum et al. 1984; Lodge 1991; Voshell 2002.)

| Category | Selected examples, common names |
|---|---|
| Waterfowl | Geese, ducks, coots |
| Fish (herbivores, omnivores) | Carp, grass carp, tilapia, roach, pinfish |
| Mammals | Manatees, dugongs, moose, muskrat, nutria |
| Reptiles | Turtles |
| Crustaceans | Crayfish, isopods |
| Gastropods | Snails |
| Insects (adults, larvae) | Beetles, caddisflies, hemipterans (as piercers) |

most well-known herbivorous fish worldwide may be the Asian grass carp (or white amur; *Ctenopharyngodon idella*). Triploid (sterile) grass carp have been introduced to many reservoirs to control nuisance macrophyte growths, but, because fertile individuals have sometimes been mistakenly evaluated as sterile, resource managers strive to carefully control introductions to avoid escapes and potential ecosystem disruptions in river ecosystems (see Chapter 14).

A few mammals are known to consume macrophytes; the best known (and most highly visible) are West Indian manatees (*Trichechus manatus*) which move between fresh- and saltwater (Stith et al. 2010). These animals weigh up to 590 kg and breed a single calf every one to two years. They are endangered, with excessive mortality occurring in Florida populations as their natural seagrass food is dying due to dense algal blooms and they starve or eat toxic macroalgae (Lapointe et al. 2015).

Many insect species also consume macrophytes—within the water, on top of the water, and along the shoreline of rivers and streams (Box 4.2; and see Chapters 5 and 6). These insects attract insectivorous predators which, in turn, attract predators from higher trophic levels. Upon plant death and senescence, the plant material is colonized by bacteria and fungi, and some of this colonized material becomes food for insect larvae. Grazers, especially waterfowl, can contribute to macrophyte dispersal by defecating undigested seeds or fragments in areas distant from the original feeding area (Lodge 1991). This activity can be ecologically harmful if the grazed plants are problematic invasive species.

---

### Box 4.2 Some macrophytes can be heavily grazed by a specific grazer!

Leaves of the yellow pond lily (*Nuphar lutea*) have high nutrient content and are grazed by larval and adult water lily leaf beetles (*Pyrrhalta nymphaea*). These grazers consume leaf material down about a quarter of the way through the leaf, leaving little trenches of excavated leaf. Grazing is heaviest on older tissues, which increases plant turnover rates through rapid production of new tissues. Like most other rooted macrophytes, water lilies mainly acquire nutrients from the sediments. Grazing of nutrient-rich leaf material promotes increased nutrient cycling from the sediments to the water column, effectively as a nutrient "pump" from the sediments. Thus, macrophyte production can enter the detrital food chain during the growing season as well as during plant senescence.

(From Wallace and O'Hop 1985.)

---

## 4.3 Summary

- Freshwater macroalgae may have various morphologies. Many of them are lumped in with periphyton; both groups are benthic algae, but periphyton grow to about 2.5 cm (1 inch) in length whereas macroalgae can be 2 m in length or longer.
- Most freshwater macroalgae are filamentous forms, along with relatively few thalloid green (streptophyte) and red (rhodophyte) taxa. The filamentous taxa are mostly benthic mat-forming cyanobacteria and green algae. Although generally beneficial, filamentous stream-inhabiting algae are infamous for causing high-biomass blooms that have many adverse ecological and economic impacts.
- Macroalgal cyanobacteria increasingly are being known as problematic, highly toxic bloom formers, and they are not well studied in comparison to other freshwater macroalgae.
- Macrophytes are generally considered within four functional groups that exploit river ecosystem habitats: emergent, floating-leaved, submersed, and free-floating.
- Macrophytes located onshore or nearshore stabilize banks with their root masses and decrease both onshore and within-stream erosion, reducing turbidity and protecting in-stream habitats from burial.
- Macrophytes serve as nearshore filters of sediments and other pollutants from stormwater runoff, and within stream and rivers macrophytes buffer currents and cause particle sedimentation.
- Macrophytes reduce water-column nutrient supplies by directly absorbing N and P from the water and sediments. In addition, the root–rhizome complex of some species enhances denitrification which results in direct loss of N from the ecosystem to the overlying air as $N_2$ gas.
- Macrophytes are directly consumed by various fauna and indirectly provide energy to the food web as detritus, as well. They also create habitat for microfauna, numerous macroinvertebrates, crustaceans, and fish. Grazers consume the associated periphyton; and the physical presence of macrophytes provides areas to hide from larger predators. In turn, smaller fish use macrophyte beds to hide from larger fish.

## References

Allan, J.D. and M.M. Castillo. 2007. *Stream Ecology: Structure and Function of Running Waters*, 2nd edition. Springer, Dordrecht.

Barko, J.W. and R.M. Smart. 1981. Sediment-based nutrition of submersed macrophytes. *Aquatic Botany* 10: 339–352.

Barko, J.W., D. Gunnison and S.R. Carpenter. 1991. Sediment interactions with submersed macrophyte growth and community dynamics. *Aquatic Botany* 41:41–65.

Belanger, T.V., D.J. Scheidt and J.R. Platko III. 1989. Effects of nutrient enrichment in the Florida Everglades. *Lake and Reservoir Management* 5:101–111.

Biggs, B.J.F. 1996. Hydraulic habitat of plants in streams. *Regulated Rivers: Research and Management* 12:131–144.

Bubier, J.L. and T.R. Moore. 1994. An ecological perspective on methane emissions from northern wetlands. *Trends in Ecology and Evolution (TREE)* 9:460–464.

Burkholder, J.M. 2009. Harmful algal blooms. Pages 264–285 in Likens, G.E (ed.), *Encyclopedia of Inland Waters*, Vol. 1. Elsevier, Oxford.

Crawford, R.M.M. (ed.). 1987. *Plant Life in Aquatic and Amphibious Habitats.* Special Publications Series of the British Ecological Society No. 5. Blackwell Scientific Publications, Boston.

Crutchfield, J.U., D.H. Schiller, D.D. Herlong and M.A. Mallin. 1992. Establishment and impact of redbelly tilapia in a vegetated cooling reservoir. *Journal of Aquatic Plant Management* 30:28–35.

Dacey, J.W.H. 1981. Pressurized ventilation in the yellow water lily. *Ecology* 62:1137–1147.

Dennis, W.M. and B.G. Isom (eds.). 1984. *Ecological Assessment of Macrophyton—Collection, Use, and Meaning of Data.* ASTM Special Technical Publication 843. ASTM, Philadelphia.

Dodds, W.K. and D.A. Gudder. 1992. The ecology of *Cladophora. Journal of Phycology* 28:415–427.

Eloranta, P. and S. Kunnas. 1979. The growth and species of the attached algae in a river system in central Finland. *Archiv für Hydrobiologie* 78:86–101.

Everitt, D.T. and J.M. Burkholder. 1991. Seasonal dynamics of macrophyte communities from a stream flowing over granite flatrock in North Carolina, USA. *Hydrobiologia* 222:159–172.

Fox, A.M. 1992. Macrophytes. Pages 216–233 in Calow, P. and G.E. Potts (eds.). *The Rivers Handbook: Hydrological and Ecological Principles.* Blackwell Scientific, Oxford.

Glime, J. M. 2017. Aquatic insects: Bryophyte roles as habitats. Pages 11-2-1–11-2-43. In: Glime, J. M. *Bryophyte Ecology. Volume 2.* Bryological Interaction. Ebook sponsored by Michigan Technological University and the International Association of Bryologists.

Graham, L.E., J.M. Graham, L.W. Wilcox and M.E. Cook. 2016. *Algae*, 3rd edition. LJLM Press, University of Wisconsin Press, Madison.

Guo, H.B. 2008. Cultivation of lotus (*Nelumbo nucifera* Gaertn. ssp. *nucifera*) and its utilization in China. *Genetic Resources and Crop Evolution* 56:323–330.

Gurbisz, C., W.M. Kemp, L.P. Sanford and R.J. Orth. 2016. Mechanisms of storm related loss and resilience in a large submersed plant bed. *Estuaries and Coasts* 39:951–966.

Hudon, C., M. De Sève and A. Cattaneo. 2014. Increasing occurrence of the benthic filamentous cyanobacterium *Lyngbya wollei*: a symptom of freshwater ecosystem degradation. *Freshwater Science* 33:606–618.

Kay, S.H. 2001. Invasive Aquatic and Wetland Plants: Field Guide. Publication UNC-SG-01-15. North Carolina Sea Grant, North Carolina State University, Raleigh.

Lapointe, B.E., L.W. Herren, D.D. Debortoli and M.A. Vogel. 2015. Evidence of sewage-driven eutrophication and harmful algal blooms in Florida's Indian River Lagoon. *Harmful Algae* 43:82–102.

Lapointe, B.E., J.M. Burkholder and K.L. Van Alstyne. 2018. Harmful macroalgal blooms in a changing world: causes, impacts, and management. Pages 515–560 in Shumway, S.E., J.M. Burkholder and S.L. Morton (eds.), *Harmful Algal Blooms and their Management: A Compendium Desk Reference.* John Wiley & Sons, Hoboken, NJ.

Larkin, D.J., A.K. Monfils, A. Boissezonc, et al. 2018. Biology, ecology, and management of starry stonewort (*Nitellopsis obtusa*; Characeae): a Red-Listed Eurasian green alga invasive in North America. *Aquatic Botany* 148:15–24.

Lembi, C.A. 2003. Control of nuisance algae. Pages 805–834 in Wehr, J. and R. Sheath (eds.), *Freshwater Algae of North America.* Academic Press, New York.

Leonard, L.A. and M.E. Luther. 1995. Flow hydrodynamics in tidal marsh canopies. *Limnology and Oceanography* 40:1474–1484.

Lipkin, Y., S. Beer, E.P.H. Best, T. Kairesalo and K. Salonen. 1986. Primary production of macrophytes: terminology, approaches and a comparison of methods. *Aquatic Botany* 26:129–142.

Lodge, D.M. 1991. Herbivory on freshwater macrophytes. *Aquatic Botany* 41:195–224.

Madsen, T.V. and E. Warncke. 1983. Velocities of currents around and within submerged aquatic vegetation. *Archiv für Hydrobiologie* 97:389–394.

McAllister, T.G., S.A. Wood and I. Hawes. 2016. The rise of toxic benthic *Phormidium* proliferations: a review of their taxonomy, distribution, toxin content and factors regulating prevalence and increased severity. *Harmful Algae* 55:282–294.

Mebane, C.A., N.S. Simon and T.R. Maret. 2013. Linking nutrient enrichment and streamflow to macrophytes in agricultural streams. *Hydrobiologia* 722:143–158.

Odum, W.E., T.J. Smith III, J.K. Hoover and C.C. McIvor. 1984. The Ecology of Tidal Freshwater Marshes of the United States East Coast: A Community Profile. Report FWS/OBS-83/17. US Fish & Wildlife Service, Washington, DC.

Patrick, R. and C. Reimer. 1966. *The Diatoms of the United States, Exclusive of Alaska and Hawaii.* Academy of Natural Sciences of Philadelphia, Philadelphia.

Peters, G.A. and J.C. Meeks. 1989. The *Azolla–Anabaena* symbiosis: basic biology. *Annual Review of Plant Physiology and Plant Molecular Biology* 40:193–210.

Petticrew, E.L. and J. Kalff. 1992. Water flow and clay retention in submerged macrophyte beds. *Canadian Journal of Fisheries and Aquatic Sciences* 49:2483–2489.

Power, M.E. 1990. Effects of fish in river food webs. *Science* 250:811–814.

Quiblier, C., W. Susanna, E.-S Isidora, H. Mark, V. Aurelie and H. Jean-François. 2013. A review of current knowledge on toxic benthic freshwater cyanobacteria—ecology, toxin production and risk management. *Water Research* 47:5464–5479.

Roberts, J. and G.G. Ganf. 1986. Annual production of *Typha orientalis* Presl. in inland Australia. *Australian Journal of Marine and Freshwater Research* 37:659–668.

Sculthorpe, C.D. 1967. *The Biology of Aquatic Vascular Plants*. St. Martin's Press, New York.

Shalygin, S., I.-S. Huang, E.H. Allen, J.M. Burkholder and P.V. Zimba. 2019. *Odorella benthonica* gen. & sp. nov. (Pleurocapsales, Cyanobacteria): an odor and prolific toxin producer isolated from a California aqueduct. *Journal of Phycology* 55:509–520.

Sheath, R.G. 1984. The biology of freshwater red algae. Pages 89–157 in Round, F.E. and D.J. Chapman (eds.), *Progress in Phycological Research Vol. 3*. Biopress, Bristol.

Sheath, R.G. and J.M. Burkholder. 1985. Characteristics of softwater streams in Rhode Island. II. Composition and seasonal dynamics of macroalgal communities. *Hydrobiologia* 128:109–118.

Sheath, R.G. and J.M. Burkholder. 1988. Stream macroalgae. Pages 53–59 in Sheath, R.G. and M.M. Harlin (eds.), *Freshwater and Marine Plants of Rhode Island*. Kendall Hunt, Dubuque, IA.

Sheath, R.G. and J.A. Hambrook. 1988. Mechanical adaptations to flow in freshwater red algae. *Journal of Phycology* 24:107–111.

Shen-Miller, J. 2002. Sacred lotus, the long-living fruits of China Antique. *Seed Science Research* 12:131–143.

Simpson, R.L., R.E. Good, M.A. Leck and D.F. Whigham. 1983. The ecology of freshwater tidal wetlands. *BioScience* 33:255–259.

Song, B., M.A. Mallin, A. Long and M.R. McIver. 2014. Factors Controlling Microbial Nitrogen Removal Efficacy in Constructed Stormwater Wetlands. Report No. 443. Water Resources Research Institute, University of North Carolina, Raleigh.

Speziale, B.J. and L.A. Dyck. 1992. *Lyngbya* infestations: comparative taxonomy of *Lyngbya wollei* comb. nov. (cyanobacteria) 1. *Journal of Phycology* 28:693–706.

Stancheva, R. and R.G. Sheath. 2016. Benthic soft-bodied algae as bioindicators of stream water quality. *Knowledge and Management of Aquatic Ecosystems* 417:15.

Stith, B.M., J.P. Reid, C.A. Langtimm, et al. 2010. Temperature inverted haloclines provide winter warm-water refugia for manatees in southwest Florida. *Estuaries and Coasts* 34:106–119.

Vis, C., A. Cattaneo and C. Hudon. 2008. Shift from chlorophytes to cyanobacteria in benthic macroalgae along a gradient of nitrate depletion. *Journal of Phycology* 44:38–44.

Voshell, J.R. 2002. *A Guide to Common Freshwater Invertebrates of North America*. The McDonald and Woodward Publishing Co., Blacksburg, VA.

Wallace, J.B. and J. O'Hop. 1985. Life on a fast pad: water lily leaf beetle impact on water lilies. *Ecology* 66:1534–1544.

Westlake, D.F. 1980. Primary production. Pages 141–246 in Lecren, E.D. and R.H. Lowe-McConnell (eds.), *The Functioning of Freshwater Ecosystems*. Cambridge University Press, Cambridge.

Westlake, D.F. 1975. Macrophytes. Pages 106–128 in Whitton, B.A. (ed.), *River Ecology*. University of California Press, Berkeley.

Wetzel, R.G. 2001. *Limnology: Lake and River Ecosystems*. 3rd edition. Academic Press, San Diego.

Wetzel, R.W. and G.E. Likens. 1991. *Limnological Analysis*, 2nd edition. Springer, New York.

Whitford, L.A. and G.L. Schumacher. 1963. Communities of algae in North Carolina streams, and their seasonal relations. *Hydrobiologia* 22:133–195.

Whitford, L.A. and G.J. Schumacher. 1964. Effect of a current on respiration and mineral uptake in *Spirogyra* and *Oedogonium*. *Ecology* 45:168–170.

Whitton, B.A. 1975. Algae. Pages 81–105 in Whitton, B.A. (ed.), *River Ecology*. Studies in Ecology 2. University of California Berkeley, California.

Wood, S.A. and J. Puddick. 2017. The abundance of toxic genotypes is a key contributor to anatoxin variability in *Phormidium*-dominated benthic mats. *Marine Drugs* 15:307.

Wood, S.A., C. Depree, L. Brown, T. McAllister and I. Hawes. 2015. Entrapped sediments as a source of phosphorus in epilithic cyanobacterial proliferations in low nutrient rivers. *PLoS ONE* 10:e0141063.

Zulkifly, S.B., J.M. Graham, E.B. Young, et al. 2013. The genus *Cladophora* Kützing (Ulvophyceae) as a globally distributed ecological engineer. *Journal of Phycology* 49:1–17.

# CHAPTER 5

# Stream and River Invertebrate Communities

The invertebrate communities in streams, rivers, and estuaries are best known as key links in the food web between primary producers and fisheries, but in fact play multiple important roles in lotic ecosystems. They serve as grazers, predators, and prey, they scavenge and transform, they recycle nutrients, and they serve as test organisms for potentially toxic substances, and benthic invertebrates serve as ecological indicators of water quality and trophic state. We begin with the zooplankton community, consisting of those microscopic and near-microscopic animals that spend much of their lives in the water column. These organisms are mobile through use of flagella and other appendages, yet are still very much at the mercy of currents.

## 5.1 Zooplankton

The zooplankton community is the trophic link between the phytoplankton community and planktivorous fish, especially the larvae of many fish species (Cramer and Marzolf 1970; Siefert 1972). Zooplankton are generally sparse in swift-moving waters but can become abundant in more lacustrine areas of lower rivers, in estuaries, and especially in reservoirs. Their abundance in lower river systems has been positively correlated with chlorophyll *a* and water temperature (Orsi and Mecum 1986). Within a given channel, the slow-flow areas such as backwaters, pools, and complex river margins (Fig. 5.1) have been shown to provide source populations for the stream in general (Walks 2007). The typical zooplankton fauna primarily contains cladocerans, copepods, rotifers, and protozoans.

Zooplankton transform primary production from phytoplankton into zooplankton biomass (secondary production). When zooplankton die or defecate, carbon is transferred to the benthos, which helps support benthic productivity (secondary or tertiary production). Zooplankton are important in the cycling of nutrients in estuarine systems. They recycle nitrogen (as organic nitrogen and ammonium) and phosphorus as excreta, and in the euphotic zone (the sunlit zone) this helps phytoplankton maintain primary productivity. This is very important in nutrient-poor lotic systems that do not receive significant nutrient inputs.

Essential taxonomic guides to this community include the classic works by Edmondson (1959) and Pennak (1978); also that of Chengalath et al. (1971), Fitzpatrick (1983), and Thorp and Covich (1991) and its several further editions; and in the estuary, valuable zooplankton guides include Smith and Johnson (1996) and Johnson and Allen (2005). Guides to sampling, biomass, and secondary production methods can be found in Edmondson and Winberg (1971), Wetzel and Likens (1991), and Harris et al. (2000). This chapter presents the basic trophic and reproductive biology of zooplankton, while more applied ecological information on lotic and lentic zooplankton is presented in Chapter 11.

### 5.1.1 Copepods (Phylum Arthropoda, Class Crustacea)

The subclass Copepoda is divided into three orders (i.e. there are three basic types of copepods).

*River Ecology*. Michael A. Mallin, Oxford University Press. © Michael A. Mallin (2023). DOI: 10.1093/oso/9780199549511.003.0006

**Fig. 5.1** Quiet backwater along the Upper Mississippi River. Such environments develop robust phytoplankton and zooplankton communities and are important nursery areas for river fishes. (Photo by the author.)

(a)                                              (b)                    (c)

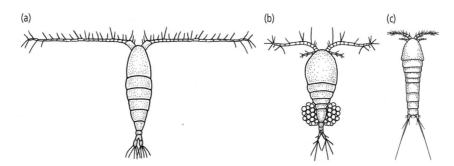

**Fig. 5.2** Stylized representations of copepods. Fig. 5.2a (left) Calanoid copepod. Fig. 5.2b (center) Cyclopoid copepod. Fig. 5.2c (right) Harpacticoid copepod. Calanoids and cyclopoids are common in plankton or reservoirs and higher-order rivers, whereas in general harpacticoids are most common in littoral or benthic zones. (Revised from Williamson 1991 except for harpacticoid which is clarified and drawn from author's photo.)

### 5.1.1.1 Calanoid Copepods

Calanoid copepods are planktonic; most species are omnivorous grazers that feed on phytoplankton or protozoans, whereas a few large species are carnivorous. They are characterized by long antennae and somewhat bullet-shaped bodies (Fig. 5.2a). Various species of *Diaptomus* (and similar genera) are common in lower rivers and reservoirs. In oligohaline rivers through polyhaline estuaries of the eastern USA the most important zooplankter is the calanoid copepod *Acartia tonsa* (Table 5.1), a very common food item found in larval and juvenile fish which usually comprises the greatest biomass of any taxa. In freshwater and estuarine systems there are usually one-to-two coexisting calanoids, but in coastal marine waters there can be several (Table 5.1).

Calanoid copepods are particle feeders that concentrate on particles in the 5- to 10-μm size range, with an upper limit of about 50 μm. For perspective, a 5-μm algal cell is about the limit of what one can identify using a light microscope. They detect food by using mechanoreceptors, which are setae on the first antennae, and chemoreceptors, also on the first antennae. Many use size-selective predation (going for the biggest bang for the buck, energywise). They can also select for more nutrient-rich particles over nutrient-poor particles. They do not filter-feed; rather, they create a current past them by flapping their appendages (which act as paddles). When an algal cell is within reach they open their second maxillae, creating a vortex. The particle is carried in, the maxillae close, and water is

**Table 5.1** Distribution of selected representative zooplankton genera based on salinity regime—lower river through to coastal ocean.

| Group | Freshwater | Estuarine | Coastal marine |
|---|---|---|---|
| Copepods | Diaptomus | Acartia | Calanus |
| | Cyclops | Parvocalanus | Labidocera |
| | Mesocyclops | Eurytemora | Centropages |
| | Tropocyclops | Oithona | Oncaea |
| Cladocerans | Bosmina | Bosmina | Penilia |
| | Daphnia | Evadne | |
| | Ceriodaphnia | Podon | |
| | Diaphanosoma | | |
| Rotifers | Keratella | Keratella | Synchaeta |
| | Brachionus | Brachionus | |
| | Synchaeta | Synchaeta | |
| | Polyarthra | | |

forced out. The particle is then eaten or rejected at the mouth (Kohl and Strickler 1981). When reproducing, many female calanoids carry a single egg sac.

### 5.1.1.2 Cyclopoid Copepods

Cyclopoid copepods are mainly planktonic, with some benthic species. They have shorter antennae than calanoids and more teardrop-shaped bodies. Common genera include *Cyclops*, *Mesocyclops*, and the smaller *Tropocyclops*; *Oithona* is a common cyclopoid in estuarine waters (Table 5.1). Cyclopoid copepods are raptorial feeders. They feed on zooplankton, other invertebrates, some algae, and even small fish (if they are big enough). This is k-selected behavior; they are always present in the plankton to some degree and eat whatever they can catch. Female cyclopoids carry two egg sacs rather than one (Fig. 5.2b).

### 5.1.1.3 Harpacticoid Copepods

Harpacticoid copepods are primarily benthic, with a few planktonic species that may bloom occasionally for short periods in estuaries. They have a wide variety of body types. Some are flattened for living among sediments, whereas others are more elongated like calanoid copepods. However, they generally have very short antennae (Fig. 5.2c). They are omnivores and some subsist on particulate organic

matter. Many species are found in the littoral zone or on or among bottom sediments.

### 5.1.1.4 Reproduction

Copepods have obligate sexual reproduction (Allan 1976). The male can be differentiated under the microscope by a modified clasping appendage and a thickened antenna. The male attaches a spermatophore (which looks like a sac-like tube) to a female at the pits located on the urosome. The spermatophore discharges into the seminal receptacle. The spermatophore can remain attached and fertilize eggs subsequently. Eggs are extruded from the female's body. Some species have egg sacs and some species cast eggs singly (like *Acartia tonsa*). The eggs develop through six naupliar stages and five copepodite stages into the adult stage (the sexually mature stage). The length of the life cycle is very species specific.

## 5.1.2. Cladocerans (Phylum Arthropoda, Class Branchiopoda, Order Cladocera)

Cladocerans are extremely common in lower rivers and in reservoirs, especially genera such as *Daphnia* and *Bosmina* (Table 5.1), and among plants in the littoral zone are chydorid cladocerans such as *Chydoris* and *Alona*. Large invasive cladocerans that have entered major US river systems from European waters include *Bythotrephes* (spiny water flea) and *Cercopagis* (fishhook waterflea) which have disrupted food chains and altered species balances (see Chapter 14, invasive species). There are only a few estuarine species (Table 5.1) and cladocerans are very rare in coastal waters. In freshwater ecosystems *Bosmina* and *Daphnia* (Fig. 5.3) in particular become abundant in spring and feed on the diatom-rich phytoplankton bloom, whereas *Diaphanosoma* are more abundant in summer. Two to four cladoceran species may co-occur in freshwater at any one time, whereas in estuaries usually only one species will appear at a time, often in "blooms" lasting for a month or two, then apparently disappear from the plankton. Because cladocerans can rapidly reproduce under food-rich conditions they are considered to be r-selected species (Allan 1976).

(a)  (b)

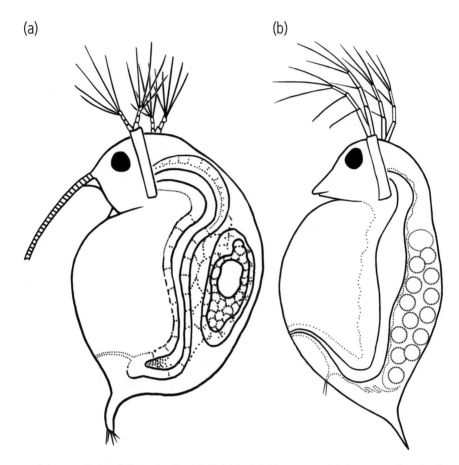

**Fig. 5.3** Common cladocerans. Fig. 5.3a (left) *Bosmina*. Fig. 5.3b (right) *Daphnia*. These genera of cladocerans are often abundant in lower rivers and reservoirs. (Clarified and drawn from author's photos.)

While feeding, these organisms utilize four to six thoracic appendages to create a feeding current. This current delivers a bolus of food to the mandibles, which are the only functional mouthparts. The upper size limit of food particles is about 50 μm. The maximum size of the particle they consume is related to the carapace length. They mostly eat algae, but some can consume bacteria-sized particles as well. In turbid situations (rivers, reservoirs, estuaries) they will consume nonliving particulate matter and glean what they can from this source (Arruda et al. 1983). Interestingly, researchers have analyzed some of these feeding methods by injecting ink into the water and using time-lapse microphotography to capture the movements of the appendages and how they create feeding currents.

### 5.1.2.1 Life History

The cladoceran life cycle (Allan 1976) depends upon food availability and climate/weather conditions. First, a resting (ephippial) egg hatches into a female. She reproduces parthenogenically (i.e. without sex) as long as favorable conditions (i.e. abundant food, proper water temperature) last. The young develop directly, without a larval stage, into miniature adults; five to six instars occur. This is r-selected behavior because it enables a species to take advantage of an algal bloom while it lasts. Under stress some males are produced, as well as some sexually reproducing females. Fertilization occurs and the resting egg is produced (diapause) and ends up in the sediments to await the return of favorable conditions.

Because of their parthenogenic reproduction, and subsequent availability of clones, cladocerans are used for water toxicity tests for industrial effluents (the *Ceriodaphnia* test is a common EPA test). A parthenogenic female is fed phytoplankton or yeast and produces a set of young, which are subsequently distributed into other vessels to reproduce. When a desired number of clones are produced, they are distributed into experimental cups or arrays of depressions to which various dilutions of the chemical in question are added. *Ceriodaphnia* survival and number of young produced are then used to assess toxicity effects. By having the same genetic material for all of your test organisms, that minimizes other influences and allows researchers to have more confidence regarding the actual pollutant effect on the test species.

### 5.1.3 Rotifers (Phylum Aschelminthes, Class Rotatoria, Order Monogonanta)

Rotifers are extremely common and diverse in the plankton of lower rivers and reservoirs, but not in rapidly flowing streams. There are only a few species in estuarine waters and only *Synchaeta* (Fig. 5.4a) is found along the continuum from fresh to oligohaline to coastal marine waters (Table 5.1). Rotifers commonly form blooms for a month or so, then that species will become scarce in the plankton. Whereas several rotifer species may coexist in freshwater situations, generally only one or two species appear in the estuarine plankton at a time (Table 5.1). Common freshwater rotifer genera include *Keratella* (Fig. 5.4b) and *Brachionus* (Fig. 5.4c), *Polyarthra*, *Hexarthra*, and *Trichocerca*. Seasonally, winter planktonic rotifer populations are low in freshwater but can be high in estuarine waters and can be abundant in any other season (Park and Marshall 2000). In rivers high densities are associated with clear, calm waters and elevated phytoplankton densities, and low abundance occurs in rapidly flowing waters with high suspended sediments and low phytoplankton (Williams 1966). Many rotifers live in the benthos and/or littoral zones among the macrophytes and associated periphyton, and some maintain a stationary location with a holdfast. A subgroup of rotifers is termed bdelloid rotifers, which may be entirely

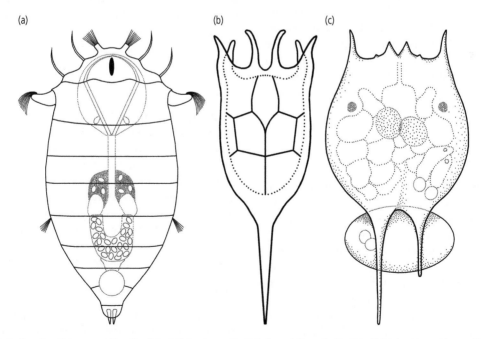

(a)    (b)    (c)

**Fig. 5.4** Fresh- to brackish-water rotifers. Fig. 5.4a (left) *Synchaeta*. Fig. 5.4b (center) *Keratella*. Fig. 5.4c (right) *Brachionus*. These rotifers are found from freshwater rivers downstream into estuaries. (Clarified and drawn from author's photos.)

parthenogenic. Rotifers use a corona of cilia for both locomotion and feeding. When feeding they create a feeding current and filter particles including phytoplankton; bacteria are consumed as well but possibly incidentally (Turner and Tester 1992). The upper size limit of food material is about 20 μm. However, a few rotifers are raptorial feeders (*Asplanchna* and *Synchaeta*). *Synchaeta* spp. include a number of lower river and estuarine species. They can eat protozoans and other small zooplankton. The nonpredatory rotifers are an important food source for larval fish (Park and Marshall 2000) and are raised as food in aquaculture facilities. Note that there have been attempts to use zooplankton, particularly rotifers, as indicator taxa (Winner 1975). However, the use of benthic invertebrates as water quality indicators is much more useful and reliable (see Section 5.2).

### 5.1.3.1 Life History

Under favorable environmental conditions, a resting (mictic, 2n) egg hatches into an amictic (1n) female (Allan 1976). Parthenogenesis then occurs, with a series of 1n females produced. Parthenogenesis continues to occur under good conditions, such as an algal bloom—again this is r-selected behavior. This can last 20–40 generations. Under environmental change or stress (starvation, temperature change, etc.) a 2n mictic female (capable of sexual reproduction) is produced. Meiosis occurs and either a 1n female or a 1n male is produced. These then undergo sexual reproduction and form the 2n mictic resting egg (diapause).

## 5.1.4. Protozoans

Protozoans are small eukaryotic organisms, of which the taxonomy is constantly in flux, i.e. there are numerous classification schemes. The term **microzooplankton** includes the various types of protozoa, including flagellates, ciliates, and amoebas. It is a loose term, and sometimes copepod nauplii or rotifers are included, as well as gastrotrichs and nematodes. Most often, however, it refers to protozoa. Protozoans are difficult to count and identify (ID) due to their small size and the fragile nature of their bodies and require special preservatives. Useful identification references include Jahn and

Jahn (1949), Thorp and Covich (1991), and Patterson (1996). The protozoans are discussed here as part of the plankton community, but they can be free-living or attached, and they are also abundant and widespread in the benthic invertebrate community as microfauna. Protozoans have long been known to be major grazers of bacteria (Azam et al. 1983), but note they are also important grazers of fecal bacteria in natural and constructed wetlands (Burtchett et al. 2017). While the discussion here is based on the ecological role of protozoans in riverine ecosystems, the reader needs to be aware that there are a number of pathogenic protozoans that may become abundant in polluted waters. Such taxa include *Cryptosporidium parvum*, *Giardia lamblia*, *Entamoeba histolytica*, and *Naegleria fowleri* (see Chapter 13).There are three major protozoan groups commonly found in the zooplankton.

### 5.1.4.1 Flagellates

Flagellates (subphylum Mastigophora) have one or more flagella; some authors consider colorless (or even pigmented) flagellated microalgae to be protozoans. A well-known biflagellated protozoan is *Bodo*. Flagellated protozoans feed heavily on bacteria (Carlough and Meyer 1991), although their feeding mechanisms are not well defined. *Bodo* uses its anterior flagella to create a feeding current to bring bacteria into its buccal cavity.

### 5.1.4.2 Ciliates

Ciliates (phylum Ciliophora) use organelles called cilia in locomotion and/or feeding. Some species are enclosed in cases (loricas); ciliates can be free-living or stalked, single or branched. Common ciliates include *Codonella*, *Paramecium*, *Strombilidium*, *Vorticella*, and the branched *Epistylis*. Ciliated protozoans use ciliary membranelles to create a feeding current to filter out particles, or else they push water through another set of cilia to filter-feed. They consume bacteria, picoplankton, and phytoplankton.

### 5.1.4.3 Amoebas

Amoebas (subphylum Sarcodina) are characterized by protoplasmic extensions by which the organisms move or obtain nutrition. They can be generally amorphous in appearance (such as *Amoeba* and *Chaos*) and very difficult to notice in plankton samples, unless inhabiting a shell (lorica) built of sand grains or other material. In freshwater systems the many species of *Difflugia* exemplify this characteristic. Estuarine and marine amoebas include foraminiferans (which have calcareous tests) and radiolarians (silica tests)—the remains of some become important portions of oceanic sediments. Some amoebas will produce pseudopods to acquire bacteria and tiny particulate matter; mucus will draw it to the body for ingestion. Note that in the wastewater treatment process, as well as stormwater runoff treatment, protozoans and other microzooplankton are important in the removal of fecal bacteria through grazing (see Chapter 13).

### 5.1.4.4 Reproduction

Protozoans reproduce primarily by fission (simple mitosis) in favorable conditions (plenty of food materials, proper temperature, etc.). There is synchronicity within a species regarding fission: it is more pronounced in the upper water column than below the euphotic zone in deep reservoirs or estuaries. Another reproductive scheme is conjugation, when a pair come together, unite, and exchange nuclear material. Meiotic and mitotic divisions occur (occurs mostly under unfavorable conditions).

## 5.2 Stream Invertebrates

Stream invertebrates cover a wide array of taxa groups, shapes, sizes, living habits, and pollution tolerances. Those organisms in some way associated with the bottom sediments are called benthic invertebrates and have been studied intently due to general ease of collection and relative fidelity to a given reach of stream or river. The array of habitats available to benthic invertebrates is vast, consisting of calm waters, turbulent waters, clear waters, turbid waters, hard surfaces, soft surfaces, bare surfaces, periphyton-covered surfaces, macrophyte vegetation, woody debris, leaf accumulations, and rocks of all sizes. Coupled with this habitat variability, the sediments may be highly heterogeneic, with different grain sizes, roots, fecal material, chemical gradients, and bacterial abundance differences,

**Table 5.2** POM size ranges. (From Cummins 1974; Maltby 1992.)

| POM | Size |
| --- | --- |
| CPOM (coarse particulate organic material) | > 1 mm across |
| FPOM (fine particulate organic material) | 50 µm–1 mm across |
| UFPOM (ultra-fine particulate organic material) | 0.5–50 µm across |

i.e. many microhabitats (Covich et al. 1999). The most basic way to categorize benthic invertebrates is by size. Macroinvertebrates are those organisms > 0.5 mm, while those that are < 0.5 mm are considered meiofauna (Cummins 1992). A very useful way to categorize benthic invertebrates is by functional groups, which in essence are feeding guilds. Invertebrates conform to the physical characteristics of the stream reach (see Chapter 6), that is to say, the current speed, depth, type of substrata (i.e. grain size), amount of hard surface available, amount of woody substrata, and the light field. As such, various benthic invertebrates shred larger particulate organic matter (POM) (Table 5.2) into smaller pieces (FPOM and UFPOM); Table 5.2), some eat (process) detritus to glean nutrition, some graze benthic microalgae and periphyton, some consume macrophyte vegetation, and some are predators on other invertebrates, zooplankton, and even small fish.

Depending upon order and species, benthic invertebrates have widely varying tolerances to pollution. Because many benthic invertebrates live for a few months or more and many are at least relatively sedentary in a given area they are useful organisms in the assessment of water bodies for pollution or trophic state. Many individual benthic invertebrate species have great value in their presence, absence, and abundance in the assessment of water quality; this is true for certain species of phytoplankton and fish as well (see review by Patrick and Palavage 1994). Going beyond individual species, numerous benthic indicator indices have been developed over the preceding century to accomplish water quality or pollution assessments. An index may be as simple as species richness (the number of different identifiable taxa in a sample) to more complicated indices based on diversity and pollution-tolerant or -intolerant species. Such indices may include evenness, species diversity, or biotic indices, depending upon computation methods or goal. Some indices are largely academic, while some are used by states, provinces, and nations to legally assess the condition of a given water body for impairment status, so restoration programs can be devised. Note that some indices are regionally specific and as such are not useful elsewhere due to different elevation, latitude, or climate. More specific information on benthic invertebrate indices is found in Washington (1984), Hauer and Lamberti (1996), and Southerland et al. (2007); note that there are fish-based biotic indices that are used for assessment purposes as well (see Chapter 7). Some useful general references that include benthic invertebrate taxonomy, physiology, and ecology are Edmondson (1959), Pennak (1978), Thorpe and Covich (1991), Merritt and Cummins (1996), and Voshell (2003); methodology on sampling and secondary productivity assessments are detailed in Wetzel and Likens (1991), Cummins (1992), and Hauer and Lamberti (1996).

The following presents stream invertebrates first by functional group, then by taxonomic classification.

## 5.2.1 Functional Groups

### 5.2.1.1 Shredders

Shredders are a feeding guild that predominates in wooded stream systems. They attack coarse particulate organic material (CPOM) (Table 5.2) and ingest leaf fragments. Some of this material is defecated partially digested, to be consumed farther downstream by other organisms. Shredders have resident gut microflora to help digest cellulose and lignin. They can exhibit preferential feeding—choosing fungal conditioned material over sterile structural material. They can also show preference between fungal species living on CPOM (Maltby 1992). Common examples of shredders include amphipods, crayfish, and stonefly larvae.

### 5.2.1.2 Collectors

Collectors feed on fine particulate organic matter (FPOM) (Table 5.2), either by filter-feeding, gathering it from surface deposits, or by "mining" the sediments (also called deposit feeding). Some collectors have resident gut flora to aid

in cellulose digestion. Common examples of collectors include bivalves (filter-feeders), some caddisflies (filter-feeders), blackfly larvae (simuliids, which can filter bacteria from the water column), mayflies (collector-gatherers), and midge larvae (chironomids) which are also collector-gatherers. Oligochaetes are aquatic worms that mine the sediments and process them for nutritious material.

Filter-feeders can process a wide variety of living (such as phytoplankton) and nonliving particulate matter. For instance, net-building caddisfly larvae produce nets with mesh-size openings ranging from 1×6 up to 403×534 µm (reviewed in Wallace et al. 1977). The different sizes of such nets depend on different species of filter-feeders and different life-stage instars, with larger meshes in swifter currents and smaller meshes in slow currents (Wallace et al. 1977).

### 5.2.1.3 Scraper-Grazers

These grazers feed on periphyton by scraping it off rocks, wood, and macrophyte stems and leaves. Other microbes, such as bacteria and protozoans, are mixed in with the algal material they scrape. Examples of scraper-grazers include snails, some mayflies, and some fish adapted to scraping periphyton in low-order streams (stonerollers are an example; see Chapter 7).

### 5.2.1.4 Chewers and Piercers

These organisms feed on living macrophyte tissue, some by using a proboscis. Examples include hemipterans (true bugs) and some chironomids (midge larvae).

### 5.2.1.5 Predators

Predatory invertebrates feed on other invertebrates, zooplankton, and even small fish. Depending upon species, they may feed in, on, or above the water. Notably, most predation occurs in daylight (sight feeding), so many insect larvae hide under rocks in daytime and come out at night to feed (some become part of the **drift community** at night). Examples of predators include Odonates (dragonfly and damselfly larvae), Coleopterans (beetles), and Hemipterans (which feed with their proboscis).

*Note*: the drift community mainly refers to organisms that hide in daylight but enter the current at night to avoid sight-feeding predators. Some organisms are dislodged during the day and consumed by fish and predatory invertebrates. The drift community also includes nonliving edible material in the current.

## 5.2.2 Taxonomic Classification

As implied above, the benthic invertebrate community is represented by many taxonomic groups. Below are some of the most common taxa groups represented, along with their phylogenetic classification (largely based on Thorp and Covitch 1991), which include some of the organisms used in water quality or trophic classification schemes.

### 5.2.2.1 Crustaceans

Crustaceans belong to the phylum Arthropoda, subphylum Crustacea, class Malacostraca. Flowing water crustaceans have a variety of feeding modes including predation, scavenging, and omnivory, but some species figure most prominently in wooded stream systems, where they perform a major role as shredders of leaves and other decaying vegetative materials. Common crustacean orders found in freshwater streams include Isopoda, Amphipoda, and Decapoda.

#### 5.2.2.1.1 Isopoda

Isopoda (sometimes called sow bugs) are commonly found in leaf packs. They are largely omnivorous and can serve as collector-gatherers, shredders, grazers, and even predators.

#### 5.2.2.1.2 Amphipoda

Amphipoda (commonly called scuds) are omnivorous, feeding as collector-gatherers, shredders, detritivores, scrapers, and predators. *Gammarus* (Fig. 5.5), *Hyalella*, and *Crangonyx* are common lotic examples of amphipods.

#### 5.2.2.1.3 Decapoda

**Decapoda** (crayfish and shrimp). Crayfish are generally opportunistic feeders, including on organic debris (shredding), and freshwater shrimp are often grazers on algae within macrophyte beds. Common crayfish include *Cambarus*, *Procambarus*, and *Orconectes* and a well-known freshwater shrimp

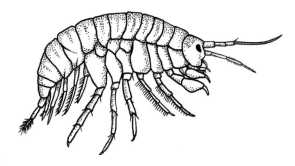

**Fig. 5.5** *Gammarus* is a common amphipod; these are largely shredders found in wooded low-order streams. (Redrawn from Voshell 2002.)

genus is *Palaemonetes* (commonly called the grass shrimp). Decapods are a major food item for fish and mammals alike. Crayfish are also raised for food in aquaculture systems in southern climes. Note that as one transitions into the estuary, crustaceans become highly abundant as collectors and predators and as economically valuable food organisms for humans.

### 5.2.2.2 Insects

Insects belong to the phylum Arthropoda, class Hexapoda, subclass Insecta. As anyone who has spent time on or near streams, rivers, and reservoirs can attest, insects are abundant and often an annoyance. However, the ones people usually see are the adults that live on or above the stream. Beneath the surface and atop or within the sediments is the insect benthic community, which is primarily composed of larval stages of numerous species. Common insect orders are discussed below.

#### 5.2.2.2.1 Plecoptera

Plecoptera (stoneflies) (Fig. 5.6a) are primarily shredders, along with some predators. This group prefers cool, well-oxygenated water, such as found in mountain or Piedmont areas where there is significant slope and mixing. This order is very pollution sensitive. *Pteronarcys*, the giant stonefly, is well known to anglers as a prey item to mimic.

#### 5.2.2.2.2 Ephemeroptera

Ephemeroptera (mayflies) are largely scrapers or collector-gatherers (Fig. 5.6b). In general, mayflies

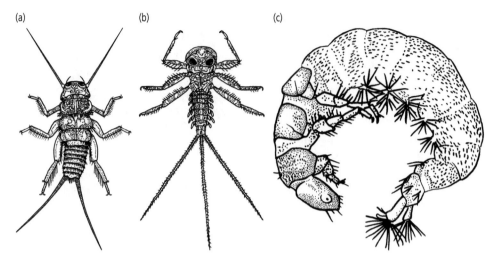

**Fig. 5.6** Aquatic insect larvae. Fig. 5.6a (left) Common stonefly Plecoptera, Perlidae. Fig. 5.6b (center) Mayfly Ephemeroptera, *Heptagenia*. Fig. 5.6c Caddisfly Plecoptera, *Hydropsyche*. (Redrawn from Giller and Malmqvist 2002.)

can live in warmer and less-oxygenated water than stoneflies. Examples include *Baetis* and *Heptagenia*. Mayflies are well known to fly fishermen who attempt to mimic various mayfly life stages for their lures.

### 5.2.2.2.3 Trichoptera

The larvae of Trichoptera (caddisflies) resemble caterpillars (Fig. 5.6c) and some species build cases to use as camouflage and protection. They have diverse feeding mechanisms as shredders, detritivores, and scrapers, but many are filter-feeders; some build nets in the current and some build tubes in the sediments and funnel water through tubes to filter. Examples include *Helicopsyche*, *Hydropsyche*, and *Cheumatopsyche*. Note that the EPT (Ephemeroptera/Plecoptera/Trichoptera) index represents the total number of different taxa from those three orders found in a given sample. This index is used as a representative of good water quality (the larger the index the better), at least in regions away from coastal lowlands where slope and water flow are naturally low.

### 5.2.2.2.4 Odonata

Odonata comprise dragonflies and damselflies. Dragonflies are commonly seen patrolling the air above water bodies looking for their prey. Beneath the water's surface the larvae of dragonflies and damselflies are predators as well. These organisms are common on Coastal Plain lowlands and can tolerate high temperatures, pollution, and fairly low DO.

### 5.2.2.2.5 Hemiptera

Hemiptera are considered the true bugs. They lack gills and larvae and adults look similar. They are largely predators on other invertebrates or piercers which feed on living plant tissue. They use their proboscis to pierce and suck fluids from food items. Examples include water boatmen (Corixidae) and water striders (Gerridae).

### 5.2.2.2.6 Coleoptera

Coleoptera are beetles. Aquatic beetles are members of a vast order and they are highly variable in shape and size. Many are predators and have mouthparts adapted for predation; but other species are scrapers, collectors, or even herbivores. Prey of predatory coleopterans range from protozoans to fish and tadpoles. Some taxa can use both surface and underwater environments. Different species have widely variable pollution tolerance, so they are useful in water quality assessments.

### 5.2.2.2.7 *Diptera*

Diptera (true flies) are a vast, highly varied group in which many feeding strategies are represented, including collector-gatherers, filter-feeders, and sediment miners. An important subgroup is the family Chironomidae (chironomids or midge larvae) which contains numerous genera (Fig. 5.7a). Well-known examples of chironomids are blood-worms, which are red-colored from a hemoglobin-like substance that promotes better oxygen uptake from the water; they can survive very low DO. These organisms are often used as indicators of poor water quality. *Chaoborus* (the phantom midge) is an **ambush predator** on zooplankton that can move vertically in the water column using hydrostatic organs. Under stratified conditions they can move to food-rich layers and feed. Other dipterans include blackfly larvae (the Simuliidae), which are filter-feeders (Fig. 5.7b) and highly abundant, especially on rocks or snags, and exploit plankton-rich running water on superstructure and rocky substrata just downstream of dams.

### 5.2.2.3 Mollusca

Molluscs are shelled organisms and are very visible members of the benthic community due to their size and protective (and sometimes elaborate) coverings. The freshwater invertebrate mollusc community consists of the following two major groups: Bivalvia and Gastropoda.

#### 5.2.2.3.1 *Bivalvia*

Class Bivalvia (clams and mussels) are filter-feeders. An example is the invasive *Corbicula fluminea*, the Asiatic clam (Fig. 5.8a), now common in US freshwaters; they have a planktonic veliger larval stage, so they are easily spread throughout river systems. A number of bivalves are invasive into the USA and discussed in more detail in Chapter 14. Various species of freshwater mussels can become abundant as filter-feeders in rivers (Wallace et al. 1977) but can be subject to localized species loss from degraded water quality. Some freshwater species have a history of commercial harvest in US freshwaters (Neves et al. 1997), and of course bivalves (oysters, clams, mussels) are valuable seafood items in estuarine and coastal marine waters.

#### 5.2.2.3.2 *Gastropoda*

Class Gastropoda (snails) have single shells which may be coiled or elongated; some shells are simple. They secrete mucus and can slide or push themselves along on it. Gastropods are mainly scrapers that use a radula to rake up periphyton on rocks, sediments, and macrophytes. They are eaten by fish, waterfowl, amphibians, and invertebrates. Many of the individual species of this group are sensitive to water quality degradation and have become locally extirpated in various streams due to excessive turbidity, hypoxia, chemical pollution, and human-caused water flow manipulations (Neves et al. 1997).

### 5.2.2.4 Oligochaeta

Oligochaetes (phylum Annelid) are aquatic segmented worms (Fig. 5.8b) and are mainly collectors/deposit feeders. Depending upon taxa, some can be found in the plankton, while most are sediment-associated where they tunnel through sediments and process them for food such as algae, bacteria, and protozoans; some will process fecal matter of other oligochaetes and invertebrates. Some oligochaetes, particularly red-colored ones, can live in low DO environments and are often

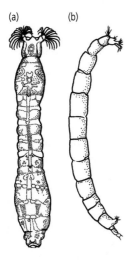

(a)    (b)

**Fig. 5.7**  Common aquatic fly larvae. Fig. 5.7a (left) Chironomid midge *Chironomus*. Fig. 5.7b (right) Simuliid blackfly; note cephalic fan-like apparatus that filters small microbiota, even bacteria, in running water. (Redrawn from Giller and Malmqvist 2002.)

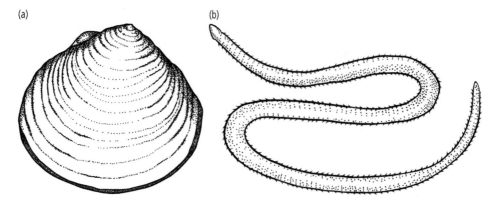

(a)                                    (b)

**Fig. 5.8** Fig. 5.8a (left) Invasive Asiatic clam *Corbicula fluminea*, now widespread throughout US waterways. Fig. 5.8b (right) Oligochaetes are extremely common and pollution-tolerant aquatic worms found from freshwater downstream into oligohaline estuaries. (Redrawn from Voshell 2002.)

used as indicators of poor water quality. Note, however, that they can be found in numerous environments, but when oligochaetes *dominate* a system it is because other taxa are excluded because they require better water quality. Examples of pollution-tolerant taxa are *Tubifex tubifex* and *Limnodrilus* spp. As salinity increases into the estuary oligochaetes are reduced, while polychaete worms (class Polychaeta) assume increasingly greater importance.

Chapter 6 takes a closer look at how the benthic invertebrate community shapes, and is shaped by, the morphology, slope, hydrology, and light regime of the river system as it moves from low-order small feeder streams through higher-order rivers and into the estuary.

Collector-gatherers are a diverse group of insect larvae and crustaceans feeding on detritus, while oligochaetes mine the sediments for nutritious particles.

- Scraper-grazers such as snails and some insect larvae scrape hard surfaces for periphyton, other invertebrates consume live macrophyte tissue, while predators such as dragonfly and beetle larvae feed on smaller animal life.
- Benthic invertebrates generally have site fidelity within given stream reaches, are relatively long-lived, and have to deal with local environmental conditions. Their presence, absence, abundance, and diversity are used in various indices to assess and describe stream ecosystem health.

## 5.3 Summary

- The lotic invertebrate community consists of zooplankton, other crustaceans, insects and their larvae, molluscs, and aquatic worms (oligochaetes).
- The zooplankton (copepods, cladocerans, rotifers, and protozoans) are abundant in lower rivers, backwater areas, reservoirs, and estuaries. They feed primarily on phytoplankton, smaller zooplankton, and bacteria.
- Shredders are crustaceans and insect larvae that process particulate matter, mostly for the associated fungi and bacteria, and are dominant organisms in wooded lower-order streams.
- Collectors include filter-feeding insects and molluscs that consume phytoplankton, bacteria, and small organic particles from the water.

## References

Allan, J.D. 1976. Life history patterns in zooplankton. *American Naturalist* 110:165–180.

Arruda, J.A., G.R. Marzolf and R.K. Faulk. 1983. The role of suspended sediments in the nutrition of zooplankton in turbid reservoirs. *Ecology* 64:1225–1235.

Azam, F., T. Fenchel, J. G. Field, J.S. Gray, L.A. Meyer-Reil and F. Thingstad. 1983. The ecological role of water-column microbes in the sea. *Marine Ecology Progress Series* 10:257–263.

Burtchett, J.M., M.A. Mallin and L.B. Cahoon. 2017. Micro-zooplankton grazing as a means of fecal bacteria removal in stormwater BMPs. *Water Science and Technology* 75:2702–2715.

Carlough, L.A. and J.L. Meyer. 1991. Bactivory by sestonic protests in a southeastern blackwater river. *Limnology and Oceanography* 36:873–883.

Chengalath, R., C.H. Fernando and M.G. George. 1971. The planktonic Rotifera of Ontario with keys to genera and species. *University of Waterloo Biology Series* 2:1–40.

Covich, A.P., M.A. Palmer and T.A. Crowl. 1999. The role of benthic invertebrate species in freshwater ecosystems. *BioScience* 49:119–127.

Cramer, J.D. and R.G. Marzolf. 1970. Selective predation on zooplankton by gizzard shad. *Transactions of the American Fisheries Society* 99:320–332.

Cummins, K.W. 1974. Structure and function of stream ecosystems. *BioScience* 24:631–641.

Cummins, K.W. 1992. Invertebrates. Chapter 11 in Calow, P. and G.E. Petts (eds.), *The Rivers Handbook: Hydrological and Ecological Principals*. Blackwell Scientific, Oxford.

Edmondson, W.T. 1959. *Ward and Whipple's Fresh-Water Biology*, 2nd edition. John Wiley & Sons, New York.

Edmondson, W.T. and G.G. Winberg. 1971. *A Manual on Methods for the Assessment of Secondary Productivity in Fresh Waters*. IBP Handbook No. 17. Blackwell Scientific Publications, Oxford.

Fitzpatrick, J.F. Jr. 1983. *How to Know the Freshwater Crustacea*. Wm. C. Brown Co., Dubuque, IA.

Giller, P.S. and B. Malmqvist. 2002. *The Biology of Streams and Rivers*. Oxford University Press, Oxford.

Harris, R.P., P.H. Wiebe, J. Lenz, H.R. Skjoldal and M. Huntley (eds.). 2000. *ICES Zooplankton Methodology Manual*. Academic Press, Cambridge, MA.

Hauer, F.R. and G.A. Lamberti (1996). *Methods in Stream Ecology*. Academic Press, Cambridge, MA.

Jahn, T. L. and F.F. Jahn. 1949. *How to Know the Protozoa*. Wm. C. Brown Co., Dubuque, IA.

Johnson, W.S. and D.M. Allen. 2005. *Zooplankton of the Atlantic and Gulf Coasts: A Guide to their Identification and Ecology*. Johns Hopkins University Press, Baltimore.

Kohl, M.A.R. and J.R. Strickler. 1981. Copepod feeding currents: food capture at low Reynold number. *Limnology and Oceanography* 26:1062–1073.

Maltby, L. 1992. Detritus processing. Chapter 15 in Calow, P. and G.E. Petts (eds.), *The Rivers Handbook: Hydrological and Ecological Principals*. Blackwell Scientific, Oxford.

Merritt, R.W. and K.W. Cummins (eds.). 1996. *An Introduction to the Aquatic Insects of North America*, 3rd edition. Kendall Hunt, Dubuque, IA.

Neves, R.J., A.E. Bogan, J.B. Williams, S.A. Ahlstedt and P.W. Hartfield. 1997. Status of aquatic mollusks in the southeastern United States: a downward spiral of diversity. Chapter 3 in Benz, G.W. and D.E. Collins (eds.), *Aqutic Fauna in Peril: The Southeastern Perspective*. Special Publication 1. Southeast Aquatic Research Institute, Lenz Desin and Communications, Decatur, GA.

Orsi, J.J. and W.L. Mecum. 1986. Zooplankton distribution and abundance in the Sacramento-San Joaquin Delta in relation to certain environmental factors. *Estuaries* 9:326–339.

Park, G.S. and H.G. Marshall. 2000. The trophic contributions of rotifers in tidal freshwater and estuarine habitats. *Estuarine, Coastal and Shelf Science* 51:729–742.

Patrick, R. and D.M. Palavage. 1994. The value of species as indicators of water quality. *Proceedings of the Academy of Natural Sciences of Philadelphia* 145:55–92.

Patterson, D.J. 1996. *Free-Living Freshwater Protozoa*. Manson Publishing, London.

Pennak, R.W. 1978. *Fresh-Water Invertebrates of the United States*, 2nd edition. Wiley, New York.

Siefert, R.E. 1972. First food of larval yellow perch, white sucker, bluegill, emerald shiner and rainbow smelt. *Transactions of the American Fisheries Society* 101:219–225.

Smith, D.L. and K.B. Johnson. 1996. *A Guide to Marine Coastal Plankton and Marine Invertebrate Larvae*, 2nd edition. Kendall Hunt, Dubuque, IA.

Southerland, M.T., G.M. Rogers, M.J. Kline, et al. 2007. Improving biological indicators to better assess the condition of streams. *Ecological Indicators* 7:751–767.

Thorp, J.H. and A.P. Covich (eds.). 1991. *Ecology and Classification of North American Invertebrates*. Academic Press, Cambridge, MA.

Turner, J.T. and P.A. Tester. 1992. Zooplankton feeding ecology: bacterivory by metazoan microzooplankton. *Journal of Experimental Marine Biology and Ecology* 160(2):149–167.

Voshell, J.R. Jr. 2003. *A Guide to Common Freshwater Invertebrates of North America*. McDonald and Woodward, Blacksburg, VA.

Walks, D.J. 2007. Persistence of plankton in flowing water. *Canadian Journal of Fisheries and Aquatic Sciences* 64:1693–1702.

Wallace, J.B., J.R. Webster and W.R. Woodall. 1977. The role of filter-feeders in flowing water. *Archiv für Hydrobiologie* 79:506–532.

Washington, H.G. 1984. Diversity, biotic and similarity indices: a review with special relevance to aquatic ecosystems. *Water Research* 18:653–649.

Wetzel, R.G. and G.E. Likens 1991. *Limnological Analysis*, 2nd edition. Springer, New York.

Williams, L.G. 1966. Dominant planktonic rotifers of major waterways of the United States. *Limnology and Oceanography* 11:83–91.

Williamson, C.E. 1991. Copepoda. Chapter 21 in Thorp, J.H. and A.P. Covich (eds.), *Ecology and Classification of North American Invertebrates*. Academic Press, Cambridge, MA.

Winner, J.M. 1975. Zooplankton. Chapter 7 in Whitton, B.A. (ed.), *River Ecology*. University of California Press, Oakland.

# Feeding the River

## Unifying Concepts

In Chapter 5 the stream invertebrate community was introduced. This chapter will discuss how that community fits into the river-stream ecosystem, as well as the heterotrophic microbiota that both process organic matter and serve as food for invertebrates and vertebrates. This chapter also emphasizes established concepts that relate autochthonous and allochthonous in-stream and riparian organic matter to lotic biotic production.

## 6.1 Feeding the Invertebrate Community

Stream invertebrates, as well as many vertebrates, are fed from sources that come from either outside of the stream (allochthonous) or that are produced from within the stream (autochthonous). The food web flows from primary producers and decomposers (Table 6.1) to grazers, shredders, and collectors, to predators and scavengers. The primary producers are discussed extensively in Chapters 3 and 4, so we will focus on the allochthonous material here.

### 6.1.1 Detritus as a Food Source: Energy from Outside of the Stream

Allochthonous energy sources (food that originates outside of the stream) include CPOM which

**Table 6.1** Food resource categories available to invertebrates (Cummins 1992).

| |
| --- |
| 1) Detritus (CPOM, FPOM, UFPOM) |
| 2) Periphyton and benthic microalgae |
| 3) Live macrophyte tissue |
| 4) Live prey (other invertebrates, zooplankton, small fish) |
| 5) Filterable microbiota (phytoplankton, bacteria, protozoans) |

constantly enters low-order streams (as leaves, twigs, etc.; Fig. 6.1a). Carbohydrates, amino acids, and phenolic substances are leached out soon after the leaf hits the water (Maltby 1992). Structural materials, such as cell walls, consist of cellulose and lignin and remain behind as CPOM (Table 5.2). Such material is often clearly most visible in streams in fall and winter as **leaf packs** (Fig. 6.1b) which are mats of decaying leaves bunched together.

CPOM is colonized by bacteria and fungi, called conditioning (Fig. 6.1c, d). Bacteria colonize edges, particularly damaged edges of leaves, and can also colonize FPOM and VFPOM. However, fungi colonize CPOM but do not colonize FPOM well; FPOM may be too small to support fungal hyphae. Some terrestrial fungi may colonize leaves before they enter the stream, while others colonize in the water (Suberkropp 1998a). Hyphomycetes are a group of fungi adapted to flowing water.

The conidia of many aquatic hyphomycetes are large, tetraradiate or sigmoid adaptations to flowing waters to increase the effectiveness with which they become trapped (Suberkropp 1998a). Peak conidial numbers occur shortly after leaf fall. Fungi can utilize plant cell walls as a carbon source (most species can degrade at least one of the components of plant cell walls). The ability of fungal hyphae to penetrate is important in processing refractory organic matter (fungi also colonize wood). Fungi use a variety of extracellular enzymes to degrade cellulose (Jenkins and Suberkropp 1995). Some will be saprophytic on other fungi and use both fungi and plant cell wall material as an energy source.

Bacteria in stream waters may be free-living or attached, suspended in the water, or associated with

*River Ecology*. Michael A. Mallin, Oxford University Press. © Michael A. Mallin (2023). DOI: 10.1093/oso/9780199549511.003.0007

**Fig. 6.1** Fig. 6.1a (top left) Fall loads vast amounts of allochthonous organic material to streams. Fig. 6.1b (top right) Leaf packs provide both shelter and food resources for various invertebrates within the stream. Figs 6.1c, d (bottom left and right) Fungi and bacteria colonize leaves and other organic material both before and once within the stream proper; bacteria will attack damaged edges such as shown in oak leaf (Fig. 6.1d), while fungi will invade large inner areas with their hyphae. (Photos by the author.)

sediments. Detrital bacteria consist of rods, cocci, and filamentous and mycelated forms, as well as myxobacteria (a group of colonial bacteria found in soil that are embedded in slime and move by gliding). Some bacteria utilize cellulose, through use of the enzyme cellulase.

What factors are important in the breakdown of detritus? There is faster breakdown of leaves with less lignin content and faster breakdown in nutrient-rich waters. Field investigations and experiments have demonstrated enhanced fungal growth in the field in nutrient-rich waters (Jenkins and Suberkropp 1995; Suberkropp and Chauvet 1995; Padgett et al. 2000). In controlled microcosm experiments, Suberkropp (1998b) found that additions of N or P alone or both together yielded enhanced

conidium production, fungal biomass as ATP, and greater leaf material weight loss compared to controls. Rosemond et al. (2015) working with leaf packs conducted two- to five-year-long experimental additions of N+P, finding enhanced C loss due to microbial activity and $CO_2$ loss to the atmosphere. The authors suggested that another impact of eutrophication was C loss from the system or enhanced stream food resource loss. Elwood et al. (1981) found that experimental additions of P to stream segments significantly increased C loss and microbial respiration. Bacterial respiration in general, in particular that of aquatic bacteria, is stimulated by inputs of P (reviewed in Mallin and Cahoon 2020). On the other side of the coin, a number of pollutants including pesticides, heavy metals, acid mine drainage, and coal ash effluent have been shown to negatively impact decomposition rates (reviewed by Maltby 1992). A series of large-scale field experiments in 100 European streams demonstrated that leaf decomposition was reduced at low nutrient concentrations, maximal at moderately high nutrient concentrations, and reduced again at very high nutrient concentrations (Woodward et al. 2012). The authors speculated that at the high end, pollution-driven factors impacted the detritivores, such as reduction of sensitive benthic species, severe oxygen depletion, or smothering.

The speed of breakdown depends upon the type of leaves and suggests a pattern of macrophyte leaves breaking down faster than non-woody terrestrial leaves, which break down faster than woody plant leaves, with pine needles most recalcitrant (Maltby 1992 and references within). Time of processing varies with structural material and the degree of recalcitrant material; as such, herbaceous leaves require 30–50 days, tree leaves 4–6 months, and resistant leaves or pine needles 1–2 years (Gregory et al. 1991).

Detrital feeders (shredders and collectors) have resident gut microflora in the hindgut that can digest cellulose and lignin. CPOM that is conditioned by microbes is more palatable to detritivores than unconditioned CPOM, the microbial content of CPOM is assimilated with greater efficiency than the structural material, and detritivores can actually discriminate between types of fungi on leaf material (reviewed by Maltby 1992).

FPOM is a mixture of autochthonous material including shredded vascular plant tissue, phytoplankton and tychoplankton, invertebrate feces, algae, and microorganisms, as well as shredded allochthonous material from outside of the stream. FPOM can be produced by the feeding of invertebrates, physical abrasion of CPOM, microbial action, scouring of periphyton from macrophytes and stream sediments, soil erosion, and flocculation of DOM (Maltby 1992; Carlough 1994). Small-sized detritus in middle- and higher-order streams consists of FPOM from upstream as well as local decomposing algae and macrophytes. This material can be nutritious to downstream invertebrates and can be harvested by collectors such as filter-feeders or collector-gatherers in pools.

## 6.1.2 Dissolved Organic Matter

DOM is another source of energy to stream invertebrates and enters the stream from various sources (Cummins 1974; Meyer 1990; Maltby 1992), as follows:

- Leached from detritus and/or released by the enzymatic activity of microbes associated with detritus, either within the stream or in riparian floodplains where it subsequently enters streams.
- Arising through the feeding activities of invertebrates on POM.
- Leaked by living algae and macrophytes.
- Released through lysis of dead cells.
- Arising from microbial activities.
- Groundwater seepage into the stream (the hyperhoeic pathway).
- Throughfall of dissolved material from the tree canopy.
- Surface runoff from natural or developed terrestrial areas.
- Anthropogenic-sourced DOM can be labile or refractory; some can be toxic to stream organisms (polycyclic aromatic hydrocarbons [PAHs], for instance; see Chapter 12).

DOM is an important source of energy for heterotrophs (microbes that do not depend on obligate photosynthesis). There are various ways in which DOM can enter the food chain. It can be directly assimilated by bacteria (Findlay et al. 1986) and heterotrophic algae (members of various algal taxa can be obligate or facultative heterotrophs; see Chapter 3). DOM can also be converted to particulate form (POM) by physicochemical processes; see Carlough (1994) for details on this process in freshwater and Camilleri and Ribi (1986) for an estuarine analog. Additionally, DOM can be adsorbed onto surfaces of something that is later eaten (leaf, FPOM, etc.); note that these particles can be extremely small, < 0.8 μm (Kerner et al. 2003). Once DOM becomes part of POM or is assimilated by bacteria or heterotrophic algae it can be eaten by a grazer and enter the food web.

### 6.1.3 Drift

The drift community refers to detritus, algae, and invertebrates carried along by the current. Adult copepods are well known to migrate upwards from the bottom at night in freshwater and estuarine ecosystems and can be a significant part of stream drift (Schram et al. 1990). Stream invertebrates move up into the water column at night to avoid sight-feeding predators (Cole 1994). Fish and predatory invertebrates feed heavily on the drift community.

### 6.1.4 Other Allochthonous Food Sources

The riparian environment provides numerous dead, living, and injured small organisms such as caterpillars, flies, earthworms, crickets, and other insects and terrestrial larvae that fall in or are washed or blown into the water and consumed by fish and larger invertebrates. These organisms are separate from organisms originating within the stream, such as invertebrate larvae, pupae, or adults. Other food materials include plant parts, living and dead, that can feed herbivores and detritivores.

## 6.2 The River Continuum Concept

The River Continuum Concept (RCC) (Vannote et al. 1980) is a useful river characterization and visualization tool/concept that was developed using rivers in the Northeast and Northwestern USA. It is applicable in many situations but not valid for lowland rivers such as southeastern blackwater rivers (see Chapter 8), other systems that lack slope in the headwaters, and nonforested systems. The major tenet of the RCC is that "Structural and functional characteristics of biotic communities along river gradients are selected to conform to the most probable position, or mean state of the physical system."

Remember from Chapter 1 our general classification of headwaters (stream orders 1–3), medium-sized rivers (orders 4–5), and large rivers (orders >5). The RCC notes that headwaters are heavily influenced by riparian vegetation (shading and organic input). With increasing stream size, there is reduced importance of terrestrial organic inputs and increased significance of autochthonous

primary production and organic matter transported from upstream.

### 6.2.1 Headwaters Streams

Headwaters streams are small, shallow streams that have maximum interaction with the landscape (Fig. 6.2a, b). They are accumulators, processors, and transporters of materials coming from the terrestrial biome, including nutrients and organic matter, both particulate and dissolved (Vannote et al. 1980). Headwaters streams in mountains, Piedmont, or otherwise hilly regions are characterized by a steep slope, rocky substratum, and narrow channel, and if in a forested region can have a dense canopy. The tree canopy, or lack thereof, can determine if a given stream reach is "powered" by photosynthesis or the heterotrophic processing of organic materials. After autumn leaf fall, during winter and early spring the lack of shade allows sufficient solar radiation to drive periphyton and benthic microalgal photosynthesis. Likewise, streams passing through meadows (Fig. 6.2c) or clearcut areas (Fig. 6.2d) will receive greater light availability for photosynthesis.

Periphyton taxa are mainly diatoms and also cyanobacteria, with lesser amounts of other taxa (see Chapter 4). This is a running water situation, so the algae need to stick to rocks and other substrata. Diatoms (mainly pennate) produce mucilage to help them adhere, while other algae have holdfasts. Bear in mind that in running water the nutrients come to the attached microflora (epilithic, epipelic, epipsammic). Periphyton are fed on by grazers such as scrapers; examples include snails and some mayflies, caddisflies, and beetles (see Chapter 5). Diatoms are easily digestible and utilized, so there is high assimilation efficiency for diatoms as food items for grazers, whereas cyanobacteria-dominated periphyton can be problematic for grazers (Burkholder et al. 2018).

A stream with a closed forest canopy in summer and fall gets much reduced solar irradiance and has subsequent reduced photosynthesis. However, it will seasonally receive large pulses of POM, i.e. leaves, twigs, pine needles, and herbaceous and shrub litter. In fall (Fig. 6.1a) the input of leaves provides energy in the form of organic matter

**Fig. 6.2** Fig. 6.2a (top left) Headwaters stream on Mt. Marcy, NY, showing interaction with forest. Fig. 6.2b (top right) Shading of Smokey Mountains stream in Tennessee. Fig. 6.2c (bottom left) Low-order North Carolina stream emerging from a shaded reach into a clearing and sunlight. Fig. 6.2d (bottom right) Periphyton growth on rocks in an unshaded first-order North Carolina stream. (Photos by the author.)

to the stream—it becomes net heterotrophic. To process this POM load, headwaters streams have a large community of shredders—organisms that process CPOM (see Chapter 5); these shredders are mainly amphipods, decapods, and various insect larvae. The shredders break down CPOM into FPOM and even UFPOM. Unused or partially processed materials are transported downstream as energy income for downstream communities (Fig. 6.3). This material is then processed by collector-gatherers, which are abundant in depositional areas (pools). Filter-feeders utilize running water areas (riffles, runs) to capture FPOM and UFPOM. An example is caddisfly larvae (such as *Hydropsyche*) which build a net/filter and periodically clean it. Another example is blackfly larvae (simuliids) which filter UPOM (including bacteria) using their

filter/tail. Additionally, the feeding, shredding, and burrowing activities of invertebrates in lower-order streams release nutrients that are then used by periphyton and macrophytes in middle reaches of the river continuum (Covich et al. 1999). Thus, in the RCC organisms upstream process material for use by organisms downstream in the continuum (Fig. 6.3).

To summarize, headwaters streams can be net autotrophic; i.e. the ratio of primary production (P)/respiration (R) is > 1, or net heterotrophic (P/R < 1), depending on the environment within the stream—if it is well lit for photosynthesis then P/R > 1 or if the reach is heavily shaded or in a depositional area then P/R < 1. Net heterotrophy also depends on season; if there is a closed canopy (summer) or heavy POM load (such as in fall) then P/R

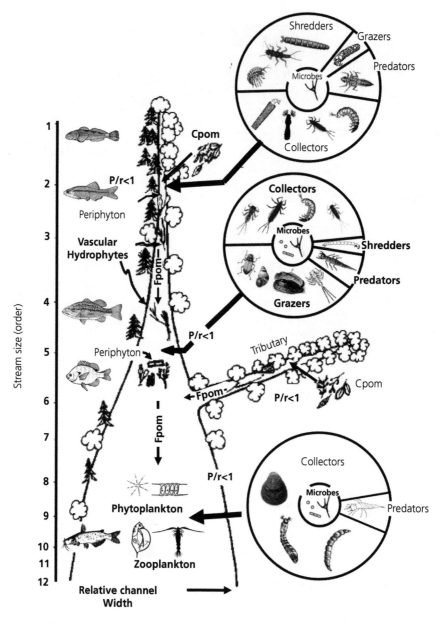

**Fig. 6.3** Schematic of the River Continuum Concept. (Modified from Vannote et al. 1980 © Canadian Science Publishing.)

< 1. If there is an open canopy in late winter or early spring this provides light for periphyton and macrophyte photosynthesis (P/R > 1).

Bear in mind that headwaters can be unstable (i.e. subject to floods); the flow variation in streams increases with decreasing drainage area (Sabo et al. 2010). Floods can add pulses of organic matter as allochthonous food sources for the stream community; conversely, floods can add loads of suspended sediments that can bury or shade out periphyton or scour invertebrate habitats. Inputs of large woody debris or boulders into a stream reach can create new habitats for organisms, as well as move and change channel subunits.

Other research (Peterson et al. 2001) found that small headwaters streams are sites of intense biological assimilation of nutrients by periphyton, bryophytes, bacteria, and fungi, as well as sites of rapid microbial nutrient transformation (spiraling). The authors suggested this elevated metabolism may be a result of high substrata surface to volume ratios in these small streams (i.e. there is usually a low volume of water relative to the stream bed and features within it, the surfaces available for processing). This situation would change dramatically under heavy rains and flooding—the nutrient spiraling length would greatly increase (see Chapter 2).

## 6.2.2 Middle Reaches of the River Continuum

According to the RCC, as the middle reaches are entered the stream is wider and the slope is less (Fig. 6.4). The upper middle reaches are generally well lit, rocky, and good habitats for periphyton (Fig. 6.3). These stretches should also be good habitats for aquatic macrophytes because the stream is still fairly shallow. Thus, this can be an autotrophic region (P/R > 1). As we know from Chapter 4, macrophytes provide good habitat for invertebrates (such as amphipods and certain hemipterans), provide substrata for periphyton (epiphytic microalgae), and provide cover and hunting grounds for fish and predatory invertebrates. Additionally, there are scrapers and other grazers to feed on the periphyton, and also filter-feeders (Wallace et al.

1977) and collectors to feed on the material coming in from upstream (FPOM, UFPOM).

As one transitions into the lower middle reaches the river becomes broader and deeper and the substrata becomes siltier from deposition as the current slows (see Chapter 1). The sediments contain a greater proportion of deposit feeders, such as clams, chironomids, and oligochaetes. There is also increasing interaction with the floodplain. Note that throughout the river continuum the variety of habitats, food material types, and feeding strategies leads to a high degree of food resource partitioning, enhancing biodiversity (Minshall et al. 1985).

## 6.2.3 Lower Reaches of the River Continuum

In the lower river reaches, i.e. the highest-order streams in the RCC, the stream is broad and deep, with well-sorted sediments. The system is likely too turbid to support periphyton or macrophytes, except for the nearshore shallows (Fig. 6.5a). The benthic invertebrate community is dominated by depositional feeders (collectors), again chironomids, oligochaetes, and clams. Depending upon turbidity and light availability there can be development of a phytoplankton community (Fig. 6.5b) that is grazed by the zooplankton community. The freshwater zooplankton community consists of copepods, cladocerans, and rotifers (see Chapter 5), all of which are an important source of food for larval and juvenile river fishes. Bear in mind this

(a)          (b)

**Fig. 6.4** Fig. 6.4a (left) Gooseberry River, Minnesota, transitioning to well-lit middle reaches. Fig. 6.4b (right) Well-lit middle reaches of Elwha River, Washington State. (Photos by the author.)

(a)  (b)

**Fig. 6.5** Fig. 6.5a (left) Turbid Lower Congaree River, South Carolina. Fig. 6.5b (right) A large regulated river like the Columbia in the Pacific Northwest can host significant phytoplankton and zooplankton communities, particularly upstream of dams. (Photos by the author.)

is a generalization and middle-to-lower reaches of some rivers (particularly coastal rivers) may be canopied, putting a seasonal as well as longitudinal autotrophic influence on the degree of phytoplankton production (Ensign et al. 2012).

Lower rivers are also key conduits for anadromous fishes (herrings, shad, striped bass, sturgeon, salmonids). Some of these taxa support important commercial and sport fisheries, while others, such as sturgeon (see Chapter 14), are endangered species.

If the lower river is turbid (Fig. 6.5a), little light is available to phytoplankton, so P/R < 1; thus the system is net heterotrophic. Note that waters that are highly colored with DOM (see Chapter 8) are subject to reduced light penetration and net heterotrophy as well (see Chapter 3, Exercise 3.1 and Fig. 3.11). If the river is less turbid but nutrient-rich, it can experience extensive algal blooms (blue-greens, greens, cryptomonads, or prymnesiophytes in the Western USA) and P/R > 1 (the river will be net autotrophic).

During floods, lower river reaches may have extensive interaction with the floodplain. These riparian forests (often called palustrine wetlands) provide food and habitat for residential and migrating fish. For example, the Amazon River has extensive floodplain interaction during high-water periods, with many fish species adapted to feeding on terrestrial vegetation (see Chapter 7). In some areas (Virginia, the lower Cape Fear basin in North Carolina) there are extensive tidal freshwater marshes,

which can have high biological diversity of both flora and fauna (Odum 1988) and are also important areas for waterfowl. Of course tributary streams, small and large, enter the main river along the continuum. Earlier it was noted that tributaries can supply phytoplankton to the river proper (see Chapter 3). Minshall et al. (1985) suggested that, depending upon the conditions within the tributary, a tributary's influence could be to either supply or dilute the main river with nutrients, phytoplankton, suspended solids, particulate organic carbon (POC), or dissolved organic carbon (DOC), altering the structure and function of the main river at least for a given distance.

### 6.2.4 Lower Rivers Merge Into Estuaries at the End of the Continuum

Estuaries, both riverine and less well-flushed sounds fed by rivers, can host large phytoplankton and zooplankton communities. Even before the estuary is reached (i.e. salinities still < 0.5 psu [practical salinity units]), some coastal rivers have extensive tidal freshwater zones that can express high phytoplankton biomass from the increased river width and increase in available solar irradiance (Ensign et al. 2012). The zooplankton of lower rivers and riverine estuaries, especially copepods (see Chapter 5), are a major food source for larval and juvenile estuarine and marine fish (Ensign et al. 2014); these estuaries contain primary nursery areas (PNAs) (see Chapter 9) for many fish species. Of course, riverine estuaries are rich commercial finfish and shellfish areas. However, estuaries can and do experience eutrophication symptoms from upstream nutrient loading (Bricker et al. 2007). Such symptoms include bottom-water hypoxia (Rabelais et al. 2001; Hagy et al. 2004), loss of seagrasses, and toxic algal blooms (Burkholder 1998; Burkholder et al. 2018).

## 6.3 River–Floodplain Interaction

### 6.3.1 The Riparian Zone

As described in Chapter 1, the **riparian zone** is the interface between terrestrial and aquatic ecosystems; its boundaries extend outward to the limit of flooding (i.e. the floodplain) and upward into the canopy of streamside vegetation (Gregory et al. 1991). In natural river ecosystems there is intense interaction between the riparian zone and the stream–river system proper. However, due to previous and ongoing small and large hydrological modifications (i.e. dams and channels; see Chapter 10) there are few remaining large rivers that can be considered to be in "natural" states. Most of our river–floodplain knowledge comes from the tropics, which in the twentieth century were less impacted by humans (but are currently being modified rapidly mainly for hydropower; see Chapter 10). One can consider the floodplain as a type of wetland.

Riparian zones, whether they are forests or marshes, serve as pollutant filters to protect the river channel from land-based human activities such as agriculture, timbering, and residential development (Naiman et al. 2005). Such areas are also ecological "edges" for terrestrial animals (they move between biomes). Additionally they serve as travel corridors for organisms such as lots of birdlife, including migrating waterfowl, and small and large mammals, and serve as dispersal areas for plants (Gregory et al. 1991; Sparks 1995; Naiman et al. 2005).

Whereas the RCC emphasizes the strong headwaters interaction with the riparian zone, the floodplain connection of middle- and higher-order river segments cannot be overestimated. The floodplain supplies POM of all sizes to the river; as noted in the RCC, particulate C is a major source of food for river biota. Large woody debris can enter the river continuum at any point (Bilby and Bisson 1998) and create logjams, debris dams, and snags (Fig. 6.6a, b) that alter currents and sediment deposition (see Chapter 8 for more on lowland river–floodplain interaction). Large woody debris is particularly abundant in the Pacific Northwest of the USA and Canada (Bilby and Bisson 1998) and can physically create pools, channels, falls, and sand or gravel bars; in the bars various plants can take root. Thus such debris can create microhabitats for periphyton, macroalgae, shredders, collectors including filter-feeders, periphyton grazers, predator invertebrates, fish, mammals, and birds. In some ecosystems, such as the increasingly rare unregulated rivers, living woody

(a)　　　　　　　　(b)

**Fig. 6.6** Fig. 6.6a (left) Logjam in stream on the Olympic Peninsula, USA. Fig. 6.6b (right) Alongshore snag in autumn, Lower Tallapoosa River, Horseshoe Bend National Military Park, AL, USA. (Photos by the author.)

debris from the floodplain can take root in bars, physically buffer currents, enhance sedimentation, and create islands (Gurnell et al. 2005). On such islands trees can grow and ponds can form, and these terrestrial and limnetic habitats lead to an increase in biodiversity.

The floodplain itself may support riparian forests, epiphytes, and macrophytes; if the floodplain hosts lakes such as oxbows (Penczak et al. 2003) or river backwaters (Fig. 5.1) these water bodies will host phytoplankton, benthic microalgae, submersed, emergent, floating leaved and free floating macrophytes. Since at some point all of these plants encounter and interact with rising river water, Wetzel and Ward (1993) note that floodplain primary productivity should be included in river productivity estimates. In addition, plants of all kinds leak DOM into water bodies, which is consumed by bacteria. Sieczko et al. (2015) note that plant-produced autochthonous DOM is a labile source of energy for consumers, whereas allochthonous DOM produced from decaying organic materials is more recalcitrant and harder to process by bacteria. Hence, the floodplain is a significant source of physical, chemical, and biological activity which helps support the river food web (Wetzel and Ward 1993), leading to the following theory on how floodplain inundation enhances river fisheries.

## 6.3.2 The Flood-Pulse Concept

The **flood-pulse concept** was based on pristine (undammed) systems. The following material was originally derived from Junk et al. (1989) and is discussed further in Bayley (1995) and Johnson et al. 1995. The concept states:

1. There is a predictable advance and retraction of water on a floodplain.
2. The biota have evolved adaptations to account for this.
3. Thus, floods are not a disturbance to the natural system—prevention of floods becomes the disturbance.
4. The regular flooding enhances productivity and maintains biotic diversity of the system.

Why would flooding increase the productivity of the floodplain? Chapter 1 discussed the timing of flooding and reasons for it in various areas of the USA; similar reasons account for flooding worldwide, depending upon season and local meteorology. During floods there is an increase in nutrients coming into the system from upstream sources; think of the Nile floodplain in pre-Aswan Dam times (Sparks 1995). Seasonal floodwaters also cause remineralization of nutrients from organic matter and soils on the floodplain. In areas where upstream agriculture is a major land use, winter–early spring flooding will wash fertilizers downstream to lower river reaches and the estuary (i.e. the freshet; see Chapter 2). Thus, riparian flooding leads to both nutrient deposits from upstream and remineralization locally. However, an important caveat to consider is the degree of denitrification that occurs on the floodplain. Ensign et al. (2008) demonstrated that regularly inundated riparian zones of tidal freshwater rivers undergo considerable denitrification; thus some portion of the nitrogen brought in or remineralized locally is lost to the system during seasonal flooding.

According to the flood-pulse concept, following flooding, both local primary production and decomposition increase on the floodplain (Fig. 6.7). While nutrient stimulation of primary production is well known, nutrients also increase rates of decomposition, as discussed in Section 6.1.1. Higher trophic level production is also stimulated. Fish yield per acre has been found to be greater in rivers with a flood pulse and floodplain than in impoundments in the same area (Bayley 1995).

Recent work in South America has demonstrated the positive impacts of flooding intensity on recruitment of fishes with a variety of reproductive strategies (de Oliviera et al. 2020). The theory suggested that elevated fish yield is associated with the seasonal inundation of the floodplain for several reasons. During floods, the rising waters cause increased expansion of the available physical habitat (called the moving littoral), which allows fish to access the terrestrial and wetland resources of the floodplain. The flooding increases the availability of invertebrates, plants, and organic matter; i.e. there are more food resources for fish. Because of the shallow and more complex habitat of the flooded terrain there is less predation on juvenile fish when they are on the floodplain—they find a refuge there, similar

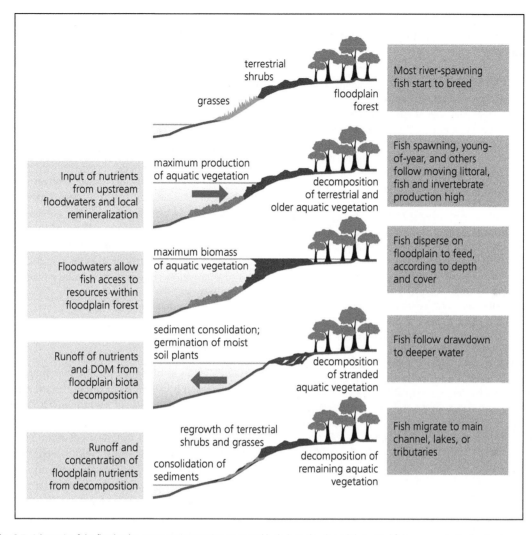

**Fig. 6.7** Schematic of the flood-pulse concept, representing an annual hydrological cycle and theoretical fish response to the flood pulse as applied to temperate regions. (Modified from Junk et al. 1989 © Canadian Science Publishing.)

to a primary nursery area. Note that the opposite occurs when there is low flow in the channel; i.e. predation by birds and large fish increases because there is less available structural habitat for young and small fish.

Unfortunately, most large river systems in the USA and Europe are dammed for various reasons. Thus, the flood pulse in these systems is greatly reduced or nonexistent. In the southeastern Coastal Plain, however, there are still some undammed rivers since the slope is too gentle for

hydropower. The basis of the flood-pulse concept was the Amazon. It should be noted that many, if not most large tropical rivers in Africa, Asia, and South America are dammed, are being dammed, or have plans for dams for hydropower, so some of this function will be lost in those ecosystems (see Chapter 10).

Small fish find a refuge on the floodplain during floods. An example of this was a large-scale inadvertent experiment caused by hurricanes in the US southeast (Box 6.1).

## Box 6.1 Flooding and fish—a coastal river example

In 1996 and 1998 Hurricanes Fran and Bonnie caused minor flooding in southeastern North Carolina, USA. However, many thousands of fish of all sizes died in the Northeast Cape Fear River from anoxia and hypoxia (see Chapter 17). These hurricanes were notable in that many CAFOs (see Chapter 13) and sewage treatment plants were damaged, leading to inputs of swine lagoon waste and raw human sewage, creating major BOD loads. There was little overtopping of river banks and the water was largely confined to the channel. Following Hurricane Bonnie, severe hypoxia prevailed in that stretch of river for nearly three weeks. The fish were essentially trapped in the channel by anoxic and hypoxic waters, causing massive mortality. However, during Hurricane Floyd in 1999 (essentially a massive flood) only 17 fish, all large, were found in the same area where earlier hurricanes had killed thousands. The small fish apparently moved onto the floodplain (inundated for weeks) where they escaped the pollution coming downriver. In contrast to Hurricanes Fran and Bonnie the waters did not become anoxic, and DO dipped below 1.0 mg/L for only a couple days before climbing back again; no major fish kills were reported from Hurricane Floyd (Mallin et al. 2002). Fish are fine with floods, people aren't—there is extensive damage to housing in low-lying areas. However, Hurricane Florence in 2018 dropped 2 ft of rain on eastern North Carolina in three days, causing many swine waste lagoons to flood and millions of gallons of human sewage to enter the rivers from power outages and flooding. A number of massive fish kills in various rivers were caused by low DO from the huge BOD load, as well as by fish strandings on flooded highways and parking lots.

## 6.4 Subsequent Theories

Other theories have been raised to address areas or processes that were not specifically dealt with in the concepts above, or to provide unification. For instance, Thorp and Delong (1994) proposed the river productivity model, which provided more emphasis on localized production and inputs in lower, unconstrained rivers. The authors suggested that during nonflood pulse periods those river segments were primarily fed by localized phytoplankton, periphyton, and macrophytes, along with direct inputs of POC and DOC from the adjoining floodplain. Finally, the river-wave concept (Humphries et al. 2014) provides an interesting and very visual take on river metabolism by envisioning sets of waves incorporating the processes discussed earlier in this chapter. The trough of the wave would be base flow, where river secondary productivity is supported by local autochthonous and allochthonous inputs. On the ascending limb of the next wave (i.e. rising flows from upstream rainfall or snowmelt) inputs from upstream predominate over local sources. The crest represents the flood stage, where the flood pulse accesses the floodplain and all its physical, chemical, and biological resources. The amplitude of the waves varies with season and meteorological events, and patchiness or disruption results from tributary inputs or river regulation.

## 6.5 Summary

- Besides algae and higher plants, the river food web is fueled by all kinds or organic matter from both within and outside of the stream: DOM and POM of all sizes.
- The RCC describes a longitudinal river visualization in which the slope, width, sediments, canopy, and shading interact to determine the type of primary producers and secondary consumers present at any given location. The RCC is valid in many locations, but there are a number of geographic exceptions where it does not or only partially applies.
- The RCC assumes mainly heterotrophic headwaters streams that interact strongly with the riparian zone and receive much POC, which is conditioned by fungi and bacteria and processed by shredders, with excess flowing downstream to collectors. In unshaded headwaters streams periphyton can be abundant and the stream net autotrophic.
- The RCC assumes that in middle reaches of the continuum the canopy opens and primary producers, such as periphyton and macrophytes, are abundant, as are their grazers. Collectors such as filter-feeders consume FPOM from upstream reaches.

- In lower rivers, the RCC predicts, light permitting, development of phytoplankton and zooplankton communities that help feed the fishery. Benthic invertebrates are dominated by filter-feeders such as bivalves and collector-gatherers including oligochaeates and chironomids at and within the well-sorted sediments.
- Large woody debris entering streams and rivers from the floodplain can channel flows, form pools and waterfalls, cause sediment deposition, and form islands. This process creates numerous microhabitats that support a diversity of algae, macrophytes, and woody plants, and both invertebrate and vertebrate animals.
- Seasonal inundation of the floodplain (the flood pulse) brings in nutrients from upstream, remineralizes nutrients from local areas, and allows small fish access to the food resources and predation refuges of this "moving littoral" over the floodplain.

## References

Bayley, P.B. 1995. Understanding large river-floodplain ecosystems. *BioScience* 45:153–158.

Bilby, R.E. and P.A. Bisson. 1998. Function and distribution of large woody debris. Chapter 13 in Naiman, R.E. and R.E. Bilby (eds.), *River Ecology and Management: Lessons from the Pacific Coastal Ecoregion*. Springer, New York.

Bricker S, B. Longstaff, W. Dennison, et al. 2007. *Effects of Nutrient Enrichment in the Nation's Estuaries: A Decade of Change. National Estuarine Eutrophication Assessment Update*. NOAA Coastal Ocean Program Decision Analysis Series No. 26. National Centers for Coastal Ocean Science, Silver Spring, MD.

Burkholder, J.M. 1998. Implications of harmful microalgal and heterotrophic dinoflagellates in management of sustainable marine fisheries. *Ecological Applications* 8:S37–S62.

Burkholder, J.M., S.E. Shumway and P.M. Glibert. 2018. Food web and ecosystem impacts of harmful algae. Chapter 7 in Shumway, S.E., J.M. Burkholder and S.L. Morton (eds.), *Harmful Algal Blooms—A Compendium Desk Reference*. John Wiley & Sons, Hoboken, NJ

Camilleri, J.C. and G. Ribi. 1986. Leaching of dissolved organic carbon (DOC) from dead leaves, formation of flakes from DOC, and feeding on flakes by crustaceans in mangroves. *Marine Biology* 91:337–344.

Carlough, L.A. 1994. Origins, structure, and trophic significance of amorphous seston in a blackwater river. *Freshwater Biology* 31:227–237.

Cole, G.A. 1994. *Textbook of Limnology*, 4th edition. Waveland Press, Long Grove, IL.

Covich, A.P., M.A. Palmer and T.A. Crowl. 1999. The role of benthic invertebrate species in freshwater ecosystems. *BioScience* 49:119–127.

Cummins, K.W. 1974. Structure and function of stream ecosystems. *BioScience* 24:631–641.

Cummins, K.W. 1992. Invertebrates. Chapter 11 in Calow, P. and G.E. Petts (eds.), *The Rivers Handbook: Hydrological and Ecological Principals*. Blackwell Scientific, Oxford.

De Oliveira, A.G., T.M. Lopes, M.A. Angulo-Valencia, et al. 2020. Relationship of freshwater fish recruitment with distinct reproductive strategies and flood attributes: a long-term view in the upper Paraná River floodplain. *Frontiers in Environmental Science* 8:577181.

Elwood, J.W., J.D. Newbold, A.F. Trimble and R.W. Stark. 1981. The limiting role of phosphorus in a woodland stream ecosystem: effects of P enrichment on leaf decomposition and primary producers. *Ecology* 62: 146–158.

Ensign, S.H., M.F. Piehler and M.W. Doyle. 2008. Riparian zone denitrification affects nitrogen flux through a tidal freshwater river. *Biogeochemistry* 91:133–150.

Ensign, S.H., M.W. Doyle and M.F. Piehler. 2012. Tidal geometry affects phytoplankton at the transition from forested streams to tidal rivers. *Freshwater Biology* 57(10):2141–2155.

Ensign, S.H., D.M. Leech and M.F. Piehler. 2014. Effects of nutrients and zooplankton on an estuary's phytoplankton: inferences from a synthesis of 30 years of data. *Ecosphere* 5(7):89.

Findlay, S., L. Carlough, M.T. Crocker, H.K. Gill, J.L. Meyer and P.J. Smith. 1986. Bacterial growth on macrophyte leachate and fate of bacterial production. *Limnology and Oceanography* 31:1335–1341.

Gregory, S.V., F.J. Swanson, W.A. McKee and K.W. Cummins. 1991. An ecosystem perspective of riparian zones. *BioScience* 41:540–551.

Gurnell, A., K. Tockner, P. Edwards and G. Petts. 2005. Effects of deposited wood on biocomplexity of river corridors. *Frontiers in Ecology and the Environment* 3: 377–382.

Hagy, J.D., W.R. Boynton, C.W. Keefe and K.V. Ward. 2004. Hypoxia in Chesapeake Bay, 1950–2001: long-term changes in relation to nutrient loading and river flow. *Estuaries* 27:634–658.

Humphries, P., H. Keckeis and B. Finlayson. 2014. The river wave concept: integrating river management. *BioScience* 64:870–882.

Jenkins, C.C. and Suberkropp, K. 1995. The influence of water chemistry on the enzymatic degradation of leaves in streams. *Freshwater Biology* 33:245–253.

Johnson, B.L., W.B. Richardson and T.J. Naimo. 1995. Past, present, and future concepts in large river ecology. *Bio-Science* 45:134–141.

Junk, W.J., P.B. Bayley and R.E. Sparks. 1989. The flood-pulse concept in river-floodplain systems. *Canadian Journal of Fisheries and Aquatic Sciences* 106:110–127.

Kerner, M., H. Hohenberg, S. Erti, M. Reckermann and A. Spitzy. 2003. Self-organization of dissolved organic matter to micelle-like microparticles in river water. *Nature* 422:150–154.

Mallin, M.A. and L.B. Cahoon. 2020. The hidden impacts of phosphorus pollution to streams and rivers. *Bio-Science*. 70:315–329.

Mallin, M.A., M.H. Posey, M.R. McIver, D.C. Parsons, S.H. Ensign and T.D. Alphin. 2002. Impacts and recovery from multiple hurricanes in a Piedmont–Coastal Plain river system. *BioScience* 52:999–1010.

Maltby, L. 1992. Detritus processing. Chapter 15 in Calow, P. and G.E. Petts (eds.), *The Rivers Handbook: Hydrological and Ecological Principals*. Blackwell Scientific, Oxford.

Meyer, J.L. 1990. A blackwater perspective on riverine ecosystems. *BioScience* 40:643–651.

Minshall, G.W., K.W. Cummins, R.C. Petersen, et al. 1985. Developments in stream ecosystem theory. *Canadian Journal of Fisheries and Aquatic Sciences* 42:1045–1055.

Naiman, R.J., H. Décamps and M.E. McClain. 2005. *Riparia: Ecology, Conservation and Management of Streamside Communities*. Elsevier, London.

Odum, W.E. 1988. Comparative ecology of tidal freshwater and salt marshes. *Annual Review of Ecological Systems* 19:147–176.

Padgett, D.E., M.A. Mallin and L.B. Cahoon. 2000. Evaluating the use of ergosterol as a bioindicator for assessing water quality. *Environmental Monitoring and Assessment* 65:547–558.

Penczak, T., G. Zięba, H. Koszaliński and A. Kruk. 2003. The importance of oxbow lakes for fish recruitment in a river system. *Archiv für Hydrobiologie* 158: 267–281.

Peterson, B.J., W.M. Wollheim, P.J. Mulholland, et al. 2001. Control of nitrogen export from watersheds by headwaters streams. *Science* 292:86–90.

Rabalais, N.N., R.E. Turner, and W.J. Wiseman Jr. 2001. Hypoxia in the Gulf of Mexico. *Journal of Environmental Quality* 30:320–329.

Rosemond, A.D., J.P. Benstead, P.M. Bumpers, et al. 2015. Experimental nutrient additions accelerate terrestrial carbon loss from stream ecosystems. *Science* 347: 1142–1145.

Sabo, J.L., J.C. Finley, T. Kenndy and D.M. Post. 2010. The role of discharge variation in scaling of drainage area and food chain length in rivers. *Science* 330:965–967.

Schram, M.D., A.V. Brown and D.C. Jackson. 1990. Diel and seasonal drift of zooplankton in a headwater stream. *American Midland Naturalist* 123:135–143.

Sieczko, A., A. Maschak and P. Peduzzi. 2015. Algal extracellular release in river-floodplain dissolved organic matter: response of extracellular enzymatic activity during a post-flood period. *Frontiers in Microbiology* 6(80):1–15.

Sparks, R.E. 1995. Need for ecosystem management of large rivers and their floodplains. *BioScience* 45:168–182.

Suberkropp, K. 1998a. Microorganisms and organic matter decomposition. Chapter 6 in Naiman, R.E. and R.E. Bilby (eds.), *River Ecology and Management: Lessons from the Pacific Coastal Ecoregion*. Springer, New York.

Suberkropp K. 1998b. Effect of dissolved nutrients on two aquatic hyphomycetes growing on leaf litter. *Mycological Research* 102:998–1002.

Suberkropp, K. and E. Chauvet. 1995. Regulation of leaf breakdown by fungi in streams: influences of water chemistry. *Ecology* 76:1433–1445.

Thorp, J.H. and M.D. Delong. 1994. The riverine productivity model: an heuristic view of carbon sources and organic processing in large river ecosystems. *OIKOS* 70(2):305–308.

Vannote, R.L., G.W. Minshall, K.W. Cummins, J.R. Sedell and C.C. Cushing. 1980. The river continuum concept. *Canadian Journal of Fisheries and Aquatic Sciences* 37: 130–137.

Wallace, J.B., J.R. Webster and W.R. Woodall. 1977. The role of filter-feeders in flowing waters. *Archiv für Hydrobiologie* 79:506–532.

Wetzel, R.E. and A.K. Ward. 1993. Primary production. Chapter 16 in Calow, P. and G.E. Petts (eds.), *The Rivers Handbook: Hydrological and Ecological Principals*. Blackwell Scientific, Oxford.

Woodward, G., M.O. Gessner, P.S. Giller, et al. 2012. Continental-scale effects of nutrient pollution on stream ecosystem functioning. *Science* 336:1438–1440.

# CHAPTER 7

# River Fishes and Other Vertebrate Communities

## 7.1 Stream and River Fisheries

Fish are the most visible and popular residents of stream and river ecosystems. They are key to multiple functions in these systems as predators and prey, recyclers of nutrients, exploiters of numerous niches, rearrangers of sediments, shapers of local vegetation through grazing, and top-down trophic controllers of selected habitats. Fish are prized by humans for sport fishing, commercial fishing, and subsistence fishing, are enjoyed aesthetically in their habitat and are also prized as subjects of fine art. The literature on fisheries science is vast, encompasses many ecosystems, and impinges on many other scientific fields. The objective of this chapter is to distill some key facets of running water fisheries to the nonspecialist; note that various aspects of river fisheries are also discussed in Chapters 6, 8, 9, and 11. The major focus will be on temperate streams, but comparisons will also be made to tropical river systems, addressed in Section 7.1.4.

### 7.1.1 Important Taxonomic Fish Groups in Running Water

In river systems many different species are able to coexist due to differences in longitudinal location along a river, preferred depths, preference for running water or pools, tolerance of lower DO, and feeding preferences. In temperate streams and rivers some of the most important riverine fish groups include those discussed below (Reeves et al. 1998; Cushing and Allan 2001; Giller and Malmqvist 2002; note also that Cushing and Allan 2001 contains excellent photographs of key illustrative species):

The largest family of freshwater fishes is the Cyprinidae, consisting of minnows and carp. Minnows are mostly small, preying on invertebrates, although some species can be quite large and prey on fish. The stoneroller *Campostoma* is actually an herbivore that feeds on algae, and carp are bottom feeders in larger streams. Common cyprinid groups are dace, shiners, chubs, and squawfish. The Cottidae are members of a vast group of sculpins that span both marine and freshwater. They are important to river ecology in that generally small-sized members of this group inhabit upper reaches of cold headwaters streams where they are adapted to feed on benthos-associated invertebrates.

The Catostomidae consist of numerous species of suckers; these are generally fishes in middle- to higher-order streams that suck or scape algae, organic matter, and invertebrates from bottom sediments. The Ictaluridae are a family of catfishes native to North America that often are the largest species in a given river. They are bottom-feeding generalists that feed on algae, invertebrates, and organic matter; larger species feed on fish, crustaceans, and frogs. Sport fishermen seek large predatory catfish out as trophy fish; channel catfish can reach up to 40 kg and flathead catfish up to 60 kg (Cushing and Allan 2001). The Percidae are northern-hemisphere fishes that include popular sport and commercial fishing targets such as yellow perch, walleye, and sauger that are found in large slow-flowing northern rivers; the smaller darters are also members of this family.

The Clupedae are herrings and shad, planktivorous fish that are found in big rivers where plankton are abundant, as well as lakes and reservoirs. Some

*River Ecology*. Michael A. Mallin, Oxford University Press. © Michael A. Mallin (2023). DOI: 10.1093/oso/9780199549511.003.0008

species are anadromous and historically were heavily fished during their upstream migrations, or runs, as food and for their roe. The Centrarchidae consist of sunfishes, crappie, and basses found in slow-moving streams and rivers as well as reservoirs. The former two groups are generally invertebrate feeders, but the basses are predators; centrarchids are often nest builders and defend them. Largemouth bass (LMB) are very popular sportfish and industries are built around sportsmen seeking out record sizes, especially in large reservoirs.

The Salmonidae (salmon and trout) contain many species that are anadromous, meaning they spawn in streamwaters, migrate to the ocean or large lakes to feed and grow, then return to the streams of their origins to spawn. Salmonids are wildly popular as sportfish, are very agile, with torpedo-shaped bodies, swim in rapid bursts in the water column, and prey on invertebrates, terrestrial insects in the drift community, and fish. Brown trout (*Salmo trutta*) characterize European waters, while cutthroat trout (*Oncorhynchus clarkia*) and brook trout (*O. mykiss*) are popular species in North American waters (Giller and Malmqvist 2002). In North America the Pacific Northwest region in particular is inextricably linked with anadromous salmonids, in terms of Native American lore, subsistence, and commercial fishing, and as objects in conflict with competing enterprises (timbering, power production) and are considered keystone species by some researchers for their broad trophic impacts (Box 7.1). Note that eels (order Anguilliformes) are catadromous and spawn in the sea, migrate as larvae to freshwater to grow, and then return to sea to spawn. Several species of lampreys also migrate from saltwater to freshwater to breed.

spawning runs upriver can be spectacular sights in free-flowing waters, but many local populations have come under great pressure from commercial fishing beginning in the nineteenth century, watershed disruptions such as logging that degrade nesting areas, and dam building for hydroelectric power, which cuts off migration routes (see Chapter 10). They build nests call redds, some of considerable size, and have thus been referred to as ecosystem engineers. The redds physically change the streambed, altering local flow patterns, increasing oxygen flux to sediments, and dislodging invertebrates from the sediments and adding them to the drift community, thus making them more available to downstream water-column consumers. Migrating salmon are an important food source for numerous species: live-caught by grizzly bears, black bears, otter, and even wolves, feeding on the most energy-rich parts of the salmon, while birds such as dippers consume young salmon. Salmon carcasses (half-eaten or otherwise) are consumed by many mammal and bird species. The carcasses also are critical as fertilizer for fluvial systems. Stable isotope analyses show that salmon-derived nutrients (which are marine in origin) move into numerous compartments of the watershed food webs. Some of the nutrients from consumed salmon are incorporated in predator and scavenger biomass. Other carcass-derived nutrients are taken up by aquatic primary producers such as periphyton, phytoplankton, and macrophytes or moved into local floodplains by deposition and water flow and taken up by forest plants of all taxonomic groups. Research has demonstrated that vegetation in smaller, less-productive watersheds benefit the most from this ocean-sourced nutrient flow. Thus, Pacific salmon are keystone organisms that have major impacts on local ecosystems physically, trophically, and in nutrient translocation and recycling.

(From Reeves et al. 1998; Schindler et al. 2003; Moore 2006; Hocking and Reynolds 2011.)

### Box 7.1 Salmon are keystone organisms in the Pacific Northwest

Pacific salmon (*Oncorhynchus* spp.) comprise a variety of species of different sizes, some of which are well known as anadromous species that spawn in freshwater streams and coastal lakes, move to the ocean to feed and grow (acquiring > 95% of their body mass there), then return to the areas of their birth to spawn. Their

Anadromous species (salmon, trout, lampreys) comprise up to 25% of the fish in rivers of the Pacific Northwest. Fisheries in that region tend to be dominated by larger species than those in eastern North America (Reeves et al. 1998). In general, fish species richness is lower in Pacific Northwestern streams than other parts of the continent, mainly due to the tectonic activity that has characterized that area (earthquakes, volcanism, and uplift separating contiguous areas) and heavy glaciation (Reeves et al. 1998).

Sturgeons are one of the most compelling groups of riverine fishes worldwide and some of the most endangered. There are 23 species of sturgeon (family Acipenseridae) worldwide (Cushing and Allan 2001) and they can grow to huge sizes. In North America the white sturgeon can reach 600 kg and famed beluga sturgeon in Russia can reach up to 1300 kg (but European sturgeons are threatened; see Section 7.1.5). Other North American sturgeons include the endangered pallid sturgeon (*Scaphirhynchus albus*) in the Mississippi River system and the threatened shovelnose sturgeon (*S. platorynchus*) in the lower Mississippi. On the US East Coast are the endangered Atlantic sturgeon (*Acipenser oxyrinchus*) and endangered shortnose sturgeon (*A. brevirostris*).

## 7.1.2 Fish Exploiting Habitats, or Making a Living

The variety of feeding strategies (Table 7.1) allows for both species diversity and exploitation of the numerous available resources. In high-relief headwaters areas, feeding on invertebrates predominates in the fish community, either associated with the bottom or as part of the drift community. According to Allan (1995) fish using rapid water are streamlined, active swimmers like salmonids, while bottom-dwelling sculpins and darters are vertically depressed, and can have a flattened ventral surface and use pectoral or pelvic fins for maneuvering. Invertebrate feeders take prey from soft bottoms or stony bottoms. Dace are bottom dwellers with subterminal mouths and feed on benthic invertebrates. Sculpins, dace, darters, adult suckers, various catfishes, and stonerollers are thus morphologically specialized for living and feeding on the bottom, whereas trout, for instance, take prey from the drift community. As mentioned earlier, the bottom-oriented stoneroller is herbivorous and scrapes rocks for periphyton.

Downstream in middle- to higher-order streams omnivores and herbivore-detritivores are common. In higher-order streams and rivers Cyprinids are abundant, and bottom-feeding suckers, carp, and most catfish feed mainly on vegetable matter and incidental invertebrates. Fish use vision, smell, taste, and touch to detect prey; in deep waters vision may be very limited due to turbidity or increased coloration from DOC. Piscivores are generally found in middle- to high-order streams. Bass are predaceous on whatever they can handle in the water column; large catfish such as flatheads and channel catfish predate upon anything they can handle largely in the lower water column.

Obligate herbivory is rather rare among temperate river fishes; in North America only about 55 out of 700 species are primarily herbivorous (Allan

**Table 7.1** Functional feeding guilds of riverine fishes. (Revised from Allan 1995.)

**Invertivores**
- Benthic invertivores feed on insect larvae and other small benthic animals. Example: sculpin—feed on insect larvae among rocks.
- General invertivores feed on terrestrial and aquatic insects and their larvae in and just above the water column, including prey in the drift community. Example: trout—feed on insects and their larvae in the drift community.

**Herbivores** feed on macrophytes, periphyton, and phytoplankton. Examples: stoneroller in upper stream reaches, grass carp, tilapia. Note: the two latter taxa have been introduced by regulatory agencies and private groups into aquatic weed-infested waters (mainly by submersed macrophytes and filamentous algae) to control the growths. Herbivory is much more widespread and specialized in tropical rivers, with specialized herbivores feeding on flowers, fruits, and seeds.

**Benthic (bottom) feeders** are often generalists, feeding on or among detritus (which is covered with fungi, bacteria, algae, protozoans, insect larvae, and worms). Their mouths may have adaptations for bottom feeding. Examples: suckers, carp, sturgeon, catfishes.

**Planktivores** feed on zooplankton—most prevalent in lower rivers and estuaries—where the water is wide, deep, and less turbulent, and there is better habitat for phytoplankton production (i.e. yielding zooplankton prey). Examples: paddlefish, shad, herring, minnows, the young of many species.

**Piscivores** eat other fish. Examples: large catfish in lower river areas, adult LMB, gar, bowfin. In the tropics various piscivorous taxa feed on selected parts of other fish, such as fins, eyes, gills, etc.

**Parasites** such as adult lampreys are parasitic on other fish.

1995). As mentioned, the stoneroller grazes algae from rocks on the streambed. In Asia the grass carp (*Ctenopharyngodon idella*) feeds on algae and macrophytes; it is imported and used for aquatic weed control in the USA, where escapees can find themselves as part of a river fish community. Various cichlids from Africa and the Middle East (tilapia) have also been imported and used for reservoir weed control in the USA (Crutchfield et al. 1992), but again, such grazers are not native to temperate waters.

Numerous riverine fish species consume zooplankton as larvae or juveniles, but only some species as adults. Along a river system plankton are only dense enough to support large fish needs in selected areas. The RCC predicts that in well-lit lower river areas, rich phytoplankton and zooplankton communities can form; likewise dense communities can form in reservoirs or otherwise behind dams. From the Clupeidae, herring and shad (with five anadromous species) are planktivores that school and strain plankton from the water (Cushing and Allan 2001). Paddlefish (Acipenseriformes) are large zooplanktivores that occur in the Mississippi River and other large systems. Unfortunately the Yangtze River's Chinese paddlefish (*Psephurus gladius*) (which grew to 7 m length) has recently been declared extinct.

Immature lamprey burrow in stream substrata and filter; some mature lampreys are parasitic on other fish as adults, sucking their blood, a type of carnivory. The parasitic sea lamprey is native to the Atlantic Ocean but successfully invaded the Great Lakes and has caused great harm to the native fishes. Finally, some fish cannot be placed in a trophic guild easily as they display flexible feeding habits and changes in feeding over a life cycle.

## 7.1.3 Longitudinal Fish Distribution

It has been long established that longitudinally, fish species richness and diversity is significantly related to increasing stream depth (Sheldon 1968). Beyond depth a variety of structural habitat features increase fish diversity and impact their distribution in streams (Gorman and Karr 1978; Diana 2004). Such features include substrata, vegetation, physical habitats such as riffles, pools, runs, bank features, and current velocity (Figs. 1.6a; 1.9b): note that some of these physical features also impact DO (see Chapter 1). As streams widen and deepen downstream, larger fish populations per unit length can be supported. Species gradually disappear, while new ones appear in the transition from headwaters to middle- and higher-order streams (Diana 2004). Bayley and Li (1992) favor a hydrological approach in conjunction with temperature and landscape properties as a template to understand fish community function and life histories and because hydrology is directly associated with physical constraints on fish habitats and also reflects geomorphology in the basin.

The RCC (see Chapter 6) provides a solid framework for understanding fish distribution patterns in many temperate running water ecosystems. Many riverine systems originate in upland areas, characterized by cool to cold water, swift currents, high DOC, rocky substrata, rapids, cascades, riffles and pools, and large allochthonous inputs of organic matter (Table 7.2). In many headwaters systems there is a close association with streamside forests. These forests supply POM to the stream which is processed by invertebrate shredders (see Chapter 6). The forest also directly supplies insects and other live food items to the stream drift community. Fish present in headwaters areas include minnows, sculpin, trout, and other taxa tolerant of cold temperatures (Cushing and Allan 2001; Diana 2004). Stenothermal refers to species that inhabit a narrow temperature range; i.e. cold stenothermal fish are limited by about 26°C (Table 7.2).

Agility in a turbulent environment favors smaller body size and higher metabolic rates. Hemoglobin delivers greater amounts of $O_2$ to active tissues in cold upland waters than to fish in higher-order, less-turbulent reaches. Substrate-oriented headwaters fish morphologies and metabolisms reflect the physical and hydrological environment (Bayley and Li 1992). Most small stream fishes are habitat specialists such as drift feeders and benthic invertivores. In headwaters streams there is a lot of input from POC from streamside forests and thus a lot of shredders in the stream to serve as prey items. Also, lots of terrestrial insects fall in and add to the drift. Fish in smaller headwaters streams tend to have a higher productivity/biomass ratio and a higher

**Table 7.2** Fish attributes and related ecological and physical characteristics in temperate rivers in relation to the RCC. (Revised from Bayley and Li 1992 and Reeves et al. 1998.)

| | Stream section | | |
|---|---|---|---|
| | Headwaters/lower orders 1st–2nd | Middle reaches 3rd–5th | Higher orders > 5th order |
| **Stream attribute** | | | |
| Current | Faster, turbulent | | Slower, laminar |
| Substratum | Coarse | Settling of cobbles and boulders | Finer, well sorted |
| Organic basis | Allochthonous leaf litter | Autochthonous periphyton, macrophytes | Mixed, phytoplankton, floodplain |
| Dominant benthos | Shredders | Grazers, collectors | Collectors |
| **Fish attribute** | | | |
| Temperature guide | Cold stenotherms (upland), eurytherms (lowlands) | Mesotherms, eurytherms | Eurytherms |
| Metabolism | High | | Low |
| (Planktonic fish in mountain or upland low-order streams have a higher metabolism to deal with swifter current and more expenditure of energy.) | | | |
| Size | Smaller | | Variable |
| Production/biomass ratio (turnover ratio) | | | |
| | Relatively high | | Relatively low |
| Lifespan | Relatively short | | Greater longevity |
| Feeding | Drift feeders, invertivores | Invertivores, piscivores, herbivores | Detritivores, planktivores, large piscivores |

Stenotherm—organism that functions within a narrow temperature range
Mesotherm—functions at moderate temperatures
Eurytherm—adaptable to a wide temperature range

intrinsic rate of increase than fish in wide, slower, downstream reaches (Bayley and Li 1992).

Headwaters of lowland streams such as in the US Coastal Plain or Midwest may not have the sharp elevation changes and swift flows that upland-originating streams do. However, they can still undergo rapid temperature changes due to weather since the waters are shallow and rapid flow changes with precipitation changes. DO can change readily due to temperature changes on storm-driven inputs of riparian swamp water. The amount of groundwater will likewise make a physical and chemical difference, again depending upon stormwater inputs but also local geology (Diana 2004).

Per the RCC, middle reaches of rivers can be open, well lit, with either soft or rocky bottoms, supporting primary producers in the form of periphyton and aquatic macrophytes (see Figs. 1.9a; 1.10b; Table 7.2). Benthic fish prey include grazers of periphyton, collector-gatherers, predatory invertebrates, and drift from upstream. Fish present in the middle reaches are more tolerant of temperature

fluctuations (eurythermal). Trout, suckers, and minnows are present; in fact middle reaches are often excellent trout streams with abundant invertebrate prey associated either with the sediments or as drift (Cushing and Allan 2001). Trout streams have great popularity among organized sport fishermen's groups, and that is a good thing as it provides outside pressure on resource agencies to focus added protection on a natural resource. The waters have to be well oxygenated for spawning in redds.

The lower reaches (higher orders) of rivers are likely to feature higher water temperatures, periodic benthic hypoxia, slower currents, finer substrata, and large channels. Fish in lower reaches are tolerant of wide temperature ranges (eurythermal) and can inhabit a wide temperature range, with warm-water fish upper thermal limits near 30°C; but note that desert fish can withstand up to 40°C (Allan 1995). Within the channel are well-sorted fine sediments, inhabited by oligochaetes and chironomids and some bivalves. In woody-influenced reaches there will also be a snag community (explored in detail in Chapter 8). At least in unfragmented rivers there will also be braided stream channels, oxbows, sloughs, backwaters, and autochthonous sources of organic matter supplied by phytoplankton and aquatic macrophytes, as well as floodplain-provided sources. The lower reaches can support a diverse community of fishes including some herbivores, detritivores, planktivores if a rich phytoplankton and a zooplankton community is present, and piscivores, with generally fewer invertebrate specialists (Bayley and Li 1992). These groups include sunfish, catfish, LMB, suckers, chub, and carp.

Fish communities are more diverse in unconstrained reaches than constrained reaches, due to increased productivity and increased habitat diversity, as discussed above (Reeves et al. 1998). Fisheries community complexity is enhanced by habitat complexity, and stream habitat stability enhances fisheries community stability (Gorman and Karr 1978).

Earlier, anadromous migrations were discussed in relation to life-stage migrations between spawning habitat and areas for growth to achieve maturity between fresh- and saltwater. Fish also migrate or travel to different habitats within a river system to find good conditions for incubation of eggs or wintering to find good habitat for survival, or undertake seasonal lateral migrations for feeding and growth (Diana 2004). Lateral habitat is important in floodplain situations; such habitats include oxbow lakes and other wetlands (Penczak et al. 2003) and the floodplain provides all types of organic matter (see Chapter 6—the flood-pulse concept). Work in South America has demonstrated that access to the floodplain and the intensity of seasonal flooding are critical to enhancing the reproductive success of fish species with a variety of reproductive strategies (de Oliviero et al. 2020).

### 7.1.4 Tropical River Fisheries

Tropical river fisheries present major contrasts to temperate river fisheries (Giller and Malmqvist 2002). The major efforts have studied the vast floodplain rivers such as the Amazon and Orinoco, some of which formed the basis for the flood-pulse concept discussed in Chapter 6. In general, there is higher diversity in tropical river systems than temperate system. For example (from Diana 2004), the fish fauna of the Amazon River (the largest river in the world by volume) has > 1000 species, the Panama Canal Zone in Central American about 456 species, compared with the Laurentian Great Lakes with about 172 species. In South America as a whole there are > 8000 freshwater fish species, but no cyprinids (minnows or carp). Five of the most important major fish taxonomic groups in South America are the Characiformes (which includes piranhas and Colossoma, or pacu), the Siluriformes (catfishes), Gymnotiformes (knifefishes), Cyprinondontiformes, and the Cichlidae. Helping to account for the greater diversity is the greater niche specialization in the tropics than among temperate fishes, including flower, seed, and fruit eating (which helps disperse seeds). Also in the tropics are high levels of detritivory, since primary productivity is so high and flooding provides all sorts of terrestrial materials to the water. The greater diversity of plant material including POM, macrophytes, and periphyton in the tropics may explain why there is a considerably greater proportion of herbivores and detritivores in South America than in North America (from Hugueny et al. 2010). Another interesting trait (niche specialization) is the ability to breathe air in lungfish, catfish, and perch—an adaptation to standing water with low DO (Diana 2004).

Massive seasonal flooding of the broad tropical floodplains in the Amazon and other South American rivers is a phenomenon that shapes fisheries diversity and has, as mentioned, informed the flood-pulse concept. During the wet season many fishes move onto the floodplain to feed and there is generally high fish growth during seasonal inundations. The floodplain plants are well adapted to seasonal flooding and can withstand flooding for long periods (Diana 2004). Flooding makes much terrestrial food available, such as terrestrial invertebrates and small vertebrates, leaves, flowers, fruits, and seeds; detritivory is widespread. Some fishes have specialized mouthparts for feeding on nuts and fruits. Feeding directly on primary producers also shortens the length of food chains (Hugueny et al. 2010), making for trophic efficiency. Further specialization includes piscivory adaptations, in which selected piscivores only eat certain parts of prey fish such as scales, eyes, or fins (Allan 1995).

## 7.1.5 Multiple Dangers to Native Fisheries

Losses of native fisheries, even to extinction, are often brought about by a combination of factors that frequently begin with overfishing, later exacerbated by habitat loss, pollution, introduction of non-native species (see Chapter 14), damming, and river fragmentation (Allan et al. 2005). Overfishing can begin with targeting a large species; with constant exploitation, both the numbers and sizes of the targeted species will decrease. As targeted species decline in size, fishermen use smaller-mesh-sized nets that target smaller-sized fish and smaller species (Allan et al. 2005). The heavily impacted larger species are also subject to further losses due to habitat loss from watershed development including floodplain clearing, channelization, dam construction, and water pollution. This process is starkly exemplified by the declines among the world's largest freshwater fish species, several of which are listed below (Allan et al. 2005; Stone 2007). The Mekong giant catfish (*Pangasianodon gigas*), a vegetarian with a record size of 2.7 m and 293 kg, is called the buffalo fish in Thailand, traditionally fished, and considered a delicacy, whose meat fetches a good price (Stone 2007). It is currently considered to be critically endangered due to overfishing, habitat loss, and changes in river flows due to upstream manipulations from dam building for hydroelectric power (see Chapter 12). Also on the Mekong is the giant freshwater stingray (*Himantura chaophraya*), which can reach sizes up to 5 m and 600 kg. It is considered endangered due to overfishing and habitat degradation. The Yangtze sturgeon (*Acipenser dabyanusm*) which ranges up to 2.5 m long is critically endangered due to overfishing and habitat loss. As mentioned earlier, the Chinese paddlefish (*Psephurus gladius*) is now considered extinct from overharvest and habitat loss. Finally, the Murray cod (*Maccullochella peelii*) in Australia's Murray–Darling river system historically could range up to 2 m and 113 kg and is considered critically endangered. This species has been overfished and its reproduction has been constrained by lack of floodplain flooding due to dam

building, and suffers from competition from the introduced Eurasian perch (*Perca fluviatilis*).

More insight into decreases in large fish production comes from two major European rivers, the Volga (the largest river in Europe) and the Danube (Schletterer et al. 2018). Historically, large anadromous fish (salmon and sturgeon) were considered flagship species in these great rivers but now suffer from reduced stocks and declining yield. For example, the anadromous Beluga sturgeon is the largest freshwater fish in the world, historically reaching up to 1300 kg. It and the smaller Russian sturgeon are both considered critically endangered.

Prior to major river fragmentation, fish productivity in these great rivers was closely related to length and height of annual flooding. River fragmentation has hit both of these rivers hard. Multiple dams have been constructed on the Volga for hydroelectric power and navigation improvement, two with fish lifts and most without any special means of moving anadromous species upstream. Note that some natural habitat remains in free-flowing sections in the headwaters and lower reaches downstream of the dams. There are also a number of introduced fish species currently reproducing in the Volga (Schletterer et al. 2018). The Danube fishery has greatly changed over recent centuries due to dam building for hydropower and flood protections, channelization for navigation, and of course pollution. In addition, anadromous species in both rivers have been overfished. The research from these major European rivers indicates that additive impacts of multiple stressors cause serious impacts on riverine fish communities (Schletterer et al. 2018).

While the declining fate of large river fish is bad in itself, individual species do not provide a full story on the health of the fish community in general. Thus, an index of biotic integrity (IBI) (pioneered by Karr 1981 and adapted by many authors) was developed that utilizes fish community attributes, or metrics, such as number of species, total abundance, extent of parasitism, presence of pollution tolerant and intolerant species, proportion of hybrid species, omnivore versus insectivore proportions, populations of top carnivores present, and number that are diseased or damaged otherwise, as well as several other parameters. These functional aspects of fish assemblages are each rated poor, fair, or good; the least impacted streams or reaches yield highest scores, while lowest scores occur in the most impacted streams or reaches. The IBI (which of course requires standardized sampling and taxonomic expertise) has been modified to many regional systems and is commonly used in assessment studies (see also Bayley and Li 1992; Diana 2004). For a given set of streams or reaches within a river, the IBI scores can be statistically compared with factors potentially impacting fisheries, such as degree of in-stream habitat loss, degree of watershed development (see Chapter 15; also Miltner et al. 2004), channelization, access to the floodplain, and degree of water pollution. A good example of the latter is excessive nutrient loading to streams.

As noted in earlier chapters, excess nutrients can lead to ecological degradation through eutrophication and heterotrophic increases in BOD. Several researchers have found that elevated stream nutrients are statistically linked to degraded stream fish communities as well. During work done in 240 wadeable streams in Wisconsin (Wang et al. 2007), fish were collected by electroshocking, and numerous habitat and water quality variables were statistically assessed for relationships with abundances of various fish groups and the IBI. Results indicated that fish variables most responsive to nutrients were salmonid abundance, percent of carnivorous fish, percent of intolerant species, and the IBI. Values for fish measures were nonlinear, highly variable at low nutrient concentrations, but at high nutrient concentrations the response variables were generally poor. Statistical nutrient thresholds beyond which fish were likely to be degraded were 0.06–0.07 mg/L for TP, 0.54–0.61 mg/L for TN, and 0.02–0.03 mg/L for ammonium. Working in Ohio streams, Miltner and Rankin (1998) found degraded IBI scores among low- and middle-order streams with elevated nutrient concentrations, with the strongest relationships in headwaters streams (see Chapter 15). Key nutrient concentrations related to degraded fisheries were 0.06 mg/L for TP and 0.61 mg/L for total inorganic nitrogen. The negative fisheries relationships with increased nutrient concentrations can be associated with the nutrient–algal bloom and hypoxia pathway, production of toxic algae, and P-stimulation of bacterial oxygen demand (see

Chapters 2, 3, and 8). Additionally, nutrient enrichment is a byproduct of industrialization, farm fertilization, industrialized livestock production, and urbanization, all of which can impact stream health in multiple ways (Wang et al. 2007).

Rivers are channelized for shipping purposes and diked for flood control (see Chapters 6, 15, and 17). These activities have the net effect of cutting off fish access to backwaters, oxbows, and secondary channels, which are important habitats for fish reproduction and growth of juveniles (nursery areas). Woody inputs provide habitat complexity to a river (see Chapter 6); when such inputs are reduced it leads to lower fish diversity (Reeves et al. 1998). Streams are also ditched for various reasons such as improving small boat access, or enhancing drainage of adjacent farmlands, or as part of irrigation schemes. Ditched streams will have unstable flows and are largely unshaded, which increases water temperatures to levels exceeding the comfort zones of many native fish species and allows for plenty of light to reach primary producers. In agriculturally impacted ditches and streams or those receiving nutrients from septic leachate or animal waste, extensive filamentous algal blooms will grow to choke channels. Nuisance blooms of filamentous algae lead to significantly reduced fish diversity (Gorman and Karr 1978). In contrast, natural streams support higher diversity than channelized streams. Naturally, structurally diverse streams buffer fish communities from extreme impacts, providing protection from solar heating and excessive algal growth by shading, pools serve as refuges in droughts, and in larger streams and rivers meanders moderate floods (Gorman and Karr 1978).

Whereas fish are the organisms most closely associated with streams and rivers by the public, other major vertebrate groups are part of lotic ecosystems. The additional groups discussed in Sections 7.2, 7.3, and 7.4 not only use these ecosystems but also in some cases have major impacts on stream physical structure and productivity.

## 7.2 Amphibians and Reptiles

Amphibians (class Amphibia) consist of the lizard-like salamanders (order Urodella) and frogs and toads (order Anura). They lay eggs in water and their early development occurs there until meta-morphosis, with time spent as larvae highly variable according to species. Salamander adults are sight-feeding carnivores, generally remaining in close contact with water, including running water. While they are usually small, there are a few large species that are significant predators. The Chinese giant salamander (*Andrias davidianus*) occurs in coldwa-ter streams in the Yangtze River basin, is considered critically endangered, and can reach sizes exceed-ing 1.5 m in length and 50 kg weight. It has a congener called the Japanese giant salamander (*A. japonicus*), endemic to Japan and of a similar size. These large salamanders consume crayfish, shrimp, worms, other amphibians, and fish. In North Amer-ica the largest salamander is the hellbender (*Crypto-branchus alleganiensis*), native to the east coast and central states where it can reach up to about 0.5 m length and prey on crayfish, worms, snails, and small fish. In the US Pacific Northwest is the Pacific giant salamander (*Dicamptodon ensatus*), reaching about 35 cm in length and found in stream head-waters (Cushing and Allan 2001). Frogs and toads as adults are insect eaters and may occur in slower-moving streams as well as ponds and lakes.

One of the key cautionary tales of current aquatic science is the major losses in amphibians that have been occurring over the past 30 years or more. Many theories have been proposed, and indeed losses and some extinctions can be attributed to habitat loss and habitat split, exploitation for vari-ous human needs, and climate change (Greenberg and Palen 2019; see Chapters 10 and 14). How-ever, losses have also occurred in pristine envi-ronments, and researchers have attributed losses in both pristine and impacted aquatic systems to a fungal disease called amphibian chitridiomyco-sis. The extent of these losses has been termed a panzootic caused by a broad amphibian-infecting species called *Batrachochytrium dendrobatidis*, and another more specific to salamanders called *B. sala-madrivorans* (O'Hanlon et al. 2018; Scheele et al. 2019). The ancestral population stems from Asia in the early twentieth century, and the spread of these pathogens has been a result of worldwide trade in amphibians (O'Hanlon et al. 2018). This disease has played a role in declines of > 500 species and 90 extinctions (Scheele et al. 2019). The

largest amphibian losses have occurred in Australia, Central America, and South America, with lesser but notable declines in Asia, Africa, Europe, and North America. The agent(s) is highly transmissible through water, and unfortunately there is no effec-tive treatment for afflicted wild amphibian popula-tions, although some species that were hit hard have shown some degree of recovery (Lips 2018; Green-berg and Palen 2019; Scheele et al. 2019). Stronger regulation in the exotic pet trade can help slow the spread of chitridiomycosis (see Chapter 14).

Reptiles (class Reptilia) consist primarily of tur-tles, snakes, and crocodilians; they are all air breath-ing, although some species spend most of their time in and near water, including slow-flowing rivers. Turtles (order Testudines) are omnivores and ubiquitous on lower rivers, sunning them-selves on banks and snags. In North America the painted turtle (*Chrysemys picta*) is very common in slow-moving waters. The American snapping turtle (*Chelydra serpentina*) is widely distributed in North America and is also found in slow-flowing waters. It is large, with a carapace reaching up to 0.5 m, and feeds on any animal life it can handle, including fish, frogs, snakes, birds, and small mammals, as well as invertebrates and vegetative matter.

Snakes (order Squamata) are all carnivores. The vast majority are terrestrial, although there are a few aquatic taxa. Nonvenomous water snakes are distributed among several genera, including Nero-dia in North America, Natrix in Europe, Sinonatrix in Asia, and Afronatrix in Africa. The water moc-casin (or cottonmouth) (*Agkistrodon piscivorous*) is a large (up to 1.5 m) North American water snake that feeds on fish and amphibians using its venom. It is fearless in its environment when encounter-ing fishermen or limnologists (Fig. 7.1a). The huge and fabled green anaconda (*Eunectes murinus*) is a South American predator found in slow-moving rivers that crushes its prey using constriction. It pre-dates on a wide variety of fish, caimans, birds, and mammals, including rather large ones (evidence of human consumption is scant). It is known to mea-sure up to 7 m in length, but is rumored up to 12 m.

Crocodilians (order Crocodilia) are carnivores and apex predators who inhabit slow-moving waters and whose prey size increases as they get larger. In North America the American alligator

(a)                                          (b)

**Fig. 7.1** Fig. 7.1a (left) Water moccasin *Agkistrodon piscivorous* approaching limnologists on the lower Cape Fear River, North Carolina. Fig. 7.1b (right) American alligator *A. mississippiensis* in the Black River, southeastern North Carolina. (Photos by the author.)

(*Alligator mississippienses*) is native to the Southeast USA but can presently be found as far north as North Carolina (Fig. 7.1b). It was formerly endangered (hunted heavily for food and leather) but has recovered due to conservation efforts. It is a top predator that can reach over 4 m in length and will consume nearly any kind of game and should be treated with extreme caution. News outlets have reported on large specimens that have been taken in Florida that contained from one to several dog collars in their guts; in 2021 in Louisiana a large alligator killed and consumed an adult human in floodwater conditions following Hurricane Ida. In recent years there have also been news stories describing killings of adult humans by large alligators in Florida and South Carolina.

Alligators are true ecological engineers. In wetland areas alligators physically create trails and wallows (ponds or "gator holes") which hold water and become drought refuges for fish and other aquatic creatures. Alligators build nest mounds or lodges; in the Florida Everglades (the River of Grass) the nest mounds are high points in the marsh and when abandoned are used by plants and animals that need dry spots to grow, bask, or lay eggs. The mounds may get colonized by trees and become the core of future islands (Day et al. 1989).

Regarding crocodiles, there is a population of American crocodiles (*Crocodylus acutus*) that is only found in a small area in southwest Florida that uses both salt- and freshwater. A river-using crocodilian in India is the gavial (*Gavailus gangeticus*). However,

some of the largest and most feared apex predators in rivers are the Nile crocodile (*Crocodylus niloticus*) in Africa and the saltwater crocodile (*C. porosus*) in Australia and Southeast Asia. In Africa the Nile crocodile will take down very large animals; videos of them killing large wildebeest at river crossings are classic nature television fare (see Section 7.3 for more on the wildebeest migrations and ecological impacts). Crocodiles are also responsible for many deaths of local fishermen, adventurers, sightseers, and others. There is a report from early limnologists working in southern Africa (Richardson and Livingstone 1962) in which they described their small wooden motorboat being attacked by a large Nile crocodile which proceeded to angrily tear the boat to pieces—they obviously were able to swim to safety with only minor injuries. In Australia the saltwater crocodile (which ventures upriver as well) is responsible for one or two deaths per year among fishermen, swimmers, or casual river visitors. In Southeast Asia crocodile deaths are believed to be much higher, although official records are less available for remote coastal fishing villages. The point is that crocodiles are top predators that take, kill, and consume nearly anything they want.

## 7.3 Mammals

Numerous mammals visit streams and rivers to drink, some to graze aquatic macrophytes (e.g. moose [*Alces alces*]), and some to fish from the bank (see Section 7.1 regarding bears consuming salmon).

Otter (*Lutra canadensis*) will hunt in and near the water to feed on fish, crabs, frogs, and crayfish, and mink (*Mustela vison*) are predators both in and outside of water, feeding on fish, shellfish, amphibians, birds, small mammals, and eggs. Muskrat (*Ondatra zibethicus macrudon*) are grazers that create streamside or marsh dens; in lower river and estuarine marshes they have been outcompeted by invasive nutria (*Myocastor coypus bonariensis*) (discussed in Chapter 14). On occasion, marine mammals such as bottlenose dolphins (*Tursipos truncatus*) will make brief forays upriver from the estuary. Caribbean or West Indian manatee (*Trichechus manatus*, order Sirenia) move between fresh- and saltwater. They weigh up to 590 kg and breed a single calf every one to two years; a variable number live in Florida, where they are well studied and also frequently in danger from errant speeding boaters in creeks and rivers, as well as algal toxins from consumption of filamentous cyanophytes when natural food such as seagrass is shaded out by algal blooms (see Chapters 3 and 4). They escape cold spells by overwintering in deep freshwater springs or power plant thermal plumes, sometimes in herds. They will also use depressions in canals as refugia, where cold freshwater covers warmer saltwater in cold snaps (Stith et al. 2010). River-using mammals are not restricted to terrestrial and aquatic biomes, however. A study of two California rivers (Jackson et al. 2020) determined that river-using bats (order Chiroptera) obtained 61–82% of their energetic demands from food (mainly emerging insects) that had been based on river algae.

Some terrestrial or semiaquatic mammals can greatly influence the physical, chemical, and biological riverine environment around them by acting as environmental engineers and physically altering the environment, or creating huge pulses of nutrients into rivers, or controlling an ecosystem as a top-down apex predator. In North America the best-known mammalian ecosystem engineer is the beaver (*Castor canadensis*) (Naiman et al. 1988; Naiman and Rogers 1997). The abundance of beaver in North America hundreds of years ago was estimated to be in the hundreds of millions, but in the 1700s and 1800s a huge market for pelts exploded in Europe and trappers drove the North American population to near extinction. Since then it has rebounded back into the millions, and their presence and activities profoundly influence freshwater ecosystems in rural areas. Beavers are herbivorous, eating woody plants and softer vegetation as well. They build dams (Fig. 7.2a), mainly on second- to fourth-order streams, creating lentic environments in which they build lodges for colonies to safely reside in. The dams change the stream from a lotic to a lentic ecosystem, altering the benthic macroinvertebrate community from shredders and scrapers to collectors such as midges and oligochaetes, and predators such as dragonfly larvae. In long-dammed sites the fish community may likewise change from lotic species to

(a)

(b)

**Fig. 7.2** Fig. 7.2a (left) Beaver dam in eastern North Carolina. Fig. 7.2b (right) Hippo (*Hippopotamus amphibius*), Kruger National Park, South Africa. (Beaver photo courtesy of Matthew R. McIver, UNC Wilmington); hippo photo by Bernard DuPont from France, CC BY-SA 2.0, https://commons.wikimedia.org/w/index.php?curid=53675121.)

a more lentic community. The dams retain sediment and associated nutrients that would otherwise go downstream, and aerobic soils become anaerobic sediments, altering biogeochemical cycling. The flooding of the nearby areas and increased hypoxia enhances soil denitrification and loss of nitrate. In rural, wooded areas beavers may construct multiple dams, greatly increasing local water retention and aquatic connectivity for wildlife. They may fell numerous trees locally as well, opening up areas for increased solar irradiance penetration and subsequent algal and duckweed blooms.

Africa has a lot of big mammals that use rivers, but none that builds dams. However, a major ecological engineer is the hippopotamus (*Hippopotamus amphibius*) (Naiman and Rogers 1997; see Pennisi 2014 for summation). Hippos are huge mammals, up to 4 tons (Fig. 7.2b), and are primarily vegetarians. However, they can be territorial and extremely violent, annually causing the deaths of many humans intruding upon their space by trampling or crushing in their massive jaws. Hippos move between the water (to rest and spend their days) and the land (to feed, mainly at night). They engineer the waterscape in two ways: their daytime wallowing in pools deepens the pools, suspending sediments, which are then moved out by currents or deposited along shore. The deeper pools thus hold more water, which is especially important during dryer periods, when the pools are important as refuges for fish and other wildlife. At night hippos move out of the water to feed on terrestrial grasses and follow well-worn hippo trails through riparian zones, or create new ones. The trails are thus kept obstruction-free and used by other wildlife. In wet periods some trails become canals, which increases the aquatic connectivity between main river channels and riparian pools for fish, amphibians, reptiles, and mammals (Naiman and Rogers 1997).

Hippos also greatly change the water chemistry downstream of their pools, sometimes in a bad way (Dutton et al. 2018). While in their watershed pools, hippos defecate—a lot, seeing that they can consume up to 40 kg of grasses in a day. In a low-rain period while the river is low this manure and its nutrients are locally confined. However, when a large rain event occurs, the washout from the pools sends copious amounts of dung-sourced nitrogen, phosphorus, and labile carbon downstream into the main river channel. This periodic nutrient supplement can support downstream algae and macrophytes, but it is also a massive BOD load, which drives down DO to severely hypoxic (<2.0 mg/L) concentrations. Researchers studying the Mara River in the Serengeti ecosystem documented 49 large rain/river flow events in three years, of which 13 produced river hypoxia and 9 led to fish kills (Dutton et al. 2018). Also entrained are elevated concentrations of ammonium, hydrogen sulfide, and methane, which contribute to the fish toxicity. Chapter 13 discusses how massive waste discharges from CAFOs likewise have caused massive BOD loads and fish kills.

Another large African mammal that feeds the river (very passively!) is the wildebeest (*Connochaetes taurinus*), which partakes in the 1.2 million animal migration in Kenya and Tanzania in east Africa, following rains. The migrating wildebeest cross rivers and experience massive deaths at river crossings primarily by drownings and partially by crocodile predation at such crossings. Researchers (Subalusky et al. 2017) studying the Mara crossings have estimated that during a 15-year period an average of 6250 wildebeest carcasses entered the river, representing 1100 tons of animal biomass. The carcasses are fed on by crocodiles, fish, and other aquatic scavengers, as well as birds such as vultures and storks. Thus, nutrients from the wildebeest remains feed local primary producers, nutrients in dung from the predators and scavengers feed primary producers well downstream, and nutrients are translocated (Vanni 2002) from the river well inland by scavenging birds. Earlier in this chapter we discussed the large nutrient subsidies that migrating salmon can leave in streams and their surroundings; think of that on a truly massive scale. Additionally, wildebeest bones slowly decay in the river (about seven years), representing a longer-term gradual nutrient supplement to the river (Subalusky et al. 2017).

Yellowstone Park in the Western USA presents an example of how feeding action from an apex predator can actually shape a riparian landscape in a top-down manner. The gray wolf (*Canis lupus*) predates upon the elk (*Cervus canadensis nelson*), which in turn can exert a powerful grazing control over stands of aspens (*Populus tremuloides*) and willows (*Salix* spp.) (Beschta and Ripple 2006; and see summary in Morell 2015). Yellowstone National Park was created in 1872 and contained wildlife in abundance; but intense poaching of elk and other ungulates led the government to station Army troops within the park in 1886 to stop the poaching. To encourage growth of the elk herd professional hunters and the troops shot numerous wolves and cougars (*Puma concolor*), driving then to local extirpation. The elk herd subsequently rose to over 10,000, grazing down vegetation including streamside woody plants such as willows so that erosion became severe and browse for other grazers scant. From the 1930s to the 1960s attempts were made to shrink the elk herd by relocations and shootings, which was unpopular and didn't really work. Such attempts were ceased and the elk herd exploded to 19,000 in the early 1990s. In 1995 wolves were re-established in the park and things rapidly changed.

A study of how these predator–prey relationships impacted river geomorphology was performed on the Gallatin River watershed within the park, using analysis of historical flows, historical photographs, and on-site surveys (Beschta and Ripple 2006). Areas analyzed and compared included a control reach well outside of the elk historical winter riparian grazing range and two reaches within elk grazing areas. Throughout the periods studied, the control (ungrazed) reach remained stable, well vegetated with willows, with a dense cover of sedges along the riparian zone, variable topography, with a number of small tributary streams entering the channel—some with beaver dams. In contrast the two reaches within the grazing range suffered major losses of willows and other browse and showed clear evidence of erosion, channel incision, channel widening, and sediment aggradation from the widening. Post-1995 photos indicated some regrowth of the willow community, following return of the wolves (Beschta and Ripple 2006). The research suggested that once the wolves were removed, the elk herd greatly increased and streamside herbivory increased as well, leading to the hydromorphic instability shown in the grazed areas, which also led to a lowering of local groundwater levels that further reduced riparian vegetation. The return of wolves has subsequently reduced

the grazing pressure from both direct predation and fear of predation causing the elk to avoid open riparian areas, leading to some recovery of the riparian willow communities.

Currently there are 11 wolf packs with about 95 individuals and a significant number of cougars active in the park as well (Morell 2015), but note that both of these predators are deeply unpopular outside the park among ranchers. However, the elk herd (and other ungulates) and their major predators (wolves, cougar, grizzly bears [*Ursus horribilis*] and coyotes [*Canis latrans*]) maintain a balance, and streamside vegetation has recovered in many areas, demonstrating the impact of a top-down effect in a terrestrial/aquatic ecosystem. Thus, while many mammals use the riverine ecosystem for activities during their lifetimes, some, either actively or passively, exert a powerful degree of control over natural lotic ecosystems.

## 7.4 Birds

Birds of various taxa groups can visit the riparian area occasionally as generalists, but some taxa are obligate riparian species that require riparian or aquatic areas for feeding, roosting, nesting, or breeding (Kelsey and West 1998). Since birds are attractive and generate great public interest, their very presence and diversity can play a role in stream and river conservation.

### 7.4.1 Feeding

Birds within river systems belong to a number of feeding guilds. Floating and diving waterbirds (dippers, ducks, grebes, cormorants) dive or duck underwater for fish, animals, or plants. Wading birds (herons, egrets, ibis) wade and feed on small fish and crustaceans. Insect eaters (swallows, wrens) catch insects in the air or pick them off of plants; while some riparian birds prefer seeds (red-winged blackbird). In riverine estuaries common groups include aerial-searching birds (gulls, terns) which dive from the air for live or dead fish or carrion. In higher-order rivers and riverine estuaries are raptors (eagles, ospreys) which catch and kill live prey in air, water, or aground (some consume carrion as well).

## 7.4.2 Distribution and Diversity

Specialist riverine birds are represented by approximately 60 species within 16 families. Asia hosts the largest number of river specialist birds; species richness is greatest in river corridors in the Himalayan Mountains, where 13 species overlap in range. Overall, riverine bird richness is highest at altitudes of 1300–1400 m within 20–40°N latitude (Buckton and Ormerod 2002). These areas generally have a lot of relief, complex riverscapes, and high primary productivity. Do river birds conform to the RCC (see Chapter 6)? Some do, but they are most evident in low-order streams, high-order streams, and riverine estuaries (Ormerod and Tyler 1993; Kelsey and West 1998; NCRS 2005; Sinha et al. 2019). In the middle orders an amalgam of various taxa groups, both passerines (songbirds) and wading birds, use the river either exclusively or periodically (Fig. 7.3).

In low-order headwaters streams, particularly in mountain or hilly terrain, the American dipper (*Cinclus mexicanus*) in North America and the white-throated dipper (*Cinclus cinclus*) in Europe are known specialists of these upper reaches. Dipper prey includes stonefly, mayfly, and caddisfly larvae and small fish, including young salmon. They are strong swimmers, with strong toes to grip submerged rocks, stout beaks to pry out hidden prey, and eyes with nictitating membranes allowing good underwater vision. The gray wagtail (*Motacilla*

*cinerea*) likes to nest near fast-moving streams and feeds on aquatic insects, molluscs, and crustaceans. Forktails are Asian passerines nesting near mountain streams and feeding on insects.

Red-winged blackbirds (*Agelaius phoeniceus*) are the quintessential marsh bird, inhabiting both freshwater marshes along middle-order streams downriver to salt marshes (Fig. 7.4a). They primarily feed on seeds, but also consume various types of insects. Marsh wrens (*Cisthorus palustris*) are marsh-breeding birds that are primarily insect eaters.

Wading birds are strongly associated with higher-order streams and riverine estuaries. They are mainly large birds (Fig. 7.4b) and thus require more space to fly and move about, which cannot be found in canopied, upland low-order streams. However, they can certainly be found in middle-order streams that are wide, working the pools and backwaters while hunting for aquatic invertebrates, crustaceans, reptiles and amphibians, small mammals, and of course fish. Some taxa (herons, egrets, ibis) exhibit colonial nesting in trees and shrubs near water, while others (rails, gallinules, coots) nest singly in marshes (according to the Natural Resources Conservation Service [NRCS 2005]).

The classic example of an obligate raptor is the bald eagle (*Haliaeetus leucocephalus*) (Fig. 7.5a), which has made a strong comeback in North America in recent decades after being near to extirpation due to habitat loss, illegal shooting, pollution, and

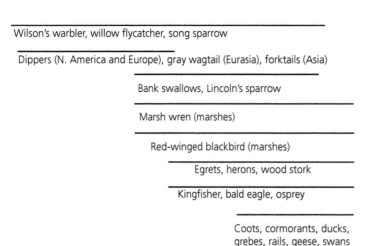

**Fig. 7.3** Birds commonly found in various river reaches, by group; note that some are river obligates while others are cosmopolitan in habitat. (Revised from Kelsey and West 1998.)

**Fig. 7.4**  Fig. 7.4a (left) Red-winged blackbird, freshwater marsh, southeastern North Carolina. Fig. 7.4b (right) Great Blue heron, Cuyahoga River, Ohio. (Photos by the author.)

**Fig. 7.5**  Fig. 7.5a (left) Bald eagle along Lake Superior in Minnesota near river inflow. Fig. 7.5b (right) Osprey nest in dead treetop along Cape Fear River, North Carolina. (Photos by the author.)

severe impacts from pesticide use especially DDT. It is primarily a fish eater but will also feed on small mammals and other prey, both alive and as carrion. Another common raptor that builds large, readily visible riverside nests is the osprey (*Pandion haliaetus*) (Fig. 7.5b).

### 7.4.3 Birds as Environmental Monitors

Sinha et al. (2019) suggest using riverine birds as biomarkers of significant landscape change due to human landscape manipulations. These authors used principal components analysis to determine that a select group of obligate river birds preferred narrow river reaches with steep slope, boulders and pebbles, channel relief, and intact streamside riparian forests, whereas other groups readily used river areas of low slope, broad channels, and adjacent agricultural land use (Sinha et al. 2019). Thus, bird communities alone may be useful to determine the degree of landscape development from the natural.

Birds are very popular with the public, and water pollution-induced damage to them can often stimulate government action. The author remembers going to limnology conferences in the 1980s and seeing awful (but effective) photos of deformed waterbird chicks from Kesterson Reservoir as a result of food-chain selenium poisoning (see Chapter 12). Metals and pesticides, especially organochlorines such as DDE (a stable metabolite of DDT), can strongly affect birds that feed at the apex of a food chain (birds of prey), which suffer mortality, thinned eggshells, and reproductive issues (Furness 1993). Such affected piscivorous birds can include cormorants, grebes, gulls, herons, kingfishers, osprey, pelicans, and terns (Ormerod and Tyler 1993). Regarding live sampling, birds can be safely captured and have their blood sampled for toxicant levels and then be released (Furness 1993). Analyses of dead waterbirds is also a useful technique, allowing researchers to determine the toxin burden in various tissues, as well as geolocating them where their corpses were found (Furness 1993). This technique has been used recently to sample per- and polyfluoroalkyl substances (PFASs) in waterbirds in the lower Cape Fear River (Robuck et al. 2021; see Chapter 12).

Can river-using birds serve as water quality monitors? The answer to this is that they can, but only to a certain degree (Ormerod and Tyler 1993). Stream and lake acidification can impact birds indirectly, by impacting their invertebrate or fish food supply (Furness 1993). The presence and absence of dippers in Wales and Scotland was quantitatively related to normal pH versus acidified waters (see studies in Ormerod and Tyler 1993). Those studies have also found dippers to have increased reproductive problems and lower body conditions in acidified waters. In organically polluted waters (see Chapter 13) invertebrates that certain bird species prey on can be severely reduced, forcing their avian predators to move, whereas other species may switch to other food sources in the same area (Ormerod and Tyler 1993). Thus, for some types of water pollution, birds are likely to be inferior as monitors compared with invertebrates.

## 7.5 Summary

- Headwaters fish are largely smaller species adapted to swift-flowing waters and cooler temperatures than fish downstream. Prey is largely invertebrates such as the abundant shredders and other insect larvae. Some fish such as trout also feed on the drift community.
- Middle-order fish may be a mixture of invertivores, herbivores, and piscivores depending upon depth, bottom structure, and vegetation. In high-order reaches and riverine estuaries detritivores may work the bottom sediments, planktivores will feed on zooplankton and phytoplankton in well-lit waters, and piscivores like LMB and large catfish will feed on smaller species.
- Tropical rivers contain much higher fish diversity than North American rivers. There are also many more herbivorous species due to massive floodplain inundations and access to flowers, fruits, seeds, and macrophytes. Piscivores are also diverse, with specialists that may eat only selected parts of other fish rather than the whole organisms.
- Amphibians are well-known residents of stream communities. Sadly, in recent decades they have been decimated by fungal diseases caused by the genus *Batrachochytrium*. The disease has apparently traveled from its native Asia worldwide via the aquarium pet trade.
- Alligators are not only top predators in southeastern North American rivers, but also ecosystem engineers, digging out wallows that are sued by fish and other aquatic organisms as refuges in drought periods. They build nest mounds in swamps that later become the core of tree islands and sites of wildlife diversity.
- Some large mammals are known ecological engineers, especially the beaver in North America. Beavers build dams that can be huge, entraining large amount of streamwater and changing the biota and soil chemistry from lotic to lentic. In African rivers hippos create deep trails that can become canals and deepen wallowing areas.
- Wildebeest, by way of mass drownings at river crossings, create large pulses of nutrients to stimulate downstream and riparian river productivity. In contrast, when storms wash out hippo wallowing areas the accumulated manure creates a massive organic BOD load that can cause river hypoxia and fish kills well downstream.
- Obligate river birds have specialist headwaters communities, with species such as dippers that feed on invertebrates at the bottom, and streamside nesting birds such as gray wagtails and forktails. Middle-order streams have a mixing of higher-order species with adventitious semiterrestrial species.
- In high-order streams and riverine estuaries wading birds are large, well-represented predators and ducks and other swimming waterbirds

are abundant. Raptors such as bald eagles and ospreys will build high riverside nests and feed on fish, mammals, and carrion from the estuary upstream into middle-order rivers.

# References

Allan, J.D. 1995. *Stream Ecology: Structure and Function of Running Waters*. Kluwer Academic Publishers, Dordrecht.

Allan, J.D., R. Abell, Z. Hogan, et al. 2005. Overfishing of inland waters. *BioScience* 55:1041–1051.

Bayley, P.B. and H.W. Li. 1992. Riverine fishes. Pages 251–281 in Calow, P. and G.E. Petts (eds.), *The Rivers Handbook: Hydrological and Ecological Principals*. Blackwell Scientific, Oxford.

Beschta, R.L. and W.J. Ripple. 2006. River channel dynamics following extirpation of wolves in northwestern Yellowstone National Park, USA. *Earth Surface Processes and Landforms* 31:1525–1539.

Buckton, S.T. and S.J. Ormerod. 2002. Global patterns of diversity among the specialist birds or riverine landscapes. *Freshwater Biology* 47:695–709.

Crutchfield, J.U., D.H. Schiller, D.D. Herlong and M.A. Mallin. 1992. Establishment and impact of two *Tilapia* species in a macrophyte-infested reservoir. *Journal of Aquatic Plant Management* 30:28–35.

Cushing, C.E. and J.D. Allan. 2001. *Streams: Their Ecology and Life*. Academic Press, Cambridge, MA.

Day, J.W., C.A.S. Hall, W.M. Kemp and A. Yanez-Arancibia. 1989. The role of wildlife in estuarine ecosystems. Chapter 11 in *Estuarine Ecology*. John Wiley & Sons, New York.

De Oliveira, A.G., T.M. Lopes, M.A. Angulo-Valencia, et al. 2020. Relationship of freshwater fish recruitment with distinct reproductive strategies and flood attributes: a long-term view in the upper Paraná River floodplain. *Frontiers in Environmental Science* 8:577181.

Diana, J.S. 2004. *Biology and Ecology of Fishes*, 2nd edition. Biological Sciences Press, Traverse City, MI.

Dutton, C.L., A.L. Subalusky, S.K. Hamilton, E.J. Rosi and D.M. Post. 2018. Organic matter loading by hippopotami causes subsidy overload resulting in downstream hypoxia and fish kills. *Nature Communications* 9:1951.

Furness, R.W. 1993. Chapter 3 in Furness, R.W. and J.J.D. Greenwood (eds.), *Birds as Monitors of Environmental Change*. Chapman and Hall, London.

Giller, P.S. and B. Malmqvist. 2002. *The Biology of Streams and Rivers*. Oxford University Press, Oxford.

Gorman, O.T. and J.R. Karr. 1978. Habitat structure and stream fish communities. *Ecology* 59:507–515.

Greenberg, D.A. and W.J. Palen. 2019. A deadly amphibian disease goes global. *Science* 363:1386–1388.

Hocking, M.D. and J.D. Reynolds. 2011. Impacts of salmon on riparian plant diversity. *Science* 331:1609–1612.

Hugueny, B., T. Oberdorff and P.A. Tedesco. 2010. Community ecology of river fishes: a large scale perspective. Pages 29–62 in Jackson, D. and K. Gido (eds.), *Community Ecology of Stream Fishes: Concepts, Approaches and Techniques*. American Fisheries Society Symposium, Bethesda.

Jackson, B.K., S.L. Stock, L.S. Harris, J.M. Szewczak, L.N. Schofield and M.A. Desrosiers. 2020. River food chains lead to riparian bats and birds in two mid-order rivers. *Ecosphere* 11:e03148.

Karr, J.R. 1981. Assessment of biotic integrity using fish communities. *Fisheries* 6:2127.

Kelsey, K.A. and S.D. West. 1998. Riparian wildlife. Chapter 10 in Naiman, R.J. and R.E. Bilby (eds.), *River Ecology and Management: Lessons from the Pacific Coastal Region*. Springer, New York.

Lips, K. 2018. The hidden biodiversity of amphibian pathogens. *Science* 360:604–605.

Miltner, R.J. and E.T. Rankin. 1998. Primary nutrients and the biotic integrity of rivers and streams. *Freshwater Biology* 40:145–158.

Miltner, R.J., D. White and C. Yoder. 2004. The biotic integrity of streams in urban and suburbanizing landscapes. *Landscape and Urban Planning* 69:87–100.

Moore, J.W. 2006. Animal ecosystem engineers in streams. *BioScience* 56:237–246.

Morell, V. 2015. Lesson from the wild lab: Yellowstone Park is a real-world laboratory of predator–prey relationships. *Science* 347:1302–1307.

Naiman, R.J. and K.H. Rogers. 1997. Large animals and system-level characteristics in river corridors. *BioScience* 47:521–529.

Naiman, R.J., C.A. Johnson and J.C. Kelley. 1988. Alteration of North American streams by beaver. *BioScience* 38:753–761.

NRCS. 2005. Wading Birds. Fish and Wildlife Habitat Management Leaflet No. 16. Wildlife Management Institute, Natural Resources Conservation Service, Washington, DC.

O'Hanlon, S.J., A. Rieux, R.A. Farrer, et al. 2018. Recent Asian origin of chytrid fungi causing global amphibian declines. *Science* 360:621–627.

Ormerod, S.J. and S.J. Tyler 1993. Birds as indicators of changes in water quality. Chapter 5 in Furness, R.W. and J.J.D. Greenwood (eds.), *Birds as Monitors of Environmental Change*. Chapman and Hall, London.

Penczak, T., G. Zięba, H. Koszaliński and A. Kruk. 2003. The importance of oxbow lakes for fish recruitment in a river system. *Archiv für Hydrobiologie* 158:267–281.

Pennisi, E. 2014. The river masters: hippos are the nutrient kingpins of Africa's waterways. *Science* 346: 802–805.

Reeves, G.H., P.A. Bisson and J.M. Dambacher. 1998. Fish communities. Pages 200–234 in Naiman, R.J. and R.E. Bilby (eds.), *River Ecology and Management: Lessons from the Pacific Coastal Ecoregion*. Springer, New York.

Richardson, J. and D. Livingstone. 1962. An attack by a Nile crocodile on a small boat. *Copeia* 1:203–204.

Robuck. A.R., J.P. McCord, M.J. Strynar, M.G. Cantwell, D.N. Wiley and R. Lohmann. 2021. Tissue-specific distribution of legacy and novel per- and polyfluoroalkyl substances in juvenile seabirds from the US Atlantic coast. *Environmental Science and Technology Letters* 8: 457–462.

Scheele, B.C., F. Pasmans, L.F. Skerratt, et al. 2019. Amphibian fungal panzootic causes catastrophic and ongoing loss of biodiversity. *Science* 363:1459–1462.

Schindler, D.E., M.D. Scheuerell, J.W. Moore, S.M. Gende, T.B. Francis and W.J. Palen. 2003. Pacific salmon and the ecology of coastal ecosystems. *Frontiers in Ecology and the Environment* 1:31–37.

Schletterer, M., V.V. Kuzovlev, Y.N. Zhenikov, et al. 2018. Fish fauna and fisheries of large European rivers: examples from the Volga and Danube. *Hydrobiologia* 814: 45–60.

Sheldon, A.L. 1968. Species diversity and longitudinal succession in stream fishes. *Ecology* 49:193–198.

Sinha, A., N. Chatterjee, S.J. Ormerod, B.S. Adhikari and R. Krishnamurthy. 2019. River birds as potential indicators of local-and catchment-scale influences on Himalayan river ecosystems. *Ecosystems and People* 15: 90–101.

Stith, B.M., J.P. Reid, C.A. Langtimm, et al. 2010. Temperature inverted haloclines provide winter warm-water refugia for manatees in southwest Florida. *Estuaries and Coasts* 34:106–119.

Stone, R. 2007. The last of the leviathans. *Science* 316:1684–1688.

Subalusky, A.L., C.L. Dutton, E.J. Rosi and D.M. Post. 2017. Annual mass drownings of the Serengeti wildebeest migration influence nutrient cycling and storage in the Mara River. *Proceedings of the National Academy of Sciences* 114:7647–7652.

Vanni, M.J. 2002. Nutrient cycling by animals in freshwater ecosystems. *Annual Review of Ecology and Systematics* 33:341–370.

Wang, L., D.M. Robertson and P.J. Garrison. 2007. Linkages between nutrients and assemblages of macroinvertebrates and fish in wadeable streams: implication to nutrient criteria development. *Environmental Management* 39:194–212.

# CHAPTER 8

# Blackwater Streams and Rivers

## 8.1 What Are Blackwater Streams and Rivers?

Blackwater streams represent the most common type of lotic system in the US southeastern Coastal Plain and are found from the New Jersey pine barrens through Virginia and south to central Florida and in areas of the Gulf Coast (Meyer 1990; Smock and Gilinsky 1992). Blackwater systems are also commonly found draining peat bog areas and forests in boreal North America, Scandinavia, and Russia and in the tropics of both hemispheres (Keskitalo and Eloranta 1999; Lewis et al. 2000). In boreal areas they are generally referred to as brownwater streams. In shallow streams the color appears brownish "tea-colored," but in deep higher-order rivers the color appears black (Fig. 8.1a). The coloration comes from soluble organic matter (primarily humic and fulvic acids) leached from terrestrial vegetation in the floodplain and exported to the streams. In these types of systems humic substances comprise anywhere from 60 to 90% of the DOM, whereas in noncolored waters the percentage is less than 50% (Keskitalo and Eloranta 1999).

The DOM that colors the water has important ecological ramifications to these systems. First, this material absorbs and attenuates light rapidly, primarily in the UV, blue, and green wavelengths (Wetzel 2001). Generally, much light attenuation occurs within the first meter of the water column, which restricts the irradiance that would otherwise be available for photosynthesis of phytoplankton and periphyton at lower depths in the water column. As shown later in this chapter, this enhances the heterotrophic nature of blackwater rivers. Second, DOM, sometimes referred to as CDOM, is a critical part of the blackwater river food web. Bacteria

and fungi consume DOM but can only readily utilize the labile portions. However, the vast majority (95%; Wetzel 2001) of DOM in these systems is in the form of high molecular weight refractory organic molecules. Bacteria utilize enzymes to help break down these refractory substances to more usable forms. This process is enhanced by photolysis by solar irradiance, particularly UV radiation, although irradiance in the PAR wavelengths can also contribute to photolysis. This process yields smaller organic fractions that consist of numerous fatty acids including acetic, formic, citric, pyruvic, malic, and levulinic that can then support bacterial growth (Wetzel 2001). Thus, even though DOM restricts light availability to the photosynthetic community, UV radiation in turn helps degrade recalcitrant organic portions of the DOM into usable substrates for bacteria, which initiates an extensive heterotrophic food web. These topics will be explored in detail below.

## 8.2 Southeastern Blackwater Stream Systems

Most of this discussion will center on the blackwater systems of the Southeastern USA, where a large amount of research has occurred in the past several decades. Coastal Plain blackwater systems are characterized by low topography, sandy sediments, extensive floodplains, and high concentrations of DOM (Meyer 1990; Smock and Gilinsky 1992). These streams and rivers have extensive interactions with the riparian swamp forests that border their channels. These riparian forests are dominated by bald cypress (*Taxodium distichum*), water tupelo (*Nyssa aquatic*), sweetgum (*Liquidamber*

*River Ecology*. Michael A. Mallin, Oxford University Press. © Michael A. Mallin (2023). DOI: 10.1093/oso/9780199549511.003.0009

(a)  (b)

**Fig. 8.1** Fig. 8.1a (left) Black River, a fifth-order Coastal Plain river. Fig. 8.1b (right) Black River entering the Piedmont-derived Cape Fear River in southeastern North Carolina. (Photos by the author.)

*styraciflua*), water oak (*Quercus nigra*), blackgum (*Nyssa biflora*), willow (*Salix* spp.), and swamp hickory (*Carya leiodermis*) (Rheinhardt et al. 1998). In scattered areas where sediments are more organic, macrophytes occur, including alligatorweed (*Alternanthera philoxeroides*), water lily (*Nuphar lutea*), and pickerel weed (*Pontederia* sp.). Blackwater streams originate on the Coastal Plain, but occasionally serve as major tributaries of rivers that originate in the Piedmont and other upland areas. Their darkly stained waters are a readily apparent contrast when they join Piedmont-derived streams (Fig. 8.1b). The organic acids lead to low pH in lower-order blackwater streams; pH subsequently increases with increasing stream order as the systems receive more alkaline groundwater inputs. For instance, in Colly Creek, a near-pristine system in southeastern North Carolina, it is not uncommon for pH levels to drop to 3 while still supporting a diverse biological community (Mallin et al. 2004). As an aside, it is notable that the low pH and high DOM in blackwater swamps make these riparian areas ideal locations for the methylation of mercury to methyl mercury (see Chapter 12), the most toxic form of that element (Tsui et al. 2020). During periods of elevated flow, blackwater rivers become even more darkly stained with increased inputs from the riparian swamps, notably increasing the attenuation of light through the water column (Meyer 1992; Mallin et al. 2002).

Coastal Plain blackwater streams also differ physically and ecologically from upland clearwater systems (Meyer 1990). Streams and rivers in mountain and Piedmont provinces pass through areas of moderate to rapid elevation decrease, with consequently significant current flow. In contrast, on the Coastal Plain the slope of headwaters streams averages about 0.1% (or a 1 m drop/km) and the slope of middle-order streams averages a mere 0.02% (Smock and Gilinsky 1992), making for slow-flowing waters. Because of the low gradient and consequent low hydrological "head," most southeastern blackwater rivers have not been impounded for hydropower purposes; thus there are still some reaches that are in relatively unmanipulated condition. There are no geologically formed falls or cascades and very few riffles—thus there is little topographic relief. The period of greatest blackwater river discharge in the Southeast USA is winter, due to low evapotranspiration. During low flow in summer some headwaters streams may dry up, except for pools, which thus become important biological refuges. During the period May 2000–April 2001 my laboratory collected monthly discharge data at locations on six second- to fourth-order streams on the North Carolina Coastal Plain (Colly Creek, Great Coharie Creek, Little Coharie Creek, Six Runs Creek, Hammond Creek, and Browns Creek). Median summer current speeds were 0.21 cm/s (*n* = 47), with a range from 41 down to 0 cm/s, or stagnant water. Because of the generally slow flow and lack of topographic relief there is little atmospheric reaeration of the water column, particularly during summer. In streams draining into lower reaches of coastal rivers, tidal movement can play a role in retarding flushing and retaining pollutants in freshwater stretches (Mallin et al. 2002; Ensign et al. 2012).

(a)                          (b)                          (c)

**Fig. 8.2** Fig. 8.2a (left) Cedar Creek in South Carolina typifies sandy-bottomed blackwater streams where woody debris provides the only solid substrata for organism use. Fig. 8.2b (middle) Debris dam in Harrisons Creek, North Carolina, a low-order blackwater stream. Fig. 8.2c (right) Snag in fifth-order Black River, North Carolina. (Photos by the author.)

The sediments are primarily well-sorted sands, with scattered patches of organic detritus that are moved downstream during spates (Fig. 8.2a). Note that in some lowland swampy areas sediments can be organic (Todd et al. 2009). In most reaches the sandy sediments and low slope are reflected by typically low turbidity, which generally averages less than 10 NTU. Because these streams are characterized by low channel roughness and shifting sandy sediments, woody material serves a critical role as stable substrates for periphyton and invertebrates. This woody material is seen mainly in the form of debris dams and snags, which are logs and tree parts in the channel. Debris dams (Fig. 8.2b) are abundant in lower-order blackwater streams, can range from spanning the width of the stream to impeding the flow in only a portion of the channel, and are formed by root masses or logs (Smock et al. 1989). Larger debris dams (including those created by beavers) increase water retention and flooding of the floodplain and lead to the creation of new stream channels. Snags, which are principally features of larger-order stream channels (Fig. 8.2c), become islands of intense trophic activity as they age that involve bacteria, periphyton, invertebrates, and fish; this will be discussed in more detail below. In relatively undisturbed blackwater stream systems woody debris abundance increases from lower-order to higher-order streams (Benke and Wallace 1990). This contrasts with upland and/or clearwater streams (and the RCC) where the strong interaction between the lowest-order streams and the riparian forests leads to greatest woody abundance in headwaters streams.

Floodplain interaction is essential to blackwater stream ecosystems. As with other types of streams it allows river biota access to the resources of the floodplain and it provides woody and non-woody organic debris inputs critical to the system (Cuffney 1988). In fact, it has been suggested that the floodplain, where extensive processing of leaf litter occurs, serves as the functional headwaters of blackwater streams by providing organic matter that supports the benthic community (Smock et al. 1985). Thus, maintenance of a healthy blackwater stream ecosystem requires avoidance of factors that decouple streams from their riparian zones. Such factors include clearcutting of riparian forests for agriculture or urbanization and cattle grazing in the riparian zone (Meyer 1990). Channel clearing removes essential woody debris and should be avoided. However, after hurricanes occur in the southeast USA, local sportsmen groups petition government agencies to clear streams and rivers of the new woody debris inputs so they can launch and operate their boats to fish (see Chapter 17). Unfortunately this removes the very habitat that provides good fishing in such ecosystems. Beaver trapping also disrupts these ecosystems by reducing water retention on floodplains and removes the habitat the dams provide for invertebrates and their predators (see Chapter 7).

A common feature in southeastern blackwater stream systems is low DO in summer months. Why would this be more common in these systems than clearwater streams? Of course, in all types of systems the warmer water temperatures in summer will lower the solubility of gases in water, but additionally, blackwater rivers drain swamps, which feature shallow water that has considerable contact with the organically rich sediments of the riparian swamp floor. Much microbial respiration occurs in this situation, reducing the DO in the swamp water (Meyer 1992). During and after rain events the water that drains into the stream channels thus already has low DO from its prolonged contact with the riparian swamp floor. Swamp and/or riparian forest drainage also carries large amounts of organic matter into the channels (Wainright et al. 1992; Dosskey and Bertsch 1994), both POM and DOM. Together these comprise total organic matter (TOM). This increases respiration rates in the stream channels by exerting a BOD as bacteria and fungi utilize the carbon within the organic matter as a substrate. BOD is a measure of the strength of the organic load in a water column as measured by the microbial community's ability to consume oxygen during respiration. Bacteria in southeastern blackwater stream systems are abundant in the water and sediments. The carbon fraction of TOM is comprised of POC and DOC; DOC is generally > 95% of TOC in the river systems (Meyer 1990). The highest concentrations of TOC occur when the river floods (water color is most intense then also), while the lowest TOC is found during drought periods. Some published DOC concentrations from Southeastern USA blackwater streams and rivers are presented in Table 8.1.

Sediment oxygen demand (SOD) can be an important regulator of water column DO as well, but SOD is not commonly calculated because it is somewhat difficult and time consuming, requiring in-situ techniques. Utley et al. (2008) performed SOD measurement at seven (primarily sandy) stream sites within three blackwater Southeastern USA river reaches within Georgia's Suwanee River basin. SOD values ranged from 0.1 to 2.3 g $O_2/m^2$/day, with an overall average of 1.36 g $O_2/m^2$/day. These results compare favorably with SOD experiments performed monthly in five tidal creeks in southeastern North Carolina (MacPherson et al. 2007) in which sediment content was mixed sand and silt with organic content ranging up to 18.3% at one

**Table 8.1** DOC data from several Southeastern USA rivers and streams, as mean (when available) or range (in mg C/L).

| Location | Mean | Range | Reference |
|---|---|---|---|
| Northeast Cape Fear River, NC (5th order) | 15.2 | | Avery et al. (2003) |
| Seven Cape Fear system streams | 12.3 | 1.3–33.5 | Mallin and Cahoon (2020) |
| Nine Georgia blackwater streams | | 11.7–62.7 | Carey et al. (2007) |
| Little River, GA (5th order) | 17.0 | 13.1–22.4 | Todd et al. (2009) |
| Ogeechee River, GA (6th order) | 12.7 | | Edwards and Meyer (1987) |
| Satilla River, GA | 20.0 | | Meyer (1992) |
| St. John's River, FL (6th order) | 20.0 | | Phlips et al. (2000) |

site. SOD in those estuarine creeks varied from 0 to 9.3 g $O_2/m^2$/day, with an overall mean of 1.58 g $O_2/m^2$/day. In comparison, Todd et al. (2009) studied SOD at two sites in southeastern Georgia's Little River watershed, at a swampy fifth-order location and a third-order location. SOD concentrations were considerable in those organic sediments with a range of 0.49–145.19 g $O_2/m^2$/day and an overall mean of 5.38 g $O_2/m^2$/day. Organic carbon content of the sediments was the principal variable correlated with SOD in that study. SOD is likely to be an important driving factor behind the naturally low summer DO levels in many blackwater systems, with its importance increasing with organic content of the sediments, especially in shallow waters where water-column BOD is comparatively less important.

Thus, southeastern blackwater streams are naturally low in DO because they are fed by waters containing already low DO, and the inputs of organic carbon, both particulate and dissolved, increases stream respiration by the abundant bacteria and fungi in the water column and sediments, which can exert a strong SOD. This has implications for water quality, because the already low DO content of these blackwater rivers makes them even more vulnerable to organic pollutant loading, increased BOD, and subsequent low DO stress in summer.

## 8.3 The Bacterial Community as a Major Driver in Blackwater Streams

Compared with many aquatic ecosystems, the microbial community plays an inordinately important role in blackwater stream ecology, not only in decomposition of POM but also in driving the food webs and controlling aspects of water quality. Research performed on Coastal Plain rivers in Georgia found that bacterial concentrations in the water column, both free-living and attached, exceeded that of most freshwater and marine systems (Edwards 1987). The bacteria consume DOM that enters the streams from riparian floodplains (Findlay et al. 1986); also, the swamps themselves are important sources of bacteria loading to the streams (Edwards 1987; Meyer 1990). The water column hosts an abundant protozoan community of microflagellates and ciliates that grazes the bacteria community (Carlough and Meyer 1989, 1991). In Georgia's sixth-order Ogeechee River the abundances of flagellates, ciliates, and bacteria are all significantly correlated with DOC concentrations (Carlough and Meyer 1989), showing the importance of bacteria as a major trophic link (more on this below). The bacterial community is not limited to the water column and sediment surface but can also be active at least 20 cm deep into the hyporheic zone (the saturated zone underlying the surface sediments) (Meyer 1990; Fuss and Smock 1996). Large amounts of organic material are stored in the hyporheos, which, like surface sediments, can be periodically flushed by spates (Fuss and Smock 1996). With the high amounts of DOC entering the streams supporting the rich heterotrophic microbial community it might be expected that planktonic metabolism in these systems is dominated by bacterial metabolism (Biddanda et al. 2001). Metabolic studies of the water column of the Ogeechee River in Georgia (Edwards and Meyer 1987) and of the various components of a Virginia blackwater stream including the water column, the sediments, the hyporheic zone, leaf litter, and woody debris (Fuss and Smock 1996) have demonstrated that these are net heterotrophic systems.

In the Ogeechee River Edwards and Meyer (1987) determined that gross primary production ranged from 0.5 to 14.0 g $O_2/m^2$/day, whereas respiration rates ranged from 3.7 to 11.75. Primary productivity was best predicted by chlorophyll *a* concentration ($r^2 = 0.50$) and negatively related to DOC ($r^2 = 0.30$), whereas the best predictor of respiration was water temperature ($r^2 = 0.27$). On average the P/R ratio in the river was only 0.25; the only period when photosynthesis exceeded respiration (P/R = 1.3) occurred following a prolonged drought (Edwards and Meyer 1987). With increasing inputs of woody material and swamp water inputs, blackwater rivers actually get more heterotrophic with increasing stream order. This contrasts with the RCC where there is increasing P/R as the systems become middle- to large-order rivers (Cuffney 1988).

## 8.4 Blackwater Stream Food Webs

Bacteria form an important base of the blackwater food web, which has two principal pathways of energy flow into larger consumers. These two pathways are the snag community (organisms associated with woody debris in the water column) and organisms associated with the sediments. The macroinvertebrate community of blackwater streams and rivers is diverse, abundant, and productive (Benke et al. 1984). The macroinvertebrate community associated with snags is dominated by filter-feeders (Fig. 8.3, including caddisfly larvae (trichopterans), blackfly larvae (simuliids), and some chironomids. In a second-order blackwater stream in South Carolina Smock et al. (1985) found that of the functional invertebrate feeding

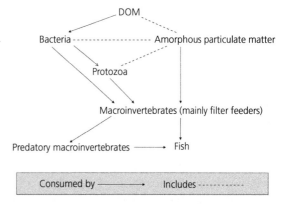

**Fig. 8.3** Trophic pathway of blackwater river snag community.

guilds (discussed in Chapter 5) annual secondary production was dominated by collector-gatherers and collector-filterers, followed by predators, with shredders and scrapers contributing little to overall secondary production. Invertebrate production per unit area was highest on snags, but overall, stream secondary production was greatest on the bottom (because of the much larger surface area), followed by snags, and mud banks with little production in leaf packs. In the limited areas where macrophytes were present, secondary production per unit area was high. In the sixth-order Satilla River, Georgia, Benke et al. (1984) found high taxonomic diversity on snags, including filter-feeding caddisfly and blackfly larvae and chironomids who themselves were consumed by invertebrate predators including hellgrammites, dragonfly, and stonefly larvae. Secondary production on snags in the Satilla River was also high.

Bacteria can be consumed directly by blackfly larvae, which are capable of filtering FPOM. Bacteria can also be consumed while associated with other food particles, such as amorphous particulate matter. These are particles that are composed of clay, inorganic detritus, and various microorganisms held together by bacterially produced mucopolysaccharide fibrils (Carlough 1994). The loci for these particles can be formed through either biotic or chemical flocculation of DOM. As these particles become larger they become sites for microbial activity and support protozoans. This material has a relatively high protein and fat content (Carlough 1994) and is consumed by the filter-feeding macroinvertebrates such as caddisfly larvae and blackfly larvae that are attached to snags and woody debris dams. Flagellated and ciliated protists within the water column can directly consume bacteria; Carlough and Meyer (1991) estimated clearance rates of 1.7–43 bacteria/flagellate/h and 25–1100 bacteria/ciliate/h. This is an example of bacteria being converted into larger food parcels which are then available to the filter-feeding invertebrate community. A variety of fishes, including bluegills, warmouth, fliers, redbreast sunfish, and pirate perch, utilize the abundant and productive invertebrate snag community as their primary food source.

The macroinvertebrate community in and on the sediments of blackwater streams and rivers includes

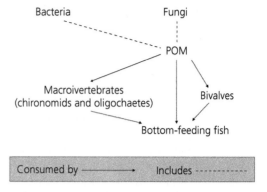

**Fig. 8.4** Blackwater river bottom trophic pathway.

collector-gatherers, such as oligochaetes and chironomids, and some filter-feeding bivalves and chironomids (Fig. 8.4). These organisms will consume bacteria and fungi that are associated with organic detritus in or on the sediments. Depending on the quantity of bacterially based carbon ingested by stream macrobenthos, the assimilation efficiency for detritus in blackwater streams may be as high as 25% (Edwards 1987). The benthic macroinvertebrate community is in turn grazed by benthic feeding fishes including spotted sucker, brown bullhead, American eel, and various catfishes. Thus, the microbial community plays a major role in supporting two important blackwater stream food webs.

## 8.5 Nutrients in Blackwater Streams

Inorganic nutrient concentrations of unimpacted blackwater streams are generally low in comparison with either upland streams or anthropogenically influenced blackwater systems (Meyer 1992; Smock and Gilinsky 1992; Mallin et al. 2004). Nutrient data from several well-studied blackwater rivers are provided in Table 8.2. Note that TN concentrations can be high, but most of that N is in the form of organic N, with much of that refractory.

Decades ago, most low-order blackwater streams were considered to be low in inorganic nutrients. However, changes in the landscape have demonstrated that the inorganic nutrient content of smaller blackwater streams can be quite high, depending upon the location and watershed sources (Table 8.3). The blackwater streams in Table 8.3 are all located in the Coastal Plain in southeastern North Carolina.

**Table 8.2** Average concentrations of nutrients and chlorophyll *a* in a selection of southeastern USA blackwater rivers, fifth order and above. Data are in µg/L.

| River (dates) | Ammonium | Nitrate | TN | Orthophosphate | TP | Chl a |
|---|---|---|---|---|---|---|
| Black, NC[1] (2015–2019) | 49 | 320 | 1119 | 37 | 116 | 1.0 |
| New, NC[1] (2004–2006) | 49 | 568 | NA | 48 | NA | 32.4 |
| Northeast Cape Fear, NC[1] (2015–2019) | 71 | 272 | 1276 | 44 | 150 | 1.0 |
| Ogeechee, GA[2] (1974–1985) | <60 | 100 | NA | NA | 50 | NA |
| Satilla, GA[2] (1968–1985) | | 100 | NA | NA | 90 | NA |
| Suwannee, FL[3] (1996–1997) | 35 | 670 | 1015 | 70 | 118 | 1.0 |

[1] Mallin laboratory unpublished data.
[2] Meyer 1992 and sources within.
[3] Bledsoe and Phlips 2000.

**Table 8.3** Concentrations of nutrients and chlorophyll *a* in a selection of southeastern North Carolina blackwater streams, first to fourth order. Data are presented as mean ± standard deviation/range and in µg/L, from unpublished Mallin laboratory data 2015–2019, except Stocking Head Creek where data are from 2013 (Mallin et al. 2015).

| Stream, order | Ammonium | Nitrate | TN | Orthophosphate | TP | Chl a |
|---|---|---|---|---|---|---|
| Panther Branch, 1st | 118±180 10–1240 | 1161±1152 10–4760 | 2056±1522 300–5560 | 47±30 5–140 | 261±170 50–1200 | 17±25 1–155 |
| Colly Creek, 2nd | 257±410 10–2390 | 27±31 10–220 | 1579±887 300–5020 | 69±–58 10–360 | 144±136 10–760 | 4±5 0–35 |
| Stocking Head Creek, 2nd | 3300±4100 0.200–10,900 | 6100±2100 1100–8400 | 8700±4100 2100–16,100 | 400±200 200–600 | 500±200 300–800 | 11±12 3–44 |
| Goshen Swamp, 3rd | 58±107 10–640 | 282±556 10–2530 | 1127±654 50–3330 | 30±18 5–90 | 146±160 10–900 | 10±26 0–175 |
| South River, 3rd | 116±144 10–620 | 128±141 10–690 | 1133±556 50–2900 | 10±8 0–50 | 80±61 10–280 | 17±26 1–145 |
| Great Coharie Creek, 4th | 57±106 10–620 | 552±558 10–2520 | 1352±632 110–3820 | 189±174 0–870 | 341±241 20–1060 | 2±1 0–6 |
| Rockfish Creek, 4th | 112±102 10–500 | 1643±793 10–8090 | 2833±2045 400–9890 | 210±298 10–1570 | 439±533 40–3020 | 3±6 0–38 |
| Six Runs Creek, 4th | 88±217 10–1690 | 887±538 10–2160 | 1717±578 500–2850 | 42±22 10–110 | 152±144 10–940 | 2±3 0–21 |

The biggest land use change in this region has been the conversion of former croplands such as tobacco to industrialized production of swine and poultry CAFOs (detailed in Chapter 13). For instance, Stocking Head Creek is a second-order stream with a watershed containing 13 swine and 11 poultry CAFOs; note that inorganic nutrient concentrations at a site in mid-watershed were extremely high (Table 8.3). Also from Table 8.3 are Goshen Swamp, Great Coharie Creek, Rockfish Creek, and Six Runs Creek, whose watersheds contain 119, 95, 74, and 153 swine CAFOs, respectively, and unknown numbers of poultry CAFOs. Inorganic nutrients, especially nitrate, are in high concentrations in those systems. Panther Branch likewise has high nitrate concentrations but is impacted by a 0.5–million gallons per day (MGD) wastewater plant outfall. Colly Creek has very low nitrate and has only six swine CAFOs in its watershed and one point source in its watershed. Thus, present-day blackwater streams can and do host high inorganic nutrient concentrations depending upon land use.

## 8.6 Phytoplankton in Blackwater Streams

Photosynthetic primary producers are also part of the blackwater stream microbial community. The phytoplankton community taxonomic composition in Coastal Plain blackwater streams has not been widely studied. A few studies indicate that the phytoplankton community is generally dominated by green algae (chlorophytes), diatoms (Bacillariophyceae), and cryptomonads (Herlong and Mallin 1985; Bledsoe and Phlips 2000; Mallin et al. 2001). Under the right environmental conditions riverine cyanobacterial blooms have been documented (Phlips et al. 2000; Ensign and Mallin 2001).

Several studies have shown that phytoplankton production in unimpacted blackwater rivers is low, which has been attributed to factors such as light limitation by water color or low inorganic nutrient concentrations (Edwards and Meyer 1987; Meyer 1992; Bledsoe and Phlips 2000).

Water color is indeed a major factor attenuating light in deep, higher-order streams. For instance, over a four-year period the median summer light attenuation coefficient $k$ for North Carolina's Northeast Cape Fear River was 3.66 m$^{-1}$ (Mallin et al. 2001), which would leave most of a 5- to 10-m-deep riverine water column below the compensation depth, i.e. net heterotrophic. Thus, water color does limit phytoplankton production in mid-sized and large blackwater rivers; note low chlorophyll $a$ concentrations in most of the rivers in Table 8.2.

In lower-order blackwater streams water-column light attenuation is less of a problem to phytoplankton than generally believed. Based on extensive stream monitoring data, the median summertime (May through August, years 2000 and 2001) depth $z$ of six second- to third-order streams on the North Carolina Coastal Plain was 0.9 m ($n = 47$). Considering the median summer surface irradiance ($I_0$) of 1515 µE/m$^2$/s for the Northeast Cape Fear River, determination of the mean irradiance ($I_x$) for a 0.9-m water column by the equation $I_x = I_0 \left[ 1 - e^{-kz} \right] / kz$ (from eqn 1.2 in Chapter 1) yields 443 µE/m$^2$/s, which is approximately the 30% irradiance level that yields near-maximum photosynthesis, $P_{max}$, in estuarine waters (Mallin and Paerl 1992). The bottom would receive approximately 56 µE/m$^2$/s$^1$, still well within the euphotic zone. First-order streams are even shallower; thus, under nonpristine (i.e. unshaded) conditions, the summer irradiance field of lower-order blackwater streams can and does provide hospitable conditions for phytoplankton growth and periodic algal bloom conditions.

Canopy shading of lower-order streams is certainly a limiting factor to phytoplankton growth in forested regions (note the low chlorophyll $a$ in forested Colly Creek, Table 8.3). However, the modern Mid-Atlantic and Southeastern Coastal Plain (Maryland, Virginia, North Carolina, and increasingly South Carolina) is being transformed by logging and agricultural activities, both traditional crop and large-scale CAFOs, which have decimated much of the forest cover. Also, many anthropogenically impacted watersheds, such as North Carolina's Cape Fear River system, have no legally required vegetated buffer zones, for either agriculture or forestry operations. Goshen Swamp (Table 8.3) suffered a clearcut in 1998, opening the canopy and increasing runoff (Ensign and Mallin 2001). Among other water quality problems this stream then hosted large algal blooms, primarily the cyanobacterium *Microcystis aeruginosa* and the euglenoids *Euglena* and *Phacus*, in summer 1998 and 1999, with chlorophyll $a$ reaching as high as 85 µg/L. In both summers the blooms then deteriorated and the subsequent BOD loads were followed by DO crashes to near anoxic levels. On North Carolina's Coastal Plain the fifth-order New River developed algal blooms exceeding 50 µg/L of chlorophyll $a$ following a large swine waste lagoon spill and chlorophyll $a$ exceeded 100 µg/L in a second-order blackwater stream in the aftermath of a swine waste lagoon leak in another watershed (detailed in Chapter 13).

More commonly, many lower-order Coastal Plain streams may pass through a landscape mosaic consisting of farm fields, CAFO sprayfields or poultry litter fields, canopied forests, and numerous road crossings (clearly visible from the air), allowing opportunity for significant solar irradiance to impact stream waters and support photosynthesis. Table 8.3 shows that some of the nutrient-rich blackwater streams on the Coastal Plain do host algal blooms and elevated chlorophyll $a$ concentrations, such as Panther Branch, Goshen Swamp, Stocking Head Creek, and South River. However, these streams are all third order or less. Blackwater streams that are fourth order or more tend to maintain low chlorophyll concentrations due to depth and light limitation regardless of nutrient loading. Even this has exceptions: Florida's nutrient-rich

St. John's River can host blooms depending upon changes in water color (Phlips et al. 2000). In a nonanthropogenically influenced system in the tropics, the Orinoco River annually hosts blue-green algal blooms as part of its hydrological cycle (Lewis et al. 2000).

## 8.7 Nutrient Loading and Other Anthropogenic Impacts

Besides driving algal bloom formation, the influence of nutrient loading on bacterioplankton and other heterotrophs is likely to be critical in regulating blackwater stream metabolism. As mentioned, because of the abundant heterotrophic community and high DOC load entering these systems, summer DO concentrations are often lower than clearwater streams, between 3 and 5 mg/L. Therefore, any anthropogenic increases in BOD will exacerbate hypoxia. Anthropogenic nutrient loading thus has consequences beyond that of algal bloom stimulation. The most abundant constituents of the heterotrophic plankton community are the naturally occurring bacteria, which can utilize both inorganic and organic nutrients (references summarized in Mallin and Cahoon 2020). Since blackwater streams are vulnerable to nutrient loading, growth stimulation of bacteria and protozoans by nutrient inputs may reduce DO to levels that stress the benthic and fish communities.

The author and colleagues conducted several series nutrient addition experiments on a variety of blackwater systems, using variable N and P additions and a collection of response variables including chlorophyll $a$, ATP, bacterial cell counts, and BOD (summarized in Mallin and Cahoon 2020). We conducted 11 nutrient addition bioassay experiments on water from two fifth-order blackwater rivers, the Black and Northeast Cape Fear Rivers (Mallin et al. 2001). Nutrient treatments were ammonium, urea, orthophosphate, glycerophosphate, and an ammonium and orthophosphate combination (N + P) as treatments, all at final addition concentrations of 1 mg/L. Nitrogen additions, either as ammonium or as urea, often stimulated phytoplankton growth, whereas phosphorus treatments did not (Table 8.4). However, glycerophosphate additions always yielded

**Table 8.4** Percentage of bioassays exhibiting significant ($p < 0.05$) responses to various nutrient treatments in two blackwater rivers, out of 11 bioassays at each site (Mallin et al. 2001).

| Variable | Black River | | Northeast Cape Fear River | |
|---|---|---|---|---|
| | Chl a | ATP | Chl a | ATP |
| Ammonium | 36 | 27 | 55 | 18 |
| Urea | 36 | 18 | 55 | 18 |
| N + P | 55 | 64 | 55 | 55 |
| Orthophosphate | 0 | 36 | 9 | 9 |
| Glycerophosphate | 0 | 92 | 0 | 36 |

significant ATP increases and orthophosphate additions occasionally did. Thus, these experiments demonstrated nitrogen limitation of the phytoplankton community and suggested that organic and inorganic phosphorus limited the growth of bacteria in these blackwater rivers.

We then tested the hypothesis that the ultimate net result of nutrient loading in blackwater streams is a significant increase in BOD in the system. We were also interested in determining the concentration at which a given nutrient yielded significant photosynthetic and/or heterotrophic responses. Thus, our 1999–2000 experiments were designed to test a range of potential nutrient concentrations, which were based on several years of monitoring these blackwater systems (Mallin et al. 2004). We tested water from two blackwater streams using varying concentrations of nitrate, urea, orthophosphate, glycerophosphate, and N + P. Colly Creek is relatively pristine, while Great Coharie Creek is an anthropogenically impacted stream (Table 8.3). Chlorophyll $a$ production, direct bacterial counts, and BOD were measured as response variables. Significant phytoplankton production over control occurred in most experiments involving N additions, regardless of whether it was in the form of nitrate or urea. Concentrations of nitrate or a urea + nitrate combination of 0.2 mg N/L or higher not only increased chlorophyll $a$ production but also significantly stimulated BOD (Mallin et al. 2004). In contrast, organic or inorganic phosphorus additions did not stimulate phytoplankton production. However, organic + inorganic P additions of 0.5–1.0 mg P/L significantly stimulated bacterial abundance and BOD over control on most occasions (Mallin et al. 2004). Additionally, 15 nutrient addition bioassay experiments were carried out in a blackwater lake (see Mallin and Cahoon 2020), with 11 showing significant stimulation by nitrate alone, while only one experiment showed stimulation by phosphate alone.

Thus, our blackwater N and P addition experiments cumulatively demonstrated that organic and inorganic nitrogen inputs stimulate phytoplankton growth, which in turn dies and decomposes in deeper, higher-order streams, becoming sources of BOD and lowering DO. The N limitation of phytoplankton production demonstrated in these Coastal Plain systems is unusual among freshwater streams, given that phytoplankton production in most freshwater systems is considered to be P limited (Paerl 1982; Hecky and Kilham 1988). P, particularly organic P, inputs directly stimulate bacterial growth, increasing BOD and potentially lowering stream DO. Thus, nutrient loading can stimulate two different microbial pathways (photosynthetic by N and heterotrophic by P), which in turn can stimulate BOD, leading to reduced DO, which is stressful to resident aquatic life (Mallin and Cahoon 2020). The low slope, slow summer flows, and naturally low summer DO conditions make Coastal Plain blackwater streams particularly susceptible to additional DO losses through BOD formation. These findings along with other regional research (Sundareshwar et al. 2003) demonstrate that ecosystem management requires consideration of nutrients limiting the heterotrophic microbial community as well as the autotrophic community.

## 8.8 Summary

- Blackwater streams are very common on the Coastal Plain of the USA from New Jersey south to Florida, and in areas of the Gulf Coast. Related systems, called brownwater streams, are common in boreal forest areas of Northern USA, Canada, Scandinavia, and Russia.
- Their dark color comes from organic acids leached from floodplain vegetation, and it attenuates light rapidly in deep blackwater rivers. These rivers are characterized by extensive floodplain interaction, which supplies DOM and POM to the stream channels.

- Blackwater rivers have unique food webs driven largely by DOM that is consumed by bacteria or repackaged into amorphous particulate matter. Bacteria are either directly consumed by blackfly larvae or consumed by the abundant protozoan community. Amorphous particulate matter and protozoans are themselves eaten by invertebrates, mainly filter–feeders.
- Debris dams and snags provide rare solid substrata within the sandy stream channels and support periphyton and an abundant, productive, and diverse invertebrate community. Blackwater stream fish in the Southeastern USA obtain an inordinate amount of their nutrition feeding on the snag community.
- The sandy sediments within blackwater rivers support a collector-gatherer invertebrate community of oligochaetes, chironomids, and some bivalves that subsist on sedimented organic matter. There is a bottom-feeding fish community that consumes these collector-gatherers, as well as other organic material on the bottom.
- Blackwater river systems diverge from the RCC in several ways:
  o Woody debris inputs increase along with increasing stream order, instead of decreasing.
  o Blackwater rivers become highly heterotrophic with increasing order, instead of showing increased autotrophy as clearwater rivers do.
  o Invertebrate scraper-grazers play an insignificant role in blackwater rivers due to lack of periphyton, whereas they are prominent in middle-order clearwater rivers.
- Algal blooms are rare in many deep blackwater streams because of the rapid light attenuation; they are also rare in shaded "pristine" lower-order streams. However, blackwater streams will support dense phytoplankton blooms as well as nuisance macrophyte blooms when impacted by nutrient loading from point sources and agricultural sources. These blooms are especially problematic where the riparian forest (and shade) has been impacted.
- Low summer DO is common to southeastern blackwater rivers, a result of inputs of POM and DOM from adjacent swamps and high bacteria concentrations in the water and sediments. This

makes blackwater streams particularly sensitive to anthropogenic organic loading.

- Experiments demonstrate that N loading stimulates algal production in blackwater systems (as opposed to P), but P inputs directly stimulate bacterial production and respiration. Thus, N and P loading may lead to lower DO through indirect and direct pathways.

- Some of these streams are not yet highly anthropogenically impacted. Their natural beauty, unique ecological character, and high invertebrate and fish production should make them candidates for strong conservation efforts.

## References

Avery, G.B., J.D. Willey and R.J. Kieber. 2003. Flux and bioavailability of Cape Fear River and rainwater dissolved organic carbon to Long Bay, southeastern United States. *Global Biogeochemical Cycles* 17:1042.

Benke, A.C. and J.B. Wallace. 1990. Wood dynamics in Coastal Plain blackwater streams. *Canadian Journal of Fisheries and Aquatic Science* 47:92–99.

Benke, A.C., T.C. Van Arsdall Jr. and D.M. Gillespie. 1984. Invertebrate productivity in a subtropical blackwater river: the importance of habitat and life history. *Ecological Monographs* 54:25–63.

Biddanda, B., M. Ogdahl and J.B. Cotner. 2001. Dominance of bacterial metabolism in oligotrophic relative to eutrophic waters. *Limnology and Oceanography* 46:730–739.

Bledsoe, E.L. and E.J. Phlips. 2000. Relationships between phytoplankton standing crop and physical, chemical, and biological gradients in the Suwannee River and plume region, U.S.A. *Estuaries* 23:458–473.

Carey, R.O., G. Vellidis, R. Lowrance and C.M. Pringle 2007. Do nutrients limit algal periphyton in small blackwater coastal plain stream? *Journal of the American Water Resources Association* 43:1183–1193.

Carlough, L.A. 1994. Origins, structure, and trophic significance of amorphous seston in a blackwater river. *Freshwater Biology* 31:227–237.

Carlough, L.A. and J.L. Meyer. 1989. Protozoans in two southeastern blackwater rivers and their importance to trophic transfer. *Limnology and Oceanography* 34:163–177.

Carlough, L.A. and J.L. Meyer. 1991. Bactivory by sestonic protests in a southeastern blackwater river. *Limnology and Oceanography* 36: 873–883.

Cuffney, T.F. 1988. Input, movement and exchange of organic matter within a subtropical coastal blackwater river-floodplain ecosystem. *Freshwater Biology* 19:305–320.

Dosskey, M.G. and P.M. Bertsch. 1994. Forest sources and pathways of organic matter transport to a blackwater stream: a hydrologic approach. *Biogeochemistry* 24:1–19.

Edwards, R.T. 1987. Sestonic bacteria as a food source for filtering invertebrates in two southeastern blackwater rivers. *Limnology and Oceanography* 32:221–234.

Edwards, R.T. and J.L. Meyer. 1987. Metabolism of a sub-tropical low gradient blackwater river. *Freshwater Biology* 17:251–263.

Ensign, S.H. and M.A. Mallin. 2001. Stream water quality following timber harvest in a Coastal Plain swamp forest. *Water Research* 35:3381–3390.

Ensign, S.H., M.W. Doyle and M.F. Piehler. 2012. Tidal geomorphology affects phytoplankton at the transition from forested streams to tidal rivers. *Freshwater Biology* 57:2141–2155.

Findlay, S., L. Carlough, M.T. Crocker, H.K. Gill, J.L. Meyer and P.J. Smith. 1986. Bacterial growth on macrophyte leachate and fate of bacterial production. *Limnology and Oceanography* 31:1335–1341.

Fuss, C.L. and L.A. Smock. 1996. Spatial and temporal variation of microbial respiration rates in a blackwater stream. *Freshwater Biology* 36:339–349.

Hecky, R.E. and P. Kilham. 1988. Nutrient limitation of phytoplankton in freshwater and marine environments: a review of recent evidence on the effects of enrichment. *Limnology and Oceanography* 33:796–822.

Herlong, D.D., and M.A. Mallin. 1985. The benthos-plankton relationship upstream and downstream of a blackwater impoundment. *Journal of Freshwater Ecology* 3:47–59.

Keskitalo, J. and P. Eloranta (eds.). 1999. *Limnology of Humic Waters.* Backhuys Publishers, Leiden.

Lewis, W.M. Jr., S.K. Hamilton, M.A. Lasi, M. Rodriguez and J.F. Saunders III. 2000. Ecological determinism on the Orinoco floodplain. *BioScience* 50:681–692.

MacPherson, T.A., M.A. Mallin and L.B. Cahoon. 2007. Biochemical and sediment oxygen demand: patterns of oxygen depletion in tidal creeks. *Hydrobiologia* 586: 235–248.

Mallin, M.A. and L.B. Cahoon. 2020. The hidden impacts of phosphorus pollution to streams and rivers. *BioScience* 70:315–329.

Mallin, M.A. and H.W. Paerl. 1992. Effects of variable irradiance on phytoplankton productivity in shallow estuaries. *Limnology and Oceanography* 37:54–62.

Mallin, M.A., L.B. Cahoon, D.C. Parsons and S.H. Ensign. 2001. Effect of nitrogen and phosphorus loading on plankton in Coastal Plain blackwater streams. *Journal of Freshwater Ecology* 16:455–466.

Mallin, M.A., M.H. Posey, M.R. McIver, D.C. Parsons, S.H. Ensign and T.D. Alphin. 2002. Impacts and recovery from multiple hurricanes in a Piedmont–Coastal Plain river system. *BioScience* 52:999–1010.

Mallin, M.A., M.R. McIver, S.H. Ensign and L.B. Cahoon. 2004. Photosynthetic and heterotrophic impacts of nutrient loading to blackwater streams. *Ecological Applications* 14:823–838.

Mallin, M.A., M.R. McIver, A.R. Robuck and A.K. Dickens. 2015. Industrial swine and poultry production causes chronic nutrient and fecal microbial stream pollution. *Water, Air and Soil Pollution* 226:407, DOI 10.1007/s11270-015-2669-y.

Meyer, J.L. 1990. A blackwater perspective on riverine ecosystems. *BioScience* 40:643–651.

Meyer, J.L. 1992. Seasonal patterns of water quality in blackwater rivers of the coastal plain, southeastern United States. Pages 249–276 in Becker, C.D. and D.A. Neitzel (eds.), *Water Quality in North American River Systems*. Batelle Press, Columbus, OH.

Paerl, H.W. 1982. Factors limiting productivity of freshwater ecosystems. *Advances in Microbial Ecology* 6:75–110.

Phlips, E.J., M. Cichra, F.J. Aldridge, J. Jembeck, J. Hendrickson and R. Brody. 2000. Light availability and variations in phytoplankton standing crops in a nutrient-rich blackwater river. *Limnology and Oceanography* 45:916–929.

Rheinhardt, R.D., M.C. Rheinhardt, M.M. Brinson and K. Fraser. 1998. Forested wetlands of low order streams in the inner Coastal Plain of North Carolina, USA. *Wetlands* 18:365–378.

Smock, L.A. and E. Gilinsky. 1992. Coastal Plain blackwater streams. Pages 271–311 in Hackney, C.T., S.M. Adams and W.H. Martin (eds.), *Biodiversity of the Southeastern United States*. John Wiley & Sons, New York.

Smock, L.A., E. Gilinsky and D.L. Stoneburner. 1985. Macroinvertebrate production in a South Carolina, United States blackwater stream. *Ecology* 66:1491–1503.

Smock, L.A., G.M. Metzler and J.E. Gladden. 1989. Role of debris dams in the structure and functioning of low-gradient headwater streams. *Ecology* 70:764–775.

Sundareshwar, P.V., J.T. Morris, E.K. Koepfler and B. Forwalt. 2003. Phosphorus limitation of coastal ecosystem processes. *Science* 299:563–565.

Todd, M.J., G. Vellidis, R.R. Lowrance and C.M. Pringle. 2009. High sediment oxygen demand within an instream swamp in southern Georgia: implications for low dissolved oxygen levels in coastal blackwater streams. *Journal of the American Water Resources Association* 45:1493–1507.

Tsui, M.T.-K., H. Uzun, A. Ruecker, et al. 2020. Concentration and isotopic composition of mercury in a blackwater river affected by extreme flooding events. *Limnology and Oceanography* 63:2158–2169.

Utley, B.C., G. Vellidis, R. Lowrance and M.C. Smith. 2008. Factors affecting sediment oxygen demand dynamics in blackwater streams of Georgia's Coastal Plain. *Journal of the American Water Resources Association* 44:742–753.

Wainright, S.C., C.A. Couch and J.L. Meyer. 1992. Fluxes of bacteria and organic matter into a blackwater river from river sediments and floodplain soils. *Freshwater Biology* 28:37–48.

Wetzel, R.G. 2001. *Limnology: Lake and River Ecosystems*, 3rd edition. Academic Press, San Diego, CA.

# CHAPTER 9

# The Ecology of Tidal Creeks

## 9.1 What Are Tidal Creeks?

A tidal creek can be defined as any natural lotic water body whose water movement is principally influenced by the tides and drains into a river, larger estuary, or the ocean. Such creeks can vary from freshwater to full-strength salinity and there does not necessarily have to be a net downstream movement of freshwater. Tidal creeks are widespread and abundant lotic ecosystems, occurring from cold latitudes to the tropics, yet their ecological relevance is undervalued as reflected by the lack of research on these systems relative to either completely freshwater streams or larger, better-known estuarine systems (such as Chesapeake Bay, the Albemarle–Pamlico Sound system, Florida Bay, San Francisco Bay). Along coastlines characterized by rapid elevation change, such as the Pacific Northwest and the northeast US and Canadian coastline, headwaters of tidal creeks are shaped by snowmelt as well as rainfall. Tidal creeks are rich areas in terms of aquatic, terrestrial, and avian wildlife and these creeks can support complex food webs (Kwak and Zedler 1997). The study of tidal creeks encompasses aspects of stream ecology, estuarine ecology, and RCC.

## 9.2 Geomorphology and Hydrology of Tidal Creeks

Tidal creeks are common in low-energy systems such as protected areas behind barrier islands along the US Atlantic Intracoastal Waterway (ICW) (Fig. 9.1a); they also commonly serve as tributaries to large estuaries or rivers (Mallin and Lewitus 2004). Tidal creeks are important areas where terrestrial and aquatic fauna move between habitats. Tidal creeks are common in salt marshes in more northern climes and mangrove swamp forests in southern climes, and are important links in the transfer of various materials (nutrients, organic matter) between terrestrial and marine biomes (Dame et al. 1991, 2000). Tidal creeks can range from first to fourth order, and larger creeks receive drainage from lower-order and intermittent tributaries (Fig. 9.1b). In the USA these types of creeks are very abundant along the US Atlantic Seaboard from New Jersey south to Florida, and along the Gulf Coast. Tidal creeks are also present, but less abundant, in high-energy systems such as the rocky intertidal of the Northeast USA and eastern Canada, the open beaches of the California coast, and the steep coastlines of the US and Canadian Pacific Northwest. Tidal creeks generally range in size from < 1 to > 10 km in length. Their depth rarely exceeds 3.0 m at high tide, and some tidal creeks contain broad intertidal sand or mud flats. In many of these systems boat passage may be limited to high tide.

## 9.3 Types of Tidal Creeks

There are three basic types of tidal creek: high-salinity creeks that do not drain uplands; mesohaline and polyhaline creeks that drain continental uplands into larger estuaries or the ocean; and fresh and oligohaline creeks that drain uplands into tidal rivers.

### 9.3.1 High-Salinity Creeks

#### 9.3.1.1 Temperate High-Salinity Tidal Creeks

High-salinity tidal creeks are considered to be polyhaline, in which salinity ranges from 25 to 35 psu, according to the Venice system (International Union

*River Ecology*. Michael A. Mallin, Oxford University Press. © Michael A. Mallin (2023). DOI: 10.1093/oso/9780199549511.003.0010

(a)                                              (b)

**Fig. 9.1**  Fig. 9.1a (left) Lower area of a tidal creek entering the Atlantic Intracoastal Waterway. Fig. 9.1b (right) Upper tidal creek areas can be freshwater, sometimes physically altered, and highly subject to anthropogenic inputs. (Photos by the author.)

of Biological Sciences [IUBS 1959]). In temperate regions such creeks are found along the sound side of barrier islands and in Coastal Plain salt marsh estuaries such as South Carolina's North Inlet. If large, high-salinity creeks will contain water continuously, but smaller-order creeks may run dry during low tide. The sediments in these creeks are often sandy near the mouth but can be muddy farther into the salt marsh. High-salinity tidal creeks host an abundance of marine life, including a highly diverse selection of nekton, consisting of resident and temporary fish and crustaceans, bivalves, and a variety of infaunal taxa (Sogard and Able 1991; Haertel-Borer et al. 2004; Harwell et al. 2011). In many cases there are oyster reefs and significant shell rubble from bivalves as well; in fact, oyster shell is often the only hard substrate in sandy bottom areas and serves as major habitat for many species of estuarine life (Lenhert and Allen 2002; Harwell et al. 2011). Nekton in general show preference for broad, shallow creeks with low flow located near uplands, with the shallows serving as a refuge from larger predators confined to deeper subtidal areas (Allen et al. 2007). Fish densities in high-salinity temperate creeks are highest from April to October (Lenhert and Allen 2002).

Importantly, most of the temperate, high-salinity tidal creeks are isolated from upland human influence (i.e. nutrient loading); thus, recycling of nutrients is key to sustaining planktonic and benthic algal communities (Lewitus et al. 2004). In contrast to upland-draining tidal creeks where nitrate dominates, ammonium from recycling is the key

inorganic N form in high-salinity creeks, with major sources of recycling being the nekton, oyster reefs, and benthic remineralization (Dame et al. 2002; Haertel-Borer et al. 2004). Nekton feeding on benthic organisms translocates nutrients from the benthos into the water column (Haertel-Borer et al. 2004).

North of the tropics this type of tidal creek will invariably be bordered by one of the *Spartina* (cordgrass) species or regional congeners. The vegetation of temperate high-salinity salt marsh tidal creeks will be detailed later in this chapter. The majority of this chapter on tidal creeks will focus on temperate creeks, so it is important to devote some time to tropical high-salinity creeks first.

### 9.3.1.2 Tropical Tidal Creeks

In more tropical areas (roughly between 25°N and 25°S latitude), mangroves generally line tidal creeks (Fig. 9.2a). Mangrove swamp forests cover vast areas of the tropics; they do so because they are adapted to salty waters and thus survive where other woody species cannot. In the southernmost USA mangrove species nearest the water are *Rhizophora mangle* (red mangrove) and *Avicennia germanis* (black mangrove); on slightly higher ground is *Laguncularia racemosa* (white mangrove) and transitioning between the true mangroves and the upland forests is buttonwood (*Conocarpus erecta*). Note that there are roughly 70 species of mangroves worldwide. Mangroves have adaptations to allow them to survive in waterlogged soils and the high salt environment (Kuenzler 1974; Dawes 1981a; Odum

(a)                                         (b)

**Fig. 9.2**  Fig. 9.2a (left) Tidal creek in mangrove forest in the Indian River Lagoon, Florida. Fig. 9.2b (right) Mangrove roots at low tide; such structure serves as substrata for algae and attached animal communities and cover for fish at high tide. (Photos by the author.)

et al. 1982). Red mangroves have prop roots for the lower part of their stems and drop roots hanging down from branches or upper stems; these roots have lenticels that allow them to get oxygen to the roots. Drop roots and prop roots support large quantities of periphytic microalgae as well as macroalgae (Fig. 9.2b). Black mangroves have pneumatophores, which are aerial roots growing up from roots buried in the soil, an adaptation that allows such roots to get air. Black mangroves and white mangroves have salt glands that extrude salt from the leaves. Both red and black mangroves extrude salt from the roots when transpiration draws water up the system, separating the salt from the freshwater at the roots by a kind of reverse osmosis. As a reproductive adaptation that utilizes water movement of the associated tidal creeks, mangroves exhibit vivipary—the seed germinates while on the tree and creates a detachable structure called a propagule. This propagule falls into the water and is carried by currents and tides elsewhere to new sites where it can thus rapidly anchor in substratum when it washes ashore.

Mangrove swamp forests are extremely productive, with productivity ranging from 1 to 7.5 g $C/m^2/day$ (Dawes 1981a). Mangroves do better when they are located streamside due to better flushing, lower salinity, higher nutrients, and lower hydrogen sulfide ($H_2S$). Along with the mangroves proper, these streamside forests support other primary producers in the form of periphyton and macroalgae (Fig. 9.2b), and the open creek

channels (Fig. 9.2a) support benthic microalgae that in turn support grazers of many species. Mangrove forests are habitat for many species—oysters, barnacles, sponges (which are filter-feeders), snails, crabs among the roots, fishes, sea turtles, alligators, rats, wildcats, monkeys, and many species of birds. The type of fauna that dominate(s) can depend on current flow. In areas of strong flow oysters, mussels, and barnacles dominate, while in quiet backwaters the most important animals are deposit feeders like crabs and snails, as well as predators.

Tidal creeks in mangrove forests have sediments that can consist of sand, shell, mud, or marl (lime-rich clay). The sediments around the roots may be more organic, as the root network serves to trap organic materials brought in by the tides and also organic materials (leaves, twigs, etc.) that are generated by the mangroves themselves and are trapped as debris among the roots. This can build up substantial layers of peat beneath mangrove forests. While seagrass beds may be located nearby in offshore areas, the creek bottoms proper are generally open, with little structure or cover available (Fig. 9.2a). Thus, the fauna associated with mangroves find a habitat refuge among the streamside roots systems (Fig. 9.2b). One study of mangrove prop roots as a habitat found 64 different fish species among 8 sample sites, with average density of $8/m^2$ (Thayer et al. 1988). These were predominantly forage species commonly found in the guts of piscivorous fish and birds in Florida Bay. However, juveniles of commercially and recreationally important species were also found (snook, gray snapper, spotted seatrout, red drum, mullet, sheepshead, barracuda), indicating that prop root habitats add value as refuge from larger predators. For some species the prop root habitat serves as a refuge in the day, while at night they venture out into the open flats or nearby seagrass beds to feed. Thus, mangroves are a nursery and feeding ground for multiple species of fish, crustaceans, birds, reptiles, and mammals. A mangrove analysis in the Gulf of California found that fisheries landings increased positively with mangrove area ($r^2 = 0.70$, $p = 0.0002$), in particular the fringe abutting the ocean or estuary (Aburto-Oropeze et al. 2008). The authors translated these data to a median valuation of US$37,500/ha of mangrove fringe.

Mangroves themselves are a large indirect source of food to the fishery. Leaves and twigs fall into the water in considerable numbers. In a southwest Florida swamp red mangroves contributed about 570 g/m$^2$/year of particulate matter (as dry weight) to the swamp (Kuenzler 1974). This organic matter is colonized and conditioned by bacteria and fungi, and it is consumed by crabs, molluscs, harpacticoid copepods, shrimp, insect larvae, small fish, and other creatures. These organisms are then preyed upon by fish, which are in turn the prey of game fish like tarpon, gray snapper, sheepshead, sea trout, and red drum. Terrestrial animals will prey upon some of these organisms and in turn can be eaten by terrestrial predators. Dead mangrove leaves leach copious quantities of DOM into the surrounding water. This leaching process does not require microbial mediation. Through physico-chemical processes this leachate forms flakes in the water, which then adhere to each other. These flakes become colonized by fungi, bacteria, and algae and are in turn eaten by copepods, isopods, amphipods, and other small organisms, providing an alternative pathway by which DOM contributes to the mangrove food web (Camilleri and Ribi 1986). This alternate food web pathway is similar to that demonstrated within southeastern blackwater rivers through formation of amorphous particulate matter (Carlough 1994; see Chapter 8).

Most of the herbivores in the mangrove swamp do not directly graze the mangroves but instead either filter-feed planktonic material and consume the organic flakes mentioned above, graze periphyton on roots or benthic microalgae in the creek channels, deposit feed, or consume mangrove detritus. On land only about 5% of the mangrove leaves are directly consumed by terrestrial biota (Kuenzler 1974). Thus, the mangrove trees in the swamp function primarily as a place for other plants (i.e. periphyton and drift macroalgae) and animal populations to live within (a habitat) rather than as a direct source of food (Kuenzler 1974).

Mangrove forests support a multitude of bird species. About half of the bird species utilize the swamp for nesting activities and the rest use it for feeding or roosting. Waders, floaters, and divers, and raptors feed on fish and invertebrates in the tidal creeks or nearby open waters. Wrens, woodpeckers, swallows, warblers, and flycatchers feed on insects in the forest. Doves and blackbirds feed on seeds outside the swamp but return for roosting or nesting. Mangrove islands host nesting colonies (rookeries) for various waterfowl such as ibises, herons, and egrets. Being surrounded by water helps protect the colonies from some of the terrestrial predators. There can be some physical harm to the forest caused by nesting colonies, but moderate amounts of excreta may benefit the primary producers (Kuenzler 1974). However, in dense rookeries the amount of guano generated may actually cause eutrophication of the nearby waters, leading to overgrowths of algae and degradation of nearby seagrass beds (Tomasko and Lapointe 1991).

Tidal creek dynamics are essential for the health of the mangrove forest and the associated biota for several reasons (Kuenzler 1974). The creeks keep the soil saturated with water, contributing to the anaerobic nature of the soil, which mangroves have adaptations to deal with. Creek tidal flow brings saltwater up the estuary, which eliminates competition from freshwater species, both plant and animal. Creek waters wet the epiphytic algae on the prop roots and drop roots, which are an important source of food for higher trophic levels. Tidal currents circulate particulate matter, including phytoplankton and resuspended periphyton, for filter-feeders such as sponges, oysters, and barnacles. Phytoplankton productivity in tidal creeks within mangroves can be comparatively high. A study in Estero Pargo, a tidal creek entering Terminos Lagoon, Mexico, found daily phytoplankton productivity ranging from 0.09 to 5.2 g $C/m^2/day$, an average of 285 g $C/m^2/year$ (Rivera-Monroy et al. 1998), with seasonal changes correlated to light and precipitation. The water movement of the creeks also leaves particulate matter which then becomes food for deposit feeders such as snail and fiddler crabs.

Mangroves, as well as salt marshes, are particularly valuable in hurricane buffering and protection of the terrestrial uplands (Spalding et al. 2014). Loss of mangroves also increases greenhouse gas production through loss of C sequestration (McLeod et al. 2011). Thus, such forests have multiple ecological and economic values and need enhanced protections.

Unfortunately, during the latter twentieth century the area of the world's mangroves was reduced by about 35% from the 1980s (Valiela et al. 2001). The loss of these important coastal habitats has been related both to the relative wealth of a country (Valiela et al. 2001), since increased wealth leads to increased development of all kinds, and to human population increase in a given region (Alongi 2002). The biggest problem was loss of mangroves from construction of housing, industrial and commercial facilities, salt flats, canals, and especially mariculture facilities (as much as 52% of mangrove losses; Valiela et al. 2001). From 2000 to 2016 the overall losses were considerably reduced (an additional 2.1% of world mangrove area was lost), but human causes accounted for 62% of the losses (Goldberg et al. 2020). Of those recent losses, 47% was due to shrimp aquaculture, rice production, and oil palm cultivation, about 12% to other human conversions such as mining and other activities, 27% through erosion, and other losses through major storm events and climate change effects (Goldberg et al. 2020).

### 9.3.2 Mesohaline and Polyhaline Temperate Tidal Creeks

A second major category of tidal creek consists of mesohaline (5.0–15.0 ppt salinity) to polyhaline (15–25 ppt salinity) tidal creeks draining continental uplands. Upland draining tidal creeks can serve as compact river continuums, dominated by freshwater in upper regions while becoming euhaline at the creek mouth. Not only does salinity distribution vary longitudinally, but also salinity distribution in many of these creeks varies tidally at any given location, with mesohaline to euhaline conditions occurring at high tide and oligohaline conditions at low tide; thus on average mesohaline conditions tend to dominate. The headwaters of tidal creeks can consist of springs, intermittent streams, urban and suburban drainage systems, or any combination of the above (Fig. 9.1b). In some cases natural drainage pathways have been "enhanced" by earth-moving equipment to considerably alter the original drainages of these systems. These creeks drain into larger estuaries such as the Atlantic ICW, Laguna Madre in Texas, Chesapeake Bay, Charleston Harbor, and the Albemarle–Pamlico Sound system, or the ocean proper (e.g. in some regions of the USA and Canadian Pacific Coast). Along the Coastal Plain or the Gulf Coast such tidal creeks may extend anywhere from less than 0.5 to 10 or more kilometers inland due to the gentle gradient and wide tidal floodplains. Along tectonic coasts, such as characterize the Pacific Northwest coast of North America, there is often a rapid transition from sea level to hilly and even mountainous terrain, with abbreviated tidal creek length, a narrow floodplain, or constrained reach (Fig. 9.3b).

Streamside vegetation ranges from wooded uplands in the fresh headwaters region, to oligohaline marsh in middle reaches, to salt marsh vegetation in the lower reaches (Fig. 9.3a). On high tides the water covers the middle marsh areas and reaches the back marsh on spring tides. Sediments in the upper creek can be organic muds, while in the lower creek well-sorted sands normally prevail. Upland-draining tidal creeks are the most studied among the three general types, probably because of the high level of human interaction with these systems. Therefore, many of the findings presented in this chapter are from these types of creeks.

### 9.2.3 Fresh and Oligohaline Tidal Creeks

The third basic category of tidal creek consists of creeks that are fresh to oligohaline (0.5–5.0 ppt salinity), found worldwide, and drain into tidal rivers such as the ones characterizing the Coastal Plain provinces of Virginia, the Carolinas, and Georgia

(a)　　　　　　　　　　　　　　　　　　　　(b)

**Fig. 9.3** Fig. 9.3a (left) Upper area of eastern Coastal Plain tidal creek with wide, marshy floodplain grading gently into forested uplands. Fig. 9.3b (right) Middle area of an Oregon tidal creek rapidly narrowing into a constrained reach in coastal foothills. (Photos by the author.)

(a)　　　　　　　　　　　　　　　　　　　　(b)

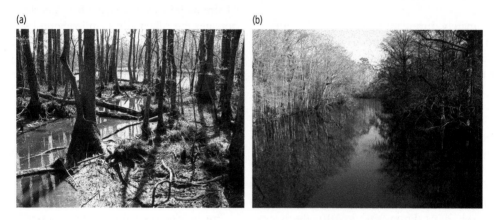

**Fig. 9.4** Fig. 9.4a (left) Shallow intermittent freshwater tidal creek draining into the Lower Black River, North Carolina. Fig. 9.4b (right) Town Creek, a third-order freshwater (blackwater) tidal creek draining to the Cape Fear River Estuary, North Carolina. (Photos by the author.)

in the USA (Fig. 9.4). These creeks are often less strongly influenced by the tides, have wooded headwaters areas, and the lower reaches may be vegetated by oligohaline or freshwater marshes or riparian swamp forest. Sediments in these systems are silt-clay, with generally high organic content (Odum 1988). These are often low-gradient streams with discharge dictated by local rainfall, as opposed to the river proper that is strongly influenced by headwaters networks well upstream. Low-order creeks of this type may be dry on low tide during periods of low river discharge. In such tidal creeks draining remote wooded areas, the major influx of anthropogenic materials (i.e. nutrients and pollutants) is likely to enter the creek from the river rather than the creek headwaters. That is a primary difference between true upland freshwater streams and

fresh to oligohaline tidal creek; whereas the river itself will influence the tidal creeks, upland rainfall and surrounding environments will primarily influence true freshwater streams.

## 9.4 Nutrient Cycling Within Tidal Creeks

Tidal creeks are interface ecosystems, connecting land to water, woodlands to marshes, uplands to lowlands, and freshwater to marine water; thus nutrient cycling is a major ecological and economic function of tidal creeks. Spatially, nutrient concentrations within a tidal creek depend upon the type of creek and its location. In temperate high-salinity creeks there are usually no major anthropogenic inputs so nutrient concentrations have little horizontal variability. In these creeks nitrate concentrations are thus typically low, and atmospheric deposition of nitrate may be a principal anthropogenic source of new nutrients. Within these creeks ammonium from recycling is the dominant form of inorganic nitrogen (see section 9.3.1.1).

Many of the mesohaline upland draining tidal creeks exhibit strong spatial nutrient variability, with highest nutrient concentrations in the uppermost, freshest stations, where there is closest proximity to human land use and subsequent anthropogenic inputs (Table 9.1). In these creeks nitrate usually dominates the inorganic N fraction and is elevated upstream due to stormwater runoff-driven inputs from fertilizers applied to lawns, gardens, landscaped areas, and golf courses (Mallin and Wheeler 2000; Mallin et al. 2004). However, orthophosphate concentrations also decline from headwaters toward the creek mouths (Mallin et al. 2004). In Californian coastal tidal creeks and rivers the percentage watershed coverage by agricultural land has been strongly correlated with nitrate and ammonium concentrations, as well as with the response-variable chlorophyll *a* (Handler et al. 2006). In creeks that drain undeveloped

watersheds, the spatial nutrient difference may be less pronounced and ammonium can play a more significant role than nitrate (MacMillin et al. 1992). In some tidal creeks inputs of nitrate-rich groundwater can occur in any area of the creek. When nearby upgradient source areas (especially golf courses) supply nitrate to groundwater, groundwater flow into springs provides inputs into tidal creeks that can have nitrate-N concentrations as high as 15 mg/L, although spring discharge is usually limited in volume (Roberts 2002). The nutrient dynamics of fresh to oligohaline tidal creeks are understudied, and nutrient concentrations and variability are likely dependent upon creek watershed development and location.

Within tidal creeks there is a transformation of inorganic nutrients into bacteria, phytoplankton, periphyton, marsh macrophytes, and woody vegetation, and into grazers and consumers. As nitrate enters continental-draining tidal creeks it is rapidly taken up by phytoplankton and benthic microalgae and tightly recycled within, with some denitrification occurring (Tobias et al. 2001; Duernberger et al. 2018). Thus, tidal creeks have strong potential

**Table 9.1** Nutrient and chlorophyll *a* concentrations in various US East Coast tidal creeks, presented as mean (and range, where available). Data are in μg/L. Note that Calico Creek had a sewage treatment plant discharge outfall entering the creek and the Cape Hatteras tidal creeks/ditches drained sandy residential areas served by septic systems.

| Creek(s) | Nitrate-N | Ammonium-N | Phosphate-P | Chl *a* |
|---|---|---|---|---|
| **Delmarva Peninsula, VA** (MacMillin et al. 1992) | | | | |
| Upper seaside creeks | 2.7 | 25.6 | 22.3 | 3.1 |
| Lower seaside creeks | 2.7 | 18.2 | 19.5 | 3.3 |
| Upper bayside creeks | 1.4 | 3.9 | 4.0 | 6.5 |
| Lower bayside creeks | 1.1 | 4.6 | 4.7 | 5.3 |
| Calico Creek, NC (Sanders et al. 1979) | 112 (42–210) | 238 (140–350) | 465 (155–1426) | 140.0 |
| **New Hanover County, NC** (Mallin et al. 2004) | | | | |
| Upper Futch Creek | 139 (4–846) | 24 (1–67) | 13 (1–233) | 8.1 (1–106) |
| Lower Futch Creek | 6 (1–35) | 20 (1–139) | 4 (1–11) | 8 (1–7) |
| Upper Howe Creek | 57 (1–190) | 36 (10–89) | 10 (1–42) | 31.3 (1–208) |
| Lower Howe Creek | 4 (1–232) | 17 (1–72) | 5 (1–19) | 2.4 (1–13) |
| Upper Hewletts Creek south branch | 107 (1–698) | 38 (1–117) | 9 (1–38) | 16.1 (2–204) |
| Upper Hewletts Creek north branch | 105 (1–582) | 47 (1–138) | 14 (1–58) | 14.3 (1–159) |
| Lower Hewletts Creek | 8 (1–49) | 16 (1–77) | 5 (1–14) | 2.2 (1–10) |
| **North Inlet, SC** (Lewitus et al. 2004) | | | | |
| Oyster Landing | 5.6 (1–19) | 57.5 (10–172) | 10.5 (4–16) | 8.9 (3–17) |
| Parsonage Creek | 7.8 (1–20) | 57.7 (9–154) | 9.0 (5–17) | 10.8 (3–30) |
| **South Bodie Island, Cape Hatteras National Seashore, NC** (Mallin and McIver 2012) | | | | |
| Five large tidal creeks/ditches | 23.0 (5–54) | 200.4 (14–748) | 130.0 (9–782) | 43.1 (6–275) |

to remove and transform anthropogenic nutrients, even extreme amounts, through uptake and microbial processing. For example, in 2005 a sewer line in Wilmington, North Carolina ruptured and sent 3,000,000 gallons of raw sewage into tidal Hewletts Creek over a 13-hour period (Mallin et al. 2007). The City of Wilmington has average nutrient concentrations in its sewage of 40.2 mg N/L for TKN (organic nitrogen + ammonium), 23.3 mg N/L for ammonium, and 5.3 mg P/L (TP). However, the nutrient load was so rapidly assimilated by the salt marsh that samples collected two days later showed only somewhat elevated TKN (< 2 mg/L), ammonia (< 1 mg/L), and TP (0.06 mg/L); additionally, the N load fueled large phytoplankton blooms in the afflicted creek (see Box 13.1).

What nutrient or nutrients typically limit phytoplankton growth within tidal creeks? This will depend upon geographic location (including soils type) as well as location within the creek itself. Along the Eastern Seaboard of the USA, as a general rule, the closer an upland-draining tidal creek area is to marine waters, the greater tendency toward nitrogen limitation. Whereas as mentioned, inorganic nutrient loading into upper creeks can be considerable, nutrient processing within the creek combined with dilution by low-nutrient marine waters tends to greatly reduce nitrate in the water column of the lower creek. A set of experiments by Mallin et al. (2004) on waters from three anthropogenically impacted upland-draining tidal creeks showed strong nitrogen limitation in the lower areas of all three creeks, but the upper creeks had a mixed response, with occasional phosphorus limitation or light limitation as well as some nitrogen limitation, all within a few linear kilometers of the creek. The phosphorus limitation presumably was caused by the greater nitrate concentrations upstream (Table 9.1), with light limitation suspected to be caused by self-shading from dense phytoplankton growth. In subtropical Ten-Mile Creek in Florida phytoplankton growth is limited by nitrogen (Lin et al. 2008). In the tropical tidal creeks and canals in south Florida, algal growth is limited primarily by phosphorus (Lapointe et al. 1990; Lapointe and Clark 1992).

## 9.5 Primary Producers Within Tidal Creeks

### 9.5.1 Macrophyte Vegetation

Along and within temperate marine and brackish tidal creeks salt and oligohaline marsh communities dominate (Table 9.2). They are distributed according to elevation and salinity tolerance. Zonation of these marshes is commonly divided into the **subtidal** (tidal creek waters or mudflats); **lower intertidal** (usually muddy substrata); **upper intertidal** (muddy substrata); **salt pannes** or **salt barrens** (above the upper intertidal, occasionally inundated, substrata more firm, can have concentrated salt through evaporation of water); and **supratidal** or **back marsh** (elevated area, very rarely inundated). Along the US East Coast lower areas of these creeks, salt marsh cordgrass (*Spartina alterniflora*) predominates, while farther up the creek as the waters become less saline the shore flora is often dominated by the emergent macrophyte black needlerush (*Juncus roemerianus*). Back marsh vegetation occasionally inundated in this manner includes herbaceous species such as salt grass (*Distichlis spicatas*), glasswort (*Salicornia virginica*), sea lavender (*Limonium carolinium* and *L. nashi*), sea oxeye (*Borrichia frutescens*), salt meadow hay (*Spartina patens*), and common reed (*Phragmites australis*). Along West Coast salt marsh tidal creeks, Pacific cordgrass (*Spartina foliosa*) and *Suaeda* spp. dominate (*Spartina alterniflora* is invasive in some West Coast areas). Woody species in East Coast back marsh areas include marsh elder (*Iva frutescens*), Eastern baccharis (*Baccharis halimifolia*), and wax myrtle (*Myrica cerifera*). In subtidal waters occasionally seagrasses occur at tidal creek mouths, while macroalgae occur attached to hard substrata or as drift algae in creeks. Eelgrass (*Zostera marina*) is a well-studied seagrass species that is cosmopolitan in the northern hemisphere. Its three-dimensional structure provides habitat for periphyton, protozoans, barnacles, polychaetes, harpacticoid copepods, snails, bivalves, and decapod crustaceans, and research has demonstrated that *Zostera* beds support significantly higher densities of fish than either macroalgal (*Ulva*) beds or unvegetated sediments (Sogard and

**Table 9.2** Major macrophyte species in and along US temperate tidal creeks and their attributes (see also Dawes 1981b; Day et al. 1989).

---

**Subtidal waters** (some taxa can use intertidal)
Eelgrass *Zostera marina*—northern hemisphere worldwide, limited by high temperatures
Manatee grass *Syringodium*—tropical, abundant in the Caribbean
Turtlegrass *Thalassia*—tropical, from Florida through northern South America
Shoalgrass *Halodule*—can live in upper intertidal
Widgeongrass *Ruppia*—tolerant of low salinity

**Subtidal to intertidal waters** (macroalgae, occurring as attached to oyster reefs or drift algae)
Sea lettuce *Ulva*—green alga
Dead man's fingers *Codium*—green alga
*Enteromorpha*—green alga
*Cladophora*—green alga, can form nuisance overgrowths (see Chapter 3)
*Ectocarpus*—brown alga
*Gracilaria*—red alga

**Lower and upper intertidal**
Smooth cordgrass *Spartina alterniflora*
Pacific cordgrass *Spartina foliosa*

**Upper intertidal**
Salt meadow hay *Spartina patens*
Black needlerush *Juncus roemerianus*—found in less saline areas

**Upper marsh/salt pannes**
Salt grass *Distichlis spicata*—has salt-extruding glands
Glasswort *Salicornia virginica*—succulent herb requiring high soil salinity
Sea lavender *Limonium carolinium*, *L. nashi*—lavender-colored small flowers
Sea oxeye *Borrichia frutescens*—large yellow flower heads
Seablite, seepweed *Suaeda* spp.—common in US West Coast salt marshes
Common reed *Phragmites australis*—once in low abundance, now considered invasive on US East Coast

**Back marsh** (generally woody species)
Marsh elder *Iva frutescens*
Wax myrtle *Myrica cerifera*
Yaupon *Ilex vomitoria*
Groundsel tree *Baccharis halimifolia*

---

Abel 1991). Seagrass itself is consumed by some species of fish, invertebrates, sea urchins, waterfowl, green turtles, and manatees.

*Spartina alterniflora* is the dominant marsh species along the US East and Gulf coasts. This species is common in the lower and upper intertidal areas and does best in areas of high tidal range such as along creek sides where there are well-drained soils. There is a 1- to 3-m-tall form (more productive) adjacent to creeks where sediments are better oxidized and porewater turnover more rapid, and a short form (0.1–0.5 m tall) a few meters away from the creek edge where sediments are more reduced and porewater does not turn over as rapidly. *Spartina alterniflora* pumps oxygen from photosynthesis or outside air down to the roots to form an oxic layer around the roots. Gas transport occurs via a series of lacunae forming a continuous gas-filled aerenchyma system from the leaves to the root tips. Tall forms have greater specific gas transport capacity from greater cross-sectional aerenchyma—this helps explain the tall form's greater nutrient uptake, better aeration, and less anaerobic root respiration frequency compared with short forms (Arenovsky and Howes 1992). In very reduced, waterlogged soils toxic products are formed by fermentation: these are released to the surrounding porewater. This is an energy-consuming process and also helps explain why plants are shorter away from the streamside. They can extrude salt from salt glands and have inrolled leaves to minimize moisture loss. Extensive root systems help retard marsh erosion. In the USA West Coast marshes the lower marsh hosts *Spartina foliosa* (Pacific cordgrass); there *S. alterniflora* is an invasive species. *Spartina patens* is found in the upper intertidal zone of the marsh. This species was historically used as graze for cattle. *Juncus roemerianus* dominates North Florida salt marshes and many areas of marsh tidal creeks along the middle Atlantic and Southeast coasts. It has cylindrical leaves to minimize moisture loss.

Primary productivity of *Spartina* is especially high, ranging from 200 to 2000 g $C/m^2$/year (summarized by Ferguson et al. 1980 and Dawes 1981b). Reasons for this are that tidal marshes are naturally fertilized twice a day by the tides, and many of these marshes are more extensive in areas of higher rainfall. It is important to note that for these species, belowground productivity is often equal to or greater than that of aboveground productivity. *Juncus* productivity ranges from about 280 to 750 g $C/m^2$/year (Ferguson et al. 1980; Dawes 1981b). *Spartina* and many of the other marsh plants are principally utilized in the food web as detritus after death and decomposition (when the associated microbial community provides the nutrition). However, the fishery habitat *Spartina* and other emergents provide when flooded cannot be underemphasized (see Section 9.7).

In marsh areas along fresh and oligohaline tidal creeks there is much higher plant diversity than in creeks draining salt marshes (Odum 1988). Emergent macrophyte species characterizing these systems along the US East Coast include wild rice (*Zizania aquatica*), arrow-arum (*Peltandra virginica*), big corgrass (*Spartina cynosiroides*), cattail (*Typha latifolia*), common reed (*Phragmites australis*), and pickerel weed (*Pontederia cordata*), with loosely attached macrophyte species such as spatterdock (*Nuphar luteum*) found along banks. Common tree species within the riparian swamp forest bordering fresh to oligohaline tidal creeks include bald cypress (*Taxodium distichum*), black gum (*Nyssa sylvatica* var. *biflora*), water gum (*Nyssa aquatica*), and Atlantic white cedar (*Chaemaecyparis thyoides*). In tropical forest areas freshwater tidal creeks drain areas vegetated by ferns, Nepa palms, and herbaceous plants (Odum 1988).

### 9.5.2 Phytoplankton

Phytoplankton communities in tidal creeks can be quite diverse, but not at the same point in time. This is so because the phytoplankton are influenced, and controlled, by several rapidly changing factors, including tidal changes (Mallin et al. 1999; Wetz et al. 2006), salinity, and rain events (Badylak et al. 2016). These creeks are dominated by dinoflagellates, diatoms, cryptomonads, and chrysophytes, with lesser abundances of green algae and even cyanobacteria in nutrient-rich fresher reaches (Mallin et al. 1999; Johnson 2005; Badylak et al. 2016). Zooplankton grazing, especially by microzooplankton, can be high and help shape phytoplankton community structure (Lewitus et al. 1998; Wetz et al. 2006). In the Hilton Head, South Carolina area, Lewitus et al. (2003) found a variety of toxic species, some of which caused fish kills, in golf course ponds which drained into tidal creeks. The well-lit waters of the tidal ponds combined with nutrient loading from golf course fertilization together made an ideal situation for these harmful algae.

In temperate tidal creeks phytoplankton biomass as chlorophyll *a* can vary from nearly undetectable to high (Table 9.1) and biomass responds to high nutrient levels (Badylak et al. 2016). Blooms can be extremely dense at times; Sanders and Kuenzler (1979) measured chlorophyll *a* as high as 280 µg/L in sewage-rich Calico Creek (Table 9.1), Baillie and Welsh (1980) measured water-column chlorophyll *a* up to 420 µg/L in a tidal creek off Long Island Sound, and Johnson (2005) measured blooms up to 55 µg/L in Hewletts Creek, North Carolina. However, such blooms are not typical and average chlorophyll *a* of temperate continental-draining tidal creeks typically ranges from approximately 2 to 6 µg/L in lower tidal creeks and from approximately 6 to 30 µg/L in upper tidal creeks (Table 9.1).

Phytoplankton productivity in upland-draining tidal creeks can be high. In a study of two creeks draining into the ICW Johnson (2005) found mean annual volumetric phytoplankton productivity of 91.2 g C/m$^3$ and mean areal productivity of 92.3 g C/m$^2$ in a creek with low levels of development, while a creek with a highly developed watershed had greater mean volumetric productivity of 246.6 g C/m$^3$ and mean areal productivity of 197.3 g C/m$^2$. Compared to other types of estuaries, the tidal creek volumetric productivity is comparatively high, while the areal productivity is more moderate due to the generally shallow waters of the tidal creeks. Johnson (2005) found that tidal creek phytoplankton productivity was strongly correlated with water temperature and solar irradiance.

### 9.5.3 Periphyton

Benthic microalgae, also called edaphic algae (Fig. 9.5a) and epiphytic microalgae (Fig. 9.5b), play a critical role in energy flow within tidal creeks. In tidal creeks their communities are primarily composed of diatoms and blue-green algae (cyanobacteria); also found are green algae, red algae, euglenophytes, and chrysophytes. Besides algae (primary producers), many other microbes live associated with them. These primary producers are very important to the estuarine food web (Sullivan and Moncreiff 1990).

Epiphytic microalgae are far more avidly consumed by most grazers than the host plants they are living on. These microalgae are grazed by nematodes, ostracods, cladocerans, rotifers, harpacticoid copepods, decapods, gastropods, amphipods, isopods, molluscs, shrimp, and fish. These grazers are in turn eaten by many species of fish and waterfowl. Grazing is readily visible within marshes lining tidal creeks in the form of the marsh periwinkle (*Littoraria irrorata*) (also called *Littorina*) grazing periphytic microalgae along the stems of *Spartina* spp. (Fig. 9.6a) and the mud snail (*Ilyanassa obsolete*) grazing benthic microalgae on the sediments (Fig. 9.6b). In two South Carolina tidal creeks Gillett et al. (2005) found that the spatial distribution of the oligochaete *Monopylephorus rubroniveus* (which accounted for 74% of oligochaete abundance in those creeks) was positively correlated with the concentration of its major food source benthic chlorophyll *a*. Benthic microalgae production is increased with nutrient inputs (Posey et al. 1999). Benthic microalgal biomass within tidal creeks declines along with increasing percentage of finer sediments as opposed to larger grain sizes (Cahoon et al. 1996); the implication is that erosion of fine sediments into such creeks may lead to lower benthic microalgal biomass and less food for the consumers.

Whereas phytoplankton in nontropical tidal creeks can be strongly seasonal, benthic microalgae are present throughout the year and assume an even greater importance in winter when phytoplankton abundance is lowest. In South Carolina tidal creeks Gillett et al. (2005) found that benthic chlorophyll *a* was significantly higher in winter than the other three seasons. At times large portions of water-column chlorophyll *a* can consist of resuspended benthic microalgae, mainly pennate

(a)                                 (b)

**Fig. 9.5** Fig. 9.5a (left) Benthic microalgae growing on sediments in a tidal creek environment. Fig. 9.5b (right) Epiphytic microalgae growing along the stems of *Spartina* streamside of a North Carolina tidal creek at low tide. (Photos by the author.)

(a)                              (b)

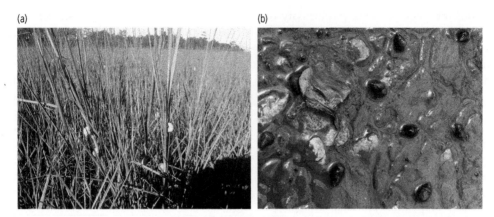

**Fig. 9.6** Fig. 9.6a (left) Marsh periwinkles *Littoraria irrorata* grazing periphyton on *Spartina* stalks. Fig. 9.6b (right) Mud snails *Ilyanassa obsoleta* grazing benthic microalgae in tidal creek salt marsh. (Marsh periwinkles photo by the author; snails photo courtesy of Dr. D.M. Sanger, South Carolina Department of Natural Resources.)

diatoms. Baillie and Welsh (1980) found that peak times for resuspension were early on the flood tide and then late on the ebb tide, when water was shallow. They attributed resuspension to convection currents caused by temperature differences between sediment and water, as well as tidal creek currents and wind action. Once suspended in the water column benthic microalgae become available to suspension feeding organisms such as shellfish. As with phytoplankton (Table 9.1), benthic microalgal biomass has been found to be greater in upper tidal creek areas than in lower creek areas (Baillie and Welsh 1980; Gillett et al. 2005).

The contribution of benthic microalgae and periphytic microalgae to total tidal creek primary production is considerable. Productivity of benthic microalgae (Fig. 9.5a) can conservatively be estimated as 25% of the overstory vascular plant production (Pomeroy 1959; Gallagher and Daiber 1974; Van Raalte et al. 1976; Sullivan and Moncrieff 1988). Epiphytic microalgae can produce at the same rate as benthic microalgae (Jones 1980), but surface area of the host plant is much greater than the benthic area so total epiphytic algal productivity is much greater than that of the host plant (Fig. 9.5b). Clearly, the overall contribution of benthic and epiphytic microalgae to the productivity of salt marshes and associated tidal creeks is considerable. In general, many of the microalgal species (especially diatoms) are more palatable to grazers than macrophyte tissue and more easily assimilated as well.

## 9.6 Tidal Creek Animal Communities

Tidal creek ecosystems contain numerous fauna utilizing living primary producers (periphyton, benthic microalgae, phytoplankton, macrophytes) as well as detritus from the above plants, upland forests, and terrestrial vegetation in general (Table 9.3). Tidal creek animals grazing periphyton include harpacticoid copepods, isopods, snails, various oligochaetes, and polychaetes; phytoplankton grazers include zooplankton, oysters, mussels, and some fish; macrophyte grazers include waterfowl, dugongs in tropical areas, mammals such as muskrats, and many insect species. Epibenthic organisms (those living on or above the sediments) utilizing detritus include crabs, shrimp, and various fish species, and infaunal species (those living within the sediments) consuming benthic microalgae, detritus, and organic matter in general include polychaetes and oligochaetes. Polychaetes in general dominate in the higher-order, higher-salinity subtidal areas of tidal creeks, while oligochaetes dominate in lower-order headwaters branches that are more subject to rapid changes in discharge (flashiness), salinity, and DO (Gillette et al. 2005; Washburn and Sanger 2011).

The eastern oyster (*Crassostrea virginica*) can alter the lower ends of tidal creeks significantly in several ways. They can be considered passive ecological engineers—they build intertidal or subtidal reefs but not deliberately. These reefs perform a number

**Table 9.3** Epibenthic and infaunal animals commonly found within US East Coast tidal creeks (Dawes 1981b; Weis and Butler 2009). Species will vary geographically, and successful invasive species may alter this list as well (see Chapter 14).

**Macrocrustaceans**
Fiddler crabs *Uca* spp.—detrital and algal feeders, live in burrows, males with enlarged claw
Salt marsh mud crabs *Sesarma reticulatum*, *Armases cinereum*—omnivores
Grass shrimp *Paleomonetes pugio*—omnivores
Blue crab *Callinectes sapidus*—carnivores, strong swimmers, valued human food item

**Molluscs (gastropods)**
Marsh periwinkles *Littoraria irrorata*—graze periphyton on marsh macrophytes
Mud snail *Ilyanassa obsolete*—graze benthic microalgae

**Molluscs (bivalves)**
Hard clams *Mercenaria mercenaria*—valued human food item
Eastern oyster *Crassostrea virginica*—passive ecological engineers, valued human food item
Ribbed mussel *Guekensia demissa*—filter-feeders, historically not eaten by humans
Blue mussels *Mytilus edulis*—valued human food item

of critical functions for these systems. Obviously they are best known to the consumer as sources of oysters for the dinner plate. However, the reef itself serves as both a source of food and a habitat for numerous organisms. An oyster reef is a solid surface in a generally soft-bottom system. Oysters, living or dead, are sites for attachment of periphytic micro- and macroalgae, and thus attract grazers of these plants. Various fishes and crustaceans utilize these reefs as habitats in which to hide and/or hunt (Harwell et al. 2011). **Rugosity** is a measure of roughness, or how well a reef is developed—well-developed and growing reefs are associated with a high degree of rugosity (Fig. 9.7a). Higher rugosity provides more habitat for mobile benthos in providing more hiding area to avoid predators and more surface areas to find periphytic food items and other prey. Also, with higher rugosity there is more space for larvae of oysters to settle and expand the reef. Oysters filter-feed, primarily on phytoplankton and resuspended benthic microalgae. They also filter and bioconcentrate bacteria; because of this, even low levels of fecal bacteria can cause closures of an area for shellfishing (in US waters the federal permissible limit of fecal coliform bacteria for shellfish waters is only 14 colony forming units [CFU]/100 ml of water).

Oysters depend on currents to bring them their food and can remove up to eight times their weight per day of seston. Studies have shown that extensive beds can significantly reduce phytoplankton and zooplankton abundance between incoming and outgoing tides or in water upstream and downstream of reefs. In a North Carolina tidal creek study (Cressman et al. 2003) there was a significant 10–25% decrease in chlorophyll *a* passing over oyster reefs and a highly variable decrease in fecal coliform bacteria by up to 45%. In experiments conducted in high-salinity tidal creeks, as well as flumes, Wetz et al. (2002) found that the presence of oyster reefs led to significantly lower chlorophyll *a* concentration than in systems without reefs. Interestingly, grazing by the oysters appeared to be selective, with significant decreases seen in photosynthetic flagellates but not in heterotrophic flagellates, ciliates, heterotrophic bacterioplankton, or cyanobacteria, with size-selection and response to chemical cues suggested as reasons for the selectivity (Wetz et al. 2002). As a measure of the regard for the oyster reef's ability to clear waters, the loss of oysters has been blamed (in part) for the increase in algal blooms (and subsequent hypoxia increases) in Chesapeake Bay. Oyster reefs as physical structures can also alter currents and cause suspended sediments to settle out of the water column (Nelson et al. 2004). Thus they clarify water by filter-feeding and by causing sedimentation of suspended solids.

Tidal creeks have high ecological value in that they attract and support abundant and diverse terrestrial animal life. Upper areas of such creeks are often characterized by terrestrial ecotones, which are zones of transition where two habitats

(a)                                                                    (b)

**Fig. 9.7** Fig. 9.7a (left) Well-developed example of an intertidal oyster reef at low tide in a tidal creek. Oyster reefs can be intertidal or subtidal depending on geographic area. Fig. 9.7b (right) Well-used game trail in ecotone between marsh and upland forest along a tidal creek. (Photos by the author.)

intermingle. Ecotones are utilized by many animal species moving between and among these habitats (Fig. 9.7b). In mid-latitudes these ecotones consist of *Spartina*-dominated salt marshes blending into *Juncus*-dominated fresh and oligohaline marshes, both of which border maritime or inland forests. Terrestrial and avian wildlife heavily utilize this "edge effect" because it gives them access to different resources found in neighboring biomes (marsh, forest, aquatic). For example, in temperate tidal creeks it is common to find well-used game trails (especially deer) wending near the upper edge of the marsh where it adjoins wooded uplands (Fig. 9.7b). The deer have access to spring water and graze in the marshes, while utilizing the woods for concealment and obtaining other food resources. Similarly other mammals such as raccoons feed on shellfish in the creek and bed down in the woods, while their predators (foxes, coyotes, etc.) will hunt the marshes and make dens in the uplands. Along rocky coastlines there is often a rapid transition from salt marsh communities to upland pine or deciduous forest in less than 1 km (Fig. 9.3b), providing even more complex animal use of various habitats.

Tidal creek areas provide habitat for a variety of birds of different functional guilds. Some species utilize adjoining woods for concealment while hunting in the shallows by either probing or aerial searching. Wading birds (herons, egrets, ibis) use shallow tidal creeks as a feeding area, and insect feeding is a key link in marsh energy transfer from

the vegetation to insects to birds (swallows, terns, wrens). Some birds nest or perch in marsh areas either in the emergent vegetation or in the woody species in the back marsh. The seed-eating redwing blackbird is the classic marsh species; also marsh wrens and clapper rails can be abundant. Ducks, grebes, and cormorants dive or duck under water for fish, animals, or plants; tidal creeks are also used as stopovers in waterfowl migrations along the Atlantic, Pacific, and Mississippi flyways. Raptors such as eagles and ospreys will use tidal creeks occasionally to catch live prey.

## 9.7 Tidal Creeks as Fisheries Habitat

Tidal creeks support some fish species as full-time residents and many more species that use the creeks either seasonally or during some stage of their life cycle (Weinstein 1979; Hettler 1989; Hoss and Thayer 1993; Ross 2003). An example of the former would be killifish or mummichog (*Fundulus heteroclitus*) which is a small-bodied generalist feeder that utilizes creek edges and the marsh surfaced at high tide. A second example is the Sheepshead minnow (*Cyprinodon variegates*) which is a mainly herbivorous resident. Common resident planktivores are Atlantic silversides (*Menida menida*) and tidewater silversides (*M. beryllina*). A variety of fishes and crustaceans that spawn offshore are carried by currents or swim into estuarine tidal creeks to feed and grow during younger stages of their lives. A commercially important schooling fish that uses the tidal creeks during a portion of its life cycle is menhaden (*Brevoortia tyrannus*). These fish feed on zooplankton as larvae and filter-feed phytoplankton and detritus as adults. They are harvested commercially as a source of fish oil for humans, as bait, and for animal feed. Other taxa that are also part-time users of estuarine tidal creeks include Atlantic croaker (*Micropogonas undulatus*), spot (*Leiostomus xanthurus*), and various flounders (*Paralichthys* spp.) which are fished commercially and for sport. Piscivorous fish of course enter tidal creeks to feed including adult flounders and bluefish (*Pomatomus saltatrix*).

Many tidal creeks and adjacent marshes are considered to be primary nursery areas (PNAs) for fisheries (Weinstein 1979; Ross 2003), a term that has both ecological meaning and legal significance in some areas. These creeks serve as nursery areas because they provide both food and protection for small and/or young fish. Streamside marshes or mangroves provide an abundance of food including microalgae, macroalgae, zooplankton, benthic invertebrates, and organic detritus. As the tide rises, the vast resources of the marshes and mangroves become available to fish and crustaceans which are able to access these resources by entering tidal creeks and the smaller rivulets that are tributary to the creek proper (Hettler 1989; Hoss and Thayer 1991; Miltner et al. 1995). Shallow creek margins appear to be areas that concentrate both fish and invertebrate prey in higher abundance relative to open channels of larger creeks (Miltner et al. 1995). Weinstein (1979) found that young fish and shellfish were particularly attracted to headwaters areas of tidal creeks, and densities decreased downstream toward larger-order creeks; several species of marine-spawned species were particularly abundant in upper creek areas. Likewise, Ross (2003) found that oligohaline creeks were particularly well utilized and that spot and croaker suffered significantly lower mortalities in such areas, especially freshwater reaches.

Tidal creek animals not only find food derived locally within such creeks but also receive potential fish food items from the terrestrial biome, such as insects and plant material washed into the creek. Sogard and Able (1991) found that salt marsh tidal creeks supported very high densities of several fish and decapod species in comparison to nearby eelgrass beds, macroalgal beds, or unvegetated estuary sites. Smaller fish can access shallower water than their larger predators can, as exemplified by intermittent or first-order tidal creeks (Hettler 1989). Additionally, vegetated marsh areas (as well as mangrove roots) provide complex structural areas in which small fish can hide from predators. The function of the PNA bears resemblance to the river floodplain function as described by the flood-pulse concept (see Chapter 6). In that concept annual flooding brings smaller fish into contact with the terrestrial food resources of the floodplain, as well as providing protection from large piscivores by shallow water and abundant cover. Many species have early life histories evolutionarily timed to maximize the benefits of floodplain flooding for their survival and growth. In an estuarine PNA the flooding is daily, and some species have life histories evolutionarily timed to swim to or be borne by currents to marshes and their creeks to maximize food and protection benefits for growth. In some states PNAs are legal designations, providing at least some measure of protection from anthropogenic impacts through the barring of point-source wastewater discharges or physically disruptive construction activities such as dredging for boat access or dock construction in these areas.

## 9.8 Economic Value of Tidal Creeks

Economically, tidal creeks have high value across a range of categories. By serving as PNAs these creeks provide an important foundation for many of the great coastal fisheries. Besides providing food and habitat for young fish and crustaceans, they are used by sport fishermen. Marine finfish such as speckled trout, red drum, and flounder are caught in estuarine creeks, as well as crustaceans, especially various species of shrimp. Fresh to oligohaline creeks are fished for largemouth bass and various sunfish and catfish. Many of these tidal creeks provide humans easy access to shellfishing beds by small boat or even by foot. Some of these areas are commercially fished, while others are used for recreational shellfishing. In some tidal creeks, beds are public and individuals can simply walk out into the creek to dig

(a)  (b)

**Fig. 9.8** Fig. 9.8a (left) Clam aquaculture at mouth of tidal creek on Virginia Eastern Shore, showing downcurrent macroalgal growth supported by excreted ammonium. Fig. 9.8b (right) Shellfish aquaculture within tidal creek on Virginia Eastern Shore. (Photos by the author.)

for clams or oysters. In others, beds are leased for harvest. Tidal creeks are also known as good water-fowling areas; many such creeks are located along the Atlantic and Mississippi flyways.

Tidal creeks in some regions have been used extensively for commercial aquaculture of various species of shellfish (Fig. 9.8). The tides bring a constant supply of feed materials, especially phytoplankton, to caged shellfish. Bivalve species most commonly raised in this manner in the eastern USA are oysters (*Crassostrea virginica*) and hard clams (*Mercenaria mercenaria*), while blue mussels (*Mytilus edulis*) are the most popular species raised in eastern Canada. Tidal creeks provide at least periodic habitat for a number of other commercially important finfish and shellfish including flounder, menhaden, blue crab (*Callinectes sapidus*), and various species of shrimp.

Real-estate values are high along tidal creeks. A major reason for this is that these systems are often open (uncanopied), creating aesthetically pleasing vistas that attract homeowners and golf course developers. A second reason is water access for boating. Many creekside single-family homes have dock structures extending over the marsh to the creek proper, and anchor boats there. In some creeks communal docks exist to serve neighborhoods. Condominium complexes may also have communal docks to serve their residents. Tidal creeks frequently serve as sites for commercial marinas. Such sites are most common in the lower reaches of these creeks where they flow into rivers or estuaries. All

of the above adds considerable monetary value to properties along or near tidal creeks. Development is most rampant along continental draining tidal creeks which have easy motor vehicle access along with the vistas. However, in some coastal areas such as Georgia, USA, individuals are building homes on hammocks (small islands) in salt marshes well away from mainland communities, and constructing bridges to these islands over high-salinity marsh creeks. Since wastewater treatment plants are not built in remote marshes, such island homes would have to rely upon septic systems for waste treatment. As detailed in Chapter 13, such on-site disposal systems function poorly in coastal areas with high water tables, and fecal pollution of neighboring waters (and shellfish) is likely to occur.

## 9.9 Anthropogenic Impacts to Tidal Creeks from Watershed Development

High-salinity salt marsh creeks are generally remote from anthropogenic pollutant sources, although some can receive pollutants from the larger sounds and estuaries if the parent systems have significant pollutant loads. But by and large it is the tidal creeks draining uplands that are the recipients of pollutants from the entire watershed that enter the creek through a network of first-order fresh and oligohaline streams, intermittent streams, and drainage ditches that often contain stormwater runoff that receives no pretreatment. Several studies of tidal creek ecosystems have shown strong statistical relationships between human development within creek watersheds (often reflected by increased watershed impervious surface coverage) and creek pollution (Table 9.4). Pollution sources are also found along immediate tidal creek shorelines. As noted in Section 9.8, tidal creek shorelines are prime real-estate for single and multifamily residential use, marinas, and golf courses. Associated clearcutting and other losses of wetland vegetation and adjacent forests reduce the buffering capacity around tidal creeks, increasing the volume and rate of stormwater inflow. The net result of these impacts is that many of these complex and productive ecosystems are anthropogenically degraded to various extents. With increasing human coastal development, existing natural tidal creeks are in danger of losing ecological value with increasing impacts.

Anthropogenic pollutants known to impact these systems include nutrients, suspended sediments, biochemical oxygen demanding materials, fecal bacteria and other microbes, pesticides and herbicides, heavy metals, petrochemicals, and other chemical contaminants. Sources of pollutants to tidal creeks are numerous. They include industrial operations both past and present (Sanger et al. 1999a, b), golf courses (Mallin and Wheeler 2000; Lewitus et al. 2003), sewage spills and leaks (Mallin et al. 2007; Tavares et al. 2008), septic systems (Lipp et al. 2001a, b; Cahoon et al. 2006), and urban and suburban runoff (Mallin et al. 2000, 2009; Holland et al. 2004; Sanger et al. 2015). All of the above sources are related to human watershed development. The relationship between watershed development and stream water quality is also discussed in detail in Chapter 15.

Elevated concentrations of fecal bacteria, viruses, and protozoans in tidal creeks are problematic to humans for two reasons. First, tidal creeks provide easy access for human contact activities such as swimming, wading, and boating (including canoeing and kayaking). Thus, in such polluted waters humans can become infected by pathogenic microbes by getting creek water in their eyes, mouth, or noses or through open wounds. Second, lower areas of tidal creeks host abundant shellfish beds that are harvested either for sport or commercially. Excessive fecal bacteria counts will lead to

closures of this resource to harvest, an economic loss as well as a local seafood delicacy loss. Of course, consuming shellfish illegally from polluted waters can lead to serious illness, and even death in some circumstances. Thus, controlling fecal microbial pollutants in tidal creeks is of great interest to coastal communities.

A number of studies have found strong positive relationships between fecal bacterial and viral indicators and the degree of watershed development (Table 9.4). Mallin et al. (2000, 2001) working in six southeastern North Carolina tidal creeks found highly significant correlations between average fecal coliform bacteria counts and watershed population ($r = 0.897$, $p = 0.039$), percentage watershed development ($r = 0.945$, $p = 0.015$), and especially percentage impervious surface coverage ($r = 0.975$, $p = 0.005$). Working in a larger set of South Carolina tidal creeks Holland et al. (2004) found a similar relationship between impervious surface coverage and fecal coliforms (Table 9.4). Sanger et al. (2015) later expanded this research to tidal creeks in three southeastern US states and came to similar conclusions using fecal coliform bacteria, *Enterococcus*, and fecal viral indicators (Table 9.4). The degree of impervious coverage where fecal bacterial contamination became problematic was as low as 10% impervious coverage. Note that lower-order tidal creeks, as opposed to larger creeks, most strongly reflected significant relationships between microbial contamination and watershed development. First-order or headwaters creeks are usually nearest to the anthropogenic sources listed above. They are also generally shallow, with weakest tidal flushing, and have fresher water (less stressful to fecal microbes).

Tidal creek nutrient cycling was discussed previously in Section 9.4. The role of watershed development in tidal creek nutrient loading is somewhat equivocal. Some studies show a statistical connection (Table 9.4), while others do not. Along the California coast the percentage watershed coverage of forest was negatively correlated with nutrient and chlorophyll *a* in tidal creeks and rivers (Handler et al. 2006). A study of tidal creeks on the Gulf Coast and the Southeast USA (Sanger et al. 2015) found significant relationships between nitrate/nitrite concentrations and

**Table 9.4** Influence of watershed impervious surface coverage on selected tidal creek ecological variables in the Southeastern USA Coastal Plain ($r^2$ = regression coefficient, $p$ = probability).

| Dependent variable | $r^2$ | $p$ | Reference and comments |
|---|---|---|---|
| **Fecal bacteria and virus abundance** | | | |
| Fecal coliforms | 0.95 | 0.005 | Mallin et al. (2001); 6 creeks, 41 stations |
| Fecal coliforms | 0.35 | 0.008 | Holland et al. (2004); 23 creeks, 1st order |
| Fecal coliforms | 0.63 | <0.05 | Sanger et al. (2015); 19 creeks, 1st order |
| *Enterococcus* | 0.35 | <0.05 | Sanger et al. (2015); 19 creeks, 1st order |
| F+ coliphages | 0.32 | <0.05 | Sanger et al. (2015); 19 creeks, 1st order |
| F− coliphages | 0.39 | <0.05 | Sanger et al. (2015); 19 creeks, 1st order |
| **Benthic macroinvertebrate impacts** | | | |
| Percentage pollution-indicative taxa | 0.34 | 0.001 | Holland et al. (2004); inverse relationship |
| Percentage pollution-sensitive taxa | 0.46 | 0.001 | Holland et al. (2004); inverse relationship |
| Percentage pollution-indicative taxa | 0.31 | <0.05 | Washburn and Sanger (2011); 19 creeks, 1st order |
| Percentage pollution-sensitive taxa | 0.27 | <0.05 | Washburn and Sanger (2011); inverse relationship |
| **Nutrient impacts** | | | |
| Ammonium | 0.31 | <0.05 | Sanger et al. (2015); 19 creeks, 1st order |
| Nitrate/nitrite | 0.30 | <0.05 | Sanger et al. (2015); 19 creeks, 1st order |
| Nitrate/nitrite | 0.25 | <0.05 | Sanger et al. (2015); 19 creeks, 2nd–3rd order |
| **Chemical impacts** | | | |
| mERMQ[1] | 0.59 | 0.0001 | Holland et al. (2004); 23 creeks, 1st order |
| mERMQ | 0.46 | <0.05 | Sanger et al. (2015); 19 creeks, 1st order |
| PAH mERMQ | 0.46 | <0.05 | Sanger et al. (2015); 19 creeks, 1st order |

[1] mERMQ compares potential cumulative effects of contaminants (see Long et al. 1995).

watershed impervious surface coverage for both lower-order (headwaters) creeks ($r^2$ = 0.51, $p <$ 0.05) and higher-order creeks ($r^2$ = 0.39, $p < 0.05$), but either weaker or no significant relationship using ammonium or phosphate concentrations. In another South Carolina tidal creek study (Van Dolah et al. 2008) focused on higher-order creeks, significant relationships between nutrients and watershed developments were generally not significant. There are some potential reasons for this disparity. First-order or headwaters creeks are usually nearest to the anthropogenic sources and generally shallow, with weakest tidal flushing. Some tidal creeks receive inputs from springs of nutrient-laden water; in some cases these inputs may come from a neighboring watershed and enter a less-developed creek (Roberts 2002). Also, tidal creek watersheds used for agriculture supply elevated nutrients to tidal creeks and rivers (Handler et al. 2006) without having urban or suburban development present. Third, because of the sporadic nature of nutrient loading, i.e. from stormwater runoff pulses and sewage spills, water-column concentration nutrient peaks can sometimes be missed by periodic sampling.

Regardless, Table 9.1 shows that the upper reaches of upland-draining tidal creeks receive the highest nutrient loads and host the densest algal blooms. Since such blooms are rare in less-developed creeks, human development, through either urbanization or agriculture, drives tidal creek nutrient loading and subsequent algal responses (Handler et al. 2006; Mallin et al. 2009). In Section 9.5.1 it was noted that such nutrient loading can lead to the presence of toxic and potentially toxic algal species. Nutrient loading also indirectly contributes to hypoxia formation in tidal creeks. Hypoxia occurs in tidal creeks, especially in upper areas, in both developed and forested creeks, and hypoxic events are most abundant in summer and fall and in warmer climates (Wenner et al. 2004). This can be a result of at least two factors. As mentioned earlier in this chapter, nutrient loading especially in upper tidal creeks can lead to phytoplankton blooms. One of the impacts of algal blooms is the creation of labile BOD as the bloom dies and decays. In several studies of tidal creeks a strong statistical correlation has been found between chlorophyll *a* and BOD (MacPherson et al. 2007; Mallin et al. 2009).

Thus, one source of hypoxia in tidal creeks is BOD from algal blooms. Additionally, in a study of several tidal creeks in North Carolina, MacPherson et al. (2007) found a strong correlation between water-column chlorophyll *a* and sediment oxygen demand. Presumably labile material sedimenting out from blooms is consumed by benthic meiofauna and bacteria, creating an oxygen demand in the sediments. Thus, nutrient loading and subsequent algal bloom responses exacerbate conditions leading to creek hypoxia. Second, work in fresh-to-oligohaline tidal creeks (Mallin et al. 2009) found statistical evidence of direct flushing of BOD materials into tidal creeks by stormwater runoff. In that study rainfall within 48 hours of sampling was positively correlated with BOD5 ($r = 0.267, p = 0.003$) and even more strongly with longer-term BOD20 ($r = 0.565, p = 0.001$). Materials carried into the creeks by stormwater runoff would include POM from yard waste and roadways, and DOC compounds. Thus, it appears that oxygen-demanding materials both are formed within the tidal creek by nutrient stimulation and algal bloom formation, and are washed into the creek by stormwater runoff.

The food webs of tidal creeks are subject to pollution from metals, pesticides, herbicides, and chemical contaminants such as PAHs and polychlorinated biphenyls (PCBs). However, since they are regularly flushed by the tides, water-column analysis of such pollutants may not be particularly revealing. As such, several studies have focused on sediment analysis to ascertain presence and impacts of such pollutants in tidal creeks. The reasons for sediment-focused research are: first, repeated inputs of such pollutants are likely to cause a buildup in the sediments, so sediment analysis provides a better long-term assessment. Second, the sediments contain infauna and support epifauna that will be in contact with such pollutants (Fig. 9.6b). Many of these benthic fauna are critical to tidal creek food webs (and fisheries by extension).

An analysis of 28 tidal creeks in the Charleston, South Carolina area revealed distinct patterns of pollutant distribution. Sanger et al. (1999a) found that sediment concentrations of the metals Cu, Pb, Zn, Cd, and Hg were significantly greater in creeks with industrialized or urbanized watersheds as opposed to forested or suburban watersheds. They also found a significant relationship between metals concentration and clay and total organic carbon content of the creek sediments. A companion study (Sanger et al. 1999b) found a similar relationship between urbanized and industrialized tidal creek watersheds and sediment concentrations of total PAHs, PCBs, and DDT. Holland et al. (2004) used the same dataset to determine the mean effects range median quotient (mERMQ) to assess the potential biologically based cumulative effect of a range of contaminants (metals, PAHs, PCBs, DDT; see Long et al. 1995). This study found a significant relationship between mERMQ concentrations and watershed impervious surface coverage for lower-order (headwaters) creeks (Table 9.4). Additionally, in a study of 15 first-order tidal creeks of varying watershed development Sanger et al. (2015) found a significant relationship between amount of impervious surface cover and mERMQ as well as PAH mERMQ which is based solely on PAH concentrations in the sediment. Finally, a study of 29 tidal creek systems in South Carolina (Van Dolah et al. 2008) found strong statistical relationships between watershed development and toxicants including PAHs, PCBs, pesticides, and metals.

Watershed development, likely through efficient delivery of pollutants, also impacts headwaters tidal creek benthic communities. Lerberg et al. (2000) found that tidal creeks with watershed impervious surface coverage > 50% were either extensively urbanized or industrialized. That study found that such creeks had high chemical pollution and had benthic communities characterized by pollution-indicative benthic species. Their benthic communities were dominated by the oligochaete *Monopylephorus rubroniveus* (in muddy sediments) and the polychaete *Laeonereis culveri* (in sandy sediments). Suburban creeks had higher benthic diversity and abundances and had fewer pollution-indicative species than creeks with forested watersheds. Several years later, Washburn and Sanger (2011) found statistically significant relationships between pollution-sensitive and pollution-tolerant groups of species depending upon amount of impervious surface cover (Table 9.4).

These studies as a whole also found that sediment pollution in first-order headwaters tidal creeks was more strongly associated with degree and type

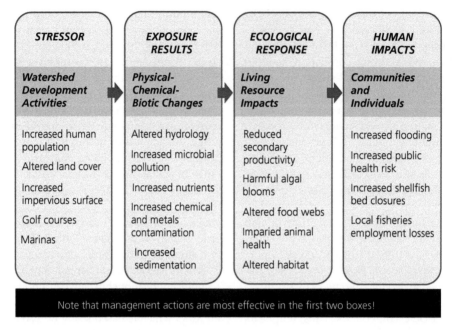

**Fig. 9.9** Conceptual model of relationships between stressors in the watershed, exposure-based changes in the tidal creek, ecological responses to physical and chemical changes, and subsequent human health and economic impacts as a result. (Modified from Sanger et al. 2015, with permission.)

of watershed development than were second-order and especially third-order creeks. Thus Holland et al. (2004) consider lower-order, headwaters tidal creeks to be sentinel systems providing early warning of ecological damage for larger creek or ecosystem pollution. In a study of 23 tidal creeks, they found that chemical signals of pollution were evident once the watershed had 10–20% impervious surface coverage, and living resources within the creeks showed degradation in the 20–30% watershed impervious surface range. Altered food webs and reductions in commercially important shrimp occurred at those levels of urbanization.

The above sets of cause-and-effect relationships can be summarized as in Fig. 9.9 (modified from Sanger et al. 2015). Actions can be taken to alter the ecological and human impacts in the latter columns. However, they are most effective in the early stages. Impervious surface coverage can be limited by governments, careful construction practices can limit hydrologic changes, and "green" golfing and "green" marinas can be mandated. Pollution, especially the non-point pollution commonly impacting tidal creeks, can be monitored

and enforced by regulatory agencies, providing political interference can be minimized. Once the resource is allowed to deteriorate significantly, it is hard and expensive to retroactively bring it back (see Chapter 16).

## 9.10 Tidal Stage and Pollutant Concentration

Finally, tidal creeks are by definition tidal, and there can be major changes in the concentrations of pollutants and other water-column constituents over a tidal cycle in such creeks. Pollutant parameter concentrations at a fixed location along a tidal creek may vary according to tidal stage. At fixed stations in a set of upland-draining North Carolina tidal creeks, salinity, turbidity, chlorophyll *a*, and fecal coliform bacteria concentration were all inversely correlated with tidal stage, and concentrations of those parameters were significantly higher at low tide compared with high tide (Mallin et al. 1999). These differences may be accounted for by several factors, including dilution by high-salinity water at high tide; the outgoing tide bringing pollutant-rich waters downstream from headwaters areas where they had been in close proximity to anthropogenic sources; and in the case of fecal coliform bacteria higher mortality in the saltier water present at high tide (Mallin et al. 1999). Similarly, in experiments in a set of high-salinity tidal creeks, Wetz et al. (2006) found that chlorophyll *a* concentrations decreased by 50% between the low and high tides, attributed potentially to tidal inputs of low chlorophyll *a* water combined with losses from grazing on the incoming tide (oysters and microzooplankton). These researchers also found that certain algal taxonomic groups (flagellates and pennate diatoms) were primarily responsible for the chlorophyll decreases. Thus, samplers need to take these changes into consideration when planning an assessment program.

## 9.11 Tidal Canals

Tidal canals are man-made systems usually built for the convenience of boaters and as "waterfront" housing (Maxted et al. 1997; Mallin and Lewitus 2004). In the USA by far the most abundant tidal canals are residential canals, which are common and abundant from the inland bays of Delaware and Maryland, within municipalities on the Outer Banks of North Carolina, in estuarine and freshwater areas in South Carolina and Georgia, and south through Florida. Houses along these tidal canals range from modest to multimillion-dollar mansions. Residential canals are especially abundant in the Florida Keys and west Florida (Gulf Coast locales). Canals are usually lined with bulkheading material, either wood, concrete, or in some cases metal sheeting. Thus, there is no littoral zone to speak of, and such associated biotic communities are therefore limited to fouling organisms. Instead, such canals usually contain numerous boat

docks with permanent or temporary anchorages for either small or large boats. Such structures may present structural habitat for fish and serve as a site for an attached algal and invertebrate community. Canals built with boaters in mind often are somewhat deeper than the estuary or river they drain into to keep sufficient water for the boats at low tide (Maxted et al. 1997; Luther et al. 2004). This depth differential, combined with the dead-end nature of many canals, makes for poor flushing of pollutants. The interception by canals of natural overland rainfall runoff may also lead to lowering of the local groundwater table.

These canals are often subject to both chemical and physical anthropogenic impacts (Lapointe and Clark 1992; Laws et al. 1994; Maxted et al. 1997). Depending on location, they can receive nutrient loads either laterally through overland stormwater runoff or groundwater inputs, or as direct discharges from boat heads. In some canal developments lawns or decorative shrubs are planted between the canal and the housing, presenting an opportunity for excess fertilizers and pesticides applied to these areas to enter the canal through stormwater runoff. Neighborhood drainage systems conveying stormwater runoff may also be directed into such residential canals through open gutters or pipes. Since canals harbor numerous boats, they are subject to inputs of petrochemicals (such as PAHs) from boat gas and oil from spillage and leaking lines and motors. Other pollutants entering tidal canals include creosote from treated lumber used for bulkheads and docks, various metals from boat paints and metal sheeting used as bulkheading, and herbicides and pesticides contained in runoff from canal-front housing areas. Additionally, many tidal canals adjoin residential communities served by septic systems; thus inorganic nitrogen is dominated by ammonia from septic leachate rather than nitrate (Tables 9.5; 9.6).

**Table 9.5** Inorganic nutrient and chlorophyll *a* concentrations in a selection of US tidal canals. Data are presented as mean (and range where available) in µg/L.

| Creek | Nitrate-N | Ammonium-N | Orthophosphate | Chl a |
|---|---|---|---|---|
| Maryland and Delaware, means of 25 tidal canals (Maxted et al. 1997) | 8.4 | 88.2 | 9.3 | 31.0 |
| Florida Keys, means of 13 canals (Lapointe et al. 1992) | 12.6 | 14.0 | 9.3 | 3.1 |
| Hawaii, Ala Wai canal (Laws et al. 1994) | (14–70) | (14–28) | (6–31) | (4–38) |

**Table 9.6** Comparison of water and sediment quality between dead-end canals and Delaware and Maryland Coastal Bay sites (as area-weighted mean concentrations ± 90% confidence interval); toxicants and metals are sediment concentrations. (Revised from Maxted et al. 1997.)

| Parameter | Coastal Bay sites | Canals |
|---|---|---|
| Dissolved oxygen (mg/L) | 6.3±0.2 | 3.8±2.0[1] |
| Nitrate (µg N/L) | 11.2±4.2 | 8.4±9.8 |
| Ammonium µg N/L) | 67.2±15.4 | 88.2±68.6 |
| Orthophosphate (µg P/L | 12.4±3.1 | 9.3±6.2 |
| Chlorophyll *a* (µg/L) | 12.2±2.0 | 25.7±7.6[1] |
| Benthic chlorophyll *a* (µg/g) | 8.1±1.4 | 31.0±16.6[1] |
| Benthic macroinvertebrates | | |
| Abundance (no./m²) | 18,724±2551 | 1917±1354[1] |
| Species richness (no./sample) | 24.2±1.2 | 3.6±2.6[1] |
| Shannon–Weiner Index | 2.73±0.1 | 0.59±0.49[1] |
| Total PAHs (ppb) | 232±92 | 2061±1103[1] |
| Total PCBs (ppb) | 2.89±1.04 | 19.8±5.5[1] |
| Chlordane (ppb) | 0.41±0.39 | 1.85±0.74[1] |
| Total DDT (ppb) | 2.15±1.09 | 3.14±2.91 |
| Endrin (ppb) | 0.04±0.02 | 0.65±0.16[1] |
| Dieldrin (ppb) | 0.13±0.07 | 1.66±1.83[1] |
| Arsenic (ppm) | 7.03±1.91 | 10.6±2.1 |
| Cadmium (ppm) | 0.14±0.05 | 0.20±0.05 |
| Copper (ppm) | 9.52±2.81 | 40.6±10.4[1] |
| Nickel (ppm) | 13.9±4.6 | 21.1±9.3 |
| Silver (ppm) | 0.05±0.02 | 0.12±0.03[1] |
| Zinc (ppm) | 64.5±16.3 | 107.9±28.9 |

[1] Significant difference between coastal bays and canals.

Fecal microbes from septic systems, stormwater runoff, or boat heads can accumulate in canals. Goyal et al. (1977) found high levels of fecal bacteria in both the water column and especially the sediments of residential tidal canals along Galveston Bay, Texas. Rainfall within 48 hours of sampling was associated with large increases in both water and sediment concentrations in that study, indicating overland runoff as a major pollution factor. In west Florida (Charlotte Harbor and Sarasota Bay) estuarine canals and bays receive fecal microbial pollution from an abundance of septic systems sited in porous soils with high water tables (Lipp et al. 2001a, b). In such coastal communities tides can influence groundwater table height and the outgoing tide actually draws polluted groundwater and associated fecal microbes into the estuarine waters (Lipp et al. 1999). The Florida Keys formerly hosted tens of thousands of septic systems and injection wells into which raw sewage was disposed. However,

the soils are karst (limestone) and very porous, and fecal microbes from human waste disposed of in this fashion moved readily through the soils into canals and coastal waters (Paul et al. 1997). These Florida Keys septic systems also served as conduits to deliver elevated concentrations of nutrients into coastal waters where they could impact sensitive seagrass beds and coral reefs (Lapointe et al. 1990). Note that in recent years many of the septic systems have been decommissioned and the residents of the Keys hooked up to centralized sewage systems in which nutrients are treated.

Maxted et al. (1997) performed an extensive study comparing dead-end canals along the Maryland and Delaware coastal bays with sites in the bays proper. Canal DO concentrations were on average 2.5 mg/L lower than the bay sites, with many canal sites failing the state DO standard (Table 9.6). Whereas there were no differences in nutrient concentrations, both water-column and benthic chlorophyll *a* were significantly greater than the bay sites, indicating rapid uptake of nutrients and good conditions for algal accumulation. The benthic community of the canals consisted of 95% annelids, whereas in the bay sites arthropods were 64%, annelids 27%, and molluscs 9% of total invertebrate numbers. Invertebrate abundance, species richness, and diversity were all significantly lower in the canals compared with the bays (Table 9.6). Total PCBs in the canal sediments were nearly seven times that of the bay sites, total PAHs were nearly nine times that of the bay sites, and in the canals average chrysene and fluorene (both are PAHs) concentrations were at levels considered unhealthful for benthic macroinvertebrates (Long et al. 1995). Most pesticides in canal sediments were significantly higher than in the bay sites, and chlordane, DDT, endrin, dieldrin, arsenic, copper, and nickel were at concentrations where adverse effects might be expected to occur in invertebrates (Table 9.6).

Dead-end canals in Delaware have been the site of large fish kills attributed to severe hypoxia and $H_2S$ buildup to toxic levels (Luther et al. 2004). Torquay Canal, adjacent to Rehoboth Bay, is about 2 m deep but has several deeper holes, and there is a 1.4-m sill at the entrance retarding circulation. Stratification accompanied by bottom-water hypoxia occurs in summer and is severe enough in the deeper holes that H$_2$S builds up to extensive concentrations. When weather events cause destratification elevated H$_2$S infuses the upper waters, especially in shallower areas, causing menhaden kills (Luther et al. 2004).

Thus, tidal canals differ from tidal creeks in that they:

- Are often dead-end systems, with consequently poor circulation.
- May be somewhat deeper than adjacent estuaries or rivers, and contain a sill, which also contributes to poor flushing and increased pollutant retention.
- Have a higher concentration of boats per linear kilometer due to densely spaced dockage behind canal-side homes.
- Have a greatly reduced littoral zone because of bulkheading.
- Have a closer proximity to direct pollutant runoff sources (lawns, gardens, streets) and lateral groundwater pollutant inputs (septic systems).

The above in sum leads to residential tidal canals being more likely than tidal creeks to suffer from stratification and hypoxia, algal blooms, elevated turbidity, fecal microbial pollution and loading, and retention of metals, pesticides, PAHs, and other chemical pollutants. While it might be argued that construction of residential canals increases the actual estuarine or riverine area of a larger, adjacent system, such canals are poor habitats for aquatic life and also serve as sources of pollutants to adjacent systems. Additionally, in areas such as the Florida Keys and the west coast of Florida numerous canals have been constructed at the expense of the natural wetland habitat, the mangrove swamp ecosystem. While regulatory agencies in the USA are required by law to protect wetlands, the poor water quality of canals requires consideration as well, with strong mitigative measure required before construction of future such systems is permitted.

Interestingly, canals at times have unexpected positive impacts to wildlife. This is best known for manatees, which in the USA are native to the Florida panhandle but constrained by cold winter weather from overwintering north of that area. Florida manatees (*Trichechus manatus*) require water

temperatures of 20°C or higher to avoid various types of stress during winter. The manatee is an endangered species that has had its problems with man, including habitat loss, eutrophication, and physical damage from careless boaters (see Chapter 7). However, it is a species that is able to overwinter successfully outside of its natural range by utilizing man-caused temperature refuges, particularly power plant thermal discharges into Florida streams. In addition to such discharge zones are passive thermal refuges (PTRs) where manatees take shelter in deeper portions of natural waters or man-made canals. One PTR of unusually high manatee abundances is in a residential development called Port of the Islands in southwest Florida (Stith et al. 2010). This is an area of extensive canal development that receives freshwater sheet flow from inland and funnels it into the estuarine canal system. As mentioned, canals have poor flushing because they are often dredged deeper than nearby estuaries and have sills that impede circulation. However, in this instance during winter a warm bottom-water layer is created by freshwater inputs flowing over the warmer, saltier water, creating a halocline and a PTR for the manatees. The magnitude of freshwater discharge controls the strength of the salinity stratification, and when such inputs are weak the halocline is lost and the PTR function deteriorates.

Canals are also built for industrial access—such as in Louisiana marshes where there is oil refining and where other chemical industries occur. Canal construction and associated spoil bank deposition contribute to wetland losses (Chesney et al. 2000; Day et al. 2000). Such canal construction can greatly change the hydrology of marsh areas, as well as increase the destructive force of hurricanes and other storms by creating a path through a natural buffer (see Chapter 17). Changes in salinity can alter plant communities and associated aquatic animal communities (Chesney et al. 2000). The presence of spoil banks from canal construction and maintenance will impede surface-building sediment inputs onto the adjacent marsh surface, and such banks will reduce marsh drainage back into the water (Day et al. 2000). This waterlogging will interact with salinity to stress the marsh vegetation. In industrial locations the presence of canals can allow pollutants leaking from ships or associated infrastructure to reach sites well away from the polluted area.

## 9.11 Summary

- There are three basic types of tidal creek: full-salinity marsh and mangrove creeks; upland-draining continental creeks with fresh headwaters and near marine lower reaches entering larger estuaries; and continental draining fresh to oligohaline tidal creeks entering rivers and estuaries. Along gently sloping shorelines such as the US East and Gulf coasts tidal creeks can stretch many kilometers inland with unconstrained reaches, while along tectonic or rocky coasts like the Pacific Northwest in northeast USA and Canada tidal creeks are shorter with much more constrained reaches, with some draining directly into the ocean.

- Full-salinity marine salt marsh and mangrove creeks are important fish and wildlife habitat and are generally far from anthropogenic sources and thus less likely to be polluted. In these creeks oyster reefs and shell and mangrove roots provide structure and habitat for algae, invertebrates, and vertebrates. Being away from upland influence, the phytoplankton and benthic microalgal communities are supported by nutrient recycling from vertebrates and invertebrates.

- Continental-draining tidal creeks have similarities to river systems, and tributaries can be designated by the Strahler stream order system. Lower-order creeks are closest to and most affected by upland pollutant sources, fresher, and flashier than higher-order reaches. Mesohaline continental tidal creeks are well studied and human influences well defined. Fresh-to-oligohaline tidal creeks draining to tidal rivers are less well studied, yet are likewise under intense anthropogenic pressure.

- Tidal creeks serve as primary nursery areas for numerous species of fish and invertebrates. Shallow areas and in-marsh areas provide food and protection from larger predators confined to deeper subtidal waters. Because of the nursery function some of these areas are provided legal protections from anthropogenic disturbances

such as dredging and point-source waste discharges.

- There are strong statistical relationships between water quality parameters and degree of human development, especially impervious surface coverage within the tidal creeks surrounding the watershed. With increasing impervious coverage there are significant increases in fecal bacteria and virus pollution, nitrate and ammonium concentrations, chemical contamination, and quality of the benthic community.

- Tidal canals are anthropogenic structures that differ from natural tidal creeks in that they are often directly adjacent to human sources of nutrients and pollutants and may be constricted by a sill. Nutrients, algal growth, metals, and chemical toxicants are concentrated and this leads to a benthic community with restricted abundance, species richness, and diversity, with some canals afflicted by fish kills.

## References

Aburto-Oropeza, E. Ezcurra, G. Danemann, V. Valdez, J. Murray and E. Sala. 2008. Mangroves in the Gulf of California increase fishery yields. *Proceedings of the National Academy of Sciences of the United States of America* 105:10456–10459.

Allen, D.M., S. Haertel-Borer, B.J. Milan, D. Bushek and R.F. Dame. 2007. Geomorphological determinants of nekton use of intertidal salt marsh creek. *Marine Ecology Progress Series* 329:57–71.

Alongi, D. 2002. Present state and future of the world's mangrove forests. *Environmental Conservation* 29: 331–349.

Arenovsky, A.L. and B.L. Howes. 1992. Lacunal allocation and gas transport capacity in the salt marsh grass *Spartina alterniflora*. *Oecologia* 90:316–322.

Badylak, S., E. Phlips, N. Dix, et al. 2016. Phytoplankton dynamics in a subtropical tidal creek: influences of rainfall and water residence time on composition and biomass. *Marine and Freshwater Research* 67:466–482.

Baillie, P.W. and B.L. Welsh. 1980. The effect of tidal resuspension on the distribution of intertidal epipelic algae in an estuary. *Estuarine and Coastal Marine Science* 10:165–180.

Cahoon, L.B., J.E. Nearhoof and C.L. Tilton. 1996. Sediment grain size effect on benthic microalgal biomass in shallow aquatic ecosystems. *Estuaries* 22:735–741.

Cahoon, L.B., J.C. Hales, E.S. Carey, S. Loucaides, K.R. Rowland and J.E. Nearhoof. 2006. Shellfish closures in southwest Brunswick County, North Carolina: septic tanks vs. storm-water runoff as fecal coliform sources. *Journal of Coastal Research* 22:319–327.

Camilleri, J.C. and G. Ribi. 1986. Leaching of dissolved organic carbon (DOC) from dead leaves, formation of flakes from DOC, and feeding on flakes by crustaceans in mangroves. *Marine Biology* 91: 337–344.

Carlough, L.A. 1994. Origins, structure, and trophic significance of amorphous seston in a blackwater river. *Freshwater Biology* 31:227–237.

Chesney, E.J., D.M. Baltz and R.G. Thomas. 2000. Louisiana estuarine and coastal fisheries and habitats: perspectives from a fish's eye view. *Ecological Applications* 10:350–366.

Cressman, K.A., M.H. Posey, M.A. Mallin, L.A. Leonard and T.D. Alphin. 2003. Effects of oyster reefs on water quality in a tidal creek estuary. *Journal of Shellfish Research* 22:753–762.

Dame, R.F., J.D. Spurrier, T.M. Williams, et al. 1991. Annual material processing by a salt marsh-estuarine basin in South Carolina. *Marine Ecology Progress Series* 72:153–166.

Dame, R., M. Alber, D. Allen, et al. 2000. Estuaries of the south Atlantic coast of North America: their geographical signatures. *Estuaries* 23:793–819.

Dame, R., D. Bushek, D. Allen, et al. 2002. Ecosystem response to bivalve density reduction: management implications. *Aquatic Ecology* 36:51–65.

Dawes, C. 1981a. Mangrove communities. Chapter 10 in *Marine Botany*. John Wiley & Sons, New York.

Dawes, C. 1981b. Salt marsh communities. Chapter 18 in *Marine Botany*. John Wiley & Sons, New York.

Day, J.W., C.A.S. Hall, W.M. Kemp and A. Yanez-Arancibia. 1989. Intertidal wetlands: salt marshes and mangrove swamps. Chapter 5 in *Estuarine Ecology*. John Wiley & Sons, New York.

Day, J.W. Jr., G.P. Shaffer, L.D. Britsch, D.J. Reed, S.R. Hawes and D. Cahoon. 2000. Pattern and process of land loss in the Mississippi Delta: a spatial and temporal analysis of wetland habitat change. *Estuaries* 23: 425–438.

Duernberger, K.A., C.R. Tobias and M.A. Mallin. 2018. Processing watershed-derived nitrogen in a southeastern USA tidal creek: an ecosystem-scale $^{15}N$ tracer study. *Limnology and Oceanography* 63:2110–2125.

Ferguson, R.L., G.W. Thayer and T.R. Rice. 1980. Marine primary producers. Chapter 2 in Verberg, F.J. and W.B. Verberg (eds.), *Functional Adaptations of Marine Organisms*. Academic Press, New York.

Gallagher, J.L. and F.C. Daiber. 1974. Primary production of edaphic algal communities in a Delaware salt marsh. *Limnology and Oceanography* 19:390–395.

Gillett, D.J., A.F. Holland and D.M. Sanger. 2005. Secondary production of a dominant oligochaete (*Monopylephorus rubroniveus*) in the tidal creeks of South Carolina and its relation to ecosystem characteristics. *Limnology and Oceanography* 50:566–577.

Goldberg, L., D. Lagomasino, N. Thomas and T. Fatoyinbo. 2020. Global declines in human-driven mangrove loss. *Global Change Biology* 26:5844–5855.

Goyal, S.M., C.P. Gerba and J.L. Melnick. 1977. Occurrence and distribution of bacterial indicators and pathogens in canal communities along the Texas coast. *Applied and Environmental Microbiology* 34:139–149.

Haertal-Borer, S.S., D.M. Allen and R.F. Dame. 2004. Fishes and shrimps are significant sources of dissolved inorganic nutrients in intertidal salt marsh creeks. *Journal of Experimental Marine Biology and Ecology* 311:79–99.

Handler, N.B., A. Paytan, C.P. Higgins, R.G. Luthy and A.B. Bocham. 2006. Human development is linked to multiple waterbody impairments along the California coast. *Estuaries and Coasts* 29:860–870.

Harwell, H.D., M.H. Posey and T.D. Alphin. 2011. Landscape aspects of oyster reefs: effects of fragmentation on habitat utilization. *Journal of Experimental Marine Biology and Ecology* 409:30–41.

Hettler, W.F. 1989. Nekton use of regularly-flooded salt-marsh cordgrass habitat in North Carolina, USA. *Marine Ecology Progress Series* 56:111–118.

Holland, A.F., D. M. Sanger, C.P. Gawle, et al. 2004. Linkages between tidal creek ecosystems and the landscape and demographic attributes of their watersheds. *Journal of Experimental Marine Biology and Ecology* 298: 151–178.

Hoss, D.E. and G.W. Thayer. 1993. The importance of habitat to the early life history of estuarine dependent fishes. *American Fisheries Society Symposium* 14:147–158.

IUBS. 1959. Symposium on the classification of brackish waters. International Union of Biological Sciences. *Archivio di Oceanographia e Limnologia* 11:243–248.

Johnson, V.L. 2005. Primary productivity by phytoplankton: temporal, spatial and tidal variability in two North Carolina tidal creeks. MS Thesis, University of North Carolina Wilmington, Wilmington.

Jones, R.C. 1980. Productivity of algal epiphytes in a Georgia salt marsh: effect of inundation frequency and implications for total marsh productivity. *Estuaries* 3: 315–317.

Kuenzler, E.J. 1974. Mangrove swamp systems. Chapter B-1, pages 346–371 in Odum, H.T., B.J. Copeland and E.A. McMahan (eds.), *Coastal Ecological Systems of the United States*. The Conservation Foundation, Washington, DC.

Kwak, T.J., and J.B. Zedler. 1997. Food web analysis of southern California coastal wetlands using multiple stable isotopes. *Oecologia* 110:262–277.

Lapointe, B.E. and M.W. Clark. 1992. Nutrient inputs from the watershed and coastal eutrophication in the Florida Keys. *Estuaries* 15:465–476.

Lapointe, B.E., J.D. O'Connell and G.S. Garrett. 1990. Nutrient couplings between on-site sewage disposal systems, groundwaters, and nearshore surface waters of the Florida Keys. *Biogeochemistry* 10:289–307.

Laws, E.A., J. Hiraoka, M. Mura, et al. 1994. Impact of land runoff on water quality in an Hawaiian estuary. *Marine Environmental Research* 38:225–241.

Lehnert, R.L. and D.M. Allen. 2002. Nekton use of subtidal oyster shell habitat in a southeastern U.S. estuary. *Estuaries* 25:1015–1024.

Lerberg, S.B., A.F. Holland and D.M. Sanger. 2000. Responses of tidal creek macrobenthic communities to the effects of watershed development. *Estuaries* 23:838–853.

Lewitus, A.J., E.T. Koepfler and J.T. Morris. 1998. Seasonal variation in the regulation of phytoplankton by nitrogen and grazing in a salt-marsh estuary. *Limnology and Oceanography* 43:636–646.

Lewitus, A.J., L.B. Schmidt, L.J. Mason, et al. 2003. Harmful algal blooms in South Carolina residential and golf course ponds. *Population and Environment* 24: 387–413.

Lewitus, A.J., T. Kawaguchi, G.R. DiTullio and J.D.M. Keesee. 2004. Iron limitation of phytoplankton in an urbanized vs. forested U.S. salt marsh estuary. *Journal of Experimental Marine Biology and Ecology* 298: 233–254.

Lin, Y., Z. He, Y. Yang, P.J. Stoffella, E.J. Phlips and C.A. Powell. 2008. Nitrogen versus phosphorus limitation of phytoplankton growth in Ten-Mile Creek, Florida, USA. *Hydrobiologia* 605:247–258.

Lipp, E.K., J.B. Rose, R. Vincent, R.C. Kurz and C. Rodriquez-Palacios. 1999. Diel Variability of Microbial Indicators of Fecal Pollution in a Tidally Influenced Canal: Charlotte Harbor, Florida. Technical Report. Southwest Florida Water Management District, Brooksville.

Lipp, E.K., S.A. Farrah and J.B. Rose. 2001a. Assessment and impact of fecal pollution and human enteric pathogens in a coastal community. *Marine Pollution Bulletin* 42:286–293.

Lipp, E.K., R. Kurz, R. Vincent, C. Rodriguez-Palacios, S.K. Farrah and J.B. Rose. 2001b. The effects of seasonal variability and weather on microbial fecal pollution and enteric pathogens in a subtropical estuary. *Estuaries* 24:266–276.

Long, E.R., D.D. McDonald, S.L. Smith and F.D. Calder. 1995. Incidence of adverse biological effects within ranges of chemical concentrations in marine and estuarine sediments. *Environmental Management* 19:81–97.

Luther, G.W. III, S. Ma, R. Trouborst, et al. 2004. The roles of anoxia, $H_2S$, and storm events in fish kills of dead-end canals of Delaware Inland Bays. *Estuaries* 27:551–560.

MacMillin, K.M., L.K. Blum and A.L. Mills. 1992. Comparison of bacterial dynamics in tidal creeks of the lower Delmarva Peninsula, Virginia, USA. *Marine Ecology Progress Series* 86:111–121.

MacPherson, T.A., M.A. Mallin and L.B. Cahoon. 2007. Biochemical and sediment oxygen demand: patterns of oxygen depletion in tidal creeks. *Hydrobiologia* 586: 235–248.

Mallin, M.A. and A.J. Lewitus. 2004. The importance of tidal creek ecosystems. *Journal of Experimental Marine Biology and Ecology* 298:145–149.

Mallin, M.A. and M.R. McIver. 2012. Pollutant impacts to Cape Hatteras National Seashore from urban runoff and septic leachate. *Marine Pollution Bulletin* 64:1356–1366.

Mallin, M.A. and T.L. Wheeler. 2000. Nutrient and fecal coliform discharge from coastal North Carolina golf courses. *Journal of Environmental Quality* 29:979–986.

Mallin, M.A., E.C. Esham, K.E. Williams and J.E. Nearhoof. 1999. Tidal stage variability of fecal coliform and chlorophyll *a* concentrations in coastal creeks. *Marine Pollution Bulletin* 38:414–422.

Mallin, M.A., K.E. Williams, E.C. Esham and R.P. Lowe. 2000. Effect of human development on bacteriological water quality in coastal watersheds. *Ecological Applications* 10:1047–1056.

Mallin, M.A., S.H. Ensign, M.R. McIver, G.C. Shank and P.K. Fowler. 2001. Demographic, landscape, and meteorological factors controlling the microbial pollution of coastal waters. *Hydrobiologia* 460:185–193.

Mallin, M.A., D.C. Parsons, V.L. Johnson, M.R. McIver and H.A. CoVan. 2004. Nutrient limitation and algal blooms in urbanizing tidal creeks. *Journal of Experimental Marine Biology and Ecology* 298:211–231.

Mallin, M.A., L.B. Cahoon, B.R. Toothman, et al. 2007. Impacts of a raw sewage spill on water and sediment quality in an urbanized estuary. *Marine Pollution Bulletin* 54:81–88.

Mallin, M.A., V.L. Johnson and S.H. Ensign. 2009. Comparative impacts of stormwater runoff on water quality of an urban, a suburban, and a rural stream. *Environmental Monitoring and Assessment* 159:475–491.

Maxted, J.R., S.B. Weisberg, J.C. Chaillou, R.A. Eskin and F.W. Kutz. 1997. The ecological condition of dead-end canals of the Delaware and Maryland coastal bays. *Estuaries* 20:319–327.

McLeod, E., G.L. Chmura, S. Bouillon, et al. 2011. A blueprint for blue carbon: toward an improved understanding of the role of vegetated coastal habitats in sequestering $CO_2$. *Frontiers in Ecology and the Environment* 9:552–560.

Miltner, R.J., S.W. Ross and M.H. Posey. 1995. Influence of food and predation on the depth distribution of juvenile spot (*Leiostomos xanthurus*) in tidal nurseries. *Canadian Journal of Fisheries and Aquatic Sciences* 52:971–982.

Nelson, K.A., L.A. Leonard, M.H. Posey, T.D. Alphin and M.A. Mallin. 2004. Transplanted oyster (*Crassostrea virginica*) beds as self-sustaining mechanisms for water quality improvement in small tidal creeks. *Journal of Experimental Marine Biology and Ecology* 298:347–368.

Odum, W.E. 1988. Comparative ecology of tidal freshwater and salt marshes. *Annual Review of Ecological Systems* 19:147–176.

Odum, W.E., C.C. McIver and T.J. Smith, III. 1982. The Ecology of the Mangroves of South Florida: a Community Profile. FWS/OBS-81/24. US Fish and Wildlife Service, Washington, DC.

Paul, J.H., J.B. Rose, S.C. Jiang, et al. 1997. Evidence for groundwater and surface marine water contamination by waste disposal wells in the Florida Keys. *Water Research* 31:1448–1454.

Pomeroy, L.R. 1959. Algal productivity in salt marshes of Georgia. *Limnology and Oceanography* 4:386–397.

Posey, M.H., T.D. Alphin, L. Cahoon, D. Lidquist and M.E. Becker. 1999. Interactive effects of nutrient additions and predation on infaunal communities. *Estuaries* 22:785–792.

Rivera-Monroy V.H., C.J. Madden, J.W. Day Jr., R.R. Twilley, F. Vera-Herrera and H. Alvarez-Guillen. 1998. Seasonal coupling of a tropical mangrove forest and an estuarine water column: enhancement of aquatic primary productivity. *Hydrobiologia* 379:41–53.

Roberts, T.L. 2002. Chemical constituents in the Pee Dee and Castle Hayne Aquifers: Porters Neck Area, New Hanover County, North Carolina. MS Thesis, University of North Carolina Wilmington, Wilmington.

Ross, S.W. 2003. The relative value of different estuarine nursery areas in North Carolina for transient juvenile marine fishes. *Fishery Bulletin* 101:384–404.

Sanders, J.G. and E.J. Kuenzler. 1979. Phytoplankton population dynamics and productivity in a sewage enriched tidal creek in North Carolina. *Estuaries* 2:87–96.

Sanger, D., A. Blair, G. DiDonato, et al. 2015. Impacts of coastal development on the ecology and human well-being of tidal creek ecosystems of the US Southeast. *Estuaries and Coasts* 38:S49–S66.

Sanger, D.M., A.F. Holland and G.I. Scott. 1999a. Tidal creek and salt marsh sediments in South Carolina coastal estuaries: I. Distribution of trace metals. *Archives of Environmental Contamination and Toxicology* 37:445–457.

Sanger, D.M., A.F. Holland and G.I. Scott. 1999b. Tidal creek and salt marsh sediments in South Carolina coastal estuaries: I. Distribution of organic contaminants. *Archives of Environmental Contamination and Toxicology* 37:458–471.

Sogard, S.M. and K.W. Able. 1991. A comparison of eelgrass, sea lettuce macroalgae, and marsh creeks as habitats for epibenthic fishes and decapods. *Estuarine, Coastal and Shelf Science* 33:501–519.

Spalding, M.D., S. Ruffo, C. Lacambra, et al. 2014. The role of ecosystems in coastal protection: adapting to climate change and coastal hazards. *Ocean and Coastal Management* 90:50–57.

Stith B.M., J.P. Reid, C.A. Langtimm, et al. 2010. Temperature inverted haloclines provide winter warm-water refugia for manatees in southwest Florida. *Estuaries and Coasts* 34:106–119.

Sullivan, M.J. and C.A. Moncrieff. 1988. Primary production of edaphic algal communities in a Mississippi salt marsh. *Journal of Phycology* 24:49–58.

Sullivan, M.J. and C.A. Moncreiff. 1990. Edaphic algae are an important component of salt marsh food webs: evidence from multiple stable isotopes analyses. *Marine Ecology Progress Series* 62:149–159.

Tavares, M.E., M.I.H. Spivey, M.R. McIver and M.A. Mallin. 2008. Testing for optical brighteners and fecal bacteria to detect sewage leaks in tidal creeks. *Journal of the North Carolina Academy of Science* 124:91–97.

Thayer, G.W., D.R. Colby and W.F. Hettler. 1988. The mangrove prop root habitat: a refuge and nursery area for fish. Pages 15–29 in *Ecologia y Conservacion del Delta de los Rios*. Usumacinta y Crijalva Memorias. INIREB-Div. Regional Tabasco, Gobierno del Estrada do Tabasco. SECUR IV. Comite Regional Conalrex. UNESCO, Paris.

Tobias, C.R., S.A. Macko, I.C. Anderson, E.A. Canuel and J.W. Harvey. 2001. Tracking the fate of a high concentration groundwater nitrate plume through a fringing marsh: a combined groundwater tracer and in situ isotope enrichment study. *Limnology and Oceanography* 46:1977–1989.

Tomasko, D.A. and B.E. Lapointe. 1991. Productivity and biomass of *Thalassia testudinum* as related to water column nutrient availability and epiphyte levels: field observations and experimental studies. *Marine Ecology Progress Series* 75:9–17.

Valiela, I., J.L. Bowen and J.K. York. 2001. Mangrove forests: one of the world's threatened major tropical environments. *BioScience* 51:807–815.

Van Dolah, R.F., G.H.M. Reikerk, D.C. Bergquist, J. Felber, D.E. Chestnut and A.F. Holland. 2008. Estuarine habitat quality reflects urbanization at large spatial scales in South Carolina's coastal zone. *Science of the Total Environment* 390:142–154.

Van Raalte, C.D., I. Valiela and J.M. Teal. 1976. Production of epibenthic salt marsh algae: light and nutrient limitation. *Limnology and Oceanography* 21:862–872.

Washburn, T. and D. Sanger. 2011. Land use effects on macrobenthic communities in southeastern United States tidal creeks. *Environmental Monitoring and Assessment* 180:177–188.

Weinstein, M.P. 1979. Shallow marsh habitats as primary nurseries for fishes and shellfish, Cape Fear River, North Carolina. *Fishery Bulletin* 77:339–357.

Weiss, J.S. and C.A. Butler. 2009. *Salt Marshes: A Natural and Unnatural History*. Rutgers University Press, New Brunswick, NJ.

Wenner, E., D. Sanger, M. Arendt, A.F. Holland and Y. Chen. 2004. Variability in dissolved oxygen and other water-quality variables within the national Estuarine Research Reserve system. *Journal of Coastal Research* 45:17–38.

Wetz, M.S., A.J. Lewitus, E.T. Koepfler and K.C. Hayes. 2002. Impact of the Eastern oyster *Crassostrea virginica* on microbial community structure in a salt marsh estuary. *Aquatic Microbial Ecology* 28:87–97.

Wetz, M.S., K.C. Hayes, A.J. Lewitus, J.L. Wolny and D.L. White. 2006. Variability in phytoplankton pigment biomass and taxonomic composition over tidal cycles in a salt marsh estuary. *Marine Ecology Progress Series* 320:109–120

CHAPTER 10

# Altering the Natural Flow

Dams and River Fragmentation

## 10.1 Dam Construction, Operation, and Flow Regulation

### 10.1.1 Importance of Dam Construction and Operation

People have been altering the natural flow of streams through the building of dams and canals for at least 3000 years (de Villiers 2000). Once agriculture began in earnest, humans built extensive irrigation systems. For example, in what is now Peru around 600 AD a warlike people called the Wari conquered and colonized a mountainous area east of present-day Lima, then proceeded to construct an expansive network of canals and aqueducts to irrigate agricultural areas on mountain terraces (Wade 2020). Their empire collapsed during a drought after some 400 years, but in the 1300s the burgeoning Incan empire used the old Wari structures to distribute water and grow crops. In the USA people constructed water wheels to power mills that ground corn and wheat to produce flour and create other products such as textiles, paper, and even whiskey (Fig. 10.1a). During the eighteenth and nineteenth centuries in what is now the Eastern USA, millponds caused by damming dotted the landscape, altering the original character of numerous streams (Walter and Merritts 2008). Besides using the power of running water to produce products, it was recognized that streams could be dammed (Fig. 10.1b) to store water that could be used to irrigate crops in times of drought, provide a steady source of drinking, cooking, and bathing water to municipalities, improve river navigation, and at the same time protect homes and crops from floods.

Eventually it was realized that the power of water could be used to produce electricity (see Chapter 12) and hydroelectric dams were built to sometimes enormous proportions. Bear in mind that use of hydroelectric power reduces the formation of greenhouse gases that are produced by coal-burning power production; i.e. it is renewable power. Regarding the latter point, Campbell and Barlow (2020) note that the recent advances in solar and wind power technology have reduced costs of those energy-generating operations to less than that of hydropower. Regarding coal or natural gas burning for power production, streams and rivers are impounded to create dependable supplies of water for industries, such as cooling water for coal and natural gas-burning and nuclear-powered electric plants (see Chapter 12). Thus, there are numerous essential uses for the impoundment of water through dam construction. However, once a river is dammed the river's hydrology, water chemistry, and biology change in many ways, both obvious and subtle. The term **river fragmentation** is used to describe some of the changes in a natural stream resulting from the building of dams and impoundments (Dynesius and Nilsson 1994).

The public often envisions dams as the huge, visually impressive concrete structures regulating rivers such as the Colorado, Columbia, Niagara, Nile, and Snake that serve as tourist attractions and are occasionally in the news (Fig. 10.2). The International Commission on Large Dams (ICOLD)

*River Ecology.* Michael A. Mallin, Oxford University Press. © Michael A. Mallin (2023). DOI: 10.1093/oso/9780199549511.003.0011

(a)  (b)

**Fig. 10.1** Fig. 10.1a (left) Harnessing water power: restored water mill works in Smoky Mountain National Park, Tennessee. Fig. 10.1b (right) Ruins of old creek dam in Cuyahoga Valley, Ohio. (Photos by the author.)

(a)  (b)

**Fig. 10.2** Fig. 10.2a (left) The 168-m-high hydroelectric Grand Coulee Dam on the Columbia River in eastern Washington State. Fig. 10.2b (right) The 220-m-high Glen Canyon Dam, confining Lake Powell on the Colorado River in northern Arizona. (Photos by the author.)

maintains a World Registry of Large Dams and considers a structural height above the foundation of 15 m or more for inclusion; the world's largest dams exceed 300 m in height. In 1940 there were 5000 large dams in the world; by 1997 there were approximately 45,000 (Black 2001). The most recent data put the number of large dams worldwide at 58,713 (ICOLD 2020). The most important reasons dams are constructed, as ranked in order, are irrigation, hydroelectric power, water supply flood control, and recreation (ICOLD 2020). Irrigation, at least on an extensive level, was evidently invented long ago by the Sumerians (de Villiers 2000). The many issues regarding irrigation will be discussed later in this chapter.

Where are the largest dams located, and why are they there? From Table 10.1 it is clear that dams are abundant in countries with large populations (requiring irrigation, power production, water storage, and flood control), nations that produce abundant crops (again, requiring irrigation), and nations with mountainous areas (to obtain high head for efficient hydroelectric power, and for flood control).

However, most dams are far smaller and can be constructed of earth, rock, or masonry of various types (Baxter 1977). Ownership of large dams is usually public (the federal government owns about 6% of the dams in the USA) or by public utilities, but over half of the many smaller dams in the USA are privately owned (Moyle 2002) by industries,

**Table 10.1** Nations hosting the greatest number of large dams (taller than 15 m) (ICOLD 2020).

| Nation | Number of large dams |
| --- | --- |
| China | 23,841 |
| USA | 9263 |
| India | 4407 |
| Japan | 3130 |
| Brazil | 1365 |
| South Korea | 1338 |
| South Africa | 1260 |
| Canada | 1156 |
| Spain | 1064 |
| Turkey | 965 |

farmers, lake associations, neighborhood associations, and other entities; some number of these have changed hands and ownership of others is unknown (Doyle et al. 2003). Particularly abundant are small dams built to create farm ponds, a steady source of water for crop irrigation. Smith et al. (2002) estimated that approximately 2.6 million small impoundments are present upon the US landscape, many for agricultural purposes.

The functional lifespan of most dams is about 60–120 years because of structural deterioration and reservoir infilling, with many dams in the USA nearing the end of their operational lives (summarized in Doyle et al. 2003). The average age of dams in the USA is > 50 years, with many considered to be hazardous (Vahedifard et al. 2017). Dam failure always has been and continues to be an important river issue that leads to loss of life, loss of homes, catastrophic property damage, and pollution (Moyle 2002). Reasons for dam failure include extreme weather events, inadequate spillway capacity leading to breaks or breaches, overtopping, structural failure of dam construction materials, instability of the foundation, internal erosion caused by seepage, cracking caused by settlement, and lack of maintenance and upkeep (Campbell and Paxson 2002; Moyle 2002; Vahedifard et al. 2017). According to the USA National Inventory of Dams, of the 88,036 state-regulated dams listed, 10,993 have a high hazard classification (i.e. loss of at least one human life is likely if the dam fails) and 10,931 have a significant hazard classification (meaning possible loss of human life and likely significant property or environmental destruction if it fails).

In recent decades new dam construction has slowed in Europe and North America due to high economic, social, and environmental costs (Graf 1999). For various reasons, including sediment infilling, public safety, lack of efficiency, or ecological issues, some dams (in the USA > 1000) are coming down (O'Conner et al. 2015). However, there are major new dam projects planned, under construction, or completed in China, India, Southeast Asia, and South America, areas of increasing populations and economic development. Much of this dam building is to provide hydroelectric power, which, on the positive side, is essentially renewable energy and results in less worldwide $CO_2$ pollution than would have occurred from coal-fired power generation. In China, massive water-related construction projects are underway and/or completed. The most famous, or infamous, is the Three Gorges Dam, discussed in Box 10.1. Other dams in China are planned, underway, or recently completed, including upstream of the Three Gorges Dam (Qui 2012). India is constructing or planning as many as 292 dams throughout the Indian Himalayas, with as many as 28 or 32 major river valleys impacted (Grumbine and Pandit 2013), with a hydroelectric power generating capacity of 126,288 megawatts (MW) planned (Bawa et al. 2010).

---

**Box 10.1  A truly massive dam in China.**

Many of the issues discussed in the text are in focus with a recent "pharaohic" scale dam project. The Three Gorges Dam on the Yangtze River in China was conceived in the 1930s, with construction beginning in 1994 and dam construction completed in 2009 at a cost of US$23–25 billion. It is 2309 m long and 185 m tall, with a generating capacity of 22,000 MW. Its major purposes were for flood control and hydroelectric power production, as well as water diversion to the arid north. At full pool the reservoir behind the dam is 660 km long; due to delays by a drought in the south resulting in releases of water downstream, the project was fully operational as of 2012. There is no passage for migratory fish, only a set of five locks for ships to traverse.

The construction of this dam and its reservoirs has had numerous social and environmental issues accompanying it. It has submerged either fully or partially 13 cities and 466 towns. It has displaced over 1.4 million people; the entire City of Kaixian (of 300,000 people) was relocated to high ground. The filling caused numerous landslides resulting in homes destroyed and fatalities; also shoreline collapses have occurred. Other geological problems including increased downstream erosion and floodplain cutoffs have occurred. Archeological sites have been lost. The dam traps vast quantities of silt that formerly contributed to marsh and island building in the river's delta. Pollution from untreated sewage and other sources becomes trapped behind the dam and algal blooms have occurred. It has caused the extinction of the river dolphin (*Lipotes vexillifer*) (the baiji) and threatened the Chinese paddlefish (*Psephurus gladius*) and Yangtze finless porpoise (*Neophocaena phocaenoides asiaorientalis*) (the jiangzhu). The dam closure caused precipitous declines in four commercially important carp species, especially their eggs and larvae. On a positive note a collaborative effort led to initiation of environmental flow releases, refined by adaptive management. As a positive response, from the years 2011 to 2015 eggs and larval counts began to show improvement to the flow manipulations. International researchers hope the adaptive management concept can be extended to the threatened species noted above, as well as reconnecting with freshwater wetland habitats

(Compiled from Mitsch et al. 2008; Stone 2008; Oster 2009; Stone 2011; Qiu 2012; Cheng et al. 2018.)

In Southeast Asia, Laos has also gone on an ambitious dam-building effort in the Mekong and its tributaries, which is predicted to severely reduce fisheries populations and impact endangered species (Stone 2016), while China has completed at least eight such facilities in the Upper Mekong, with more planned (Grumbine and Xu 2011; Grumbine et al. 2012; Kondolf et al. 2022). Cambodia has constructed six hydropower dams on the Upper Mekong, with plans (currently on hold) for up to 11 more in middle and lower reaches (Winemiller et al. 2016). While hydroelectric power facilities reduce greenhouse gas dependence, the consequences of all this dam building to the Mekong ecosystem cannot be overestimated. Impacts to the fishery are discussed below. Dams collect sediments, which are essential to building and maintenance of river

deltas—especially in the case of the Mekong where massive rice cultivation occurs. Researchers (Kondolf et al. 2022) estimate that between the massive sediment trapping by dams, sand mining in river bottoms, and sea level rise (see Chapter 17) anywhere from 10 to 90% of the Mekong Delta will be under water by 2100, a massive loss to food production. Finally, although the vast majority of European rivers have long been regulated, there are a few free-flowing rivers in the western Balkans that contain high biological diversity, especially of trout (Schiffman 2022). Unfortunately, a number of dams for hydropower are planned or under way, and local scientists and conservationists are attempting to save these free-flowing systems.

In South America, once a region of relatively pristine rivers and used by ecologists as a benchmark of such, numerous new dams are under construction or proposed, especially in Brazil, with many in areas of high fish biodiversity (Winemiller et al. 2016); and even Patagonia, a remote region known for its biodiversity, has up to seven dams planned (Vince 2010). The Amazon basin, where much research on pristine rivers had formerly been performed, has approximately 158 large dams either being constructed or in operation, with hundreds more planned (Flecker et al. 2022). Several large dams are under construction or planned along the Upper Nile River in Sudan and Ethiopia, which pose serious threats to numerous Neolithic and medieval archeological sites, along with water quantity conflicts with Egypt (Lawler 2012). Note that the numbers of dams planned (as indicated above) is constantly changing according to changes in governments, finances and international treaties—or lack thereof.

These recent and upcoming dams are invariably constructed to maximize energy "bang-for-the-buck," with little if any consideration of the numerous ecological impacts. It does not have to be this way, as argued by Flecker et al. (2022). When planning for power-generation dam location and design particularly in the Amazon basin, these researchers have proposed a "multiobjective optimization framework" using five river ecosystem services (or issues)—flow regulation, connectivity, sediment transport, fish diversity, and greenhouse gas emissions—to consider, and trade off, to minimize ecological damages while accruing no loss to proposed power generation. Flecker et al. (2022) rightly acknowledge that a basinwide rather than political boundary approach is critical, which means international buy-in—no small feat.

Most large river systems in the USA have been dammed, primarily for various reasons including municipal water supply, power generation, and agricultural irrigation. Thus, the flood pulse in these systems is greatly reduced or nonexistent. In the southeastern Coastal Plain there are still some undammed rivers since the slope is too gentle for hydropower. The basis of the flood-pulse concept was the Amazon, where, as earlier noted, a number of large tropical rivers are now being dammed for hydropower, so some of this function will be lost in these ecosystems (Vince 2010; Flecker et al. 2022).

## 10.1.2 Negative Effects of Damming and Flow Regulation

Clearly, the Earth's burgeoning population needs a certain number of large dams for the reasons listed above. Major questions remain as to whether they are all needed, or can alternatives be found? First we will examine some of the negative ecological effects of dams—as introduction to this material, a famous cautionary tale is presented in Box 10.2 that involves a number of impacts from construction of a major dam project.

---

### Box 10.2 The Aswan High Dam.

The Aswan High Dam story is the classic example of what can occur when a major dam is constructed on a river that had served millions of people over several millennia. Historically, the floodplain of the Nile River had served as the breadbasket of Egypt, with farmers depending upon the river for irrigation and a supply of nutrient-rich sediment that covered the fields following floods. The floods were controlled by rainfall well upstream in Ethiopia. Not only did the farmed fields receive sediment deposits, but also the Nile delta did as well.

Floods based on rainfall are of course notoriously unpredictable, and destructive floods occurred, along with periods of drought. To help control flooding and maintain a steady supply of irrigation water, a dam was completed on the Nile near Aswan in 1902. This 54-m-high, 1,900-m-long dam allowed for passage of sediment

downstream, but some peak floods could still overflow it. Thus, the Aswan High Dam was constructed upstream of the old dam and began impounding water in 1964; this dam was huge at 111 m tall, 3,600 m long, 40 m wide on top, and 980 m wide at the bottom. Besides regulating the floods and supplying irrigation water, the dam creates 2.1 gigawatts of hydroelectric power. The vast Lake Nasser was created behind the High Dam, displacing some 90,000 Nubians who formerly lived there.

A human health problem that arose post dam construction was increased epidemics of the disease bilharzia (schistosomiasis) following the change from seasonal irrigation water to perennial availability of water in the abundant area canals. This disease is spread by parasitic worms released from freshwater snails that then enter human hosts, and the now-perennial availability of water in canals greatly increased the contact time between humans and worms. The dam's presence has led to a chain of ecological, hydrological, and geological issues. The dam completely cut off the supply of sediment to the floodplain downstream as well as to the delta. Whereas the sediment load had supplied fertilizers to agricultural fields previously, now the farmers had to depend upon manufactured fertilizers, with use of phosphate fertilizers doubling and use of nitrogen fertilizers tripling from pre-dam years. The removal of sediment supply to the Nile Delta has led to greatly increased rates of land subsidence. Coupled with expected increases in sea level rise from climate change, this is leading to tragic consequences for the millions of people living in or depending upon the delta for sustenance. Due to the loss of nutrients reaching the Mediterranean from the sediment load, the once-massive sardine harvest plunged more than 80%, a situation that continued for 15 years. However, in the late 1980s and early 1990s the fishery in general began to improve. This improvement has been attributed variously to better fishing techniques, more nutrients from the increased use of applied fertilizers, or more nutrients from human sewage from the vast population entering the Nile and subsequently the Mediterranean. Dam building has not ceased in this region, however. Upstream of Egypt, Sudanese officials have constructed a new dam and planned a number of other dams on the Nile and its tributaries that threaten numerous archeological treasures.

(Compiled from de Villiers 2000; Nixon 2004; Bohannon 2010; Lawler 2012.)

---

Before discussing the ecological impacts of dams and diversions, it is important to understand the interaction between the river and its riparian zone in a natural situation. Recall from Chapter 6 the flood-pulse concept (Junk et al. 1989; Bayley 1995; Johnson et al. 1995). The concept states that there is a predictable seasonal advance and retraction of water on a floodplain to which the biota have evolved movement patterns. The regular flooding enhances productivity and maintains biotic diversity of the system by increasing nutrient availability, increasing access to floodplain food availability. The flooding causes expansion of the physical habitat laterally into the floodplain, which allows fish to feed and escape larger predatory fish in this physical refuge. Accordingly, floods are not a disturbance to the natural system; prevention of floods becomes the disturbance to the natural system. Fisheries yields as catch per unit effort are greater in floodplains than in lakes unconnected to the river (Bayley 1995). Besides the positive impacts to riverine productivity the flood pulse brings, an unimpacted floodplain is an area of high physical and biological diversity that is put at risk by water-control structures.

### 10.1.2.1 Interrupting the Flood Pulse

The flood-pulse concept showed how the biota of an unaltered river are adapted to and depend upon the flooding of the floodplain, which is important to healthy fish production. When a dam is constructed this regular (and irregular) pulsing is diminished or removed altogether. Since numerous organisms are adapted to the natural discharge and water levels (Dynesius and Nilsson 1994) their reproduction cycles and magnitudes are disrupted, sometimes in a major way. With access to the floodplain denied, small fish no longer have a refuge from predation, nor do they have access to the food sources of the floodplain. Inputs of riverborne nutrients to the floodplain are reduced, as is the amount of remineralization of organic nutrients on the floodplain. Upstream-sourced nutrients are no longer processed on the floodplain, but are carried downstream to the lower river and estuary, to support potentially harmful algal blooms.

For example, at risk is the Mekong River flood pulse that feeds Cambodia's Tonlé Sap Lake, a critical element in a massive freshwater fishery that yields about 2.2 million metric tons/year (Grumbine et al. 2012). The Laotian government plans multiple hydropower projects on the Lower

Mekong River that would disrupt historical flow patterns and the ecosystems and fisheries depending on them, as well as individual threatened species such as the Mekong giant catfish (Stone 2011).

### 10.1.2.2 Loss of River Corridor Function

The use of rivers for upstream and downstream movement by anadromous and catadromous fish is the most well-known example of rivers as movement corridors. However, numerous mammals, birds, and invertebrates also move upstream and downstream either during a life cycle or simply to forage. Once a dam is constructed the ability of a given river to serve as a corridor for aquatic and terrestrial wildlife is reduced. With small dams this wildlife corridor function may be reduced only incrementally, but large dams make effective barriers. These impacts collectively lead to localized species extinctions or fragmentation of populations.

To account, at least in part, for the disruption of anadromous fish movement dam operators have conceived various schemes and alterations such as building fish ladders (Fig. 10.3a), trapping and transferring fish upstream through boat locks (Fig. 10.3b), and constructing various types of fishways around the dam.

### 10.1.2.3 Loss of Pollutant Filtration Function

One of the ecological roles of riparian wetlands is to serve as a filter for human-impacted waters flowing downstream and for impacted waters entering the river laterally from urbanized or agricultural areas. Extensive riparian wetlands assimilate nutrients and convert them to vegetation and eventually higher biota. Wetland areas in the riparian zone enhance denitrification. Suspended sediments collect in riparian wetlands and form new substrata for vegetation growth. Metals are adsorbed by organic soils, and chemical toxins are diluted or microbially degraded in wetlands. The riparian zone also functions to reduce downstream flooding. However, following dam construction the downstream riparian zone thins, with removal or reduction of the flood pulse. With dam construction, particularly larger dams, less river water reaches the riparian zone,

(a)    (b)

**Fig. 10.3**  Fig. 10.3a (left) Fish ladder associated with dam on North Umpqua River, Oregon. Fig. 10.3b (right) Anadromous fish have been transferred upstream in locks, with limited success. (Photos by the author.)

wetlands dry up, and these pollution reduction and flood prevention functions diminish.

### 10.1.2.4 Loss of Physical Habitat Creation Downstream

The normal flooding process in a river continually destroys older habitat by washing it away, but also creates new habitat by the deposition of sediments on newly built bars. Such sediment is the source of new habitat for pioneering plant taxa such as cottonwood trees, and for invertebrates and vertebrates as well. In unaltered rivers, the channel tends to migrate over the floodplain in broad unconstrained reaches. The natural aquatic habitats in unaltered floodplains include the main channel, dead-end side channels, abandoned meanders (oxbow lakes), tributaries of various sizes, springs, marshes and other wetlands, and temporary pools. Such habitats will have differing algal, macrophyte, and woody vegetation associated with them and ecotones (transitional zones) connecting terrestrial areas with such varied aquatic habitats (Ward et al. 1999). The result is a mosaic of rich ecological complexity, which supports high ecological diversity. Due to storms and droughts there is also periodic disturbance that further enhances organism diversity and productivity, especially of successional species (Gregory et al. 1991; Ward et al. 1999). Oxbows serve as nursery areas for fish, with extensive movement from the channel back and forth to the oxbow habitats (Penczak et al. 2003).

Construction of a dam reduces or removes the flow of sediment downstream. This reduction of sediments and water causes habitat destruction and less creation of new habitat, impacting organism productivity. Dams greatly reduce downstream channel migration over the floodplain by factors of three to six, resulting in less new habitat creation, such as pioneering forest on newly built bars, and less creation of aquatic habitat, such as oxbows and backwaters (Gregory et al. 1991; Shields et al. 2000). This also results in loss of biodiversity (fish, birds, other taxa) that immigrate into or use such habitats. Dams reduce the frequency of storm or drought disturbance on the downstream environment, which will lower biodiversity especially of pioneering species. The reduction of the flood pulse causes riparian habitats to dry up. Overall, river flow regulation by dams, levees, and other channel stabilization structures tends to reduce habitat complexity, diversity, and fish harvest (reviewed in Shields et al. 2000).

Not only is physical habitat lost but also access to other aquatic habitat. Construction of upstream storage reservoirs on the Danube and Missouri Rivers trapped large amounts of sediment, allowing the rivers below the dams to scour away and deepen the channels, reducing the access of the main channel to the floodplain, with subsequent loss of wildlife habitat (Sparks 1995). Thus, a dam can reduce the flood pulse both by regulating floodwaters and by deepening the downstream channel.

These impacts collectively lead to localized species extinctions or fragmentation of populations.

A large-scale satellite-based study (Dethier et al. 2022) found that dams made major impacts on sediment fluxes to coastal areas. In the global north there have been major sediment decreases to coastal waters due to dam building from the mid-twentieth century on. This has negative implications for marsh building in lower estuaries, especially in light of global warming and sea level rise. In the southern hemisphere there was increased sediment discharge due to deforestation, mining, and palm oil plantations. However, note that with the current massive focus on hydropower dam building in the global south that sediment discharge may greatly decrease.

### 10.1.2.5 Dam Releases Alter Natural Flow Regimes

Water releases from impoundments are done for anthropogenic reasons, such as providing hydroelectric power, providing water downstream during droughts, and releasing excess water building up within the reservoir. However, the natural river biota is adapted to seasonal variations in flow. For example, spawning success of anadromous fishes (such as salmon, sturgeon, striped bass, and shad) depends on proper meteorological/hydrological conditions for maximum survival. In the Roanoke River system of Virginia and North Carolina, releases from upstream dams altered natural flow regimes, impacting striped bass production. Rulifson and Manooch (1990) determined that recruitment of this species was poor when downstream flow was either high or very low, but best at low to moderate flows, which happened to best mimic average flows of the pre-dam conditions. Flow-related factors that impact migratory species include seasonal timing, location of spawning, daily and hourly spawning activity patterns, larval transport and feeding, abundance of zooplankton for larval feeding, and location of nursery grounds (Rulifson and Manooch 1990.

Not only does alteration of the river corridor function and major changes in flow impact productivity of migratory fishes, but in the case of salmon there may be watershed-scale ecological productivity consequences in terms of nutrient translocation.

In the Pacific Northwest, following upstream migration and spawning, salmon carcasses become major sources of nutrients to some lotic upstream ecosystems, i.e. translocation of nutrients (Schindler et al. 2003). With decreased passage of migrating salmon, less fertilization can mean lower primary and secondary production that can impact higher trophic levels.

Hydroelectric power dams do not produce the toxic chemicals associated with fossil fuel generation (covered in Chapter 12). However, large hydropower dams, as noted, have a number of negative ecological impacts to rivers. With smaller hydropower facilities, such impacts are lessened. However, even with small dams, water flows are altered downstream because discharges change throughout the day depending upon electrical power needs, a process called hydropeaking. Research by Kennedy et al. (2016) found that such flow alterations lead to drying and large mortalities of attached insect larvae (filter-feeders). Because such larvae are critical food sources to higher animal life, such flow alterations can impact fisheries populations.

### 10.1.2.6 Freshwater Dams Can Impact on Marine Fisheries

Dams and water control upstream can impact marine fisheries as well as freshwater (summarized by Drinkwater and Frank 1994). If upstream consumption reduces river discharge it may lead to increased pollutant concentrations entering the estuary and more rapid sedimentation deposition and estuarine toxins buildup. Water usage upstream can impact interactions between estuarine-spawning organisms and critical flows needed for oxygenation or salinity requirements. If too much water is released during spring or summer the added nutrient pulse can increase estuarine algal bloom frequency and related hypoxia issues. Again, with less floodplain interaction nutrient removal through uptake or denitrification is reduced and the excess nutrients can stimulate harmful algal blooms downstream.

As noted, vast amount of formerly suspended sediments are stored behind large dams—this deprives estuarine marshes of substrata needed to grow new marsh upon and keep pace with rising seas and subsiding marshes (Box 10.1 and Box 10.2). Not only is sediment storage by large dams important, but the numerous small dammed ponds collectively retain more suspended sediments that large impoundments do (Smith et al. 2002; Jackson and Pringle 2010).

### 10.1.2.7 Dams Can Degrade or Alter Water Quality

The presence of a dam and the water releases associated with it can considerably alter the natural water temperature regimes (Baxter 1977). The reservoir can stratify, with warmer water above and cool (or cold) water below. Dam releases from the hypolimnion can send colder water downstream, altering biotic communities. If water coming over a spillway or through a turbine is supersaturated with gases, it can cause potentially fatal gas bubble disease in downstream fish (Baxter 1977).

The presence of dams can lead to downstream eutrophication not only by allowing more nutrient movement but also by altering nutrient ratios. The impact of dams constructed on the Danube River and other tributaries on Black Sea deterioration is an example. Excessive consumption of reservoir water for irrigation, industry, and municipalities, combined with temporal changes in river discharge due to power production usage reduced marsh flooding and altered natural salinities, impacting the fisheries (Drinkwater and Frank 1994). Nutrient concentrations increased from discharges of agricultural runoff. At the same time N and P increased due to runoff and Si discharges decreased due to retention behind the dams (Humborg et al. 1997). The retention of Si behind dams can be due to Si associated with clay suspended sediments settling out or Si assimilated within diatoms settling out. With elevated N and P driving diatom blooms behind dams, this would accelerate the uptake of Si and subsequent settling upon cell death (Humborg et al. 2000). The reduction in Si:N:P ratios altered phytoplankton composition in the Black Sea from diatom-dominated to coccolithophores and flagellates, and the N and P increases led to excessive blooms and BOD-driven anoxic conditions, severely damaging the fishery (Drinkwater and Frank 1994; Humborg et al. 1997). Mechanistic models predict that the proliferation of large dams will increasingly

reduce Si and P relative to N, increasing P limitation in some coastal waters and, again, "starving" Si-limited diatoms, the critical base of many coastal food chains (Maavara et al. 2020).

### 10.1.2.8 Dams and Fault Lines Don't Mix

New reservoirs can lead to earthquakes (Baxter 1977). For example, in China the reservoir behind the Zipingpu Dam (which is located 500 m from the Beichuan Fault) was filled in 2006 with several million tons of water. A massive earthquake occurred in May 2009 in Sichuan Province, killing 80,000 people. The dam was located just 5.5 km from the quake's epicenter (Kerr and Stone 2009; Qiu 2012). Reservoir-triggered quakes have occurred in India and elsewhere as well. Localized landslides and collapses have also been associated with large dam construction (Box 10.2).

### 10.1.2.9 Dams Can Impact Human Health

There are occasional health and disease problems associated with dams, particularly in developing nations (Baxter 1977). While quality of life may improve in various ways due to dam functioning, incidences of waterborne diseases can increase in villages near dam sites; incidence of disease falls the farther the village is from the dam site (Tetteh et al. 2004). Such diseases include malaria, schistosomiasis (Box 10.1), infectious hepatitis, skin diseases, and diarrheal diseases, as well as ailments resulting from cyanobacterial toxins thriving in quiescent waters within impoundments (Black 2001; Tetteh et al. 2004). The flooding of boreal forest soils by damming and reservoir creation causes the release of mercury (Hg) into the overlying water. From there, biomagnification up the food chain can lead to Hg contamination of fish and higher predators (Drinkwater and Frank 1994).

### 10.1.2.10 Dams Can Have Negative Socio-Economic Impacts

There are numerous socio-economic costs to local communities in addition to ecological problems resulting from construction of large dams. It has been estimated that 40–80 million people have been displaced from their homes by dam projects (see Black 2001), and that is old data. Such individuals face issues with finding new employment or trades, loss or reduction in income, and emotional upheaval for families uprooted from their traditional home areas, particularly among rural populations in developing nations (Box 10.2). Regarding heritage, archeological sites are buried and lost under massive amounts of water (Folger 2008; Lawler 2012).

### 10.1.2.11 Dams Are Expensive to Build

Most dam projects have cost overruns, and dam repair costs, especially for large dams, can be high (Black 2001). Some dams come to the end of a useful existence and upkeep or upgrades become more expensive than removal of the dam (Doyle et al. 2003).

### 10.1.2.12 Dams Can Lead to Conflicts

Dams and water consumption upstream can lead to serious international conflicts because of the effects downstream. Rivers crossing borders have always led to upstream–downstream user conflicts, especially concerning major withdrawals. Upstream users may keep too much water for themselves, or change the natural flow patterns which will impact agriculture, wildlife, and other activities to downstream users. For example, in Southeast Asia some downstream nations blame China's operation of several hydroelectric dams on the Lancang Rivers, which becomes the Mekong River south of China, for many impacts. These include lowering the Mekong depth so that tour boat and cargo boat operations are greatly reduced, changing flow patterns so that fish catch downstream is reduced, causing a loss of sediment downstream that leads to bank erosion, and possibly impacting spawning of the Mekong giant catfish (Stone 2011). Of course, drought periods will exacerbate such conflicts. More on this matter is presented below regarding interbasin transfer.

### 10.1.2.13 Associated Structures

Campbell and Barlow (2020) make a solid case that when controlling impacts, it is important not to ignore the collective ecological impacts of the numerous smaller but abundant structures on floodplains and along river tributaries that retain or alter water flows. These include irrigation works, flood control devices, and small weirs.

## 10.1.3 Dam Removal

### 10.1.3.1 Positive Impacts

In recent years dam removal has become more frequent, although primarily for small dams. Sometimes removal occurs because a hydroelectric dam has become outdated and no longer needed; other times dam removal is necessary as a safety issue due to dam aging and deterioration (Doyle et al. 2003). Sometimes dam removal occurs to counter some of the negative ecological (especially fisheries) effects discussed above. Positive effects of dam removal on the river ecosystem include the following:

- Return of the flood pulse, which allows seasonal flooding to inundate the historic floodplain. This would allow young and otherwise smaller fish access to the food resources of the floodplain, as well as protection from predators in the complex habitat. This would in effect be an expanded nursery area for many species, which would increase fish production and other associated wildlife. Data demonstrate that higher fish yields are found in river floodplain water bodies than in the channel proper (Bayley 1995).
- Improvement or restoration of anadromous fish passage: with the dam gone, such fishes can move unimpeded upstream to spawning areas. There have been a number of success stories regarding anadromous fish repopulation of formerly inaccessible river reaches, two examples of which follow. In 1998 the Quaker Neck Dam was removed from North Carolina's Neuse River, restoring access to more than 120 km of main channel spawning habitat and 1488 km of tributary spawning habitat to anadromous fishes (Burdick and Hightower 2006). An egg and larval fish study performed in 2003 and 2004 demonstrated considerable anadromous fish spawning activity well upstream of the old dam site, especially for American shad and striped bass (Burdick and Hightower 2006). In 1999 the Edwards Dam on the Kennebec River in Maine was removed. This resulted in alewife, Atlantic salmon, striped bass, and American shad rapidly returning to former upstream areas (Doyle et al. 2003; Postel and Richter 2003).

- Dam removal will return the river corridor function to nonanadromous organisms. This would allow restoration of a more natural river discharge periodicity which would help flow-dependent freshwater and marine fish and invertebrates.
- Following dam removal, rivers appear to recover rather rapidly, at least hydrologically and geophysically, to something approaching pre-dam state (O'Conner et al. 2015). In an example of biological improvement, Tuckerman and Zawiski (2007) reported on the impacts of dam removal in the upper regions of the infamous Cuyahoga River in northern Ohio. Rather than mandate expensive waste discharge tightening, local stakeholders decided to remove the Munro Falls dam and alter the historic Kent dam to allow for natural flow around it. Within two years significant increases in DO were documented, as well as notable improvements to the river fishery.
- Dam removal as part of river restoration schemes would help the recreational desirability of a river (i.e. ecotourism)—though whitewater rafters/kayakers may prefer dam-generated rapids.

### 10.1.3.2 Potential Negative Impacts

Dam removals need careful consideration of multiple factors. In some cases dam removal can cause new problems to the river ecosystem. Unfortunately, while there are good data regarding the effects, good and bad, of small dam removal, there are too little data available on impacts of large dam removal (Doyle et al. 2003).

- Since dams considered for removal are likely to have been in place decades or even centuries, they have been accumulating sediments behind them, as well as whatever is associated with those sediments. Sediments eroded from rural areas may have legacy pesticides associated with them, as well as nutrients from fertilizers. Sediments eroded from construction areas may contain petrochemicals and metals. Sediments sourced from urban areas may contain PAHs, metals, PCBs, and nutrients from fertilizers. As such, the fate of toxicants that are built up in the sediments behind dams may be a problem to wildlife (Stanley and Doyle 2003). For example, the dumping of PCBs into the Hudson River (see Chapter 12) caused a buildup of these toxins behind dams, and dredging activities had to concentrate there to remove contaminated sediments. The Ft. Edward Dam was removed on the Hudson in 1973 and PCB-laden sediments were carried to downstream reaches (Doyle et al. 2003). Along the Dan River in North Carolina and Virginia, following the accidental discharge of stored power-plant ash pond wastes in 2014, sediment contamination was heaviest behind dams (see Chapter 12), so again dredging had to be targeted in those areas. Even in areas where acute toxin spills have not occurred, the buildup of pollutants from upstream erosion and stormwater runoff can occur behind dams (Doyle et al. 2003). Thus, such areas require sediment toxicity testing before dam removal should be undertaken.
- Dam removal may release stored nutrients from sediments behind the dams and lead to eutrophication symptoms downstream, such as blooms of filamentous algae (Stanley and Doyle 2003). Phosphorus and ammonium adhere to particulate matter including suspended sediments (see Chapter 2) and will accumulate behind dams. If a dam is removed nutrients associated with stored sediments will be carried well downstream, even into estuarine waters, potentially stimulating algal blooms.
- The release of stored sediment alone will have consequences, even if such sediments have low pollutant concentrations. For instance, following dam removal, the rapid increase in downstream sediment loading could severely impact endangered species, such as mussels, many species of which require clean waters for survival and propagation (Doyle et al. 2003). If the released sediments cloud the water and reduce macrophyte photosynthesis, or physically cover macrophyte beds, fish habitat is lost. If a large dam storing massive amounts of sediments is breached, downstream ship channels can be affected, requiring additional dredging. The removal of the aforementioned Ft. Edward Dam

led to sediment infilling of downstream canals (Doyle et al. 2003).

- If the dam in question supports a large reservoir upstream of it, removal of the dam and subsequent dewatering of the reservoir will rapidly change the aquatic environment, resulting in stranded fish, washout of lentic species, and massive local mortality (Stanley and Doyle 2003).

- There is no guaranteeing that, once a dam is removed, the river downstream will return to its original configuration over the floodplain (Doyle et al. 2003), although rapid recovery has been noted in some situations (O'Connell et al. 2015). Over the years the dam was in operation the upper watershed may have changed dramatically due to anthropogenic development, or the downstream environment may have been altered substantially.

- Non-native species are continually being introduced into riverine environments (see Chapter 14). If such species are introduced into upper or even headwaters stretches, or from upstream tributaries, then they will spread downstream over the dam. However, if such species including fish or snails are introduced into lower rivers, they will need to travel upstream to spread into new habitats. The presence of a dam may serve as a barrier to keep undesirable introduced species from entering new stretches of river. In some circumstances levees can protect sensitive native species that live or reproduce off-channel in riparian wetlands by protecting them from interaction of invasive species in the main channel (Jackson and Pringle 2010). Besides animals, riparian areas are corridors for dispersal of plants (Gregory et al. 1991). If undesirable invasive plant species enter a watershed, a dam may provide an effective barrier to further spread up- or downstream.

## 10.1.4 Dam Alteration

Dam alteration can improve ecosystem function without going through the process of dam removal and its potential negative consequences. Installation of "rock arch ramps" is a type of dam alteration that was developed in the US Upper Midwest to improve fish passage for anadromous species on low-head dams. This process consists of piling up rock rubble to create a slope, with passageways between rock "vanes" to allow fish movement, and created "pools" behind large boulders to allow fish to rest on the way up. This concept has recently been transferred to the US Southeast (Fig. 10.4), where anadromous species include striped bass, shad, and endangered sturgeons. For example, in the Cape Fear River a system of three lock and dam structures (named Lock and Dam 1, 2, and 3) was built between 1915 and 1935 to aid commercial boat passage upstream. This led to a drastic loss in anadromous species that could no longer pass upstream; 50% of the anadromous fish were blocked at Lock and Dam 1 alone. The

(a)
(b)

**Fig. 10.4** Fig. 10.4a (left) Fishermen at Lock and Dam #1 on Cape Fear River before ramp construction. Fig. 10.4b (right) Lock and Dam #1 after construction of the rock arch ramp to improve fish passage. (Photos by the author.)

locks were subsequently modified so the lockmaster could "lock" anadromous fish through (Fig. 10.3b)—which helped somewhat but still only aided a fraction of the fish. In 2009 a rock arch ramp was begun. Before such construction occurred it was found that water cascading over the 11-ft dam scoured a hole 40 ft deep at the base of the dam—a destabilizing process. Workers had to dump 16,000 cubic yards of rock into the scour hole to stabilize the lower dam area first. The rock arch, with a 20° slope, was completed in 2012 that allows fish to migrate upward across the face of the dam and over it.

If a dam has stored large amounts of sediment over the years, the downstream river can be sediment-starved, with consequences as discussed in Section 10.1.2. The stored sediments also clearly shorten a dam's working life by infilling. Stored sediments can be dredged and trucked out downstream or elsewhere (Brainard 2011). Or a dam can be altered to permit sediment passage. A method called sluicing can funnel sediment past a dam through pipes in a deliberate manner so as not to disrupt downstream ecosystems (Brainard 2011).

## 10.2 Flow Diversion and Consumption

Flow regulation is the alteration of natural river seasonal and directional flow patterns that comes about from reservoir operation, as well as interbasin transfer and irrigation. As of 1994 it was estimated that about 60% of the world's streamflow was regulated and about 79% of total water discharge in North America, Europe, and the old Soviet Union was regulated (Dynessius and Nilsson 1994). Shumilova et al. (2018) calls the largest of these efforts water transfer megaprojects, with of future planned projects 42 are for agricultural purposes, 13 for hydroelectric power, and others for mining, navigation, and even ecosystem restoration. As a rather extreme recent example of flow alteration and regulation, in 2021 and early 2022 China did extensive water engineering to host the 2022 winter Olympic Games, building reservoirs and vast pipeline systems to carry water from distant locations to the sites near Beijing, where it was made into snow for artificially carved ski runs, as well as planting entire forests and covering them with artificial snow to create a winter ambience (Myers et al. 2022).

### 10.2.1 Interbasin Transfer

Interbasin transfer is the act of transferring water from one river system to another, or through a man-made diversion to the sea. This occurs worldwide, and in the USA in multiple locations. Generally, a burgeoning community requires more water for municipal uses and increases pumping from the most convenient river. However, wastewater from that community may be released from treatment systems into a different river, resulting in a net loss to the source water river. This has a number of implications. This process of course leaves less water for downstream consumers, which may become a serious issue should prolonged drought occur. Concerning water quality, downstream point source dischargers will continue to discharge the same amounts of waste into the now-shrunken river, and pollutant concentrations will become magnified with less water for dilution. Lower flow volume will result in lower river discharge, which causes suspended sediments to settle out sooner, decreasing light attenuation. In nutrient-rich rivers the clearer water combined with the elevated nutrients may stimulate nuisance or toxic algal blooms to occur. Also, as was discussed regarding dam construction, the decrease in suspended sediments carried downstream will leave less incoming sediment available for marsh building in the estuary. Regardless, interbasin transfer continues to be contentious between cities, states, and even nations.

A massive-scale interbasin transfer is underway in China. There the government has the goal of transferring 45 billion $m^3$ of water annually from south China north to the Beijing area to relieve water shortages in the north. This will utilize tunnels, canals, and blasting and pump stations. There is a high monetary cost to the project (US$77 billion) and numerous environmental risks, and such a transfer does not solve existing water pollution problems. However, China has embarked on a water conservation program to reduce usage in irrigated agriculture (Liu and Yang 2012).

Sometimes river diversions are done for sound ecological purposes. For instance, in 1991 a portion of the Mississippi River was diverted to reduce Gulf of Mexico wetland losses caused by river armoring. This was the Caernarvon Freshwater Diversion (CFD), diverted into the formerly isolated Breton Sound Estuary. Results of follow-up studies indicated that river diversion appears to be an appropriate and useful managerial tool that, while building marshes through sediment loading, has the ecological benefits of increasing production of key fisheries species and creating wetland nursery habitat (de Mutsert and Cowan 2012).

## 10.2.2 Water Consumption as Flow Loss

Water is consumed (i.e. removed from its source without replenishment) through daily human uses such as drinking, cooking, bathing, and cleaning; industrial processing of many kinds; and crop irrigation. Irrigation water that is not immediately used by crops is lost or "consumed" through evaporation from soils or irrigation canals or through transpiration by crops. Irrigation accounts for approximately 90% of global water consumption, with India, Pakistan, China, and the USA accounting for 36, 15, 13, and 8% of global irrigation consumption, respectively (West et al. 2014). Water consumption also occurs when a dam is built and the reservoir behind it is filling; such an action can be exceedingly threatening to downstream river users in a drought (Bagla 2010). Irrigation consumption alone may reduce large portions of the natural flow and disrupt the downstream ecosystem. Currently there are major problems related to irrigation in the western USA, and these problems continue to get worse as the climate changes (Pederson et al. 2011). To grow the vast crops in California's Central Valley, water is conveyed hundreds of kilometers through the California Aqueduct and other canals to irrigate an otherwise arid region (Fig. 10.5a). Water is diverted through the All-American Canal from the Colorado River to irrigate California's Imperial Valley (Fig. 10.5b; and see below). Interestingly, the Imperial Valley then sells

surplus water (available through implementation of irrigation conservation techniques) to San Diego (De Villiers 2000).

The story of the Colorado River in the arid Western USA exemplifies many of the cross-state problems, cross-border problems, improbable urban development scenarios, and imprudent water-distribution decisions that have occurred worldwide where irrigation water is scarce, including massive river diversions to support agriculture and municipalities well outside of its watershed. The mainstem Colorado arises in the mountains of its namesake state, then flows southwesterly through Utah, Arizona, Nevada, and California, before crossing the border into Mexico, where it, or what is left of it, enters the Gulf of California. Major tributaries enter it from Wyoming and New Mexico along the way. It was completely wild and untamed in 1869 when Major John Wesley Powell led the first scientific and exploratory expedition down the Green and Colorado Rivers through the Grand Canyon and followed up with a second expedition in 1871–1872. Major Powell was a renaissance man—a soldier, scientist, linguist, author, and explorer who later headed the United States Geological Survey for 13 years. In the Civil War he was a Union artillery commander who lost an arm to a Confederate bullet at Shiloh, yet returned to active combat duty for several more battles. Always possessed of a questing mind, Major Powell was said have foraged for geological specimens and fossils in the trenches during the Siege of Vicksburg. After the war he was a college geology professor in Illinois before undertaking several rugged western expeditions for scientific purposes, despite having a single arm. He took time to learn local Native American tribal languages and befriended them. As a scientist, he reported that much of the West was too arid to be developed, and what was to be developed should be done on a watershed approach (clearly well ahead of his time). However, then as now, politicians often sponsor municipal and agricultural development for reasons having nothing to do with science or logic. The latter attitude has led to the improbable scenario of major western cities including Las Vegas, Phoenix, Tucson, and San Diego growing ever larger in what is basically desert

(a)                                                         (b)

**Fig. 10.5** Fig. 10.5a (left) The California Aqueduct brings water hundreds of kilometers south from northern rivers through arid lands to irrigate the Central Valley's agriculture and supply water to southern California urban and industrial areas. Fig. 10.5b (right) Irrigation from the Colorado River makes California's arid Imperial Valley crop-rich, with sharp contrast with adjoining unirrigated land. (Photos by the author.)

terrain. The Glen Canyon Dam (Fig. 10.2b) retains Lake Powell (named for Major Powell, who had originally named Glen Canyon for its geological features) and the Hoover Dam retains Lake Mead. Lake Powell, upstream of the Grand Canyon, and Lake Mead, downstream of the Grand Canyon, are the two largest reservoirs in the USA and supply water to 25 million people (Folger 2008). The Colorado watershed is divided into an upper basin of Colorado, Wyoming, New Mexico, and Utah, and the lower basin of Arizona, California, and Nevada— each basin is supposed to receive half of the annual flow (Folger 2008). A small amount is doled out to Mexico, by treaty; the decrease from historically much larger flows has shrunk the marshes and backwaters of the estuarine delta (Blinn and Poff 2005). Regarding these huge western reservoirs, Graf (1999) coined the term "hydrologic colonialism" to refer to exporting huge volumes of water for use elsewhere and generating large amounts of hydroelectricity for use elsewhere, while retaining the environmental costs in the local river basin. A prime example is the Imperial Valley in southern California, which produces vast amounts of vegetables that feed the USA particularly when winter comes to northern states. The Colorado River feeds the Imperial Valley through the 82-mile All-American Canal, making this area the single largest user of the Colorado River (Folger 2008).

There are 40 large-flow regulation structures along the Colorado, retaining and diverting massive amounts of water (Blinn and Poff 2005). The dams and diversions have radically changed the native flora and fauna of the system. Water temperatures are altered by dam releases, sediment is retained and thus new habitat is not created, and water quality has changed through inputs of runoff and evaporation from the reservoirs. Of the 49 native fish species in the system, 4 are extinct and 40 more are considered threatened or endangered due to habitat changes and competition from introduced non-native species, and non-native plant species invade the altered habitats of the riparian areas (Blinn and Poff 2005). On a positive note, federal agencies perform periodic high-flow dam releases on the Glen Canyon Dam in order to supply sediment to the downstream areas to create new habitat and mimic, for a time, natural flows (Cornwall 2017). Unfortunately in 2021 and 2022 Lakes Mead and Powell faced unprecedented drops in water level due to climate-change-induced droughts (see Chapter 17).

Besides the American West, Australia, North Africa, and the Middle East all have major ecological issues related to supporting agriculture in an arid climate. Probably the best-known international example of a major ecological disruption caused by irrigation consumption of river water is the Aral Sea story (Box 10.3).

## Box 10.3 The Aral Sea—ecological disaster and partial recovery.

The Aral Sea was once the fourth largest lake in the world, covering 67,000 km$^2$. Located in the East Asia section of the old Soviet Union where the republics of Kazakhstan (in the north) and Uzbekistan (in the south) meet, this sea was fed by two rivers, the Syr Darya from the north and the Amu Darya from the south. In the 1950s Soviet planners decided that using the rivers that feed the Aral Sea for crop irrigation would be more economically useful that letting them flow into the sea. They embarked on a massive water diversion project, and as such they increased irrigated land in the Aral basin by 50%, turning it into a highly profitable agricultural region primarily for cotton and wheat production. However, beginning in the late 1950s into the late 1990s, the Aral Sea's surface area dropped 40%, volume dropped 60%, and salinity tripled. With the decrease in volume and area, all 24 native fish species disappeared. At the base of the food chain, cladoceran zooplankton dwindled from 12 species to 2. Additionally, many land animal species were locally extirpated. Fisheries commercial catch went from 44,000 tons in the 1950s to zero in the 1990s (causing a loss of 60,000 jobs in the fishing industry). This caused approximately 100,000 people to leave the area seeking new work (displaced). The low flows into the former sea concentrated salts, herbicides, pesticides, metals, and other toxins. Residents of the area saw a dramatic increase in incidents of typhoid fever, tuberculosis, and hepatitis, and esophageal cancer greatly increased (from toxic dust storms from the lake bed), with infant mortality quadrupling.

However, this degradation in recent years has been at least partially reversed. In 2001 the World Bank and the Kazakhstan government put major funding into reconstructing canals and waterworks and allowing more flow into the northern part of the Sea. In 2005 a 13-km dam, or possibly more appropriately dike, was completed across the Kazakhstan portion of the lake bed to capture the water flowing into the northern part of the sea from the Syr Darya. Additionally, the poor-quality, Soviet-era canals and locks were repaired and improved to reduce seepage loss, improving crop irrigation along the river and increasing the incoming river discharge volume at the same time. As a result, the northern portion of the Aral Sea has seen a strong recovery, with the water level increasing by 2 m and the lake surface by 900 km$^2$ and salinity decreasing considerably. Eleven species of fish are now living in the northern portion (mostly stocked from hatcheries), fish biomass within the lake has more than quadrupled, and the fishery has strongly returned. As to the base of the fishery food chain, the zooplankton have also rebounded. Riparian wetlands have reestablished, with terrestrial wildlife increasing as a result. However, the larger southern portion of the Aral Sea is still polluted and shrinking. That portion of the Aral Sea is in Uzbekistan, where the focus is upon agriculture and the Amu Darya is still used for that purpose alone. Additionally, that nation is considering exploring for oil in the former lake bed.

(Compiled from de Villiers 2000; Kamalov 2003; Usmanova 2003; Conant 2006; Pala 2011.)

A recent and nontraditional irrigation consumption activity is the biofuels (such as ethanol) industry (Service 2009). This business is in a state of flux, with (in the USA) expansion of such business depending largely upon congressional mandates rather than efficiency of use. Corn is well established as an ethanol source and requires significant irrigation (as well as fertilization; see Chapter 2). Other potential biofuel crops such as switchgrass have been proposed to be grown on more marginal lands than those used for traditional food crops, and irrigation of such fuel crops may add an increasing and significant demand on water supplies. Note that in the USA cost-effective biological processing of grasses and vegetation other than corn has proven to be elusive. Whereas in Brazil sugarcane is well established as a biofuels crop and is water-intensive (West et al. 2014), projections of water use to support non-corn-based biofuel endeavors in the USA are at present highly speculative.

Another part of the water consumption story in regards to fuel supply needs concerns extraction of fossil fuels (although most water used in these processes is from local groundwater rather than area rivers). Huge quantities of water are used to extract usable fossil fuel from tar and oil sands, as well as in hydraulic fracturing (fracking) to obtain natural gas deep underground (Ritter 2011). These processes result in heavily polluted process water that requires treatment before release into the environment or even recycling (see Chapter 12). Removal of groundwater for these purposes may lower local groundwater tables, potentially impacting stream discharge and resident biota. In northern Alberta, the tar sands (or bituminous sands) are mined with the aid of water pumped in from the Athabasca River or from wells. After mining, it is too polluted to be released back into the river, and 85% of it is recycled for further use in the oil sands operations or pumped into underground wells (Ritter 2011).

Finally, rivers do not recognize human borders but flow where they will. As discussed, river diversions lead to border conflicts. River water overconsumption, whether enhanced by a dam or not, also leads to conflicts when an upstream nation consumes excessive amounts of river water before the river enters a downstream nation. Presently such conflicts are occurring between India and Pakistan, where rampant hydroelectric dam building in upper rivers has led to fears over water security in downstream Pakistan (Bagla 2010). Another cross-border issue is in East Asia. Located in Kazakhstan, Lake Balkhash, covering 16,000 km², is the 13th largest lake in the world. However, in recent years its principal feeder river, the Ili, has been carrying less and less water into Kazakhstan from China. This river runs through Xinjiang Province, where China has greatly increased the local population and tripled the amount of land irrigated by the Ili (Stone 2012). Dam building in the Upper Nile is currently creating great tensions among the user nations.

## 10.3 Natural Flow Impacts Through Riparian Modification

### 10.3.1 Levees and Armored Shorelines

Throughout history, floods have been terrifying and costly natural disasters. The most famous US flood incident was the Johnstown (Pennsylvania) flood of May 31, 1889, in which the South Fork Dam gave way in a major storm and 2209 people died, with 1600 homes destroyed. Floods worldwide have had severe impacts on humans in terms of loss of life, property damage, and crop damage. As humans developed the technology (or use of slave labor) to protect themselves and their property from floods, levees were born. Riverbanks that are physically modified may become decoupled from the floodplain; that is to say, river floodplain interaction

becomes limited or even nonexistent. This can and has been accomplished by large-scale use of levees, of which an excellent example is the Mississippi River. In pre-industrial times (i.e. "natural" conditions) the Mississippi flood pulse would spread out over and into the floodplain wetlands, which would accept, absorb, and cleanse waters. Flood control of this great waterway resulted in extensive armored shorelines along middle and lower river sections and reducing meander belts. Perversely, confining the waters has the destructive effect of increasing flood stages (Myers and White 1993). During severe floods, the confined flood waters then breach levees and expand over residential or agriculture areas. An example is the Mississippi River floods of 1993 which impacted nine states, breaching or overtopping 1000 different levees and causing some $12 billion in damages (Myers and White 1993; see Chapter 17).

Regarding agriculture, just as dams prevent nutrient-rich silt from coming downstream to fertilize floodplains (i.e. the Aswan High Dam; Box 10.2), levees prevent the replenishment of nutrient-rich silt over the floodplain, leading to increased fertilizer use. Additionally, just as dams cut off sediment supplies from upstream areas to estuarine marshes, levees cut off sediment supplies that would normally enter the river laterally from riparian areas (Sparks 1995).

### 10.3.2 Canal Building

Water quality aspects of tidal canals are discussed in detail in Chapter 9. Canals are sometimes built to connect waters from two watersheds or connect freshwater sources with marine or estuarine areas. In the latter situation this of course leads to changes in salinity to natural habitats, affecting native biota and sediment geochemistry. In freshwater areas the connections provided by canals, while making boating or shipping more convenient, allow for easier spread of non-native species of aquatic plants and animals from one watershed to another. For example, Jackson and Pringle (2010) noted that the Erie Canal and St. Lawrence Seaway were instrumental in the spread of some of the harmful invasives entering North America's Great

Lakes. Canal building (often combined with wetland/marsh removal) alters natural flows as well. In southern Louisiana a network of nearly 13,000 km of canals and channels has been built for shipping and to provide access for workboats to service the oil and chemical industries. Such created waterways have opened pathways for hurricane-induced storm surges, such as occurred in Hurricane Katrina. Normally, marsh wetlands buffer storm surges, but man-made channels, combined with wetland loss, intensify storm surge waters. The canals reroute sediment delivery that is critical in marsh building, which combined with sediment delivery blockage from the levee system results in marsh subsidence and further loss of protection from major storms (Kerr 2012).

### 10.3.3 Flow Alteration Through Floodplain Alteration

Altering the floodplain itself is physical riparian land transformation. Such alterations can result in extensive changes to natural flow patterns and/or normal ecological functions. Such floodplain alterations are caused by timber harvesting, grazing on the floodplain, urbanization, road building, and agricultural activities (Allan and Flecker 1993). Floodplain alterations can also lead to the phenomenon known as habitat split, where anthropogenic activity on land severely separates habitat areas for organisms requiring both terrestrial and aquatic environments to complete their life cycles (Becker et al. 2007).

Riparian forest loss causes decreased organism diversity through reduction of woody debris (snags) in streams and reduction of habitat complexity (the former floodplain). The loss of snags will impact the fishery through loss of the snag invertebrate community, as well as loss of habitat for use as cover. Creation of stream log jams will be reduced as well as the low-flow microhabitats in the lee of such natural structures.

With anthropogenic-driven loss of riparian ground cover (and the streamside buffer zone) there are increases in stream sedimentation and siltation. This can result in streamflow changes by bank slumping into the stream proper, creation

of mid-stream bars, burying of riffles under silt, and loss of the associated microhabitats associated with such riffles, including aerated fish spawning sites. On a larger scale the loss of floodplain forests increases ecosystem physical instability. The floodplain forest absorbs excess rainfall and intercepts runoff, absorbing the water or transpiring it back to the atmosphere. With the forest gone the excess runoff entering streams and rivers is greatly magnified, causing flooding. For example, timber harvest along Himalayan streams has led to increased flooding in India and Bangladesh (Allan and Flecker 1993).

In a category of its own is freshwater loss due to climate change. Humans are undoubtedly responsible for injection of large amounts of greenhouse gases to the atmosphere and the consequent greenhouse effect (see Chapter 17). This planetary warming causes loss of glaciers (Overpeck and Udall 2010; Pederson et al. 2011; Gardner et al. 2013; Marzeion et al. 2014), which are major freshwater sources in mountain ranges worldwide. As the glaciers disappear less source water is available to feed stream flow, ecosystems change, rivers dry, and conflicts over diminished water supplies heat up (Bawa et al. 2010). Not only is freshwater likely to become less available, but also with ice melt, productive river deltas could be submerged by rising sea levels (Grumbine et al. 2012).

## 10.4 Decision-Making Regarding River Fragmentation and River Recovery

Riverine ecosystems are complex physically, chemically, and biologically and have multiple economic and human health issues associated with them. Regardless of whether decisions need to be made regarding removal of dams, performing interbasin transfer, increasing upstream irrigation or municipal consumption, or allowing major floodplain alterations, a decision should not be made until all ramifications are fully examined. A major part of this decision-making process is "What entities get how much water? And how much does the river need to support its natural wildlife?" Throughout this chapter we have discussed consumptive uses of a river's water, such as irrigation for agriculture, municipal water for drinking, bathing, cooking, washing, etc., cooling water for steam-generated power, sufficient current for hydroelectric power generation, and other industrial uses for water— all of these uses have advocates. Likewise, maintaining a healthy fishery, supporting endangered species, supporting anadromous fishes, maintaining or restoring a river's access to the floodplain (and all the ensuing benefits), riverine recreation such as ecotourism, kayaking, swimming, fishing, etc. also have advocates. One key is bringing these disparate groups (i.e. stakeholders) together to consider all of a river's uses.

One obvious but underutilized effort to shed some clarity on such matters is to assemble a group of scientists (biologists, geologists, hydrologists, geographers, etc.) to determine how a river functions ecologically during different seasons and under different hydrological regimes (high versus low water) to provide proper sediment and other instream habitat, support larval transport, distribute nutrients and organic matter, maintain ability to dilute and process pollutants, and so on. With the aid of modelers, this provides a scientific basis to first understand a river's needs before performing major alterations. This is an ideal approach, considered to be holistic in nature, and practiced in South Africa and Australia (Richter et al. 2006). While this approach is logical, some entity of course has to provide the funding for such a study—a tricky thing in itself.

When a holistic approach is not feasible, a program known as adaptive management is recommended by the Nature Conservancy and others (Richter et al. 2006). It is a five-step process as follows: 1) convene an orientation meeting for planning and organization purposes with key stakeholders; 2) perform a literature search relevant to the local flow-dependent data and ecological processes; 3) convene a workshop to develop objectives, identify information gaps, and make initial flow recommendations; 4) implement initial flow recommendations to test their appropriateness; and 5) monitor system response and conduct further research as needed. This adaptive process can be considered ongoing and the latter three processes can be altered and retested as needed (Richter et al. 2006).

To give us an idea of the potential complexity of such decision-making, there are numerous biologically related functions that require ecological sustainable flows (Postel and Richter 2003; Richter et al. 2006). To give us an idea of the potential complexity of such decision-making, there are numerous biologically-related functions that require ecological sustainable flows (Postel and Richter 2003; Richter et al. 2006), some of these are listed in Table 10.2.

The ecological data required for a full assessment will not be available in many circumstances. When river-specific information is scant, another approach that has been used is to determine hydrological regimes based on other rivers of a similar class. Streams that bear similarities based on hydrological regime, physiographic area, climate, and biota can be appropriately classified, with reference streams determined to provide a basis against which flow modifications can be statistically assessed (detailed in Arthington et al. 2006).

## 10.5 Using Less Water: Conservation Techniques

In early 2015, then Governor Jerry Brown ordered the first mandatory water restrictions for the state of California (agriculture exempted). Years of drought, exacerbated by a human-induced changing climate (Barnett et al. 2008), had led to lower snowfall in the mountains and subsequently less water for rivers in much of the American West. Major western reservoirs on rivers such as the Colorado River continue to sink to new lows with continuing increases in demand but no increase in supply. Profligate use of water from rivers by municipalities and agricultural areas often located well outside of the watershed have led to overuse in a number of western US states, leading to absurd phenomena such as well-watered golf courses in the desert (Fig. 10.6), extensive ornamental plantings in arid areas, and irrigation of water-loving crop species. Of course one factor promoting such overuse has been that water is often subsidized by government entities and is provided to the public well below what a true market-driven price would actually be (de Villiers 2000). In communities where water use is

**Table 10.2** Selected ecological functions performed by different flow levels. Amended, adapted from and reduced from Box 2-1 in Postel and Richter (2003).

| Flow level | Ecological functions |
|---|---|
| Low (base) flows | Provide adequate habitat space for aquatic taxa |
| | Maintain suitable water temperature, chemistry, and DO |
| | Maintain water table in floodplain and appropriate soil moisture for native flora |
| | Provide drinking water for terrestrial fauna |
| | Keep fish and amphibian eggs suspended |
| | Enable fish to move to feeding and spawning areas |
| | Support organisms that live in saturated sediments |
| Higher flows | Shape physical nature of channels including riffles, pools, runs |
| | Determine size of and sort streambed substrata (sand, silt, cobbles, boulders, etc.) |
| | Prevent riparian vegetation from clogging the channel |
| | Aerate eggs in spawning grounds |
| | Flush accumulated organic waste and pollutants |
| | Dilute and flush algal blooms in nutrient-rich rivers |
| Flood conditions | Provide prolonged access to floodplain (flood pulse) |
| | Provide migration and spawning cues for fish |
| | Recharge water table on floodplain |
| | Provide disturbances to keep biodiversity high |
| | Add new sediment to create new habitats |
| | Drive new movement of river channel over floodplain by creating new meanders and cutting off oxbows for new lacustrine habitats |
| | Distribute seeds and fruits of aquatic and terrestrial plants |

(a)    (b)

**Fig. 10.6** Fig. 10.6a (right) Greens where there should be no green. A watered golf course in arid New Mexico. Fig. 10.6b (right) Golf course in Phoenix, southern Arizona. (Photos by the author.)

unmetered people have little incentive to conserve if use does not lead to increases in water prices (Thompson 1999).

Infrastructure is the key to efficient municipal water delivery from the riverine or groundwater source. Aging or poorly constructed pipes and pumps leak and such leakage can lose 10–20% of water before it reaches its ultimate destinations

(Thompson 1999). Likewise, as seen with the Aral Sea example (Box 10.3), this holds true for water slated for irrigation purposes. Much irrigation water is lost between the river and the cropland through leaking pipes, unlined canals, and evaporation from open reservoirs and canals (de Villiers 2000; Grant et al. 2012). Regarding crop irrigation, there is a hierarchy in water efficiency depending on the method, with flooding the least efficient, followed by spray irrigation, with drip irrigation—essentially delivering water to plant roots—being the most efficient (Thompson 1999). Reductions in agricultural consumption also include laser-based leveling of fields to reduce runoff, precision timing of watering to provide optimum watering when it is most needed, recovery of excess water at the field tailwaters for reuse, and storage of excess water in underground reservoirs to avoid evaporative losses until needed (Thompson 1999; de Villiers 2000). Also, some crops (such as almonds) require much more water than others (although they may not be as lucrative to a farmer), and researchers continually work to develop crop varieties that can thrive on lower water levels. Israel is a world leader in irrigation efficiency.

Institutions, businesses, and individual homeowners all can reduce water consumption considerably by similar techniques. Regarding indoor usage, the installation of low-flow toilets and waterless urinals provides a huge reduction in municipal water use, without sacrificing comfort or sanitation. Installation of low-flow showerheads and use of more efficient washing machines are also water-saving options (Thompson 1999). Gray water is the water that has been used for domestic laundry, dishwashing, and bathing. In some areas it has been used for toilet flushing and yard irrigation, resulting in substantial municipal water use decreases (Grant

et al. 2012). Instead of wasting rainfall by allowing it all to enter the stormwater system, rain falling on rooftops can be intercepted and directed into cisterns for large facilities (Fig. 10.7a) and rain barrels for individual homes (see demonstration project in Fig. 10.7b). The stored rain can then be used to water gardens and ornamentals. While Las Vegas is much maligned for its (former) extravagant uses of water in displays, it at present is under strict water reuse regulations for homes, casinos, and other businesses (Folger 2008). A lush green lawn in the American West, while previously a much-loved affectation, is becoming a thing of the past in diverse areas including southern California, Nevada, New Mexico, and Arizona. Xeriscaping, which is the use of native ornamental plants requiring little water, combined with attractive rock gardens is replacing the standard lawn.

Municipal wastewater treatment produces copious amounts of treated wastewater that is usually piped into a stream or river. Provided that the treatment level is sufficient and disinfection is performed, this "used" water can be reused for watering non-food-crop-vegetated areas such as golf courses, common areas, and lawns. Treated wastewater can be suitable for industrial cooling (Grant et al. 2012). The term "purple pipe" refers to such reusable water. Some municipalities produce drinkable water after tertiary wastewater treatment, although a considerable psychological barrier

(a)

(b)

**Fig. 10.7** Fig. 10.7a (left) One of several cisterns located at Wrightsville Beach, North Carolina, fire station; roof runoff is used for washing fire trucks and watering shrubbery and lawns. Fig. 10.7b (right) Rain barrels are useful for individual homes or other structures to capture rain for watering purposes. (Photos by the author.)

would have to be overcome before drinking recycled sewage water becomes widespread. The above water conservation techniques are most effective when built into the municipal or agricultural infrastructure to begin with, rather than as later revisions (Postel and Richter 2003). There is good opportunity to apply such measures in developing nations where growing human populations must be served.

## 10.6 Summary

- Dams are built to impound stream waters for irrigation, hydroelectric power (as renewable energy, which can be reused downstream and reduces $CO_2$ pollution from coal burning), industrial cooling water (such as coal- and nuclear-powered power plants [see Chapter 12]), and flood control.

- Dams have significant impacts on downstream ecosystems by cutting off the flood pulse, reducing downstream disturbance frequency, disrupting fish and wildlife passage, reducing sediment supply needed for new habitat creation, lowering the downstream channel, altering water temperature, and reducing downstream water quality.

- Dam construction and operation have other nonecological impacts as well. These include geological disruptions, archeological losses, human health effects, human population displacement, and cost overruns.

- Dam removal can alleviate some of the above impacts. However, ecological issues requiring consideration include polluted sediment stored behind dams, impacts of high turbidity and suspended sediments on downstream communities, eutrophication impacts of stored nutrient release downstream, and more open passage for invasive species with dam removal.

- Interbasin transfer is pumping significant volumes of river water from one watershed into another for irrigation or municipal use. This can lead to conflicts between nations, states, or communities. In some cases such as the American West, vast amounts of water are shipped to distant agricultural or municipal areas, with consequences to downstream river and estuarine ecology.

- River water consumption for irrigation and industrial and municipal use can severely impact aquatic ecosystems and also lead to border conflicts. Such consumption can be alleviated in part by irrigation conservation techniques (laser field leveling, drip irrigation, timed irrigation, less thirsty crops) and municipal conservation (low-flow showers, low-flush toilets, waterless urinals, xeriscaping instead of lawns, gray water recycling).

## References

Allan, J.D. and A.S. Flecker. 1993. Biodiversity conservation in running waters. *BioScience* 43:32–43.

Arthington, A.H., S.E. Bunn, N.L. Poff and R.J. Naiman. 2006. The challenge of providing environmental flow rules to sustain river ecosystems. *Ecological Applications* 16:1311–1318.

Bagla, P. 2010. Along the Indus River, saber rattling over water security. *Science* 328:12226–1227.

Barnett, T.P., D.W. Pierce, H.G. Hidalgo, et al. 2008. Human-induced changes in the hydrology of the western United States. *Science* 319:1080–1082.

Bawa, K.S., L.P. Koh, T.M. Lee, et al. 2010. China, India and the environment. *Science* 327:1457–1459.

Baxter, R.M. 1977. Environmental effects of dams and impoundments. *Annual Review of Ecology and Systematics* 8:255–283.

Bayley, P.B. 1995. Understanding large river-floodplain ecosystems. *BioScience* 45:153–158.

Becker, C.G., C.R. Fonseca, C.F.B. Haddad, R.F. Batista and P.I. Prado. 2007. Habitat split and the global decline of amphibians. *Science* 318:1775–1777.

Black, H. 2001. Dam-building decisions: a new field of fairness. *Environmental Health Perspectives* 109:A80–A83.

Blinn, D.W. and N.L. Poff. 2005 Colorado River basin. Pages 483–526 in Benke, A.C. and C.E. Cushing (eds.), *Rivers of North America*. Elsevier, Amsterdam.

Bohannon, J. 2010. The Nile Delta's sinking future. *Science* 327:1444–1447.

Brainard, J. 2011. Countdown for the Conowingo. *Chesapeake Quarterly* 10:8–11.

Burdick, S.M. and J. E. Hightower. 2006. Distribution of spawning activity by anadromous fishes in an Atlantic slope drainage after removal of a low-head dam. *Transactions of the American Fisheries Society* 135:1290–1300.

Campbell, D. and G. Paxson. 2002. Flood passage for earth dams. *Lake Line* 22:26–28.

Campbell, I. and C. Barlow. 2020. Hydropower development and the loss of fisheries in the Mekong River Basin. *Frontiers in Environmental Science* 8:566509.

Cheng, L., J.J. Opperman, D. Tickner, R. Speed, Q. Guo and D. Chen. 2018. Managing the Three Gorges Dam to implement environmental flows in the Yangtze River. *Frontiers in Environmental Science* 6:64.

Conant, E. 2006. Return of the Areal Sea. Discover Magazine September:54–58.

Cornwall, W. 2017. U.S.–Mexico water pact aims for a greener Colorado delta. *Science* 357:635.

De Mutsert, K. and J.H. Cowan Jr. 2012. A before-after-control-impact analysis of the effects of a Mississippi River freshwater diversion on estuarine nekton in Louisiana, USA. *Estuaries and Coasts* 35:1232–1248.

De Villiers, M. 2000. *Water: The Fate of Our Most Precious Resource.* Houghton Mifflin, Boston.

Dethier, E.N., C.E. Renshaw and F.J. Magilligan. 2022. Rapid changes to global river suspended sediment flux by humans. *Science* 376:1447–1452.

Doyle, M.W., J.M. Harbor and E.H. Stanley. 2003. Toward policies and decision making for dam removal. *Environmental Management* 31:453–465.

Drinkwater, K.F. and K.T. Frank. 1994. Effect of river regulation and diversion on marine fish and invertebrates. *Aquatic Conservation: Freshwater and Marine Ecosystems* 4:135–151.

Dynesius, M. and C. Nilsson. 1994. Fragmentation and flow regulation of river systems in the northern third of the world. *Science* 266:753–762.

Flecker, A.S., U. Shi, R.M. Almeida, et al. 2022. Reducing adverse impacts of Amazon hydropower expansion. *Science* 375:753–760.

Folger, T. 2008. Requiem for a river. *onEarth*, spring issue, 24–35.

Gardner, A.S., G. Moholdt, J.G. Cogley, et al. 2013. A reconciled estimate of glacier contributions or sea level rise: 2003 to 2009. *Science* 340:852–857.

Graf, W.L. 1999. Dam nation: a geographic census of American dams and their large-scale hydrologic impacts. *Water Resources Research* 35:1305–1311.

Grant, S.B., J.-D. Saphores, D.L. Feldman, et al. 2012. Taking the "waste" out of "wastewater" for human water security and ecosystem sustainability. *Science* 337:681–686.

Gregory, S.V., F.J. Swanson, W.A. McKee and K.W. Cummins. 1991. An ecosystem perspective of riparian zones. *BioScience* 41:540–551.

Grumbine, R.E. and M.K. Pandit. 2013. Threats from India's Himalaya dams. *Science* 339:36–37.

Grumbine, R.E. and J. Xu. 2011. Mekong hydropower development. *Science* 332:178–179.

Grumbine, R.E., J. Dore and J. Xu. 2012. Mekong hydropower: drivers of change and governance challenges. *Frontiers in Ecology and the Environment* 10:91–98.

Humborg, C., V. Ittekkot, A. Cociasu and B.V. Bodungen. 1997. Effect of the Danube River dam on Black Sea biogeochemistry and ecosystem structure. *Nature* 386:385–388.

Humborg, C., D.J. Conley, L. Rahm, F. Wulff, A. Cociasu and V. Ittekkot. 2000. Silicon retention in river basins: far reaching effects on biogeochemistry and aquatic food webs in coastal marine environments. *AMBIO* 29:45–50.

ICOLD. 2020. The International Commission on Large Dams. http://www.icold-cigb.org/.

Jackson, C.R. and C.M. Pringle. 2010. Ecological benefits of reduced hydrologic connectivity in intensely developed landscapes. *BioScience* 60:37–46.

Johnson, B.L., W.B. Richardson and T.J. Naimo. 1995. Past, present, and future concepts in large river ecology. *BioScience* 45:134–141.

Junk, W.J., P.B. Bayley and R.E. Sparks. 1989. The flood-pulse concept in river-floodplain systems. *Canadian Journal of Fisheries and Aquatic Sciences* 106:110–127.

Kamalov, Y. 2003. The Aral Sea: problems, legends, solutions. *Water Science and Technology* 48:225–231.

Kennedy, T.A., J.D. Muehlbauer, C.B. Yackulic, et al. 2016. Flow management for hydropower extirpates aquatic insects, undermining river food webs. *BioScience* 66:561–575.

Kerr, R.A. 2012. Rebuilding wetlands by managing the muddy Mississippi. *Science* 335:520–523.

Kerr, R.A. and R. Stone. 2009. A human trigger for the great quake of Sichuan? *Science* 323:322.

Kondolf, G.M., R.J.P. Schmitt, P.A. Carling, et al. 2022. Save the Mekong from drowning. *Science* 376:583–585.

Lawler, A. 2012. Dams along Sudanese Nile threaten ancient sites. *Science* 336:967–968.

Liu, J. and W. Yang. 2012. Water sustainability for China and beyond. *Science* 337:649–650.

Maavara, T., Z. Akbarzadeh and P. Van Cappellen. 2020. Global dam-driven changes to riverine N:P:Si ratios delivered to the coastal ocean. *Geophysical Research Letters* 47:e2020GL088288.

Marzeion, B., J.G. Cogley, K. Richter and D. Parkes. 2014. Attribution of global glacier mass loss to anthropogenic and natural causes. *Science* 345:919–921.

Mitsch, W.J., J. Lu, X. Yuan, W. He and L. Zheng. 2008. Optimizing ecosystem services in China. *Science* 322:528.

Moyle, J. 2002. The need for dam safety standards and regulations. *Lake Line* 22:15–18.

Myers, M.F. and G.F. White. 1993. The challenge of the Mississippi flood. *Environment* 35:6–35.

Myers, S.L., K. Bradsher and T. Panja. 2022. The games that Xi built, entirely on his terms. The New York Times January 23.

Nixon, S.W. 2004. The artificial Nile. *American Scientist* 92:158–165.

O'Conner, J.E., J.J. Duda and G.E. Grant. 2015. 1,000 dams down and counting; dam removals are reconnecting rivers in the United States. *Science* 348:496–497.

Oster, S. 2009. Why Chinese dam is forging another exodus. *Wall Street Journal* November 6.

Overpeck, J. and B. Udall. 2010. Dry times ahead. *Science* 328:1642–1643.

Pala, C. 2011. In northern Aral Sea, rebound comes with a big catch. *Science* 334:303.

Pederson, G.T., S.T. Gray, C.A. Woodhouse, et al. 2011. The unusual nature of recent snowpack declines in the North American cordillera. *Science* 333:332–335.

Penczak, T., G. Zieba, H. Koszalinski and A. Kruk. 2003. The importance of oxbow lakes for fish recruitment in a river system. *Archiv für Hydrobiologie* 158:267–281.

Postel, S. and B. Richter. 2003. *Rivers for Life*. Island Press, Washington, DC.

Qiu, J. 2012. Trouble on the Yangtze. *Science* 336:288–291.

Richter, B.D., A.T. Warner, J.L. Meyer and K. Lutz. 2006. A collaborative and adaptive process for developing environmental flow recommendations. *River Research and Applications* 22:297–318.

Ritter, S.K. 2011. Water for oil. *Chemical and Engineering News* September 5.

Rulifson, R.A. and C.S. Manooch III. 1990. Recruitment of juvenile striped bass in the Roanoke River, North Carolina, as related to reservoir discharge. *North American Journal of Fisheries Management* 10:397–407.

Schiffman, R. 2022. In the Balkans, researchers mobilize to protect a river. *Science* 377:13–14.

Schindler, D.E., M.D. Scheuerell, J.W. Moore, S.M. Gende, T.B. Francis and W.J. Palen. 2003. Pacific salmon and the ecology of coastal ecosystems. *Frontiers in Ecology and the Environment* 1:31–37.

Service, R. 2009. Another biofuels drawback: the demand for irrigation. *Science* 326:516–517.

Shields, F.D. Jr., A. Simon and L.J. Steffen. 2000. Reservoir effects on downstream river channel migration. *Environmental Conservation* 27:54–66.

Shumilova, O., K. Tockner, M. Thieme, A. Koska and C. Zarfl. 2018. Global water transfer megaprojects: a potential solution for the water-food-energy nexus? *Frontiers in Environmental Science* 6:150.

Smith, S.V., W.H. Renwick, J.D. Bartley and R.W. Buddemeier. 2002. Distribution and significance of small, artificial water bodies across the United States landscape. *The Science of the Total Environment* 299:21–36.

Sparks, R.E. 1995. Need for ecosystem management of large rivers and their floodplains. *BioScience* 45:168–182.

Stanley, E.H. and M.W. Doyle. 2003. Trading off: the ecological effects of dam removal. *Frontiers in Ecology and the Environment* 1:15–22.

Stone, R. 2008. Three Gorges Dam: into the unknown. *Science* 321:628–632.

Stone, R. 2011. Mayhem on the Mekong. *Science* 333:814–818.

Stone, R. 2012. For China and Kazakhstan, no meeting of the minds on water. *Science* 337:405–407.

Stone, R. 2016. Dam-building threatens Mekong fisheries. *Science* 354:1084–1085.

Tetteh, I.K., E. Frempong and E. Awnah. 2004. An analysis of the environmental health impact of the Barakese Dam in Kumasi, Ghana. *Journal of Environmental Management* 72:189–194.

Thompson, S.A. 1999. *Water Use, Management, and Planning in the United States*. Academic Press, San Diego.

Tuckerman, S. and B. Zawiski. 2007. Case studies of dam removal and TMDLs: process and results. *Journal of Great Lakes Research* 33(special issue 2):103–116.

Usmanova, R.M. 2003. Aral Sea and sustainable development. *Water Science and Technology* 47:41–47.

Vahedifard, F., A. AghaKouchak, E. Ragno, S. Shahrokhabadi and I. Mallakpour. 2017. Lessons from the Oroville Dam. *Science* 355:1139–1140.

Vince, G. 2010. Dams for Patagonia. *Science* 329:382–385.

Wade, L. 2020. Engineering an empire. *Science* 368:235–239.

Walter, R.C. and D.J. Merritts. 2008. Natural streams and the legacy of water-powered mills. *Science* 319:299–304.

Ward, J.V., K. Tockner and F. Schiemer. 1999. Biodiversity of floodplain river ecosystems: ecotones and connectivity. *Regulated Rivers: Research & Management* 15:125–139.

West, P.C., J.S. Gerber, P.M. Engstrom, et al. 2014. Leverage points for improving global food security and the environment. *Science* 345:325–328.

Winemiller, K.O., P.B. McIntyre, L. Castello, et al. 2016. Balancing hydropower and biodiversity in the Amazon, Congo and Mekong. *Science* 351:128–129.

# CHAPTER 11

# Reservoir Limnology

Reservoirs have characteristics of both rivers and lakes, and thus reservoir limnology lies at the intersection of river ecology and lake limnology. There are a number of useful limnology textbooks and reference volumes that include aspects of reservoir research, including Horne and Goldman (1994), Wetzel (2001), Dodds and Whiles (2010), and Cole and Weihe (2016). In general, European, North American, and some South American reservoirs are well studied, whereas many Asian, African, and South American reservoirs are not, and as noted in Chapter 10, more and even larger ones are coming on-line. The objective of this chapter is to introduce and discuss those limnological concepts that are most relevant to reservoir ecology, draw comparisons among rivers, reservoirs, and lakes, and discuss unique features of reservoir ecosystems.

Reservoirs have been in use for a long time. Evidence of infrastructure for water movement and storage nearly 3000 years old comes from Asia and the Middle East (de Villiers 2000). Reservoir construction has been ongoing for hundreds of years in Europe, with two Spanish reservoirs, Cornalbo and Prosperpina, dating from the second century and other large reservoirs subsequently constructed in the seventeenth and eighteenth centuries in Germany, Spain, France, and the UK (Leonard and Crouzet 1999). In the US Southwest, the native Anasazi people lived at Mesa Verde from about 550 to 1300 AD. Rainfall was never abundant in the arid southwest and these people built a number of reservoirs either in canyons or upon mesas to collect and store rainwater, with these reservoirs operational for 150–350 years, depending upon the system (Wright 2008). Drought and warfare caused the collapse of this civilization by 1300 AD. The oldest existing dam in the USA was built in 1677 (Mill

Pond Dam, Connecticut; Graf 1999). In both the USA and Europe a proliferation of reservoir building occurred throughout the 1900s, with the rush of dam building curtailing in the USA in the 1980s (Graf 1999). The twentieth-century profusion of new large water bodies led to reservoirs becoming systems of interest to limnologists over half a century ago (Wright 1958). A number of helpful reviews of reservoir physical, chemical, and biotic characteristics have been produced, including Baxter (1977), Soballe et al. (1992), and a multichapter volume edited by Thornton et al. (1990).

## 11.1 Geographical and Physical Considerations

When a high-order stream or the flow from a series of lower-order streams is dammed, a lake-like, or **lentic**, environment is created. This body of water is called a **reservoir**; a less-popular term for such a water body is **impoundment**. Reservoirs are built for a variety of reasons, including flood control, municipal water supply, irrigation of crops, hydroelectric power, and to provide a source of industrial and power plant cooling water. Reservoirs are rarely built for recreational purposes; this is normally a secondary reason used to help convince regulatory agencies and municipalities that they should allow (or fund) reservoir construction. Regarding hydroelectric power generation, **run-of-the-river reservoirs** are systems with rather short retention times whose purpose is to direct flows for hydroelectric projects; these systems are sometimes arranged in series along a river (Baxter 1977).

Reservoir surface area and depths vary greatly, depending on location (deep valley, gentle sloping

*River Ecology*. Michael A. Mallin, Oxford University Press. © Michael A. Mallin (2023). DOI: 10.1093/oso/9780199549511.003.0012

plain) and purpose (local use, regional use, even distant uses such as in the US Southwest). Some dams impound truly massive volumes of water. Table 11.1 lists some of the largest reservoirs on Earth; note that Canada and Russia are well represented, along with several African nations.

Reservoirs are predominantly located in regions where natural lakes are few in number (Thornton 1990a). Where do lakes naturally occur? Natural lakes are created by a number of means, but three major forces stand out in particular. In glaciated regions (such as much of Canada, northern Europe, the Upper US Midwest, and parts of New England) lakes have been created by damming by debris deposited by a glacier's terminal moraine, gouging, and subsequent withdrawal of glaciers along river valleys, and melting of large blocks of ice buried in the soil, leaving kettle lakes or smaller prairie potholes. In areas of karst topography (such as Florida and the Yucatan region of Mexico) lakes can be created by dissolution of the calcium carbonate ($CaCO_3$) substrata—on a small scale think in terms of the sinkholes that periodically swallow car dealerships in Florida. Movement of the Earth's crust, i.e. tectonic forces, creates large lakes in areas where the Earth is active, especially the vast lakes formed in Africa's Great Rift Valley. On a smaller scale there are many other types of lakes and means of formation, including the shallow aeolian lakes created by wind patterns in some regions such as areas of the US Great Plains, the bay lakes on the Coastal Plain of the Carolinas that may have been created by meteor showers (although much stranger theories have been proposed as well), and the oxbow lakes discussed in Chapter 1. But vast areas of land are well away from glaciated regions, karst topography, and tectonic activity and have fertile soils amenable to crop growth. Here humans build reservoirs to store water for the uses noted above (and in Chapter 10). Reservoirs are abundant in the Central, Southeastern, Southwestern, and Far Western USA and some are located in the Pacific Northwest and New England (Thornton 1990a). In Europe most reservoirs have been constructed in Spain, the UK, and central Europe (Table 11.2). As a comparison, the National Inventory of Dams map of major dams compiled by the US Geological Survey includes 8100 dams > 8 m tall in the USA proper and its territories.

Reservoirs in mountain areas (in the USA most famously the Tennessee Valley Authority system and the Pacific Northwest) are generally used for the production of hydroelectric power, because the steep slope of the terrain produces higher energy yield. A subset of such systems are called pumped-storage reservoirs, in which some of the generated electricity is used to pump water up to a higher level to be stored and available for hydroelectric power generation at a later time (Baxter 1977). Reservoirs in the Central, Far West, Midwest, and Lowland areas of the USA are generally built to

**Table 11.1** The largest reservoirs on Earth by water storage volume ($m^3$), according to ICOLD.

| Dam name | Storage volume ($m^3$) | Nation |
|---|---|---|
| Kariba | 180,600,000 | Zimbabwe/Zambia |
| Bratsk | 169,000,000 | Russia (Russian Federation) |
| Aswan High | 162,000,000 | Egypt |
| Akosombo | 150,000,000 | Ghana |
| Daniel Johnson (Manic 5) | 141,851,350 | Canada |
| Guri | 135,000,000 | Venezuela |
| W.A.C. Bennett | 74,300,000 | Canada |
| Krasnoyarsk | 73,300,000 | Russia (Russian Federation) |
| Zeya | 68,400,000 | Russia (Russian Federation) |
| Hidase | 63,000,000 | Ethiopia |
| Robert-Bourassa | 61,715,000 | Canada |
| La Grande-3 | 60,020,000 | Canada |
| Ust-Ilim | 59,300,000 | Russia (Russian Federation) |
| Cutarm Creek | 58,595,982 | Canada |
| Boguchany | 58,200,000 | Russia (Russian Federation) |

**Table 11.2** European nations containing the greatest number of large (dam > 15 m) reservoirs (Leonard and Crouzet 1999). Note such reservoirs are most abundant in nonglaciated regions.

| Country | Number of reservoirs |
| --- | --- |
| Spain | 849 |
| France | 521 |
| UK | 517 |
| Italy | 425 |
| Norway | 316 |
| Germany | 260 |

produce a dependable water supply for municipalities and agriculture, or for industrial usage. Where along a drainage basin a reservoir is built is important in a number of ways. Reservoirs that are built on a higher-order (more than fifth-order) stream are called main-stem reservoirs (Fig. 11.1), while reservoirs built on lower-order streams are sometimes referred to as tributary reservoirs (Soballe et al. 1992). Generally, water residence time is shorter in main-stem reservoirs and longer in tributary reservoirs. This impacts reservoir plant growth in that turbidity is higher in reservoirs of short residence time, shading out phytoplankton, benthic microalgae, and macrophytes, while longer residence time allows for turbidity settling, greater light penetration, and autotrophic community development (Soballe and Kimmel 1987). In contrast to the US West, many southeastern US reservoirs are built higher up in the drainage basin on lower-order streams. These reservoirs in the southeastern US differ in a number of ways from reservoirs built in other regions of North America; this will be explored later in this chapter.

### 11.1.1 Reservoir Shape and Depth Considerations

Most reservoirs fundamentally differ from natural lakes in several ways (Table 11.3). Lakes tend to have catchments that surround the lake, whereas catchments for reservoirs are principally located

**Fig. 11.1** Very linear reservoir located in the arid US Southwest fed by a high-order stream. (Photo by the author.)

**Table 11.3** Differences between lakes and reservoirs, broadly generalized

|  | Lakes | Reservoirs |
|---|---|---|
| Physical shoreline development | Low | Complex |
| Deepest area | Middle | Near dam |
| Thalweg | No | Yes |
| Water residence time | Long | Short to medium |
| Turbidity | Low | Moderate to high |
| Nutrient loading | Low to moderate | Moderate to high |
| Mixing | Wind or seasonal | Current, wind, seasonal |
| Depth control | Rain, drought | Human use, rain, drought |
| Botanical shoreline development | Mature | Low or transitional |
| Species diversity | Moderate | Moderate to high |
| Endemic species | Yes | No |
| Invasive species | Less subject to | Highly subject to |
| Eutrophication | Less subject to | More subject to |

upstream of the reservoir (Fig. 11.1); thus many reservoirs are fed by high-order streams, whereas lakes may have a number of low-order streams that enter them. Most lakes are deepest in the middle, whereas reservoirs are generally deepest near the dam. For these reasons, lakes tend to be more round, while reservoirs are more dendritic or they can be very lengthy along the axis of the main feeder river. The relationship can be described by the shoreline development index ($D_L$):

$$D_L = L/2(\pi A)^{0.5},$$

where L is the circumference of the lake (shoreline length) and A is the lake's surface area.

A perfect circle has an index value of 1.0. This is most approximated in nature by lakes that form in ancient volcano calderas, like Crater Lake in Oregon (Fig. 11.2a). Dodds and Whiles (2010) computed the $D_L$ of Crater Lake as 1.6. The dendritic nature of reservoirs increases $D_L$, in some cases greatly (Fig. 11.2b). Lake Powell, on the Colorado River, consumed a lot of canyons during its filling, and thus it is very dendritic. This large reservoir has a shoreline of approximately 3219 km and a surface area of about 653.2 km$^2$ (Hueftle 1995); thus we get a $D_L$ of about 35.5. Soballe et al. (1992) noted that median $D_L$ among a set of US Army Corps of Engineers reservoirs was 12.0, whereas most natural lakes have a $D_L$ < 3.0.

The water level of lakes historically has been controlled by natural occurrences such as droughts, rainfall patterns, and snowmelt. While these factors also play a role in reservoir water depth, humans control the ultimate water levels based on needs for power generation, flood control, irrigation, and municipal water supply. Human activities over recent decades also impact reservoir depth through climate change. In summer 2021 some of the great reservoirs in the western US, such as Lake Powell and Lake Mead on the Colorado River, were at record lows due to drought, low snowmelt, and continually increasing human withdrawal (Stokestad 2021). Climate change is a major operator here (see Chapter 17); with increased temperatures there is also increased evapotranspiration, and the aforementioned loss of snow cover decreases reflection of solar radiation, compounding water loss in systems such as the Colorado (Milly and Dunne 2020). In reservoirs built for cooling water it is critical to maintain high volumes of water within the system so that cooling water is always available. In reservoirs built for hydroelectric power, such as run-of-the-river reservoirs, it is of interest to produce flows that produce electrical power in response to periods when use is highest, such as day rather than night (called hydropeaking; see Poff and Schmidt 2016). Other reservoirs are built for storage of municipal water or irrigation water, wherein levels may fluctuate seasonally more rapidly than in natural lakes and volumes may change dramatically due to use patterns. This results in a less stable shoreline, or **littoral** region; thus there is normally less of a littoral macrophyte community and associated littoral

(a)                                                      (b)

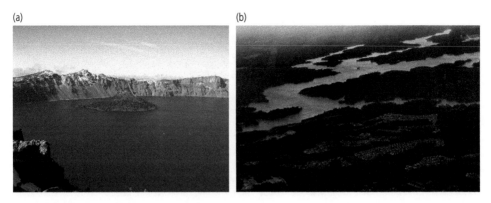

**Fig. 11.2**  Fig. 11.2a (left) Very round Crater Lake, Oregon, as seen from a nearby mountain slope. Fig. 11.2b (right) An evening view of very dendritic Lake Norman, a North Carolina Piedmont reservoir. (Photos by the author.)

(a)                                                      (b)

**Fig. 11.3**  Fig. 11.3a (left) Poorly developed reservoir shorelines due to human water withdrawals; Lake Lanier, Georgia, the City of Atlanta water supply. Fig. 11.3b (right) Lost Creek Reservoir, Oregon. (Photos by the author.)

animal community in reservoirs (Fig. 11.3). Finally, lakes can normally be divided into a littoral zone and a lentic or **lacustrine** zone away from shore, whereas reservoirs often express horizontally zonation depending on flow regime. Some researchers consider reservoirs as occupying an intermediate position between rivers and lakes in terms of physical, nutrient, and biological characteristics, primarily a result of water residence time (Soballe and Kimmel 1987).

### 11.1.2  Reservoir Horizontal Zonation

A typical main-stem reservoir can be divided into three functional zones. The headwaters, which is the area of primary river input, is called the **riverine zone** (Fig. 11.4a). Regarding mixing, water

entering a reservoir in the riverine zone will often be of different temperature than that residing within. Since temperature affects density, this may cause density currents that may persist well into the reservoir. Colder water will move along the bottom, warm water will slide over the surface, or a current may flow at some intermediate depth; greater mixing will occur at a wider and deeper point along the reservoir axis, sometimes called the plunge point (Soballe et al. 1992).

The riverine zone receives a heavy suspended sediment load that includes coarse inorganic particles, CPOM such as twigs and other plant-derived materials, and large amounts of silts and clays, leading to high turbidity levels in this zone (Fig. 11.4); depending on current velocity, turbidity currents may extend well into the reservoir (Baxter

(a)                                    (b)

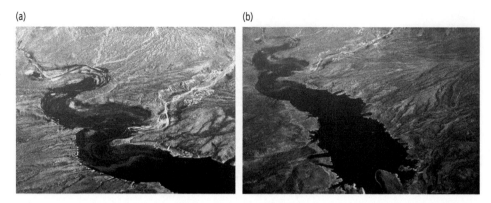

**Fig. 11.4** Fig. 11.4a (right) Reservoir near Phoenix, Arizona; close-up photo of riverine zone showing turbid water and sediment deposition caused by decreasing water velocity. Fig. 11.4b (right) Same reservoir showing riverine, transition, and lacustrine zones. (Photos by the author.)

1977). The lack of wetland and macrophyte development in the river input area exacerbates this suspended sediment loading, which is particularly apparent in western and southwestern reservoirs. Sediment input and deposition becomes a problem as reservoirs age, because sizable amounts of their original water storage capacity become depleted due to sediment buildup. By far the greatest amount of sedimentation occurs in the upper portions of reservoirs where larger suspended particles (sands, coarse silt, and CPOM) tend to settle out as the river current slows (Thornton 1990b). Deposited sediments may form a noticeable **delta** near the river input (Fig. 11.4a) (Baxter 1977); on a small scale such a delta is often visible in receiving areas of a stormwater detention pond draining urban or suburban construction sites. Finer silts and clay particles remain in suspension and settle out farther into the reservoir as currents continue to slow. As discussed (see Chapter 2), phosphate, ammonium, and organic nutrients are often bound to clays, and thus reservoir nutrient concentrations are highest in the riverine zone. High suspended sediment loads, coupled with the nutrients bound to finer sediments, often result in greater nutrient loading to reservoirs than to lakes (Thornton 1990b). However, the high turbidity levels cause light limitation of phytoplankton productivity (Kimmel et al. 1990). In addition to light limitation, the incoming current leads to short water retention time in this zone that further retards development of the

phytoplankton community. Even though riverine or stream-originated phytoplankton communities may be sparse entering the reservoir, a zooplankton community can develop in the lower portion of the riverine zone if the detrital particles entering the system have adsorbed sufficient DOM to render them suitably nutritious (Arruda et al. 1983).

The middle portion of the reservoir is called the **transition zone** (Fig. 11.5) where the slower current allows for settling out of smaller suspended particles, allowing for increased penetration of solar irradiance. This decrease in turbidity can be abrupt (see Fig. 11.4b). However, there are still plenty of nutrients available from inputs of dissolved nutrients and desorption from suspended or settled particles. Therefore, phytoplankton productivity is high (Fig. 11.5) and the zooplankton community is well developed in response (Kimmel et al. 1990). Note that this occurrence is analogous to a turbid but nutrient-rich riverine estuary entering the stratified ocean; this forms a **front**, where suspended sediments settle out, nutrient-rich waters are exposed to high light, and a chlorophyll maximum forms. The chlorophyll maximum along a front attracts invertebrate grazers, small plankton and invertebrate feeding fish, and large predatory fish. Sport fishermen know this well and troll or cast along fronts, which are commonly known as tide lines. Back to reservoirs; since littoral macrophyte development is often limited in reservoirs (Figs 11.3 and 11.4) phytoplankton are usually the dominant primary producers. There is a

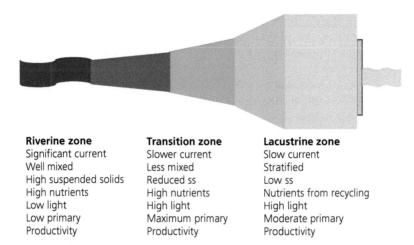

**Riverine zone**
Significant current
Well mixed
High suspended solids
High nutrients
Low light
Low primary
Productivity

**Transition zone**
Slower current
Less mixed
Reduced ss
High nutrients
High light
Maximum primary
Productivity

**Lacustrine zone**
Slow current
Stratified
Low ss
Nutrients from recycling
High light
Moderate primary
Productivity

**Fig. 11.5** Stylized depiction of horizontal reservoir zonation showing brown, sediment, and nutrient-rich river waters entering reservoir (riverine zone); current slows and suspended solids settle out, allowing enhanced light transmission and high nutrients to maximize phytoplankton production (green). Note this transition can be abrupt (a front). Productivity becomes more nutrient-limited downstream into the lacustrine zone. Nearest the dam (blue) there is high light transmission but nutrients for phytoplankton are only available from recycling. (Modified from Kimmel and Groeger 1984.)

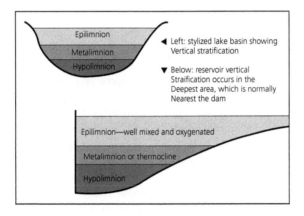

**Fig. 11.6** Depictions of vertical stratification and associated water layers in stylized lakes and reservoirs. (Figure by the author.)

considerable amount of settling of living and non-living FPOM brought in by the river, as well as sedimentation of phytoplankton produced in the transition zone. Microbial decomposition of this material can cause hypoxic or anoxic conditions in the **hypolimnion**, or near-bottom layer of the reservoir (Fig. 11.6). This can lead to further desorption of phosphorus from the bottom sediments, making this nutrient available to the phytoplankton.

Downstream, the reservoir deepens and widens, becoming lake-like, and is referred to as the **lacustrine zone** (Fig. 11.5) Here, the finest particles tend to settle out, turbidity is lowest, and solar irradiance penetration deepest. The turbidity zonation in reservoirs is clearly visible from the air as one flies over the central and western portions of the USA (Fig. 11.4). In the lacustrine zone nutrients may become limiting to phytoplankton growth because of uptake in the transition zone and settling of suspended sediment-sorbed nutrients farther up the reservoir, and nutrient recycling between the zooplankton and phytoplankton communities is important (Kimmel et al. 1990). These functional horizontal zones are not static, but shift depending on water residence time in the reservoir. During droughts the lacustrine zone may extend upstream nearly to the stream input. Conversely, during periods of elevated rainfall and river flow the riverine zone may reach far downstream into the lower reservoir, virtually eliminating the lacustrine zone. Human manipulation of reservoir outflows for water supply, irrigation, and power generation will also affect horizontal zonation.

### 11.1.3 Reservoir Mixing

Aside from the horizontal zonation described above, there is vertical water column zonation that may or may not be similar to that of lakes, depending upon reservoir design and usage. The vertical zonation of natural lakes is greatly affected by water temperature, leading to temperature stratification (Fig. 11.6). In summer the upper layer heats up, and because warm water is lighter than colder water it sits above the cooler water in a lens. The thickness of the upper lens, called the **epilimnion**, depends upon solar heating, which is controlled by the length of the warm season, latitude, rainfall, and drought. Below the epilimnion lies a transitional zone referred to as the **thermocline**, where the most rapid temperature change occurs (Fig. 11.6). This zone of variable thickness is also called the **metalimnion**. The lowest layer of the lake is the zone of coldest water, called the **hypolimnion**. The metalimnion is a strong barrier to vertical mixing during summer, which impacts organism distribution. An important consideration is the impact on DO. The hypolimnion receives the organic debris from the layers above it, such as remains of algal blooms, decaying bodies of zooplankton, invertebrates, and fish, and organic debris from the floodplain. Thus, it is a zone where BOD can be high. With lack of exposure to oxygenated surface waters, the respiration caused by the BOD load can lead to hypoxia or even anoxia in the hypolimnion and bottom sediments, excluding sensitive species of invertebrates and fish. Reservoirs also experience vertical stratification, but often more locally. Since reservoirs are generally deepest near the dam, that is where the strongest stratification is usually found (Fig. 11.6). Arms of dendritic reservoirs may be shallow and not experience stratification.

The mixing of reservoirs is in part controlled by the same factors that impact lake mixing, but reservoir design, morphology, and human operations add additional considerations. Upper layers of lakes can mix as a result of wind activity. Wind blowing across a large lake can cause mixing down to several meters by creating Langmuir cells, which are near-surface cells that are either aligned with the wind or 0–20° at an angle from the wind. Such mixing will also impact large reservoirs, i.e. those with a lot of open water, referred to as fetch. Other sporadic mixing can arise from **seiches**, which are oscillations of the upper layers caused by wind forces causing a piling up of water, forcing the downwind side of a lake to rise and subsequently fall. Hurricanes cause seiches to occur in lower rivers and large estuaries by changes in barometric pressure forcing similar oscillations. Again, large reservoirs may experience such weather-induced seiches. However, seasonal mixing is a well-known phenomenon that occurs in lakes, referred to as lake turnover, where water temperatures in the epilimnion and hypolimnion equilibrate, allowing for complete water-column mixing. This typically occurs in spring and fall in temperate lakes and some reservoirs. Systems that turn over twice a year are called dimictic systems, systems turning over once a year are monomictic, and frequently mixed systems are polymictic. Systems that are permanently stratified or else rarely mix are referred to as meromictic.

So what additional factors impact mixing in reservoirs? Shape is one consideration. A long, narrow reservoir (i.e. Fig. 11.1) may not be impacted by Langmuir circulation or internal seiches due to the lateral lack of open-water fetch. Reservoirs that are highly dendritic (Fig. 11.2b) may have rather shallow upper arms that do not stratify; rather, temperatures throughout the water column may be similar enough to permit mixing in summer or winter when stratification affects deeper water bodies or deeper portions of the reservoir. These factors are essentially results of the design considerations of the reservoir builders (usually a municipality, regional authority, or public utility).

Another human-caused impact to reservoir circulation is the depth at which water is released from the dam. In some cases excess water flows over the dam (Fig. 11.7a) or through valves in

(a)        (b)

(c)        (d)

**Fig. 11.7** Fig. 11.7a (top left) Release of water over the top of a dam. Fig. 11.7b (top right) Hypolimnetic release from John Martin Reservoir, Colorado. Fig. 11.7c (bottom left) Lake Rogers, North Carolina, showing release over the top of a dam but adjoining emergency spillway. Fig. 11.7d (bottom right) Release over a spillway. (Photos by the author.)

the upper dam, with little impact on stratification. Upper dam release also results in more accumulation of sediments, organic material, and nutrients behind the dam (Kennedy 2001). In other circumstances there may be a hypolimnetic release of waters to the receiving stream below the dam (Fig. 11.7b). Drawing out the colder hypolimnetic water may impact mixing and position of the thermocline and thickness of the epilimnion and hypolimnion. The depth where DO concentrations change most rapidly (located at the metalimnion) is referred to as the **oxycline**. Water flowing over a dam or down a spillway (Fig. 11.7c, d) or out through upper drainage valves will be better oxygenated entering the tailwaters than that coming out from hypolimnetic releases.

Finally, water residence time (WRT) within a reservoir can vary far more than within a natural lake for a variety of reasons. WRT in a reservoir is basically the reservoir volume divided by the discharge (see Leopold and Wohlman 1957 and Walks 2007 for more on reservoir and river water residence times):

$$\mathrm{WRT} = \text{length} \times \text{width} \times \text{depth} \left( \mathrm{km}^3 \right) /$$

$$\text{discharge} \left( \mathrm{km}^3 / \mathrm{day} \right)$$

Run-of-the-river reservoirs have short WRTs and can be quite turbid (Kennedy 2001). High flushing will exacerbate the nutrient load that the associated reservoir is subject to (Jones et al. 2008a) regardless of watershed land use. If a reservoir is primarily built for flood control, its outflow will be manipulated for that reason above all others. If it is built for water supply or irrigation and it is near full pool, extensive releases can occur; conversely, during drought considerable consumption can occur.

Reservoirs can be subject to seasonally variable water residence time, with much shorter WRT during high inflow periods such as winter precipitation or spring snowmelt and longer WRT in summer. Reservoirs that are more likely to have longer WRT are tributary reservoirs in general (these can be deep and are fed by smaller streams rather than main stems), southeastern reservoirs (where snowmelt is usually not an issue), and industrial cooling reservoirs (which seek to maintain a consistent level for industrial or power plant cooling needs). As will be discussed in the next two sections, WRT is an important determinant of phytoplankton and zooplankton community abundance and composition.

## 11.2 Feeding the Reservoir Biotic Communities

Whereas most riverine food webs are supported by the benthic algal and macroinvertebrate communities and/or allochthonous materials from the floodplain, most reservoir food webs are plankton-based. Reservoir waters are deeper and wider and have (usually) lower velocity than rivers, so where turbidity permits, rich phytoplankton communities can develop. The phytoplankton communities are fed by nutrient loading, which can be considerable in many reservoirs. For instance, reservoirs used for agricultural irrigation often in turn receive fertilizer-rich runoff from planted fields. The amount of agricultural land in a reservoir's drainage basin has been strongly correlated with reservoir nutrient concentrations in widely varying geographic regions (Bott et al. 2006; Jones et al. 2008a). Reservoirs that provide water to nearby municipalities in turn are subject to receiving urban and suburban stormwater runoff from developments. This is exemplified by the North Carolina Piedmont region, where several large reservoirs supply drinking water to nearby cities (Touchette et al. 2007).

### 11.2.1 Reservoir Phytoplankton

Many phytoplankton taxonomic groups are found in reservoirs, although in numbers and biomass a few major groups dominate. Diatoms are present year-round, both centric and pennate species, but this group dominates the flora in spring (see below). Cyanobacteria (blue-green algae) are the most abundant taxa group in summer, although in reservoirs in warmer climates they can dominate for much of the year (Touchette et al. 2007). Chlorophytes (green algae) are present year-round and are often most abundant in late spring and fall. Chrysophytes occur year-round, usually in low to moderate numbers, although some species can form problematic blooms. Cryptomonads are flagellated species, some of which are heterotrophs that are usually present in low number but can bloom at times. Euglenophytes are also commonly found in reservoirs—some of these taxa are myxotrophic as well.

Many temperate reservoirs have seasonal phytoplankton productivity and taxa group cycling that are similar to lakes. Such cycling is discussed in detail in Horne and Goldman (1994) and Wetzel (2001) and is generalized as follows. In winter, planktonic productivity, like terrestrial primary productivity, is at a minimum. In northern climes such as Alaska, Canada, the northern USA, and central Europe many reservoirs (as well as lakes) are iced over, sometimes to considerable depth. High-altitude reservoirs in mountainous zones are likewise subject to ice cover. Ice is less dense than water and floats over it, so water freezes first at the surface and the freezing works downward. The ice and snow covering greatly reduces light availability for phytoplankton and other aquatic plants. In less northern areas (such as the US Mid-Atlantic Piedmont region) reservoirs will not ice-over completely but shallow upper arms may see an ice covering. In addition to ice and snow reducing light penetration the short daylengths and reduced solar irradiance levels contribute to plant growth limitation. In winter the water column is relatively stable, zooplankton abundance and grazing is low, and the phytoplankton community has low biomass consisting largely of small flagellated cryptomonads, chrysophytes, dinoflagellates, and green algae, the species of which are adapted to low temperature and low light conditions.

As spring comes on things change dramatically. Increasing water temperatures melt ice coverings. Increased solar irradiation intensity and depth penetration, and increasing daylength provide a stimulus to phytoplankton and aquatic macrophytes. At the same time, nutrient (N and P) loading to reservoirs and lakes increases, providing additional stimulus. This nutrient increase occurs because with snowmelt comes a surge of runoff. The runoff carries with it nutrients remineralized from organic matter on the landscape (animal manure, decaying vegetation, nutrients associated with suspended sediments, etc.), which enters streams and rivers and subsequently reservoirs and lakes. Additionally, agricultural fields are fertilized and some portion of this fertilizer exits the fields in stormwater runoff and makes its way to reservoirs and lakes. During this period reservoir and lake waters are well mixed due to spring overturn, bringing up Si from the sediments and maintaining diatoms in the euphotic zone. This results in a spring phytoplankton bloom. While a number of taxa groups increase at this time, the most prominent group of spring bloomers are the diatoms. Diatoms are known to be strong competitors for N and P when Si is abundant (Sommer 1988); Si enters water bodies in runoff from lithogenic sources. Additionally, in early spring grazing pressure by zooplankton is still low.

As spring moves into summer the situation changes considerably. Nutrient inputs to the reservoir are curtailed because there is less farm field fertilization and increased interception of runoff on the vegetated landscape. As indicated above, thermal stratification sets in and water-column mixing is reduced or essentially nonexistent in strongly stratified situations. With the onset of stratification less motile phytoplankton such as diatoms settle out of the water column. The reduced nutrient availability and grazing by a now-abundant population of zooplankton cause the demise of spring blooms. Thus, recycling of nutrients by zooplankton in the upper water column becomes increasingly important in maintaining phytoplankton productivity in the euphotic zone. In early summer there may be a "clear-water" phase caused by the grazing and lack of new nutrients. Cyanobacterial blooms commonly occur in mid-to-late summer and these organisms have adaptations to maintain dominance in summer (see reviews by Paerl 1988 and Burkholder 2002). Heterocyte-bearing genera such as *Anabaena* are able to fix N from the atmosphere, giving a competitive advantage (see Chapter 3). Some cyanobacterial taxa can regulate their depth in the water column by creation and

destruction of gas vesicles (contained within gas vacuoles)—rising to the surface in daylight and sinking into more nutrient-rich hypolimnetic waters in the evening. Surface scum-forming cyanophytes can intercept solar irradiance before it reaches other taxa groups within the water column, giving a competitive advantage. Cyanobacteria have UV-resistant carotenoid pigments to protect them at the surface from harsh sunlight (see Chapter 3). Notably, cyanobacteria make generally poor zooplankton food, because of either toxigenic properties or mucilaginous sheaths that deter effective grazing (Arnold 1971).

With fall, cyanobacterial dominance generally ends and chlorophytes, diatoms, and various flagellates are more prominent. As water temperatures decrease there may be a fall overturn, allowing for remixing of hypolimnetic nutrients into the upper water column. Daylength grows shorter and solar irradiance decreases, however, so blooms are less common. Eventually the onset of winter reduces reservoir phytoplankton productivity to minimal levels. It is also important to remember that members of several phytoplankton taxa groups, such as dinoflagellates, cryptomonads, and euglenophytes, can supplement photosynthesis or subsist wholly through heterotrophy (consuming DOC compounds) and phagotrophy (engulfing particulate prey by various means). These organisms can maintain a presence in the water column when light or nutrients are limiting for photosynthetic phytoplankton.

Current velocity, which controls water residence time also structures the phytoplankton community (Fig. 11.8). With slow currents the phytoplankton community has time to develop and the reduced current allows for sedimentation of suspended solids, reducing turbidity and permitting increased light transmission. With rapid currents the phytoplankton community is suppressed, although the increased velocity entrains benthic diatoms into the plankton as tychoplankton (Beaver et al. 2013, 2015); such diatoms can survive at low light levels (Sommer 1988). With elevated current, phytoplankton productivity is likely to be limited by reduced light transmission from increased turbidity (Fig. 11.8), while in periods of slow flow (and reduced turbidity) nutrient limitation is more likely to occur (Koch et al. 2004).

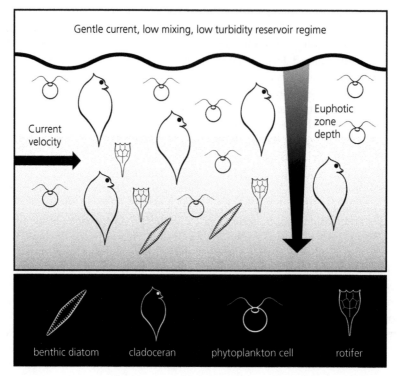

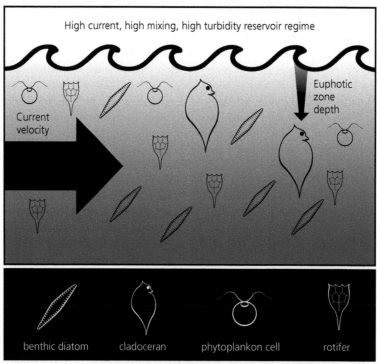

**Fig. 11.8** Top: depiction of a reservoir plankton community response to a gentle flow situation; water column is well lit, permitting good phytoplankton production and growth of crustacean zooplankton. Bottom: depiction of a reservoir plankton community response to a rapid system discharge (i.e. low WRT); phytoplankton are light-limited by turbidity, but tychoplankton (such as pennate diatoms) are swept into the water column by high current, and the current suppresses cladoceran and copepod production, although rotifers do well (Koch et al. 2004; Walks 2007, Beaver et al. 2013; 2015). (Figure by the author.)

Periphyton can be important in some reservoir situations. Benthic microalgae in reservoirs are usually limited to a rather narrow band around the shoreline, unless the reservoir is very shallow. Bear in mind that reservoirs are often turbid, so sufficient light is limiting in waters of appreciable depth. If the reservoir level is relatively stable a macrophyte community may develop around the shoreline or in shallow reservoir arms. Macrophytes offer substrata for epiphytic microalgae, which in turn are prey for littoral zooplankton and other invertebrate grazers. Macrophytes additionally offer small fish habitat, both as cover from predators and as areas in which to feed.

## 11.2.2 Reservoir Zooplankton

The zooplankton community is the trophic link between the phytoplankton community and planktivorous fish, especially the larvae of many fish species (Cramer and Marzolf 1970; Siefert 1972; Marzolf 1990). The typical reservoir zooplankton fauna primarily contains cladocerans, copepods, (primarily calanoids and cyclopoids), and rotifers (Table 11.4; see also Chapter 5). A typical reservoir taxonomic assemblage generally consists of three to six cladoceran species, one or two calanoid copepod species, two to five cyclopoid copepod species, and five to fifteen planktonic rotifer species, as well as some number of protozoan taxa. The littoral zone will also have its own fauna, primarily consisting of chydorid cladocerans and bdelloid

**Table 11.4** List of common planktonic reservoir zooplankton genera divided by suborder; note that genus and species names will vary geographically.

| Cladocera | Copepoda | Rotifera |
|---|---|---|
| Bosmina | Cyclops | Asplanchna |
| Ceriodaphnia | Diaptomus | Brachionus |
| Daphnia | Mesocyclops | Conochilus |
| Diaphanosoma | Tropocyclops | Conochiloides |
| Holopedium | | Filinia |
| Leptodora | | Hexarthra |
| Moina | | Kellicottia |
| | | Keratella |
| | | Platyias |
| | | Polyarthra |
| | | Synchaeta |

rotifers and harpacticoid copepods. Cladocerans use their appendages to create feeding currents and are typically herbivorous on phytoplankton and whatever nutritious detritus is present. Cladocerans can reproduce parthenogenically to take advantage of a significant food source while it is available (r-selected behavior), such as a reservoir algal bloom. It is notable that many reservoirs have also acquired invasive zooplankters, especially large predatory cladocerans such as *Bythotrephes longimanus*, *Cercopagus pengoi*, and others that can outcompete native species such as *Daphnia lumholzii* (see Chapter 14). In many circumstances several reservoirs may be built on the same river system, facilitating spread of invasive pelagic species.

Calanoid copepods are typically grazers of phytoplankton and protozoans and are largely suspension feeders that create a feeding current (DeMott and Watson 1991), although some species can be predaceous. Reservoirs often contain two or even more species of the calanoid copepod *Diaptomus* (or its congeners) which may be separated temporally but may overlap during some periods. Diaptomids in temperate reservoirs tend toward spring and fall peaks and winter minima. Cyclopoid copepods do not create feeding currents and are raptorial feeders (De Mott and Watson 1991) and largely predaceous upon smaller zooplankton as well as some phytoplankton; some predaceous cyclopoid copepods can consume larval fish. *Mesocyclops* (typically *M. edax*) is a predaceous cyclopoid copepod and may be found throughout the year, as is the much smaller cyclopoid copepod *Tropocyclops* (often *T. prasinus*). Copepods have an extraordinary sensory system concerning potential food items. Their first antennae contain sensory hairs called mechanoreceptors that pick up hydrodynamic disturbances caused by potential prey items. Mechanoreception appears to be the primary means that copepods detect prey items at a distance, with calanoids cuing in on larger phytoplankton particles and cyclopoids cuing in on motile prey that create disturbances in the water (DeMott and Watson 1991). In closer proximity copepods also can use both contact and distance chemoreceptors to discriminate food quality or toxicity (reviewed by Price 1988). Copepods will reject food at the mouth if it is toxic or otherwise unpalatable or difficult to process, but can also orient filaments of benign algae at the mouth to con- sume them (Price 1988; DeMott and Moxter 1991). Useful information on specific feeding behaviors of zooplankton can be found in Allan (1976), Koehl and Strickler (1981), and Davis (1984).

Rotifers are abundant and diverse in reservoirs. Most rotifers can reproduce parthenogenically and, like cladocerans, can thus reproduce rapidly to take advantage of algal blooms. Most rotifers feed by producing feeding currents using their ciliated coro- nas and normally consume small phytoplankton; bacteria are consumed incidentally as well (Turner and Tester 1992). A set of hardened structures called trophi are located below the cilia and grind food. A few rotifer genera, particularly *Asplanchna* and *Syn- chaeta*, are predaceous. The smallest grazers found in reservoirs are protozoans, which are typically members of three functional groups (flagellates, ciliates, and amoebae). As noted previously (see Chapter 3), some flagellates are heterotrophic or myxotrophic and thus considered protozoans by various researchers, while other researchers con- sider them algae. Protozoans feed by a variety of means, including engulfing prey, creating feeding currents using ciliary rings, and collecting small particles on pseudopodia or in protoplasmic nets. They consume bacteria and small phytoplankton (nanoplankton and picoplankton) as well as other minute particles. Because protozoans vary widely in size, in reservoirs the method of sampling will determine the number of taxa collected; the ciliates *Codonella*, *Difflugia*, and *Vorticella* are periodically caught in small mesh-sized zooplankton net hauls. Most protozoans are only quantifiable by collecting whole water samples and concentrating them. Some protozoans, such as species of *Epistylis*, are parasitic upon fish and crustaceans in reservoirs and rivers.

Seasonally, in temperate reservoirs the lowest zooplankton abundances occur in winter. In spring there is often a bloom of cladocerans, particularly daphnids, which graze the spring phytoplankton bloom, and diatoms are generally considered a good food for grazers (Porter 1977; Crumpton and Wet- zel 1982; Willen 1991). Often two species of *Daphnia* co-occur, with one usually larger than the other and able to consume larger food items. A lesser fall peak of daphnids also generally occurs. *Ceriodaphnia* occurs variably in spring and fall and can also bloom in summer, usually in somewhat lesser abundances

than *Daphnia*. *Ceriodaphnia* is commonly used as a test organism for toxicity of industrial effluents. Its ability to reproduce parthenogenically yields genetic copies that provide excellent replicates for experimental tests of varying concentrations of potential toxins. *Bosmina* occurs generally from spring through fall and *Diaphanosoma* is typically a summer species. *Holopedium* is a summer clado- ceran enclosed in a gelatinous matrix and when it blooms it is well known (and thoroughly despised) for clogging up the nets of fishermen and limnolo- gists. Warm water temperatures are often positively correlated with the abundance of larger zooplank- ton in reservoirs (Beaver et al. 2013, 2015). In some reservoirs low water temperatures have been found to favor rotifers (Beaver et al. 2013, 2015); in contrast rotifers have been correlated with warmer water in the lower Missouri River (Havel et al. 2009), and in Southeastern USA reservoirs rotifer seasonality can be quite variable (Mallin 1986). Clearly, factors other than water temperature (WRT, food resources, predation) are likely involved in river and reservoir rotifer population seasonality.

WRT within a reservoir is an important factor determining zooplankton composition, seasonality, and abundance, as well as phytoplankton commu- nity structure and which factors limit phytoplank- ton productivity. During periods of low WRT crus- tacean zooplankton are generally scarce in the main channel water column (Walks 2007; Beaver 2013, 2015) (Fig. 11.8). High flushing can cause zooplank- ton losses through advection out of the system and can cause physical damage to the organisms. Also, the species with a slower growth rate (such as larger crustaceans) will require a longer residence time to successfully reproduce, and periods of rapid flush- ing will not permit that at times (Walks 2007). In con- trast, organisms such as rotifers have a much more rapid growth rate (Walks 2007) and can succeed in periods of reduced WRT; this has been verified in field studies (Havel et al. 2009; Beaver 2013, 2015).

### 11.2.3 Reservoir Benthos

The benthic invertebrate community of reservoirs can differ considerably from that of rivers, depend- ing upon WRT. Streams and upper rivers are fed largely by periphyton and detritus (i.e. consumed by shredders, scraper-grazers, and filter-feeders).

In a run-of-the-river reservoir with a short WRT many of these organisms will continue to play a role in the benthos. However, in reservoirs with moderate to long WRTs only the zones adjacent to the main feeder river, or the feeder stream function in this manner. In deeper reservoir waters the benthic community is fed by plankton-based detritus (sedimenting phytoplankton or dead zooplankton) which is consumed at the sediments by collector-gatherers, especially chironomid larvae and oligochaete worms. Filter-feeding bivalves can be found but are limited by solid substrata to attach to (and may become problematic in certain situations; see Section 11.8). Other larval invertebrates will predate upon zooplankton in the water column; the ambush predator *Chaoborus* (called the "phantom midge") can vertically migrate and is a good example of this strategy. In shallow reservoir arms where aquatic vegetation is abundant invertebrates adapted to piercing and feeding upon vegetation, such as hemipterans, may be abundant. Predatory coleopterans and odonates that feed on other invertebrates will always be present as well. In reservoirs with stable shorelines various invertebrate larvae as well as terrestrial insects are common among littoral vegetation. However, in reservoirs that have unstable shorelines due to water level fluctuations (Fig. 11.3) littoral invertebrate communities may be depauperate.

Bivalves are common in reservoirs, but unfortunately in the USA reservoir bivalve communities are presently dominated by invasive species. The Asiatic clam (*Corbicula fluminea*) was introduced into US waters in 1980 and is currently well established throughout most major waterways. The zebra mussel (*Dreissena polymorpha*) was established in the USA around 1988 and is expanding its range rapidly at present. In massive Lake Mead, a major reservoir on the Colorado River, the quagga mussel (*Dreissena bugensis*) is now a major presence due to the abundant rocky substrata that promotes settling (Beaver et al. 2010). These invasive species and others are discussed in Chapter 14.

### 11.2.4  Reservoir Fish

An overarching fact to bear in mind when discussing reservoir fisheries is that very few reservoirs

that were built for municipal or agricultural water supply or for industrial cooling water have a naturally occurring community of fishes exclusively derived from the feeder river or streams. Since a side value of such reservoirs is to provide the public with recreation, especially fishing opportunities, such reservoirs are invariably stocked with warmwater game and forage fish by managers. In some cases, especially reservoirs in the Southwestern USA, game-fish stocking has been deliberate but nonofficial (see Chapter 14). Outside of mountain hydroelectric power reservoirs, reservoirs are often built in warm climates (where glaciers did not impact the landscape). Thus, the fish community in nonmountain reservoirs is typically dominated by warmwater rather than coldwater species. Because of the lake–reservoir differences described above, the fish community must be able to deal with shifting water levels, a less-developed littoral zone, and turbidity that can at times be severe enough to impact feeding. Reservoirs are often stocked with forage and game fish as well (Table 11.5), sometimes with non-native species (see Chapter 14).

Within the reservoir environment fish are distributed according to habitat preference (O'Brien 1990), with pelagic (i.e. open-water) areas used by walleye, crappies, yellowfin perch, gizzard, and threadfin shad. Littoral fishes include many sunfishes and minnows, and bottom species include carp, carpsuckers, buffaloes, and various catfishes.

Hydroelectric power plant reservoirs built in mountainous areas can host coldwater fish species such as salmonids and trout. As mentioned in Chapter 10, dam building usually has severe impacts on these migratory species and special dam adaptations and fish transportation are required to maintain their populations.

Most reservoir fish species feed on zooplankton as larvae (Cramer and Marzolf 1970; Siefert 1972) and many switch to larger invertebrates as they grow into juveniles. Some species, notably gizzard and threadfin shad, feed selectively on zooplankton as larvae and later by pumping water through their gill rakers and straining out zooplankton and large phytoplankton (Cramer and Marzolf 1970). This feeding mode may aid in their success in turbid reservoirs, where sight-feeding may be limited. Various minnows and silversides consume large

**Table 11.5** Typical US warmwater reservoir fish, including species often stocked. (From Noble 1981; O'Brien 1990.)

| Sport fish | Forage fish |
|---|---|
| Largemouth bass *Micropterus salmoides* | Gizzard shad *Dorosoma cepedianum* |
| White bass *Morone chrysops* | Threadfin shad *D. petenense* |
| Striped bass *M. saxatilis* | Bluegill *Lepomis macrochirus* |
| Walleye *Sander vitreum* | Golden shiner *Notemigonis crysoleucas* |
| Flathead catfish *Pylodictus olivaris* | Fathead minnow *Pimephales promelas* |
| Channel catfish *Ictalurus puctatus* | Alewife *Alosa pseudoharengus* |
| Crappie *Pomoxis* spp. | Silversides order Atheriniformes |

zooplankton as adults by sight-feeding in the shallows (Noble 1981). Littoral fishes such as sunfishes consume benthic invertebrates as well as terrestrial insects that enter the water nearshore. Some reservoirs have been stocked with herbivorous grass carp or tilapia to control excessive vegetation (Leslie et al. 1987); however, such fish can become too successful and decimate macrophytes to an alarming degree (Crutchfield et al. 1992). Piscivores such as largemouth and white bass, walleye, and crappie feed heavily on gizzard and threadfin shad as well as various sunfishes. Some bottom feeders such as large catfish are piscivorous as well, while others such as carp are generalists that consume whatever they can find that is palatable. Humans, in turn, often prefer piscivores such as largemouth bass and walleye as sport fish, but also fish for sunfish and catfishes.

Many studies have concluded that complex habitat in the form of littoral zone macrophyte vegetation is a benefit to the fishery in general. Such habitat provides nesting areas for various species including largemouth bass and black crappies, and provides a rich feeding area for littoral fishes seeking invertebrates among the vegetation. The vegetation additionally provides cover for young or small fish from larger predators. However, as noted earlier, many reservoirs experience frequent drawdowns according to human usage and drought, so such reservoirs have an undeveloped shoreline (Fig. 11.3) that is inhospitable to littoral macrophytes, benthic invertebrates, and littoral fish. Notably, reservoir drawdown has been used as a management technique to restructure fish populations as well (Noble 1981). Lake Okeechobee is a natural lake in southern Florida, but its water levels are highly variable due to both rainfall and human manipulation through canals and water regulation structures. Research conducted over several years (Havens et al. 2005) found that largemouth bass populations fared poorly under two scenarios: 1) when there was prolonged deep water that reduced the biomass and extent of submersed aquatic vegetation, and 2) when waters were shallow and vascular SAV coverage was low. In contrast, bass had high recruitment when there were moderate water depths and a large and diverse community of vascular SAV, including *Vallisenaria*, *Hydrilla*, *Potamogeton*, *Eleocharis*, and *Paspalidium*. Human development along reservoir shorelines may cause loss of the natural vegetated littoral zone, even in reservoirs that have stable shorelines. As such, Purcell et al. (2013) analyzed fish abundance and diversity in relation to levels of shoreline development in an Alabama reservoir. They found fish abundance and diversity was highest along undeveloped shorelines, second highest along riprap areas, and poorest along bulkheaded areas.

Because reservoirs are so widely used for sport, and even commercial fishing, it is of interest to fishery managers to try to enhance fisheries. As such, a number of efforts have been made toward predicting fish biomass in reservoirs using morphological, chemical, and biotic parameters. The morphoedaphic index (MEI) has been used for decades to predict fish yield and production in lakes and reservoirs:

$$MEI = \text{total dissolved solids (mg/L)} /$$
$$\text{mean depth (m)}.$$

A number of studies have demonstrated good correlations between the MEI and fish production (Ryder et al. 1974; Oglesby 1977). Elevated dissolved solids combined with generally shallow depth provide elevated fish production, whereas lower dissolved solids and deeper water lead to less-productive systems. It has been suggested that in shallower systems the littoral zone is larger, nutrients are recycled more rapidly there, and primary productivity is higher, leading to greater secondary and tertiary production (Hanson and Leggett 1982). There is generally good correlation between total dissolved solids (TDS) and TP in reservoirs, and this is a likely reason for the link to fish fertility (Hanson and Leggett 1982). In a set of southern Appalachian reservoirs there were strong correlations determined between fish standing stock and TP concentrations, with the strongest relationship between planktivores and P, but weakest between piscivores and P (Yurk and Ney 1989). Michaletz et al. (2012) also found only a weak relationship between the piscivorous largemouth bass and P in a set of Missouri reservoirs.

Finally, trophy fishing in reservoirs, particularly for largemouth bass, is a cottage industry in itself. Tournaments featuring prizes are commonly held on some of the larger Southeastern USA dendritic reservoirs and fishermen pay large sums for expensive bass boats and gear for the chance to nab a record fish. Research has determined that growth of sport fishes is related to lake fertility (as P) except in the case of hypereutrophic systems and their inherent poor water quality, whereas increased suspended solids concentrations were generally negatively associated with sport fish growth and size structure (Michaletz et al. 2012). Again, having good coverage of a diverse community of littoral coverage of submersed and emergent vegetation is important in generating good recruitment of largemouth bass (Havens et al. 2005).

## 11.3 Ecological Impacts of Reservoir Turbidity

Due to construction, drainage areas, and geographic location, reservoirs tend to be more turbid than natural lakes, with the riverine zone most affected (Fig. 11.3a). The amount of sediments delivered to main-stem reservoirs leads to infilling, giving them a shorter geological lifespan than natural lakes

(Jones and Knowlton 2005). From Chapter 1 it has been established that suspended particles, particularly clays, adsorb numerous substances such as phosphorus, bacteria, metals, and DOM. Experiments and field observations have revealed that turbidity has profound impacts on reservoir biota in many ways, including organism distribution, productivity, and trophic interactions. Probably the most obvious is the deleterious impact turbidity has on reducing solar irradiance and the subsequent reduction of photosynthesis in phytoplankton, benthic microalgae, periphyton, and submersed aquatic vegetation (Wehr and Descy 1998; Koch et al. 2004); additionally turbidity reduces the efficiency by which chlorophyll *a* production responds to the commonly limiting nutrient P (Jones and Knowlton 2005). Regarding higher trophic levels, sight-feeding fishes suffer from reduced prey visibility in turbid waters (Gardner 1981; O'Brien 1990).

As noted (see Chapter 2), clay-suspended particles adsorb P; such particles can also serve as a source of P for phytoplankton in reservoir situations (depending on the situation). Research by Cuker et al. (1990) used limnocorral experiments (large plastic in-situ enclosures) adding montmorillonite clay (0.01–2.0 µm), which causes more turbidity and stays in suspension longer than larger kaolinitic clay (0.1–5.0 µm). They found that experimental additions of montmorillonite clay caused greater suppression of primary productivity, chlorophyll *a*, and algal abundance than kaolinite additions. However, additions of P (as in eutrophic situations) can partially overcome the suppression of algal productivity and growth in clay addition treatments. The P fertilization promoted clearing of kaolinitic turbidity, but was less effective with montmorillonite clay turbidity. Clearing of turbidity by P additions is apparently a result of enhanced co-flocculation and subsequent settling of the suspended clay particles (Burkholder et al. 1998). Regarding phytoplankton species effects, experimental clay turbidity additions have led to increases in cyanobacteria relative to other taxa (Cuker et al. 1990; Burkholder et al. 1998). To investigate species-specific impacts of clay and P introductions in a reservoir situation, Burkholder (1992) combined limnocorral experiments with track light microscope autoradiography and additions of radiolabeled $^{33}PO_4^{3-}$.

Her experiments revealed a number of profound turbidity impacts to phytoplankton beyond simple suppression or photosynthesis (Box 11.1).

> ### Box 11.1  Phytoplankton, nutrients, and clay turbidity interactions.
>
> Clay additions in reservoir limnocorrals caused dinoflagellates to aggregate onto clay particles and/or form cysts, eliminating most free-swimming cells from the plankton. Dinoflagellates that encyst or form temporary cysts can survive the suspended sediments inputs as they are protected from direct contact with the clay particle and can reemerge under better conditions (an illustration of how the temporary cyst mechanism enhances dinoflagellate survival). In response to suspended sediment additions colonies of the most abundant cyanobacterial taxon (*Merismopedia*) fragmented and colony size decreased, enabling more cells to survive free and reestablish the population later. However, colonial algae which did not fragment easily and had sheaths were strongly adsorbed by clay and settled into the sediments. P was taken up by free-swimming dinoflagellates; they aggregated with clay and *Merismopedia* as well as clay particles alone. The suspended sediment loading stimulated uptake of P among all the algal taxa analyzed, at a higher rate than before the clay treatment. Thus, additions of suspended sediment-bound P serve as an opportunity for algal cells to increase P reserves to survive low-nutrient conditions. Burkholder (1992) noted that there were many dinoflagellate-sourced cytoplasmic nets and peduncles present in the clay treatments, so heterotrophy is a strategy for survival under turbidity loading for dinoflagellates. Another conclusion reached was that the planktonic ecosystem can be resilient to intermittent turbidity loading in that the community regained much of its former structure seven days after the experimental loading events. Thus, periodic turbidity pulses into reservoirs can have immediate major impacts on all trophic levels. In the case of zooplankton, changing of food sources can augment a disrupted food supply, whereas phytoplankton have physical and behavioral adaptations that can at least partially compensate for turbidity pulse disruption.

Regarding animal distribution, Matthews (1984) analyzed how turbid inflows into reservoirs influence larval shad and drum behavior. Larval fish are sight-feeders—they respond to turbid inflows by concentrating in the upper water layers where there is more light; however, larger zooplankton normally vertically migrate downward in daytime to avoid sight-feeding predators. Thus, the larval fish that concentrate in the upper layer in order to see prey are actually food-stressed; gut content analysis showed little or no food in their digestive tracts. As another example, Cuker (1993) used further reservoir limnocorral experiments including fish additions (sunfish, bass) to examine trophic interactions with clay turbidity additions. Under control conditions, i.e. no clay additions, when sunfish were added the phantom midge larva *Chaoborus* was reduced, freeing zooplankton from predation from this invertebrate. However, with clay turbidity, added visual predation on *Chaoborus* from fish was reduced, so more *Chaoborus* survived, leading to more predation on zooplankton. The rotifer community was similar in all treatments; likely they showed selective feeding on algal cells rather than inorganic particles. Cuker (1993) suggested that turbidity from suspended clay (or dense phytoplankton blooms) creates a refuge from predation for *Chaoborus* in shallow reservoirs, like darkness does.

Thus, turbidity inputs influence the zooplankton community through top-down (vertebrate or large invertebrate grazing) food chain effects. Turbidity inputs also influence zooplankton through bottom-up (food source) impacts. Research by Cuker and Hudson (1992) found that additions of kaolinite had little effect on zooplankton, but montmorillonite caused strong numerical reductions of crustacean zooplankton and rotifers (*Diaptomus* and *Diaphanosoma* were most tolerant to clay additions). The reductions were likely caused by losses in the phytoplankton (the zooplankton food base) through clay-induced light limitation and P sequestration. Clay also likely interfered with zooplankton feeding, filling them up with clay instead of algae. Arruda et al. (1983) found that zooplankton such as daphnids can ingest small suspended sediment particles; however, ingestion rates of phytoplankton prey and incorporation of the prey items were depressed with higher amounts of suspended sediments. These authors suggested that clay particles, particularly small ones with large surface/volume ratios, can adsorb DOM, and daphnids can survive and grow on such amended particles.

## 11.4 Dam Tailwaters

The regulated outflow from a reservoir is released below the dam, into an area known as the **tailwaters**. While this is a lotic, rather than lentic, ecosystem, its biotic community is very different from the river or stream upstream or well downstream of the reservoir. The outflow from the dam contains an abundance of phytoplankton and zooplankton of all sizes, and the plankton from the dam release can persist for several kilometers downstream before dissipating (Armitage and Capper 1976; Novotny and Hoyt 1982). Regarding zooplankton, abundances of copepods and cladocerans diminish first, with rotifers maintaining significant concentrations farther downstream (Novotny and Hoyt 1982; Herlong and Mallin 1985; Havel et al. 2009). Losses of zooplankton downstream from dams can occur from predation by fish or larger invertebrates, physical destruction through turbulence and substrata contact, and rapid transit times that are too short to allow for successful reproduction (Armitage and Capper 1976); Herlong and Mallin 1985; Walks 2007). The dam releases structure the benthic invertebrate and fish communities of the tailwaters.

For example, upstream of Robinson Impoundment, South Carolina, phytoplankton community densities in Black Creek were very low (Table 11.6) and zooplankton abundance approximately 500/m³ (mostly rotifers and copepod nauplii). In the immediate tailwaters, phytoplankton densities were approximately 4000/ml and zooplankton densities 150 times greater than in the inflowing stream, with abundant adult copepods and cladocerans. Upstream of the impoundment the macroinvertebrate snag community was diverse (Table 11.6) and dominated by the filter-feeders *Simulium* and

*Tanytarsus*, utilizing UFPOM such as bacteria and detritus. In the immediate tailwaters there was decreased macroinvertebrate diversity, with the community dominated by filter-feeding caddisflies *Hydropsyche* and *Neureclipsis crepecularis*, utilizing the larger FPOM in the form of the outpouring reservoir plankton. Two and a half kilometers downstream of the dam there was a 76% decrease in zooplankton, with an 83% decrease in crustacean zooplankton and 60% decrease in rotifers. The benthic community became more diverse and capable of utilizing UFPOM, paralleling the decrease of larger zooplankton and retention of small zooplankton and phytoplankton. The benthic community at this point exhibited a trend toward returning to the diverse food sources it utilized upstream of the impoundment in the blackwater stream. High zooplankton and macroinvertebrate abundance in tailwaters leads to elevated numbers of consumers in the form of finfish and the larger piscivorous fish that consume them. Fishermen are aware of this and dam tailwaters are often considered "hotholes" for sport fishermen.

## 11.5 Southeastern Reservoirs

Many of the reservoirs in the Piedmont and Upper Coastal Plain regions of the Southeastern USA differ from reservoirs outside of these areas in a number of ways. Land slopes are generally more gradual the closer to the ocean one gets; thus dams are often constructed in the upper reaches of catchments where slope is steeper and the reservoir can hold more water with less surface area affected. Instead of a single inflowing channel there are often several low-order streams that enter southeastern reservoirs, causing a dendritic shape (Fig. 11.2b). Thus, the horizontal biotic and abiotic zonation common in

**Table 11.6** Mean total phytoplankton and zooplankton and various benthic invertebrate functional group abundances upstream (US) of Robinson Impoundment, just below the dam in the tailwaters (TW), and 2.5 km downstream of the dam (Herlong and Mallin 1985).

| Site | Phytoplankton (no./ml) | Zooplankton (no./m³) | Filter-feeders (no./m²) | Collectors (no./m²) | Predators (no./m²) | Scrapers (no./m²) |
|------|------|------|------|------|------|------|
| US | 230 | 536 | 505 | 217 | 126 | 45 |
| TW | 3935 | 79,916 | 741 | 82 | 3 | 2 |
| 2.5 km | 3233 | 19,238 | 337 | 159 | 89 | 51 |

other types of reservoirs can be absent from many southeastern reservoirs. Shorelines are irregular and characterized by coves, islands, and embayments; i.e. they are morphologically complex (Soballe et al. 1992). Causeways for automobile traffic are often constructed across various arms of the reservoirs, further increasing habitat heterogeneity. Because of the many microenvironments in and tributaries to these systems there can be considerable spatial heterogeneity in the flora and fauna within southeastern reservoirs (Mallin 1986). However, because shallow coves characterize these systems, nuisance aquatic macrophyte growths can be problematic (Crutchfield et al. 1992), particularly in combination with the long growing season in the southeast. Such invasive macrophytes are easily spread from reservoir to reservoir through small boat bilges and attached to propellers and anchors (see Chapter 14).

## 11.6  New Reservoirs

Aside from the impacts mentioned in Chapter 10, construction of a new reservoir dramatically changes the trophic ecology of the river ecosystem. Engineers determine the area of the immediate catchment that will be filled and bulldozers go to work clearing the trees and other standing terrestrial vegetation. Some of the trees may be removed as lumber if they have such value, whereas others are burned (creating greenhouse gases). Regardless, a lot of woody and nonwoody organic material is left on the area that is to be flooded. Once the dam is closed there is a major decrease in stream flow and the formerly lotic system assumes lentic characteristics. Whereas in the stream system the major autotrophs may have been periphyton and rooted aquatic macrophytes, the increased water residence time allows for the development of the phytoplankton community. The sources of nutrients for the phytoplankton community are incoming stream waters, as well as leachate from newly submerged soils and the decay of submerged organic matter.

Because there is so much submerged organic material, there is a major increase in the abundance and activity of organisms involved in the decay process, particularly bacteria and fungi. The abundance of microbial heterotrophs can cause hypoxic conditions, particularly in the deeper areas of the new reservoir. These microbes also serve as food for protozoans, some species of benthic macroinvertebrates, and planktonic and benthic zooplankton. In shallow, less turbid areas the submerged timber serves as substrata for periphyton and their associated benthic grazers. Thus, many new habitats are formed for benthic and littoral species. The new zooplankton community is dominated by r-selected species (mainly rotifers and cladocerans) that consume the readily available and filterable bacteria and phytoplankton. As noted earlier, these species rapidly reproduce by parthenogenesis (asexual reproduction) to take advantage of the abundant food. With the cessation of stream flow, lentic fauna replace the lotic benthic macroinvertebrate fauna that characterized the stream. Filter-feeding caddisfly larvae that are current-dependent for food supply are replaced by a variety of collector-gatherers and predators characteristic of quiescent waters. In deeper areas, chironomids and oligochaetes that can survive low DO environments predominate (Hergenrader and Lessig 1980). Some new reservoirs may contain standing trees. These trees, along with fallen trees, provide hard surface area for the growth of periphyton, a key food source for many invertebrates (Kimmel et al. 1990). The physical presence of standing and fallen timber serves as habitat for chironomids and bivalves, among other invertebrates, and also provides cover for young fish to avoid predators (Baxter 1977).

The concomitant increases in heterotrophic microbes, phytoplankton, zooplankton, and benthic grazers is termed **trophic upsurge** and culminates in high fish productivity. This trophic upsurge is well known to sport fishermen who avidly seek out new reservoirs for a productive fishing experience, and typically lasts from 5 to 20 years. The high fish production is a result of the abundance of planktonic and benthic food material, the availability of cover and habitat for young fish among the inundated vegetation, and the increase in fish spawning sites among the flooded terrestrial vegetation. Finally, fish and wildlife agencies, and power companies who use the reservoir, will usually stock both sport and forage fish to stimulate sport fishing interest (and generate favorable publicity) in the new reservoir. New reservoirs also become habitat for resident and migrating waterfowl (Baxter 1977), which brings in waterfowl hunters, providing good public relations for the reservoir operators and this subset of stakeholders.

## 11.7 Eutrophication and Reservoirs

Riverine eutrophication has been discussed previously in relation to nutrients (see Chapter 2) and phytoplankton (see Chapter 3). Reservoirs can be highly subject to eutrophication (Whittier et al. 2002; Touchette et al. 2007). Reservoirs built for irrigation often contain significant agricultural lands in their watersheds and their tributary streams thus receive nutrient-rich runoff from fertilized fields (Jones et al. 2008a). Water supply reservoirs built near cities are subject to urban and suburban runoff. Reservoir chlorophyll *a* concentrations are often highly correlated with concentrations of TN and TP (Whittier et al. 2002; Jones and Knowlton 2005; Touchette et al. 2007; Burkholder et al. 2022). Eutrophication in reservoirs usually results in excessive phytoplankton growth (often of cyanobacteria), leading to hypoxia issues, taste and odor problems for water supplies, or blooms of toxic algae. In some shallow reservoirs, macrophyte growth can also be exacerbated by nutrient loading.

### 11.7.1 Trophic Classification of Reservoirs

Classification of reservoirs and lakes by trophic state has great utility, not only in giving researchers a rather instantaneous picture of the system, but also in providing a basis for determining what systems should be restored (or simply improved) and how well such a process worked if attempted. Classification has been frequently attempted using a number of methods. Commonly, researchers have determined mean concentrations of TN, TP, and chlorophyll *a* that place reservoirs (and lakes) into trophic categories, and in some cases subcategories (Table 11.7).

Most classification does require some degree of subjectivity, due to issues with geography, hydrology, system morphology, etc. Carlson (1977) proposed a trophic state index (TSI) that avoided the use of the trophic state terminology as above, and instead used a numerical value (from 0 to 100) to categorize lakes and reservoirs based on either Secchi depth (SD), TP concentration, or chlorophyll *a*, as in many inland waters these parameters are statistically interrelated (Table 11.8). He considered a new trophic state as being created with each new doubling of chlorophyll *a*.

Such a numerical index has particular value when comparing a number of lakes and reservoirs to each other, and also providing a means to make comparisons over time for individual systems if the system is either becoming more eutrophic or improving due to positive management actions. Carlson (1977) noted that problems with the index could arise when using SD if the system is highly colored by DOM or has high inorganic turbidity, and with using TP under N-limited conditions. A variety of other classification schemes have been devised over the years as well.

**Table 11.7** Trophic states of lakes and reservoirs based on average concentrations and ranges of nutrients and chlorophyll *a*; note that liberties were taken interpreting Wetzel's (2001) Table 15–13 in that he further subdivided into additional categories such as ultraoligotrophic, oligomesotrophic, mesoeutrophic, and hypereutrophic.

| Trophic state | TN (μg/L) | TP (μg/L) | Chl *a* (μg/L) | Reference |
|---|---|---|---|---|
| Oligotrophic | < 600 | < 10 | < 3 | Wetzel (2001) |
| | < 500 | < 25 | < 8 | Dodds et al. (1998) |
| Mesotrophic | 500–1100 | 10–30 | 2–15 | Wetzel (2001) |
| | 500–1260 | 25–71 | 8–25 | Dodds et al. (1998) |
| Eutrophic | > 1100 | > 30 | > 15 | Wetzel (2001) |
| | > 1260 | > 71 | > 25 | Dodds et al. (1998) |

Wetzel (2001) based on opinions of numerous authors.
Dodds et al. (1998) statistical analysis based on 403 lakes and reservoirs.

**Table 11.8** TSI and associated parameters. (From Carlson 1977.)

| TSI | SD (m) | TP (µg/L) | Chl a (µg/L) |
|-----|--------|-----------|--------------|
| 0   | 64     | 0.75      | 0.04         |
| 10  | 32     | 1.50      | 0.12         |
| 20  | 16     | 3.00      | 0.34         |
| 30  | 8      | 6.00      | 0.94         |
| 40  | 4      | 12.0      | 2.60         |
| 50  | 2      | 24.0      | 6.40         |
| 60  | 1      | 48.0      | 20.0         |
| 70  | 0.5    | 96.0      | 56.0         |
| 80  | 0.25   | 192.0     | 154.0        |
| 90  | 0.12   | 384.0     | 427.0        |
| 100 | 0.062  | 768.0     | 1183.0       |

## 11.7.2 Eutrophication and Hypoxia

As in rivers and estuaries, decaying algal blooms in reservoirs become sources of labile BOD, which can contribute to reservoir hypoxia in bottom waters; occasionally such hypoxic zones can also reach vertically well up in the water column nearly to surface waters (Burkholder et al. 2022). Hypoxia is exacerbated by temperature stratification and is likely to be found in the deepest waters nearest the dam (Fig. 11.6). As noted previously, strongly hypoxic waters will exclude most species of fish and degrade the benthic invertebrate community. Regarding the benthic community, eutrophication-induced hypoxia drives the invertebrate community to one dominated by *Tubifex* worms and chironomids, which are tolerant of low DO (Hergenrader and Lessig 1980). The benthos are a major trophic link to fishes, and thus large hypoxic areas will impact the health of the fishery. The impact of hypoxia may not be confined to the reservoir hypolimnion, however. With reservoir dams that are designed to release hypolimnetic water downstream from the dam, this can present a stressful situation where hypoxic waters can impact the receiving stream for an extended period. This will stress or drive away less-tolerant fish taxa and kill sensitive benthic invertebrates in the tailwaters. Since, as described in Section 11.4, tailwaters are often prime targeted places for fishermen, such hypoxic releases can prove problematic. Other water quality issues with dam releases are discussed in Chapter 10.

## 11.7.3 Reservoir Toxic Algal Blooms

Toxic algal blooms have long been associated with reservoirs, particularly when the reservoirs have been situated in agricultural watersheds that provide elevated nutrient loading to the reservoir. Commonly, cyanobacterial blooms cause death and illness among domesticated or wild animals that drink or swim in lakes or reservoirs containing such blooms (Repavich et al. 1990). Even human deaths have been recorded. In a Brazilian dialysis center in 1996, 76 patients receiving hemodialysis treatments died after intravenous exposure to the cyanotoxin microcystin, at concentrations of 19.5 µg/L (Carmichael et al. 2001). The toxin had apparently entered the center's water supply from its water source, a reservoir that was afflicted with microcystin-producing cyanobacterial blooms. Nonfatal effects on humans of water contact with cyanobacterial blooms include allergic reactions, skin irritation, and gastrointestinal disorders (Falconer 1989). Cyanobacterial toxicity or the presence of the toxin microcystin has been detected in many reservoirs in surveys from diverse geographical areas (Repavich et al. 1990; Touchette et al. 2007; Graham and Jones 2009), and was correlated with chlorophyll *a*, TN, and TP in a set of 177 Missouri reservoirs (Graham and Jones 2009). Since cyanobacteria make up large portions of the phytoplankton community in many reservoir systems (Touchette et al. 2007; Jones et al. 2008a, b) the potential for cyanobacteria-derived problematic levels of toxin production will likely increase as reservoirs age and become more eutrophic. In municipal water supply reservoirs cyanobacterial blooms can cause taste and odor problems that the water treatment plant must spend additional taxpayer dollars to remove before distribution to the public.

Another toxic algal taxonomic group that has become problematic in US southwestern and southeastern reservoirs within the past few decades is the Haptophytes, of the class Prymnesiophyceae (see Chapter 3). These are minute single cells with two flagella and a third flagella-like structure called a haptonema, with different morphology than the two flagella; they are photosynthetic organisms. Some haptophytes are toxic and cause large fish kills, especially *Prymnesium parvum* in fresh-to-brackish reservoirs. This species is called "golden algae" and has been causing blooms and large fish kills in US reservoirs since 1985 (Roelke et al. 2011). This species is generally believed to have invaded the USA (Texas) from ship ballast waters.

## 11.8 Reservoirs and Non-Native Species

Reservoirs are constructed where there was no lake previously and provide new lentic habitats for numerous non-native plant and animal species. In a large regional comparison within the US Northeast, reservoirs were found to contain more exotic species than lakes (Whittier et al. 2002). Reservoirs built to supply cooling water to steam electric plants or for other industrial uses are often planned with a public recreational component, usually as part of the deal with state or local agencies or municipalities to build the system. Since this involves construction of boat landings and subsequent heavy use by fishermen or other recreational users, there is an elevated risk of invasive species entering the reservoir through boat bilges or attached to boat propellers or trailers (accidental introductions; see Chapter 14). Shallow arms of reservoirs are especially favorable environments for invasive aquatic macrophytes, which can crowd out native vegetation and impede boat traffic. Invasive molluscs prefer areas where there is a steady current, and water intake pipes can become clogged, with additional operating costs for clearing the intake systems to the utility or industry (see Chapter 14 for more on costs from invasive species).

In many areas reservoirs provide habitat in series on major waterways which permits non-native species to reach new ranges, sometimes with spectacular success. This has led to establishment of invasive invertebrates throughout entire major watersheds. Examples of such extensive and rapid spreading include the Asian clam *Corbicula* and the zebra and quagga mussels in the US East and Midwest, and the recent major western expansion of the quagga mussel into the vast Lake Mead ecosystem on the Colorado River (Beaver et al. 2010). Regarding the American West, as noted earlier, many large reservoirs have been constructed in rather arid terrain. The native fish populations that originally entered these new lentic systems have been significantly transformed through the deliberate introductions of game fish that anglers brought from lakes and reservoirs in the eastern USA (Rahel 2000; see more on this in Chapter 14).

## 11.9 Summary

- Reservoirs are typically located in areas where natural lakes are few. They are built for flood control, irrigation, municipal water supply, and hydroelectric power generation.
- Reservoirs typically have more irregular shorelines than natural lakes, are deepest at the dam rather than the middle, and have shorter WRT and less-developed shorelines due to human water use demands.
- Reservoirs are usually more turbid, more nutrient-loaded, and more eutrophic than natural lakes. Nuisance and toxic algal blooms are common in reservoirs.
- Many reservoirs have horizontal zonation with a riverine, transitional, and lacustrine zone. However, reservoirs with a relatively long WRT have seasonal biotic and productivity patterns similar to natural lakes.
- Short WRTs cause elevated turbidity, lower phytoplankton productivity, a sparser crustacean zooplankton community, but increased meroplanktonic algae and rotifer abundances.
- Reservoirs in the southeastern USA differ from most other reservoirs in that they are more dendritic, have more complex plankton communities, are fed by lower-order streams, and are built farther up in the watershed.
- Reservoirs can have excellent fishing due to stocking of sport and forage fishes, but are more subject to invasive plant, invertebrate, and fish species than natural lakes.
- The tailwaters below dams receive high numbers of zooplankton and phytoplankton from the reservoir lacustrine zone and support a large number of benthic consumers and fish that prey on them. This adds up to a well-known sport fishery.

# References

Allan, J.D. 1976. Life history patterns in zooplankton. *The American Naturalist* 110:165–180.

Armitage, P.D. and M.H. Capper. 1976. The number, biomass and transport downstream of microcrustaceans and *Hydra* from Cow Green Reservoir (Upper Teesdale). *Freshwater Biology* 6:425–432.

Arnold, D.E. 1971. Ingestion, assimilation, survival, and reproduction by *Daphnia pulex* fed seven species of blue-green algae. *Limnology and Oceanography* 16:906–920.

Arruda, J.A., G.R. Marzolf and R.K. Faulk. 1983. The role of suspended sediments in the nutrition of zooplankton in turbid reservoirs. *Ecology* 64:1225–1235.

Baxter, R.M. 1977. Environmental effects of dams and impoundments. *Annual Review of Ecology and Systematics* 8:255–283.

Beaver, J.R., T.E. Tietjen, B.J. Blasius-Wert, et al. 2010. Persistence of *Daphnia* in the epilimnion of Lake Mead, Arizona–Nevada, during extreme drought and expansion of invasive quagga mussels (2000-2009). *Lake and Reservoir Management* 26:273–282.

Beaver, J.R., D.E. Jensen, D.A. Casamatta, et al. 2013. Response of phytoplankton and zooplankton communities in six reservoirs of the middle Missouri River (USA) to drought conditions and a major flood event. *Hydrobiologia* 705:173–189.

Beaver, J.R., K.C. Scotese, E.E. Manis, S.T.J. Juul, J. Carroll and T.R. Renicker. 2015. Variation in water residence time is the primary determinant of phytoplankton and zooplankton composition in a Pacific Northwest reservoir ecosystem (Lower Snake River, USA). *River Systems* 21:261–275.

Bott, T.L., D.S. Montgomery, D.B. Arscott and C.L. Dow. 2006. Primary productivity in receiving reservoirs: links to influent streams. *Journal of the North American Benthological Society* 25:1045–1061.

Burkholder, J.M. 1992. Phytoplankton and episodic suspended sediment loading: phosphate partitioning and mechanisms for survival. *Limnology and Oceanography* 37:974–988.

Burkholder. J.M. 2002. Cyanobacteria. Pages 952–982 in Bitton, G. (ed.), *Encyclopedia of Environmental Microbiology*. Wiley, New York.

Burkholder, J.M., L.M. Larsen, H.B. Glasgow, K.M. Mason, P. Gama and J.E. Parsons. 1998. Influence of sediment and phosphorus loading on phytoplankton communities in an urban Piedmont reservoir. *Journal of Lake and Reservoir Management* 14:110–121.

Burkholder, J.M., C.A. Kinder, D.A. Dickey, et al. 2022. Classic indicators and diel dissolved oxygen versus trend analysis in assessing eutrophication of potable-water reservoirs. *Ecological Applications* 32:e2541.

Carlson, R.E. 1977. A trophic state index for lakes. *Limnology and Oceanography* 22:361–369.

Carmichael, W.W., S.M.F.O. Azevedo, J.S. An, et al. 2001. Human fatalities from cyanobacteria: chemical and biological evidence for cyanotoxins. *Environmental Health Perspectives* 109:663–668.

Cole, G.A. and P.E. Weihe. 2016. *Textbook of Limnology*, 5th edition. Waveland Press, Long Grove, IL.

Cramer, J.D. and G.R. Marzolf. 1970. Selective predation on zooplankton by gizzard shad. *Transactions of the American Fisheries Society* 99:320–332.

Crumpton, W.G. and R.G. Wetzel. 1982. Effects of differential growth and mortality in the seasonal succession of phytoplankton in populations in Lawrence Lake, Michigan. *Ecology* 63:1729–1739.

Crutchfield, J.U., D.H. Schiller, D.D. Herlong and M.A. Mallin. 1992. Establishment and impact of two *Tilapia* species in a macrophyte-infested reservoir. *Journal of Aquatic Plant Management* 30:28–35.

Cuker, B.E. 1993. Suspended clays alter trophic interactions in the plankton. *Ecology* 74:944–953.

Cuker, B.E. and L. Hudson Jr. 1992. Type of suspended clay influences zooplankton response to phosphorus loading. *Limnology and Oceanography* 37:566–576.

Cuker, B.E., P.T. Gama and J.M. Burkholder. 1990. Type of suspended clay influences lake productivity and phytoplankton community response to phosphorus loading. *Limnology and Oceanography* 35:830–839.

Davis, C.C. 1984. Planktonic copepoda (including Monstrilloida). Chapter 4 in Steidinger, K.A. and L.M. Walker (eds.), *Marine Plankton Life Cycle Strategies*. CRC Press, Boca Raton, FL.

DeMott, W.R. and F. Moxter 1991. Foraging on cyanobacteria by copepods: responses to chemical defenses and resource abundance. *Ecology* 72:1820–1834.

DeMott, W.R. and M.D. Watson. 1991. Remote detection of algae by copepods: responses to algal size, odors and motility. *Journal of Plankton Research* 13:1203–1222.

De Villiers, M. 2000. *Water: The Fate of our Most Precious Resource*. Houghton Mifflin, Boston.

Dodds, W. and M. Whiles. 2010. *Freshwater Ecology: Concepts & Environmental Applications of Limnology*, 2nd edition. Elsevier, Amsterdam.

Dodds, W.K., J.R. Jones and E.B. Welch. 1998. Suggested classification of stream trophic state: distributions of temperate stream types by chlorophyll, total nitrogen, and phosphorus. *Water Research* 32:1455–1462.

Falconer, I. 1989. Effects on human health of some toxic cyanobacteria (blue-green algae) in reservoirs. *Toxicity Assessment: An International Journal* 4:174–184.

Gardner, M.B. 1981. Effects of turbidity on feeding rates and selectivity of bluegills. *Transactions of the American Fisheries Society* 110:446–450.

Graf, W.L. 1999. Dam nation: a geographic census of American dams and their large-scale hydrologic impacts. *Water Resources Research* 35:1305–1311.

Graham, J.L. and J.R. Jones. 2009. Microcystin in Missouri reservoirs. *Lake and Reservoir Management* 25:253–263.

Hanson, J.M. and W.C. Leggett. 1982. Empirical prediction of fish biomass and yield. *Canadian Journal of Fisheries and Aquatic Sciences* 39:257–263.

Havel, J.E., K.A. Medley, K.D. Dickerson, et al. 2009. Effect of main-stem dams on zooplankton communities of the Missouri River (USA). *Hydrobiologia* 628:121–135.

Havens, K.E., D. Fox, S. Gornak and C. Hanlon. 2005. Aquatic vegetation and largemouth bass population responses to water-level variations in Lake Okeechobee, Florida (USA). *Hydrobiologia* 539:225–237.

Hergenrader, G.L. and D.C. Lessig. 1980. Eutrophication of the Salt Valley Reservoirs, 1968-73 III. The macroinvertebrate community: its development, composition, and change in response to eutrophication. *Hydrobiologia* 75:7–25.

Herlong, D.D. and M.A. Mallin. 1985. The benthos–plankton relationship upstream and downstream of a blackwater impoundment. *Journal of Freshwater Ecology* 3:47–59.

Horne, A.J. and C.R. Goldman. 1994. *Limnology*, 2nd edition. McGraw-Hill, New York.

Hueftle, S. 1995. Lake Powell, the future of a reservoir. *Lakeline* 15:20–23.

Jones, J.R. and M.F. Knowlton. 2005. Chlorophyll response to nutrients and non-algal seston in Missouri reservoirs and oxbow lakes. *Lake and Reservoir Management* 21:361–371.

Jones, J.R., M.F. Knowlton and D.V. Obrecht. 2008a. Role of land cover and hydrology in determining nutrients in mid-continent reservoirs: implications for nutrient criteria and management. *Lake and Reservoir Management* 24:1–9.

Jones, J.R., D.V. Obrecht, B.D. Perkins, et al. 2008b. Nutrients, seston, transparency of Missouri reservoirs and oxbow lakes: an analysis of regional limnology. *Lake and Reservoir Management* 24:155–180.

Kennedy, R.H. 2001. Considerations for establishing nutrient criteria for reservoirs. *Lake and Reservoir Management* 17:175–187.

Kimmel, B.L. and A.W. Groeger. 1984. Factors controlling primary production in lakes and reservoirs: a perspective. *Lake and Reservoir Management* 1:277–281.

Kimmel, B.L., O.T. Lind and L.J. Paulson. 1990. Reservoir primary production. Pages 133–194 in Thornton, K.W., B.L. Kimmel and F.E. Payne (eds.), *Reservoir Limnology: Ecological Perspectives*. John Wiley & Sons, New York.

Koch, R.W., D.L. Guelda and P.A. Bukaveckas. 2004. Phytoplankton growth in the Ohio, Cumberland and Tennessee Rivers, USA: inter-site differences in light and nutrient limitation. *Aquatic Ecology* 38:17–26.

Koehl, M.A.R. and J.R. Strickler. 1981. Copepod feeding currents: food capture at low Reynolds number. *Limnology and Oceanography* 26:1062–1073.

Leonard, J. and P. Crouzet. 1999. Lakes and Reservoirs in the EEA Area. Topic Report No. 1. European Environment Agency, Copenhagen.

Leopold, L.B. and M.G. Wolman. 1957. River Channel Patterns: Braided, Meandering and Straight. Geological Survey Professional Paper 282-B. US Geological Survey, United States Department of the Interior, Baltimore, MD.

Leslie, A.J., J.M. Van Dyke, R.S. Hestand and B.Z. Thompson. 1987. Management of aquatic plants in multi-use lakes with grass carp *Ctenopharingodon idella*. *Lake and Reservoir Management* 3(1):266–276.

Mallin, M.A. 1986. Zooplankton community comparisons among five southeastern United States power plant reservoirs. *Journal of the Elisha Mitchell Scientific Society* 102:25–34.

Marzolf, G.R. 1990. Reservoirs as environments for zooplankton. Pages 195–208 in Thornton, K.W., B.L. Kimmel and F.E. Payne (eds.), *Reservoir Limnology: Ecological Perspectives*. John Wiley & Sons, New York.

Matthews, W.J. 1984. Influence of turbid inflows on vertical distribution of larval shad and freshwater drum. *Transactions of the American Fisheries Society* 113:192–198.

Michaletz, P.H., D.V. Obrecht and J.R. Jones. 2012. Influence of environmental variables and species interactions on sport fish communities in small Missouri impoundments. *North American Journal of Fisheries Management* 32:1146–1159.

Milly, P.C.D. and K.A. Dunne. 2020. Colorado River flow dwindles as warming-driven loss of reflective snow energizes evaporation. *Science* 367:1252–1255.

Noble, R.L. 1981. Management of forage fishes in impoundments of the southern United States. *Transactions of the American Fisheries Society* 110:738–750.

Novotny, J.F. and R.D. Hoyt. 1982. Seasonal zooplankton concentrations in Barren River Lake and tailwater, Kentucky. *Journal of Freshwater Ecology* 1:651–662.

O'Brien, W.J. 1990. Perspectives on fish in reservoir ecosystems. Pages 209–226 in Thornton, K.W., B.L. Kimmel and F.E. Payne (eds.), *Reservoir Limnology: Ecological Perspectives*. John Wiley & Sons, New York.

Oglesby, R.T. 1977. Relationships of fish yield to lake phytoplankton standing crop, production, and morphoedaphic factors. *Journal of the Fisheries Research Board of Canada* 34:2271–2279.

Paerl, H.W. 1988. Nuisance phytoplankton blooms in coastal, estuarine and inland waters. *Limnology and Oceanography* 33:823–847.

Poff, N.L. and J.C. Schmidt. 2016. How dams can go with the flow. *Science* 353:1099–1100.

Porter, K.G. 1977. The plant–animal interface in freshwater ecosystems. *American Naturalist* 65:159–170.

Price, H.J. 1988. Feeding mechanisms in marine and freshwater zooplankton. *Bulletin of Marine Science* 43: 327–343.

Purcell, T.R., D.R. DeVries and R.A. Wright. 2013. The relationship between shoreline development and resident fish communities in a Southeastern US reservoir. *Lake and Reservoir Management* 29:270–278.

Rahel, F.J. 2000. Homogenization of fish faunas across the United States. *Science* 288:854–856.

Repavich, W.M., W.C. Sonzogni, J.H. Standridge, R.E. Wedepohl and L.F. Meissner. 1990. Cyanobacteria (blue-green algae) in Wisconsin waters: acute and chronic toxicity. *Water Research* 24:225–231.

Roelke, D.L., J.P. Grover, B.W. Brooks, et al. 2011. A decade of fish-killing *Prymnesium parvum* blooms in Texas: roles of inflow and salinity. *Journal of Plankton Research* 33:243–253.

Ryder, R.A., S.R. Kerr, K.H. Lolvrus and H.A. Regieri. 1974. The morphoedaphic index, a fish yield estimator—review and evaluation. *Journal of the Fisheries Research Board of Canada* 31:663–688.

Siefert, R.E. 1972. First food of larval yellow perch, white sucker, bluegill, emerald shiner and rainbow smelt. *Transactions of the American Fisheries Society* 101:219–225.

Soballe, D.M. and B.L. Kimmel. 1987. A large-scale comparison of factors influencing phytoplankton abundance in rivers, lakes, and impoundments. *Ecology* 68:1943–1954.

Soballe, D.M., B.L. Kimmel, R.H. Kennedy and R.F. Gaugush. 1992. Reservoirs. Chapter 11 in Hackney, C.T., S.M. Adams and W.H. Martin (eds.), *Biodiversity of the Southeastern United stats: Aquatic Communities*. John Wiley & Sons, New York.

Sommer, U. 1988. Growth and survival strategies of planktonic diatoms. Pages 227–260 in Sandgren, C.D. (ed.), *Growth and Reproductive Strategies of Freshwater Phytoplankton*. Cambridge University Press, Cambridge.

Stokestad, E. 2021. A voice for the river. *Science* 373:17–21.

Thornton, K.W. 1990a. Perspectives on reservoir limnology. Pages 1–14 in Thornton, K.W., B.L. Kimmel and F.E. Payne (eds.), *Reservoir Limnology: Ecological Perspectives*. John Wiley & Sons, New York.

Thornton, K.W. 1990b. Sedimentary processes. Pages 43–70 in Thornton, K.W., B.L. Kimmel and F.E. Payne (eds.), *Reservoir Limnology: Ecological Perspectives*. John Wiley & Sons, New York.

Thornton, K.W., B.L. Kimmel and F.E. Payne (eds.). 1990. *Reservoir Limnology: Ecological Perspectives*. John Wiley & Sons, New York.

Touchette, B.W., J.M. Burkholder, E.H. Hannon, et al. 2007. Eutrophication and cyanobacterial blooms in run-of-river impoundments in North Carolina, U.S.A. *Lake and Reservoir Management* 23:179–192.

Turner, J.T. and P.A. Tester. 1992. Zooplankton feeding ecology: bacterivory by metazoan microzooplankton. *Journal of Experimental Marine Biology and Ecology* 160:149–167.

Walks, D.J. 2007. Persistence of plankton in flowing waters. *Canadian Journal of Fisheries and Aquatic Sciences* 64:1693–1702.

Wehr, J.D. and J.-P. Descy. 1998. Use of phytoplankton in large river management. *Journal of Phycology* 34:741–749.

Wetzel, R.G. 2001. *Limnology: Lake and River Ecosystems*, 3rd edition. Academic Press, San Diego.

Whittier, T.R., D.P. Larsen, S.A. Peterson and T.M. Kincaid. 2002. A comparison of impoundments and natural drainage lakes in the Northeast USA. *Hydrobiologia* 270:157–171.

Willen, E. 1991. Planktonic diatoms—an ecological review. *Algological Studies* 62:69–106.

Wright, J.C. 1958. The limnology of Canyon Ferry Reservoir. I. Phytoplankton–zooplankton relationships in the euphotic zone during September and October, 1956. *Limnology and Oceanography* 3:150–159.

Wright, K.R. 2008. Ancestral Puebloan water handling. *Lakeline* 28(4):23–28.

Yurk, J.J. and J.J. Ney. 1989. Phosphorus–fish community biomass relationships in southern Appalachian reservoirs; can lakes be too clean for fish? *Lake and Reservoir Management* 5:83–90.

# CHAPTER 12

# Industrial Pollution of Streams and Rivers

Chemical pollution of streams and rivers is a vast subject, primarily caused by industries but also due to agriculture and other more diverse sources. Power generation plants are a good place to start, releasing chemical pollutants but also causing physical damage to the environment.

## 12.1 Power Plant Effects

Harnessing the power of water has been a major use of streams and rivers for centuries, largely for grinding and milling to produce foodstuffs (Walter and Merritts 2008). Early in the twentieth century it was discovered that the power of water could be exploited to create electrical power on a massive scale, termed hydroelectric power, sometimes shortened to hydropower. Briefly, this process consists of funneling the gravity-driven power of falling water at a dam to turn turbines of various types. The spinning of the turbines creates a magnetic field that produces electricity. The water used in such endeavors is not actually consumed, or chemically treated. Hydropower is considered to be renewable energy, as the same water in a river can be used by multiple hydroelectric plants along its length. Environmental impacts of hydropower primarily are dam-related and consist of impacts on biodiversity, anadromous species passage, floodplain inundation reduction, and riparian zone alteration, as discussed in Chapter 11, although hydropower dams with little storage capacity will have reduced impacts compared with large dams.

Of course, the distribution of hydroelectric power generation is limited by sufficient available geographical slope to make construction worthwhile;

i.e. it is most popular in or downstream of mountainous regions. By far the largest amount of electrical power is produced by fossil-fuel-based generation. In this process, a fuel (mainly coal, natural gas, and occasionally oil) is burned to produce heat to boil water and produce steam, which is used to spin turbines. These are referred to as steam electric plants. Nuclear fission is also used to generate heat and steam in some plants, but nuclear power has been less popular than it was in the past, before the accidents at Three-Mile Island in the USA and Chernobyl in Russia. The Fukishima reactor accident in Japan, caused by a tsunami, has further reduced the popularity of nuclear power plant construction (although power generation by such facilities does reduce reliance on greenhouse-gas-producing fossil fuels).

It has been well publicized in recent years how fossil-fuel-burning power plants (Fig. 12.1a) produce massive amounts of greenhouse gases, contributing to global climate change, sea level rise, and ocean acidification. These facilities can also be ecologically disruptive and polluting to rivers, reservoirs, and estuaries in a number of ways, some of which are quite insidious and little known to the public (or the scientific community at large). Ecological disruption to local water bodies from steam generation occurs largely because steam electric plants require vast amounts of water for cooling. Cooling can be accomplished by one of two methods. The first is called once-through cooling and it consists of large amounts of water being pumped from a river or reservoir into the plant to help cool the system, with the water subsequently being released to its source (or another water body). As

*River Ecology*. Michael A. Mallin, Oxford University Press. © Michael A. Mallin (2023). DOI: 10.1093/oso/9780199549511.003.0013

(a)                                                                              (b)

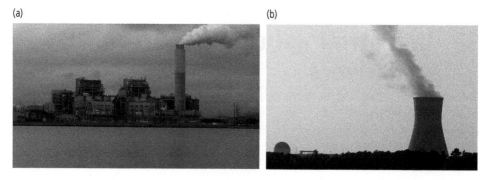

**Figure 12.1**   Fig. 12.1a (left) Coal-fired steam electrical generating plant near Asheville, North Carolina, located on Lake Julian, a constructed cooling pond. Fig. 12.1b (right) Cooling tower serving a nuclear power plant; the tower allows for recycling of cooling water, but evaporative losses can impact reservoir volume during drought periods. (Photos by the author.)

mentioned in the Chapters 10 and 11, many dams are built and reservoirs created to provide a dependable source of cooling water for nuclear or fossil-fuel power plants. Note that the cooling water released from nuclear power plants is nonradioactive. Water can also be recycled within the plant grounds and reused for cooling, which reduces the amount of outside cooling water used. When flying near or driving by a power plant, you can occasionally see large towers next to the plant, either rectangular or vase-shaped, with clouds billowing from them. The clouds consist of steam, not smoke. The structures are cooling towers (Fig. 12.1b) and they are used to reduce the temperature of the cooling water run through the plant, so that it may be recycled or safely released. Note that with the use of cooling towers there is a lot of evaporation from the site (Fig. 12.1b). This loss of local water can become ecologically significant during a summer drought, when excessive amounts of water may have to be drawn down from the reservoir, or operations may have to be reduced to use less water. The use of large quantities of cooling water also leads to a number of other physical, biological, and chemical impacts to the aquatic environment.

### 12.1.1   Impingement and Entrainment

A cooling system usually includes an intake canal from the water source to the plant, from where the cooling water is pumped into the system (Fig. 12.2). The pumping of cooling water into a plant pulls in large quantities of resident biota with the water,

including phytoplankton, zooplankton, meroplankton, and fish. This process is called **entrainment**. Intake canals and pipes are screened to prevent large living and nonliving objects from entering the cooling system (Fig. 12.3). Large quantities of phytoplankton, zooplankton, and fish larvae can be killed daily by passage through power plant cooling systems. This is the trade-off between use of cooling towers and the consequent evaporative loss of water (Fig. 12.1b) versus the expanded use of outside water (once-through cooling) with associated entrainment impacts to biota. Entrainment can kill or damage organisms by exposure to high temperatures or physical abrasion. The amount of mortality depends on how high the water temperatures get and how long the passage takes. Researchers performed a set of experiments in an apparatus designed to mimic mechanical, thermal, chemical (from chlorination), and pressure stresses that estuarine zooplankton or larval crustaceans might experience in passage (Bamber and Seaby 2004). Effects varied by species, with mechanical stress impacting only lobster larvae, chlorine causing significant mortality to copepods and shrimp larvae, pressure impacting copepods, and thermal stress impacting all species, especially synergistically with other stressors. Thermal mortality in the various species was estimated to occur between 34 and 37°C.

Little can be done about loss of plankton due to their small size. As mentioned, plant water intakes are screened to prevent larger fish and other organisms from entering the cooling system. However, the strong current generated by the water pumps

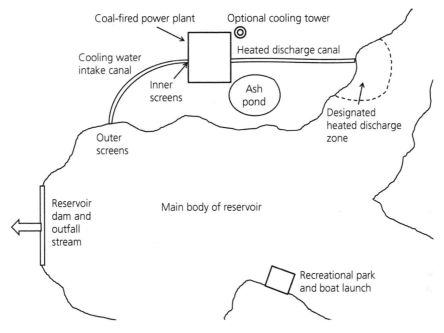

**Figure 12.2** Stylized depiction of a coal-fired power plant drawing cooling water from a reservoir (sometimes referred to as a cooling pond). Plans for company-funded public recreational facilities are often required by state agencies before reservoir construction is approved. (Figure by the author.)

forces fish and other organisms against the screens at high pressure, often resulting in their death, a process called **impingement**. In the 1960s this process was brought to the attention of the public by environmental activists, when thousands of striped bass killed by impingement by the Consolidated Edison Indian Point Plant on the Hudson River were photographed by wildlife officials. Large reductions in the mortality rate of finfish have been achieved by various intake modifications. One such modification (used by Duke Energy's Brunswick Nuclear Power Plant on North Carolina's Cape Fear River) is the use of rotating screens, which continually drop impinged fish into a receiving channel, from which they are released into a nearby marsh well away from the riverside intake area.

### 12.1.2 Thermal Pollution

Thermal stress impacts river and reservoir biota in several ways. Organisms that become entrained and enter the cooling system are subject to temperatures that may be lethal. Many power plants that are built on reservoirs are permitted to raise the water temperature in a designated mixing zone (Fig. 12.2) by a certain amount, set by government regulators. Temperatures within this mixing zone, particularly in summer, can prove lethal to sessile organisms and drive away or kill motile organisms. This zone can also serve as a thermal barrier to fish, so that they are unable to cross it to enter another segment of the reservoir. Water temperatures in mixing zones can reach 40°C or more during summer in reservoirs in the southeastern USA. Thermal tolerances of fish vary by species and can also vary within species according to geographic location and acclimation time. In general, many species are adversely affected by temperatures of 35°C or greater and suffer acute mortality at temperatures exceeding 40°C (Beitinger et al. 2000; Souchon and Tissot 2012). Importantly, larval fish often are less able to deal with thermal stress than adults (Souchon and Tissot 2012), which has implications for entrainment through power plant cooling systems. Many species of zooplankton become stressed in water temperatures exceeding 30°C and experience

**Figure 12.3** Outer screen protecting cooling water intake for Brunswick Nuclear Power Plant, a once-through cooling system drawing water from Cape Fear River Estuary; heated water is discharged into the Atlantic Ocean. (Photo by the author.)

mortality at temperatures between 35 and 40°C (Goss and Bunting 1976; Galkovskaja 1987; Mallin and Partin 1989).

While heated discharges are problematic in warmer months, during winter months mixing zones can serve as thermal refuges—areas of enhanced primary and secondary productivity. Such areas can attract or stimulate the growth of biota normally not present during winter, and even stimulate introduced species. This can lead to ecosystem-level effects. For example, Lake Julian is a power plant cooling reservoir in the North Carolina mountains near Asheville (Fig. 12.1a). In 1965 the North Carolina Wildlife Resources Commission introduced the blue tilapia (*Oreochromis aurea*), an African cichlid, into the reservoir. Whereas it normally could not survive winter in the North Carolina mountains, thermal loading from the power plant allows it to maintain a viable population year-round. In addition to supporting tilapia, though, the thermal discharge has the effect of stimulating

cool-weather growth of various species of filamentous green (*Spirogyra* and *Mougeotia*) and blue-green (*Oscillatoria*) algal growth. The blooms, in turn, are held in check by intense grazing pressure from the tilapia, which can digest normally unpalatable algae by very acidic (pH of approximately 2.0) stomach acids (Mallin 1986). Another example of ecosystem-level thermal effects occurred in South Carolina's Par Pond, a cooling water reservoir associated with a nuclear power steam electric plant (Janssen and Giesy 1984). In winter, thermally tolerant zooplankton became abundant in the hot water discharge pond that flows into Par Pond proper. These zooplankton blooms attracted large numbers of feeding blueback herring into the mixing zone, where excessive temperatures (> 33°C) incapacitated or killed many herring. Additionally, the thermal discharge killed bluegill venturing in. The abundance of dead and dying fish attracted largemouth bass to foray into this "hot spot" (although for a limited time) for a winter feast.

In terms of human health, thermal discharges into reservoirs can provide an attractive environment for the pathogenic amoebae *Naeglaria fowleri*. This microorganism enters swimmers through their nasal passages and moves into and multiplies in the nervous system, causing primary amoebic meningoencephalitis—which leads to cerebral hemorrhaging. While contracting this disease is rare, it is almost always fatal; of over 148 known cases reported since 1978, there have been only four survivors. *Naeglaria fowleri* thrives in naturally warm or artificially warmed lakes and reservoirs and is most likely to occur in the water column during summer. Stevens et al. (1977) surveyed a selection of control reservoirs versus reservoirs and rivers with power plant cooling water discharges, finding toxic strains of *N. fowleri* in several of the thermally enhanced systems but none in control reservoirs. In Europe another survey found *N. fowleri* present in thermal outfalls from a number of non-power plant factories (de Jonckheere and van de Voorde 1977). Thus, the presence of thermal outfalls enhances the abundance of this pathogen. As to bacterial pathogens, one study of cooling towers concluded that the potential for encountering aerosol-borne fecal bacteria was low (Adams et al. 1978).

### 12.1.3 Chemical Pollution

Power plants also discharge many chemicals into receiving rivers and reservoirs, such as byproducts of coal ash, metals leached from cooling systems, and cleaning compounds. As mentioned, radioactive materials are not discharged into receiving waters from nuclear power plants except during very rare catastrophic events, and will not be discussed here. Most chemical discharges are chronic outputs from fossil-fuel plants. While occasional massive releases of chemical pollutants from power plants will make news headlines (however briefly) the majority of citizens are unaware of chronic chemical discharges and their potential effects. Fossil-fuel burning is a dirty process, particularly when coal is used, and unwanted secondary effects can occur in the environment. Byproducts from coal burning include the toxic metals mercury, selenium, cadmium, and arsenic (Schmitt 1998), which are left in the ash residue. Other chemicals can be problematic depending upon the situation. To prevent long-distance airborne transport of ash and its associated pollutants, electrostatic precipitators are usually used to clear the ash from the system. The most common means of ash removal historically has been through wet-ash systems, where ash is wetted and pumped outside to be stored on-site in coal ash piles or ash ponds. Ash ponds should be lined to prevent leakage, but most are not, particularly older ones. Leaching into groundwater from unlined coal ash ponds occurs and threatens nearby water bodies and in some cases human drinking water supplies. States generally allow groundwater contamination in the immediate vicinity of the ash ponds (compliance boundary). About 300 coal ash ponds are believed to exist in the USA. Waste coal ash is sometimes recycled into road beds and concrete. The walls of ash ponds are also subject to breakage, resulting in occasional spectacular large-scale pollution incidents (see Chapter 17 regarding hurricane damage to a riverside ash pond).

A major ash pond pollution episode occurred on December 22, 2008 near Harriman, Tennessee, where the earthen wall containing a 40-acre coal ash pond for the Kingston Fossil Plant, operated by Tennessee Valley Authority (TVA), gave way and 5.4 million cubic yards (> 1 billion gallons or 3.8 billion L) of coal ash waste polluted 300 acres of land and entered the Emory River. In 2009 samples from the river showed arsenic contamination > 100 times the state arsenic standard. Three houses were destroyed and 12 damaged, and the operator, TVA, was fined $11.5 million for violations of water pollution and solid waste disposal laws. Later the utility agreed to settle property damages with 800 property owners for $27.8 million. Costs for cleanup and restoration combined were about $1.2 billion, and TVA agreed to a 30-year monitoring program at the site and is also converting its other waste ponds to dry-ash handling. Stream research showed that in areas of restricted flow, arsenic accumulated to unsafe levels in surface water and was diluted to low levels downstream in the Emory and Clinch Rivers, while arsenic and mercury accumulated to potentially toxic concentrations in the downstream river sediments (Ruhl et al. 2009). Further studies (Otter et al. 2012) demonstrated elevated levels of arsenic and selenium in largemouth bass, bluegill, white crappie, and redear sunfish from sites near the Kingston spill, with isotopic studies suggesting incorporation of the pollutants into the food web.

The incident had a large human health impact as well. In November 2018 a federal court ruled that the company that TVA hired to clean up the ash spill failed to provide protective measures for its workers on-site; according to various news sources, since the incident approximately 50 men have died and about 400 others became ill from exposure (various cancers and other ailments) during the long cleanup period. Court-ordered mediation has thus far failed to produce a settlement.

In February 2014 a major spill occurred at a North Carolina ash pond owned by Duke Energy from a 48-inch pipe that may have leaked for up to seven days. Approximately 39,000 tons of coal ash slurry spilled into the Dan River and travelled at least 70 miles downstream, coating the bottom. The US EPA required Duke Energy to dredge 2500 tons of ash from the river bed from an area abutting a dam, where it had accumulated, as well as another 500 tons from other river areas and from settling tanks in drinking water treatment plants. Beyond the short- and long-term impacts of such ash pond accidents, power plant discharges and drainages have caused chronic chemical damage to the aquatic community as well (Box 12.1).

---

### Box 12.1. Power Plant Chemical Pollution

Fossil-fuel-fired steam electric plants can cause chronic chemical pollution to nearby waterways, such as cooling reservoirs adjacent to the plant. Selenium is an example of a toxic byproduct of coal burning that can enter the environment through airborne ash or local runoff. In the 1970s and 1980s major selenium contamination problems occurred in two North Carolina power plant cooling reservoirs, Hyco Reservoir (then operated by the Carolina Power & Light Company) and Belews Lake (operated by Duke Power). These plants used wet-ash handling systems. Selenium entered the reservoirs from ash pile runoff and was biomagnified up the food chain. In Belews Lake 16 species of fish suffered severe reproductive failure and were eliminated from the reservoir. Species impacted by selenium toxicity in Hyco Reservoir included bluegill (*Lepomis macrochirus*), warmouth (*Lepomis gulosus*), redear sunfish (*Lepomis microlophus*),

black crappie (*Pomoxis nigromaculatus*), and largemouth bass (*Micropterus salmoides*). In the case of Hyco Reservoir, long-term monitoring programs and extensive laboratory experiments were carried out, funded by the power company. Affected reservoir fish were characterized by elevated selenium in liver and muscle tissue. Symptoms of exposure included outright mortality, reproductive failure, mouth deformities and skeletal deformities, cataracts, larval edema, and other maladies (note that the author of this book was employed as a biologist for the power company at that time and frequently saw the deformities on fish while sampling Hyco Reservoir; see Fig. 12.4). While selenium concentrations were also elevated in reservoir plankton and benthic invertebrates, only the fish suffered significant mortalities, attesting to the strong biomagnification pathway. Controlled experiments demonstrated that many of the symptoms seen in fish collected from the field could be mimicked by selenium addition, with larval survival or impacts caused by parental selenium exposure or water contact. The State of North Carolina subsequently reduced the selenium water quality standard from 10 to 5 µg/L, and fish consumption advisories were issued for all fish species in Hyco Reservoir fish. To comply with the new standard the Carolina Power & Light Company installed a $48 million dry fly ash handling system. In such systems fly ash removed by electrostatic precipitators is moved via forced air into concrete silos, from which it is either removed to a landfill or recycled for use in concrete. In response, during the late 1980s and throughout the 1990s selenium concentrations decreased significantly in plankton, benthic invertebrate, and fish tissue, and strong recoveries of species such as largemouth bass, sunfish, and crappie were documented in Hyco Reservoir. Duke Power also went to a dry-ash handling system for Belews Lake, but Belews Lake had been so severely impacted that restocking was required to produce a viable fish community

(Compiled from Lemly 1985; Gillespie and Baumann 1986; Woock et al. 1987; Crutchfield 2000; Lemly 2014; see also Section 12.3.5.)

---

Another type of chronic chemical pollution occurred in South Carolina's Lake Robinson, a reservoir fed by Black Creek, a naturally acidic blackwater stream. The reservoir was constructed to serve as a cooling pond for a nuclear power plant owned by the Carolina Power & Light Company. High concentrations of copper and zinc were found in the reservoir biota, causing deformities in sunfish and other finfish. The metals source was the brass tubing used in the cooling system. The acidic blackwater leached the copper and zinc from the tubes, allowing it to enter the food chain. The tubing subsequently was replaced with stainless steel, copper and zinc concentrations rapidly decreased, and deformities stopped. As the tubing was replaced, the reservoir phytoplankton community also underwent a shift in dominance, from diatoms to chlorophytes.

Thus, steam-generating power plants cause thermal injury and physical damage to nekton and plankton entrained in cooling waters. Thermal discharges of cooling waters have a variety of ecological impacts to reservoir- and river-receiving waters. Fossil-fuel-burning plants, especially coal-burning, produce chronic and occasionally acute pollution of waterways by various metals, which will be examined in more detail in the following section. Power plants discharge large amounts of environmentally harmful chemicals into the atmosphere through smokestack emissions as well (see subsequent material on acid precipitation in this chapter). Currently there is major emphasis on phasing out coal-fired power plants in favor of increased solar and wind power generation, although coal-fired generation remains strong in certain nations, particularly China and India.

## 12.2 Fossil-Fuel Extraction and Mining for Metals

### 12.2.1 Oil Extraction

Our discussion of chemical pollution to waterways now turns to problems associated with extraction of fossil fuels and metals from the Earth. We begin with oil. Oil has many uses, but by far its largest use is in transportation, in part as a lubricant but mainly when it is refined to gasoline to power cars and trucks, airplanes, and boats. Traditional **oil extraction** through wells offshore can lead to massive-scale marine oil spills. Oil spills can cause large quantities of pollutants to enter water bodies. When considering marine systems, the size and widespread impacts of some of these incidents have been truly astonishing; names that easily come to mind are the Deepwater Horizon spill in the Gulf of Mexico, the Exxon Valdez incident in Alaska, and

the Amoco Cadiz in France. Much of what we have learned regarding the physiological and ecological impacts of oil spills on aquatic biota has come about as a result of such major spills. However, significant spills in freshwater ecosystems occur all too frequently, generally as a result of pipeline leaks and breaks, and are usually underreported in the news media. However, the transport of oil and its derivatives on land by pipeline or vehicle also leads to spills that injure or even devastate aquatic ecosystems. A few of these are discussed below.

One of the largest, if not *the* largest, freshwater oil spill in the USA occurred in July 2010 when an Enbridge Energy pipe burst and approximately 1,000,000 gallons of heavy crude oil flowed through Talmadge Creek and then into the Kalamazoo River in Michigan. The oil originated in the Athabasca region of Canada. The spill polluted 56 km of the Kalamazoo River and the cleanup involved dredging of sediments as well as shoreline and floodplain cleanup, with costs believed to be approximately $765 million. The US Department of Transportation fined Enbridge $3.7 million for numerous violations associated with the spill. In June 2012 portions of the river were reopened. In July 2011 an ExxonMobil pipeline broke during a flood and spilled 42,000 gallons of oil into the Yellowstone River, with the oil spreading more than 100 km downstream. News reports indicated that at one point more than 1000 ExxonMobil contractors were involved in the cleanup, with only 1% of the spilled oil actually recovered. Exxon stated that the spill cost them about $135 million. In March 2013 the Department of Transportation fined ExxonMobil Corp. $1.7 million for pipeline safety violations. For a third example, a train derailment in 2018 led to a 230,000-gallon oil spill into the Rock River in northwest Iowa. While these are large incidents, numerous smaller oil spills and leaks occur regularly from transport systems to pollute fresh and estuarine waters.

Fish kills and oiled birds, mammals, and turtles are spill damages obvious to the public. However, more subtle impacts to biota are often not measured with long-term investigations except in the case of massive marine spills. PAHs are toxic components of crude oil and are carcinogens; such impacts, along with reproductive failures among biota, are more subtle effects. Of course, other impacts from oil spills to ecosystems occur from cleanup operations themselves. Crews use pressure hoses to flush out oiled shorelines, trample through habitats, scrape oiled areas using shovels and heavy equipment, and so on. Thus human use of fossil fuels for transportation as well as power generation imparts widespread impacts to water resources.

## 12.2.2. Tar Sand Oil Extraction

As traditional resources of oil shrink, other forms of oil are being avidly sought after. One of these sources is oil locked in **tar sands** (also called oil sands) which are essentially hydrocarbon-infused sandstone. By far the largest known site is in the Athabasca region of Alberta, Canada, where a major extraction operation is ongoing. This area contains an estimated 173 billion barrels of recoverable oil, and Canada is now a major supplier of oil to the USA. The product, called bitumen, is not pumped out of the ground but rather it is mined. Much of the bitumen lies at least 60 m below the surface, which is stripped away, including the overlying boreal forest. The bitumen that is dug out is processed by high-temperature and scalding water to produce the crude, which is then shipped to refineries in the USA. The process uses huge amounts of water, and the remaining liquid waste is a mixture of water, oil, sand, mercury, arsenic, and other toxins. Large amounts of greenhouse gases are released as well. Another, less-damaging process is to use a steam treatment deep underground to extract the product. The source of process water is the adjacent Athabasca River, which empties into Lake Athabasca 150 km to the north. The waste slurry (called tailings water) is pumped into vast ponds along the river banks, where slow dewatering occurs by various means. The ponds unfortunately attract flocks of migratory birds, which can suffer from various ailments and stresses associated with contact with the oil in the ponds. An analysis by Timoney and Ronconi (2010) found that whereas the industry self-reported 65 bird deaths annually, the actual number as derived using scientific analysis of raw company data (provided through a freedom of information request) ranged from 458 to 5029 birds annually, representing 43 species.

The proximity of the ponds and mines to the Athabasca River allows for seepage of toxins derived from the bitumen, extracted product, and operations into the river itself. Kelly et al. (2010) demonstrated that the tar sands industry was responsible for the release of elevated concentrations of EPA priority pollutants downstream in the Athabasca River and its delta, as well as a site near Fort Chipewyan in Lake Athabasca proper. The pollutants enter the aquatic environment both via airborne deposition from ash and as leakage or discharge from the tailings ponds. Sediment cores from area lakes found that PAHs increased significantly following the onset of bitumen development compared to previously, with concentrations 2.5–23 times greater than pre-1960 levels (Kurek et al. 2013;). Numerous news articles have reported on deformed fish caught in Lake Athabasca and cancer clusters among the Native Americans who live in Fort Chipewyan and traditionally used the lake for commercial and subsistence fishing (e.g. Harkinson 2008). However, definitive analyses on the histopathology of the deformed fish as well as investigations of potential carcinogenic effects on local residents have yet to be published in the scientific literature. Extracting oil for tar sands is an expensive process, however, and only economically viable when world market prices for oil are high.

## 12.2.3 Natural Gas Extraction

Natural gas is becoming the fossil fuel of choice to fire steam electrical generating plants. It is far less polluting than oil or coal, it produces fewer greenhouse gases, and advances in technology have made extraction much more cost effective. Hydraulic fracturing, commonly known as **fracking**, is the process by which gas trapped in rock layers far beneath the Earth's surface can be accessed. In this process, wells are drilled vertically first and then horizontal shafts are drilled into the shale layers. Next a mixture of water and chemicals is injected into the seams to fracture open the rock and release the gas. The injection water contains a variety of chemicals to accelerate the process; many of these chemical mixtures are considered proprietary information by fracking companies and the use of this water is controversial for that reason. A congressional investigation indicated that over 750 different chemicals have been used in fracking fluids, with 29 being known or suspected carcinogens (Entrekin et al. 2011). Three to five million gallons of water per gas well are usually needed, and sand or ceramic proppant is needed to prop open the shale fractures (Bomgardner 2012). Some of the process water, along with groundwater, comes back out of the shaft and is brackish, contains high levels of dissolved solids, and is chemical-laden (Entrekin et al. 2011). Standard wastewater treatment plants are unable to process it; thus it must be treated by other means for use in recycling (Bomgardner 2012) or else injected as wastewater into injection wells. The latter practice in some geological formations such as areas of the Midwest (van der Elst et al. 2013) and Oklahoma (Hand 2014) has led to geological instability and earthquakes.

Wastewater from the gas wells is not normally dumped into streams and rivers, but on occasion this has happened. For instance, during a 2007 incident in Kentucky, fracking fluids overflowed retention pits and flowed into Acorn Fork Creek in the Cumberland River watershed, lowering stream pH from 7.5 to 5.6 and raising conductivity from 200 to 35,000 µS/cm (Papoulias and Velasco 2013). The spill caused a fish and invertebrate kill, including deaths to federally threatened blackside dace (*Chrosomus cumberlandensis*) and impacting over 2.7 km of the stream. Follow-up histopathological investigations of fish in the impacted area found that compared to local control areas, the fish in the impacted waters had tissue injury and stress indicators consistent with exposure to low pH and toxic metals levels (Papoulias and Velasco 2013). A spatial analysis of the Fayetteville and Marcellus shale formations found that gas wells were sited on average 300 m from streams, and many were within 100 m of streams (Entrekin et al. 2011). The on-site construction, increased truck traffic, and road and bridge building associated with preparing new sites led to significant increases in turbidity in nearby streams; the same spatial study found that stream turbidity was strongly correlated ($r = 0.91$, $p = 0.003$) with area well density (Entrekin et al. 2011). Thus, impacts to groundwater from fracking fluids, accidental discharge to streams, and surface activity associated with well construction and operation are all potentially damaging to surface waters. Fracking techniques are also used to access light crude oil that is bound up in permeable shale formations, mostly in Montana, North Dakota, and Texas. Again, large amounts of water mixed with chemicals are injected; little information on surface water impacts is presently available.

### 12.2.4 Mining for Metals and Coal

**Surface mining** for various metals has caused much stream pollution. This can occur from local runoff of mining waste into streams, pollution by ore processing, leakage from waste ponds, and land alteration activities associated with servicing the mine sites; smelting and refining of metals also generates airborne metals pollution, and has for millennia (Hong et al. 1996). Ores are usually mixtures of metals, so in the various processes involved in accessing the target metals a number of toxic metals such as mercury, cadmium, and arsenic can be released and pollute waterways. Mining sulfide minerals such as copper, nickel, lead, zinc, silver, and especially coal can cause sulfur to be oxidized to sulfur dioxide ($SO_2$), which enters the atmosphere to be converted to sulfuric acid, i.e. acid rain (Schmitt 1998), which will pollute streams and lakes far downwind (see below).

**Coal mining** has been and continues to be a source of water pollution. Traditional coal mining involved digging shafts into the Earth in coal-rich regions, mining the seams, and extracting the ores. This generates mining waste that is subject to leaching by rainfall, creating polluted runoff including acidic waters. Acid mine drainage is caused when buried pyrite minerals in the coal mine wastes (the overburden) are exposed to air and water and thus oxidized, creating sulfuric acid. Acid mine drainage lowers stream pH well below 6, down to as low as 2. The low pH alone will severely stress stream biota and reduce biodiversity. However, the acidic drainage will also leach toxic concentrations of copper, aluminum, and zinc from the rocks and soils (Schmitt 1998), further damaging biota, including rare and endemic species.

A large-scale and very environmentally damaging coal mining practice is **mountaintop mining**.

Despite far less reliance on coal for power production, this practice still occurs in areas of West Virginia, Virginia, and Kentucky. To get at rich coal seams buried up to 200 m deep underground, overlying forests are clearcut, topsoil is removed, explosives are used to break up overburden, and the overburden is removed by massive metal buckets on chains called draglines. The overburden is dumped into valleys, burying natural streams and all their biota. This practice has converted at least 1.1 million ha of forest to surface mine and buried well over 2000 km of natural streams (Bernhardt and Palmer 2011). Natural drainage is further altered by earth compaction by heavy equipment, increasing surface runoff. The small valleys that suffer burial often contain headwaters streams, some of which are intermittent or ephemeral. Such habitats can be rich in biotic diversity and harbor rare and/or endemic species (Bernhardt and Palmer 2011). Drainage from the rubble pollutes downstream higher-order streams with metals such as selenium, sulfate, calcium, and magnesium and it raises stream pH and total dissolved solids concentrations (Palmer et al. 2010). A geographic information system (GIS)-based analysis concluded that this practice of mountaintop removal has led to chemical impacts on some 22% of West Virginia streams (Bernhardt et al. 2012). That analysis concluded that when upper watershed mine area coverage reached 5.4%, significant chemical pollution was evident, and deleterious impacts on stream macroinvertebrates occurred at even lower areal mine coverage of the watershed. The pollutants resulting from off-site drainage into downstream systems also damage local fisheries. For instance, in the Guyandotte River basin a study of fish assemblages in reference sites versus sites exposed to mountaintop mining drainages found higher taxonomic and functional diversity in the reference sites, with differences attributed to elevated selenium and conductivity in the impacted sites (Hitt and Chambers 2014). The mine operators are legally responsible for rehabilitating the mined areas by covering affected areas with dirt, regrading and recontouring, planting vegetation, and creating running watercourses along contour pathways. However, there has been no documented evidence of successful recoveries of biodiversity and other natural ecosystem functions from such "rehabilitated" areas (Palmer et al. 2010). As mentioned, coal mining in the USA and some of Europe has dramatically decreased in recent years, although it continues elsewhere.

## 12.3 Aquatic Metals Pollution

Metals pollution began about 5000 years ago—since humans began mining and refining metals, reaching a peak (for copper and lead) at the height of the Roman Empire that would not again be reached until the Industrial Revolution (Hong et al. 1994, 1996). Elevated concentrations in deep ice cores demonstrate that pollution from airborne metals can occur far from the source as well as near mining or smelting operations. Numerous industries produce or use numerous metals and chemicals that end up in waterways. Some of these also have breakdown products that cause environmental damage. Many types of industries use rivers for water supplies and waste disposal. The types of known and potential chemical contaminants are too numerous to list; therefore I will describe several of the better studied aquatic chemical pollutants, their effects on humans and biota, and their sources. An important concept to keep in mind is that of **bioaccumulation**. Bioaccumulation occurs when a metal or toxic chemical becomes more highly concentrated as it moves up the food chain. How this occurs is that a contaminant enters a watercourse through direct discharge, through stormwater runoff, or as airborne deposition. It can be taken up from the water in low concentrations by algae, which are eaten in large amounts by zooplankton, which concentrate it more. Small fish consume large quantities of zooplankton, concentrating the contaminant more; predatory fish eat many small fish, and the largest predators (including birds and mammals that eat fish) concentrate the contaminant the most. Large predatory sport fish therefore pose the biggest danger to human consumers. Fish that are herbivorous will have the lowest toxin concentrations and may be the safest to eat. Many metals and toxic chemical compounds are bioaccumulated, but some are not. Generally, in water bodies metals will be most highly concentrated in sediments and organism tissues, particularly those of higher predators. Some of these polluting metals are discussed in the following sections.

## 12.3.1 Mercury

One of the most notorious polluting metals impacting the biota of streams, rivers, lakes, reservoirs, and estuaries, as well as human health is **mercury** (Hg). There are natural sources of mercury, such as weathering of rock and discharge from volcanoes; however, mercury has many anthropogenic sources. It is or has been a component of paints, pesticides, batteries, vaccines, thermometers, and fluorescent light bulbs and it was formerly used in the pulp and paper industries as a fungicide. It enters the environment in wastewater treatment plant effluent, urban stormwater runoff, and as waste from small-scale (artisanal) gold mining; through discharges from medical waste incinerators and chemical manufacturing including cement; however, coal-fired electric utilities are the largest single source (US EPA 2000a; Evers et al. 2007).

Elemental Hg can readily become airborne and pollute far from its source. Mercury has an approximate 0.5- to 1-year residence time in the atmosphere (Krabbenhoft and Sunderland 2013). When inorganic mercury enters the aquatic system, a portion of it can become methylated by bacteria in the sediments to its most toxic form, **methylmercury** (MeHg). Methylation occurs most readily in blackwater rivers, streams, and floodplain wetlands that have high DOC concentrations, along with low pH and appropriate bacteria (Wasserman et al. 2003). Documented areas where there are particularly high levels of mercury in fish and other organisms include the blackwater river floodplains of the Southeastern USA; lake, reservoir, and river floodplain areas in New England and northeastern Canada; the Adirondack Mountains; the Florida Everglades; and the Amazon River watershed (Evers et al. 2007; Peterson et al. 2007; Chalmers et al. 2011). New reservoirs are subject to elevated mercury from initial saturation of soils outfluxing mercury to the water. Also the decompositional environment in the hypolimnion and bottom sediments encourages bacterial methylation of mercury. As discussed in Chapter 11, reservoirs can undergo major water-level fluctuations depending

upon use. Such water-level fluctuations in reservoirs encourage mercury methylation in the littoral zones—transitioning oxidation and reducing states promote bacterial methylation (Evers et al. 2007).

Mercury has a rich history as a dangerous pollutant. In previous centuries mercury was used in the hat-making process in England. Its use caused a neurological disorder among hat-makers called **Mad Hatter's disease**. Lewis Carroll is believed to have based his Alice in Wonderland character "the Mad Hatter" on this unfortunate malady characteristic of his time period. The first major aquatic mercury pollution incident was documented from Japan in the twentieth century and became known as the Minamata Disaster. In 1932 an industrial plant producing acetaldehyde began discharging MeHg waste into Minimata Bay, which subsequently made its way into fish and shellfish. By the early 1950s there were severe nervous disorders and deaths among farm animals, cats, and waterfowl that ate fish from the bay, followed in 1956 by human cases characterized by numbness, dementia, disorientation, and death. In addition to becoming ill from consuming contaminated fish, some victims were affected congenitally by being born of mothers who ate the fish, suffering severe effects. Early on, it was estimated that among humans there were 111 confirmed cases and 41 deaths (Irukayama 1966); however, over time up to 2000 deaths were attributed to the poisoning (Wasserman et al. 2003) and numerous more people were affected by severe symptoms (potentially up to 100,000), with many still seeking financial compensation (Normile 2013). A wastewater treatment system installed at the plant did not remove the mercury. The discharge was stopped in 1968 when the plant stopped making acetaldehyde, ending the discharge of mercury. The government did not close the bay to fishing until 1975, after which the bay was dredged and reopened in 1997 (Normile 2013).

Regarding human health, mercury is readily taken up by tissues and can enter the brain but unfortunately is only slowly eliminated from the body. It is known to be harmful to humans even at low doses and is principally known to cause neurological effects. As in the Minimata Disaster, mercury can be absorbed by humans through the consumption of contaminated fish, and the EPA

**Table 12.1** Human health standards (non-cancer health risk; figures in ppm or μg/g as wet wt) for metals and selected organic pollutants in fish tissue (US EPA 2000b, 2004), based on consuming four fish meals per month.

Arsenic (inorganic): 0.35–0.70; cancer risk 0.008–0.016
Cadmium: 0.35–0.70
Mercury: 0.12–0.23
Selenium: 5.9–12.0
Total DDT: 0.059–0.12; cancer risk 0.035–0.069
Dieldrin: 0.059–0.12; cancer risk 0.00073–0.0015
Endosulfan: 7.0–14.0
Lindane: 0.35–0.70; cancer risk 009–0.018
Total PAHs: 0.0016–0.012 (cancer risk)
Total PCBs: 0.023–0.047; cancer risk 0.0059–0.012

has developed consumption guidelines to protect human health (Table 12.1). Pregnant women should not eat high mercury content fish species and men and nonpregnant women should eat no more than one meal/week of these species. As well, mercury can be absorbed through the skin and from vapors from combusted mercury that are breathed in.

As mentioned, in artisanal gold mining mercury is used and released to the environment; however, it is present in significant concentrations in some Amazon soil layers. The release of large quantities of mercury to the larger environment in this region is exacerbated by clearcutting, soil erosion into streams and rivers, and burning of forest areas—creating airborne mercury (Wasserman et al. 2003). A study of villagers living along the Rio Tapojos found that (according to hair samples) villagers who ate a lot of river fish had significantly higher mercury levels than those who did not (Lebel et al. 1997). That same study showed that piscivorous fish had higher mercury body burdens than herbivorous fish. Other studies in Amazonia have shown excessive levels of mercury in humans (Silva-Forsberg et al. 1999) as well as in sediments, aquatic macrophytes, and fish (Martinelli et al. 1988). In affected Amazon Indian villages researchers observed dose-related nervous system problems in young adults when fish diets were high in mercury—sources were waste from gold mining and leachate from soils, exacerbated by land-disturbing activities. In a study of two villages 200 km downstream of a gold-mining area, hair samples of villagers showed comparatively high concentrations of mercury (Lebel et al. 1996). Higher

levels of hair mercury content were related to several neurophysiological issues including reduced color discrimination capacity, near visual contrast sensitivity, peripheral vision field profiles, and, in women, reduced manual dexterity.

Regarding wildlife, mercury is bioaccumulated through the food chain, meaning that it becomes more highly concentrated in higher predators (Peterson et al. 2007). High mercury levels can be toxic to phytoplankton, zooplankton, benthos, and fish, as well as higher-level consumers like mammals and birds (Scheuhammer et al. 2007). Ecologically critical tissue levels of mercury are 0.16 μg/g for piscivorous fish, 3.0 for loons, and 30 for river otter and mink (Evers et al. 2007). Mercury accumulation is not just confined to aquatic communities of mercury-polluted waters; research has found that in riparian areas terrestrial bird species that feed on invertebrates (especially predaceous spiders) can bioaccumulate elevated mercury body burdens through feeding activities (Cristol et al. 2008). Eggs of aquatic bird species collected at the delta where the Peace and Athabasca Rivers enter Canada's Lake Athabasca contain elevated mercury (Hebert 2019). This area is downstream from the massive oil sands operations (see Section 12.2.2) that release mercury into the environment, and mercury body burden was positively related to river discharge and sedimentation.

A western US fish study found that body burden positively related to fish length, and to piscivores as opposed to herbivores and generalists (Peterson et al. 2007). Western streams have pH values > 6 and low DOC; thus methylation would be low. Mercury body burden was primarily related to wind depositional patterns; deposition in swampy blackwater areas leads to methylation. A USGS study showed a similar distribution of total mercury in fish tissue; again, this was lowest in mountain and western areas and highest in swampy lowlands (Larsen et al. 2013). Another analysis found only a weak relationship between body burden and length but strong relationships with species and trophic status, primarily piscivores (Sackett et al. 2009). That study also found positive relationships between body burden and ecoregion (Coastal Plain), land use (agriculture), and low pH. An analysis of trends in fish tissue concentrations using data from 1969 to 2005 showed that concentrations of mercury in tissue were decreasing in many areas of the USA but increasing in southeastern states (Chalmers et al. 2011). However, in a study of predator and bottom-feeding fish from 500 lakes from 2000 to 2003, mercury was found in all fish samples, with mercury in fillets exceeding the EPA 300 µg/kg fish tissue consumption criterion in 49% of the lakes sampled (Stahl et al. 2009). In freshwater, fish that are often on consumption advisory lists for mercury include predators like largemouth bass, chain pickerel, bowfin, certain catfish, and walleye. Fish usually considered safe for consumption include trout, bluegill, and various farm-raised fish.

In recent years there has been positive movement toward reducing mercury in the environment. In 2012 the EPA issued new regulations requiring fossil-fuel power plants to reduce emissions of mercury (as well as arsenic, cyanide, and other toxins). On January 19, 2013 delegates from over 100 countries at the Minimata Convention in Geneva agreed to limit mercury production as of 2017. The treaty mandated phasing out mercury in some batteries, fluorescent lamps, cosmetics, and medical devices and set limits on mercury emissions from coal-fired power plants, cement factories, and waste incineration; also strategies must be devised for reduction of mercury due to small-scale gold-mining activities, a growing source (Erikson 2013). The latter is likely to continue to be challenging, as much small-scale gold mining in Amazonia is unofficial with little or no concern regarding stream pollution. Signatories have subsequently met several times (2018–2022) to discuss goals and problems and further refine tactics.

## 12.3.2 Arsenic

Due to its devastating health and ecological effects mercury is well studied and well publicized in the media. However, other metals can be significant pollutants impacting humans and aquatic life both. **Arsenic** (As) is a potentially toxic metal that can be widespread in water bodies. Large doses of arsenic are acutely toxic to humans and animals. Humans and other mammals can ingest arsenic from eating contaminated fish and shellfish; however, arsenic is not bioaccumulated in the food chain. Chronic impacts on human health include skin lesions and gastrointestinal, cardiovascular, hematological, liver, and kidney disorders (US EPA 2000b). Sources of arsenic include fossil-fuel power plant emissions, hazardous waste site leachate, wood preservatives, pesticides and herbicides, and mining/smelting operations (Schmitt 1998; US EPA 2000a). Thus arsenic can find its way into streams and other water bodies through anthropogenic discharges, runoff, and airborne deposition. It is also a naturally occurring element in various soils and rocks worldwide and can be leached into water bodies at toxic levels. As such, archeologists suspect that arsenic poisoning through consumption of tainted river water has been responsible for human mortalities (among the Chinchorro people in South America) as far back as 6800 years ago (Pringle 2009). Arsenic contamination of drinking water sources remains a problem today. For example, in Bangladesh, river and stream water supplies are often contaminated with pathogenic fecal microbes, rendering consumption of untreated water hazardous. In response, numerous wells were dug with international financial contributions to allow access to groundwater. However, groundwater, especially from upper aquifers, is often naturally loaded with arsenic, and human toxicity has occurred among the largely poor residents unknowingly drinking contaminated water.

## 12.3.3 Cadmium

**Cadmium** (Cd) sources include waste from pigment works, printing, lead mines, chemical industries, paints, batteries, pesticides and herbicides, and sewage plant discharges (US EPA 2000a). Cadmium can be toxic to organisms and does biomagnify in the food chain. It can act synergistically with other metals to increase toxicity. It is known to cause kidney damage, neurotoxicity, and bone disorders in humans (US EPA 2000b). In studies of cadmium toxicity to fish, several authors (summarized in Weis and Weis 1989) have noted cardiac malformations, retarded growth, skeletal malformation, and increased mortality following exposure to various concentrations of cadmium.

A major Japanese riverine cadmium pollution incident occurred over several decades in the mid-twentieth century, although cadmium was not identified as the pollutant until the 1960s (Kaji 2012). The Kamioka mine was located in the headwaters of the Junzu River, and the water downstream was used for irrigation. Downstream residents were afflicted by kidney lesions, severe back and joint pains, decalcification of bones, and many fatalities; the disease was termed itai-itai (ouch-ouch) disease. Estimates of the number of affected people variously range from 195 up to 400 (primarily women), with as many as 200 fatalities potentially attributed to the disease. In 1972, after a major pollution lawsuit, the effluent was controlled and cadmium concentrations in the river eventually returned to background concentrations (Kaji 2012). Cadmium continues to be involved in major river pollution incidents. China had a large cadmium spill from a smelter in 2005 that polluted large sections of the Bei River; the majority of the cadmium was stopped by a dam downstream. The government solution to reduce the danger was to drain large quantities of water from nearby reservoirs into the river to dilute the cadmium. China also had a massive spill of benzene from a chemical plant explosion into a northeastern river, which subsequently flowed into Russia.

### 12.3.4 Lead

Another metal that has a long history of human toxicity is **lead** (Pb). This metal has had numerous uses, including in pipes, solder, paints, and battery cases; it formerly was a gasoline additive, and lead smelters use furnaces to remove and recycle lead from scrap. In humans it causes metabolic and neurophysiologic disorders, especially in children (anemia, low IQ). During Roman Imperial times it was used as lining in water pipes and wealthy families used lead in flatware; thus the preponderance of lead-caused mental disorders, especially within royal families, and it is believed to have played a role in the fall of the Roman Empire (Hong et al. 1994). In previous decades lead toxicity caused extensive mortality among waterfowl (ducks, geese, and swans), which will ingest lead shotgun pellets and sinkers mistakenly as small stones or grit for their gizzards, thus becoming poisoned. In the USA alone at one point approximately 1.5–3

million waterfowl deaths annually were attributed to lead poisoning (De Francisco et al. 2003). Lead shot is now banned for waterfowling in the USA and waterfowl shot is now made from stainless steel, bismuth, or tungsten. Research has shown that fish experimentally dosed with lead experienced reduced hatching success, cardiac malfunctions, and especially skeletal malformations (Weis and Weis 1989). Phaseout of lead from gasoline began in many nations in the 1980s, and the United Nations led a successful program beginning in 2002 to eliminate lead from gasoline globally. In the USA lead in reservoir sediments decreased from 1975 to 1997 (Larsen et al. 2013). Note that multistate toxicology research in the USA has found that eagles are still suffering from an unusually high degree of lead poisoning, potentially due to winter scavenging from the remains of terrestrial game animals shot during the hunting season by lead ammunition (Slabe et al. 2022).

### 12.3.5 Selenium

**Selenium** (Se) is a potentially toxic metal that becomes highly magnified in the food chain. For example, in Belews Lake (Box 12.1) selenium in periphyton was biomagnified 500 times that of water concentrations, while selenium in fish tissue was biomagnified 4000 times that of the reservoir water. Chronic exposure to selenium in humans in food and water causes symptoms including skin disorders, hair and nail loss, limb pain, convulsions, and paralysis (US EPA 2000b). Taxa groups lower in the food chain (phytoplankton, zooplankton, invertebrates) require far higher body burdens of selenium before ill effects are seen, compared with fish and birds (Lemly 1993). Elevated selenium in fish tissue causes decreased larval survival and deformations (Fig. 12.4) including edema, spinal and mouth deformations, and eye cataracts (Gillespie and Bauman 1986; Woock et al. 1987). Fish tissue concentrations of selenium exceeding 12 µg/g have been suggested by Lemly (1993) to be detrimental to freshwater fish health and reproduction, while reservoir water concentrations of 10–20 µg/L can be toxic concentrations for fish reproduction and survival (Lemly 1985). Selenium tends to magnify strongly in fish gonads, so reproductive impairment occurs

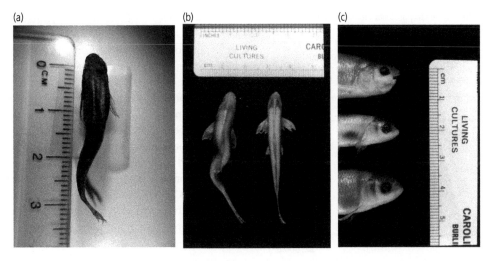

**Figure 12.4** Fish with deformities found in reservoirs polluted by selenium. (Left) Mosquitofish (*Gambusia affinis*) with scoliosis from Sutton Reservoir, North Carolina, 2013. (Center) Red shiner (*Notropis lutrensis*) with scoliosis compared with normal specimen from Belews Lake, NC, 1980. (Right) Red shiners from Belews Lake: top individual has a deformed mouth which cannot close, along with other deformities; middle individual has an underbite lower jaw and shortened head; bottom individual is normal. (Photos and explanations courtesy of Dr. A. Dennis Lemly, Wake Forest University.)

at much lower concentrations than those required to directly impact adult health (Lemly 1985).

As mentioned earlier (Box 12.1), a major source of selenium is emissions from fossil-fuel combustion and leachate from ash ponds associated with coal-fired power plants (Woock et al. 1987; Lemly 1993; Crutchfield 2000). Other sources of selenium pollution include cement manufacture; selenium can also be leached from coal and shale-oil mining sites (Lemly 1993), including the notorious "mountaintop removal" mining practiced in Appalachia (Palmer et al. 2010). Selenium is abundant in certain soils, mostly in the western USA. One of the most well-known cases of reservoir contamination by excessive selenium is from Kesterson Reservoir in California. This reservoir was constructed to accept saline agricultural drainage from part of the San Joaquin Valley and was opened in 1980. However, over the next several years approximately 9000 kg of selenium was leached from farm soils by extensive irrigation and entered the reservoir (Quinn and Vorster 1998). What generated headlines in the news media were the graphic photos of deformed waterfowl chicks that began appearing following reservoir opening. The waterfowl using the reservoir ate selenium-contaminated fish and suffered selenium teratogenicity leading to malformed embryos and nesting failure. Massive fish kills and reproductive

failures in reservoir drainages also occurred. The reservoir was closed in 1986 due to the toxicity. Rangers have had to drive off visiting waterfowl flocks with gunshots to prevent further exposure.

### 12.3.6 Sediment Pollution by Metals and Other Substances

Metals and other pollutants tend to concentrate in aquatic sediments, especially organic sediments. This provides the greatest exposure to benthic organisms, particularly invertebrates. As benthic invertebrates are major fish food items, biomagnification up the food chain into higher predators (and sought-after sport fish) is an obvious problem. Some fish scavenge whatever they can from the bottom, providing additional exposure. Researchers have thus developed guidelines (Table 12.2), based on numerous laboratory and field studies, that regulatory agencies and researchers can reference when assessing streams, rivers, reservoirs, and lakes for pollutant exposure and potential remediation.

## 12.4 Chemical Pollution

Industrial pollution of waterways has a long and notorious history harking back to the Industrial Revolution. Wastes from tanneries, dye-works,

**Table 12.2** Sediment quality guidelines for freshwater for selected metals and organic pollutant concentrations potentially harmful to aquatic life (adapted from MacDonald et al. 2000). TEC = threshold effect concentration, below which harmful effects on aquatic communities are unlikely to be observed. PEC = probable effect concentration, above which harmful effects are likely to be observed.

| Pollutant | TEC | PEC |
|---|---|---|
| **Metal (mg/kg dry wt)** | | |
| Arsenic (As) | 9.79 | 33.0 |
| Cadmium (Cd) | 0.99 | 4.98 |
| Chromium (Cr) | 43.4 | 111.0 |
| Copper (Cu) | 31.6 | 149.0 |
| Lead (Pb) | 35.8 | 128.0 |
| Mercury (Hg) | 0.18 | 1.06 |
| Nickel (Ni) | 22.7 | 48.6 |
| Zinc (Zn) | 121.0 | 459.0 |
| Total PAHs ($\mu$g/kg dry wt) | 1610.0 | 22,800.0 |
| **Chemical pollutants ($\mu$g/kg dry wt)** | | |
| Anthracene | 57.2 | 845.0 |
| Benzo(a)pyrene | 150.0 | 1450.0 |
| Chrysene | 166.0 | 1290.0 |
| Fluoranthene | 423.0 | 2230.0 |
| Pyrene | 195.0 | 1520.0 |
| Total PCBs ($\mu$g/kg dry wt) | 59.8 | 676.0 |
| **Organochlorine pesticides ($\mu$g/kg dry wt)** | | |
| Chlordane | 3.24 | 17.6 |
| Dieldrin | 1.90 | 61.8 |
| Endrin | 2.22 | 207.0 |
| Lindane | 2.37 | 4.99 |
| Total DDTs | 4.16 | 572.0 |

textiles, breweries, the paper industry, and other industries of the time were released untreated into the nearest stream or river (Billen et al. 1999) and caused (along with raw sewage inputs) the severe pollution that once afflicted some of the great rivers of Europe, such as the Thames, the Seine, and the Rhine. In the New World, rapid industrialization in the nineteenth and twentieth centuries likewise led to extensive stream pollution, some of which gained infamy in the public eye such as the Cuyahoga River, Elizabeth River, Toms River, Love Canal, and Gowanus Canal (some of which were associated with human cancer clusters). While the US Clean Water Act and other assorted legislation in the early 1970s made massive strides in water pollution abatement, chemical pollution of our waterways continues, often in more subtle forms.

Over many decades thousands of synthetic organic compounds have been developed, many of which were developed as pesticides, herbicides, and fungicides. Other synthetic organics have industrial uses and others occur as byproducts of industrial processes or breakdown products of other chemicals. However, many if not most of these compounds have not been tested for carcinogenicity or other effects; thus, official safety standards for water, sediments, and body burden have not been established for many organic compounds (Huff 2011). The cost of analysis for these chemicals in environmental samples is high and therefore there are large spatial and temporal gaps in sampling for them.

### 12.4.1 Pesticides and Herbicides

Improving agricultural yields has been the major driving force behind the development of **pesticides**, including chlorinated hydrocarbons (DDT, aldrin, dieldrin, chlordane, lindane, Kepone, endosulfan); these compounds are also called organochlorine pesticides. These pesticides were developed in the mid-twentieth century and saw their widest use up until the 1970s. Such products can be persistent in the environment (2 to > 30 years). Toxicity to fish and invertebrates is often high and these compounds also have sublethal effects including neurotoxicity, teratogenicity, impaired reproduction and learning behavior, and slow reflexes (Köhler and Triebskorn 2013). Toxicity to water-associated birds by organochlorine pesticides is well known (Furness 1993). Some act as photosynthetic inhibitors and reduce algal growth rates (DeLorenzo et al. 2001). Organochlorine pesticides are not easily metabolized or excreted and bioaccumulate up the food chain (US EPA 2000a). Another group of pesticides are the organophosphates (including malathion and parathion, among others). Other pesticides include carbamate and synthetic pyrethroid compounds, and neonicotinoids. While the latter pesticides degrade more rapidly and accumulate less than chlorinated hydrocarbons, they have still been implicated in kills of fish, invertebrates, and birds, and also impact endangered species (Schmitt 1998; Köhler and Triebskorn 2013; Erikson 2022a). Unfortunately, insects develop resistance to pesticides through genetic changes that increase their efficiency for detoxification via oxidation, conjugation of hydrophilic compounds, and excretion of insecticides (Heckle 2012). Resistance leads to continual development of more effective pesticides, whose long-term direct and indirect impacts may be poorly quantified. Pesticides break down in the environment either through biological transformations, usually via microbial mediation, or abiotically as chemical or photochemical reactions—such as UV degradation (Fenner et al. 2013). Presently, to be licensed pesticides must have a short breakdown period, but in the past this issue was not addressed, leading to the long-term presence in aquatic animals and sediments that is still problematic.

Fish and wildlife kills, reproductive failures, and deformed young tend to garner the most public and scientific attention (see Rachel Carson's eye-opening 1962 book *Silent Spring*, for instance); however, pesticides and herbicides impact water resources in many more subtle ways. Pesticides can impact microorganisms such as phytoplankton, zooplankton, and bacteria as well (DeLorenzo et al. 2001). Reducing aquatic photosynthesis by phytoplankton or impacting some species more than others can reverberate up the food chain, altering biodiversity, affecting species competition, reducing prey for higher organisms, and lowering system productivity. Additionally, nontargeted microorganisms can alter or degrade pesticides, sometimes into byproducts that are more toxic or disruptive to the ecosystem than the original pesticide, and interactions between pesticides and other environmental factors such as nutrient loading are understudied (Schmitt 1998; Delorenzo et al. 2001; Fenner et al. 2013; Köhler and Triebskorn 2013). Some of the more well-known pesticides considered to be aquatic contaminants are discussed below. While the impacts of the persistent organochlorine pesticides are best known, it is important to keep in mind that other classes of pesticides including organophosphates have been implicated in severe animal and human impacts as well (Mascarelli 2013).

The best-known or perhaps most notorious pesticide is **DDT**—its breakdown products DDD and DDE are also toxic. This compound is not water soluble but is soluble in organic solvents. It decomposes slowly and is still found in significant concentrations in fish and wildlife in the USA. It is toxic to

humans at high doses and sublethal doses can be accumulative and cause liver damage. DDT affects calcium metabolism in birds, causing thin eggshells, deformation of chicks, and reduced reproduction; the severe impacts to the brown pelican in California in the 1950s were a highly visible result of widespread DDT use that led to public outcry regarding indiscriminant pesticide use. Sadly, regulatory action often needs to be stimulated by such highly publicized environmental catastrophes. DDT was banned in 1972 in the USA but is still used illegally, and is used legally for mosquito control in African and Asian nations plagued by malaria. DDT in reservoir sediments declined from 1965 to 1997 (summarized in Larsen et al. 2013).

A well-known pesticide pollution incident that affected an entire river and estuary is that of **Kepone**, an organochlorine pesticide patented in 1952 which was manufactured in Virginia from 1966 to 1975. The companies that manufactured it were Allied Chemical and later LSP Inc., and they released their effluent into the James River. Over 1.5 million kg of Kepone were produced in that period and polluted the entire tidal river and estuary. Several Kepone workers were hospitalized with a neurological disorder the workers called the "Kepone shakes." As a result of the human effects extensive environmental sampling was carried out in the river and estuary, with elevated Kepone concentrations found in finfish and oysters. In response, authorities closed 98 miles of the tidal river and estuary for finfish and shellfish harvest; the company ceased manufacturing Kepone in 1975 and it was banned in 1977.

Laboratory studies showed that marine life could purge themselves of the toxin in two to seven weeks in clean water (Huggett and Bender 1980), and the James was reopened to oystering. Numerous tests were conducted on various fish species to determine concentrations causing mortality or injury. Kepone was also detected in waterfowl using the James, raptors in and even outside the watershed, and rats in adjoining uplands (Huggett and Bender 1980). Although the manufacture and discharge had ceased, there was a reservoir of Kepone in the sediments. The immense cost of cleaning the river (dredging), plus the biological danger were so high that it was decided to let nature take its course and bury the Kepone under the sediments. The river was reopened to fishing after 13 years (1988).

**Dieldrin** is a formerly used organochlorine pesticide that still occurs in significant concentrations in aquatic animals today. It was developed and used to control soil-dwelling pests and for termite control. However, it was responsible for massive fish kills and contributing to the deaths of birds and mammals as well (summarized in Schmitt 1998). A production phase-out began in 1974 and it was banned in 1987. Animal studies showed liver toxicity, neurotoxicity, and reproductive disorders (Weis and Weis 1989). Observed effects on exposed humans include liver toxicity and neurotoxicity, and it is listed as a probable carcinogen (US EPA 2000b).

Herbicides also impart toxicity and aquatic ecosystem disruption; some are no longer in use, but others are widely used in agriculture as well as urban situations (Schmitt 1998; Delorenzo et al. 2001). **Atrazine** is banned in some countries, but is a commonly used herbicide; the US EPA permits up to 3 ppb in drinking water. Atrazine is one of the most extensively used herbicides in the USA and accounts for 60% of the of the total pesticide volume applied to crops each year (DeLorenzo et al. 2001). Original tests of atrazine tended to show little toxicity except at very high doses; this pesticide has a short half-life and shows little bioaccumulation, and thus it was considered safe for use. However, experiments with frogs showed that exposures as low as 0.1 ppb caused 20% of the frogs to have multiple sex organs or both male and female sex organs; males had a 10 times decrease in testosterone levels at atrazine exposures of 25 ppb (Hayes et al. 2002). Those exposure levels are well below what can be found in US streams. The author was attacked by the pesticide industry. Atrazine has also been shown to reduce spawning and reproduction in fish, as well as cause tissue abnormalities at levels below the EPA benchmarks for aquatic life of 65 μg/L (Tillitt et al. 2010). Regardless, in 2020 the EPA reapproved the used of atrazine with some additional safety-oriented application restrictions; note that as of 2022 the EPA proposed further restrictions on atrazine based on a 3.4 μg/L level of concern, a tightening of the 15 μg/L level adapted in 2020.

Glyphosate is a widely used herbicide (the active ingredient in Roundup) and manufacturers have been the target of many lawsuits, particularly as a potential cause of human cancers and impacts on endangered species (Erikson 2022b). Bear in mind that this brief overview has featured largely those pesticides with known, well-reported side effects on aquatic life. Other pesticides are not discussed here due to ongoing research or litigation— regarding pesticides, the regulatory authorities are always playing "catch-up" to what is already on the market.

### 12.4.2 Polychlorinated Biphenyls

Regarding nonpesticide compounds, a widely distributed and very toxic group of chemicals in our waterways are **PCBs**. Manufacture of these chemicals began in 1929 and some 635 million kg were produced in the USA alone before their manufacture was banned in 1979 (Schmitt 1998). Because of their high stability and resistance to heat and pressure these chemicals found extensive use in transformers, capacitors, heat exchangers, hydraulic systems, vacuum pumps, lubricating oils, and paints. They enter the environment through landfills, sewage effluent, industrial dumping, and incineration of waste—PCBs are widely dispersed in the atmosphere and contaminate water and sediments worldwide. Despite their stability, polychlorinated compounds including PCBs are biodegraded by a variety of microbial and enzymatic mechanisms in the environment (Wackett and Robinson 2020).

PCBs accumulate in body fats and biomagnify in the food chain. The very stability and resistance characteristics that made these compounds desirable industrially make them very persistent compounds in the environment. Their concentrations in water and sediments have decreased since PCB manufacture ceased, but they are still found in elevated levels in invertebrates, fish, and mammals, and the reservoir of PCBs in aquatic sediments allows for their reentry into the water column upon sediment stirring by storms, boats, etc. Since PCBs bind to sediments, bottom fish and invertebrates get the highest exposure. In a study of predator and bottom-feeding fish from 500 lakes from 2000 to 2003, PCBs were found in all fish samples, with total PCBs exceeding a 12-ppb human health risk-based consumption standard in 17% of the lakes sampled (Stahl et al. 2009).

PCBs have caused mortality, liver damage, cancer, skin lesions, and reproductive damage (larval or hatchling deformities, poor hatchling success) in numerous species of fish including lake trout and coho and chinook salmon; birds such as night herons, ring-billed gulls, ospreys, bald eagles, terns, and cormorants; and mammals such as mink, river otters, and seals (summarized in Schmitt 1998; Colborn and Thayer 2000). Fish (as prey items) with concentrations of 0.01–0.15 µg PCBs/g have been found to cause toxic effects in their mammal consumers, while liver concentrations of 6.6–11.0 µg/g (geometric mean 8.7 µg/g) caused physiological damage to those organisms (Kamman et al. 2000). In humans, studies conducted near Lake Michigan implicated consumption of PCB-laden fish by mothers to infant maladies including anemia, susceptibility to disease, smaller body weight and smaller head circumference, and behavioral defects. PCBs are considered probable human carcinogens (US EPA 2000a). Dumping of PCBs into a major waterway has led to one of the most damaging and long-running incidents of industrial pollution in US history (see Box 12.2).

---

**Box 12.2 The Hudson River PCB pollution story.**

The story of PCB contamination of the Hudson River is an example of industrial pollution that demonstrates how lack of knowledge of a heavily used chemical's biological effects, combined with careless handling, caused pollution on a massive scale that was and is extremely costly monetarily to both private industry and the public. Two large General Electric (GE) factories at Hudson Falls and Fort Edward, New York State, began in 1947 using PCBs in capacitors they manufactured, and over 30 years they dumped over 1.1 million lb (0.5 million kg) of waste PCBs into the Hudson River until 1977, when PCBs were banned. GE could have made efforts to treat the waste PCBs, but opted not to due to the added expense. The PCBs entered the food chain, with significant concentrations accumulating in microzooplankton, macrozooplankton, benthos, and fish, and the PCBs especially found a reservoir in the sediments. Since the dumping ceased the sediments became the main source of PCBs to the water

column and biota. Most species of fish that are desirable to the fisherman and consumer have body burdens of > 5 mg/kg (the limit for edible fish). Commercial fishing was and still is closed (except for shad, which spend most of their time offshore), and there are human consumption advisories for increased cancer risk currently in place. Catch and release only sport fishing is permitted in affected areas. Such bans obviously are a financial burden to commercial and sport fishermen and the associated businesses supporting those pastimes. Additionally individuals who practice subsistence fishing are barred from consumption, or a take a health and legal risk if they do catch and consume fish.

The Hudson is the lengthiest EPA Superfund site in the USA, with a 306-km reach from Hudson Falls through New York Harbor contaminated by PCBs. After long court battles the EPA in February 2002 required GE to spend $500 million to dredge PCB-polluted sediments along a 40-mile stretch of river; the plan including Phases I and II was approved. GE expected to spend at least $460 million to remove 2.0 million $m^3$ of contaminated sediments, concentrating efforts on "hot spots" where PCBs accumulated, in hope of removing approximately 68,000 kg of PCBs. GE previously spent $200 million cleaning up PCBs in the vicinity of the capacitor plants, and of course millions of dollars more on lawyers, consulting firms, and public relations to fight the EPA. Dredging to remove the PCBs began in 2011 and was completed as of late 2015. The company said it removed 140,000 kg of PCBs within 1.2 million $m^3$ of sediment, at a total project cost of $1.6 billion. Note that PCBs are degraded by microbes in the environment, so natural loss also occurs in the river sediments (Wackett and Robinson 2020).

(Compiled from Brown et al. 1985; Cronin and Kennedy 1997; Schmidt 2001; Claudio 2002.)

---

## 12.4.3 Dioxins

**Dioxins** are polychlorinated dibenzo-para-dioxins (PCDDs), of which there are about 75 compounds. Dioxins occur as trace impurities in manufactured chemicals and industrial wastes, unintended byproducts of manufacturing processes involving chlorine or waste burning, and the pulp and paper industry. They enter the environment through hazardous waste dumping and industrial discharge, are airborne and waterborne, and bioaccumulate in the food chain (US EPA 2000a). Besides the chemical industry, a major source of dioxins is primary metals processing, which is the smelting and refining of metals from ores or scrap metals (Hogue 2012).

Dioxins are very stable, accumulate in fat, resist breakdown, are not readily excreted, and can cause severe or lethal effects even in small doses. Some are extremely toxic. The most potent dioxin compound is 2,3,7,8-tetrachlorodibenzo-*p*-dioxin (TCDD). TCDD has been linked to cancer, immune system effects, nerve damage, and birth defects. It was a component of herbicide (the infamous Agent Orange) used heavily by the US military in Vietnam to clear jungle and destroy North Vietnamese crops. Its legacy was and is the disfigurement of Vietnamese people exposed to it (soldiers and peasants) and birth defects, cancer, and other illness to the Vietnamese and to American troops exposed to it. Exposure of dioxins to aquatic organisms, birds, and mammals results in mortality or reproductive, mutagenic, teratogenic, and carcinogenic effects; in fact TCDD is the most potent animal carcinogen evaluated by the EPA (US EPA 2000a). An EPA study of fish chemical residues detected TCDD at 70% of 388 sites (US EPA 2000a). There are no existing regulations for contamination and protection of wildlife and aquatic organisms, but in 2012 the EPA established safe levels of human exposure to TCDD (Hogue 2012).

In 2004 Viktor Yushchenko, the Ukrainian presidential candidate, was poisoned by dioxin (TCDD) and suffered severe cramps, ulcers, and facial disfigurement called chloracne. Yushchenko believed he was poisoned at a dinner with the heads of the Ukrainian secret police. He survived, recovered, won the election, and served as President of Ukraine from 2005 to 2010.

### 12.4.4 Polycyclic Aromatic Hydrocarbons

Important pollutants of urban streams, lakes, and reservoirs are **PAHs**, which are organic compounds that have a fixed ring structure of two or more benzene rings. PAHs are an urban pollutant. Sources are the burning of gasoline, oil, coal, and wood involved with power generation, heat generation, or automotive transportation; various industrial processes, foundries, waste incineration, and urban runoff carrying PAHs originating from asphalt, automobile tires, exhaust emissions, oil sands processing, and oil and gasoline spills. PAHs were also sourced from creosote, formerly a widely used wood preservative used in dock construction. PAHs accumulate in aquatic animals from water, food, and sediments—bivalves in particular are at risk. They may be absorbed through the lungs, stomach, or skin. PAHs appear to bioaccumulate in fish and crustaceans if sediments are the source (US EPA 2000a), although they are biodegraded by aerobic and anaerobic bacteria in the environment (Wackett and Robinson 2020). Laboratory animal studies show adverse effects on the cardiovascular, respiratory, gastrointestinal, and immune and central nervous systems and PAHs are believed to be carcinogenic to humans (Schmitt 1998; US EPA 2000a). PAHs trended upward in urban lakes from 1975 to 2000 (Larsen et al. 2013). In a study of 10 US reservoirs and lakes (Van Metre et al. 2000) PAHs showed an increase in sediment concentration in all 10 systems. The data also indicated that the sources of PAHs to the water had changed over time from uncombusted PAHs (such as from oil seeps and petroleum spills) to combusted fossil fuels from vehicle exhaust, heating, and power generation. The authors considered these data a strong signal of increasing urbanization, especially involving auto usage.

### 12.4.5 Xenoestrogens

A widespread, but as yet poorly understood group of pollutants are **endocrine-disrupting compounds (EDCs)**, also called **xenoestrogens** (foreign estrogens). Chemical compounds known or suspected of expressing endocrine disruption include synthetic hormones used in human contraceptives; natural hormones used to promote livestock growth; organochlorine pesticides including DDT, endosulfan, Kepone, and atrazine; metals such as mercury and cadmium; industrial chemicals such as PCBs, PAHs, and dioxins; and breakdown products of plastics and detergents (Davis and Bradlow 1995; Colborn and Thayer 2000; Czarny et al. 2017). EDCs are introduced into wildlife or humans from the environment, and consumption of contaminated fish is a significant route of exposure. EDCs mimic the action of estrogen produced in cells or alter the hormone's production or transport. Some xenoe-strogens (synthetic ones found in certain pesticides, drugs, fuels, and plastics) can amplify estrogen's effects. Exposure can be passed on to offspring from mothers who have consumed EDC-contaminated fish, among both wildlife and humans (Colborn and Thayer 2000). Too much estrogen is known to be a risk factor in breast cancer; research suggests that xenoestrogens also lead to breast cancer (Davis and Bradlow 1995).

Synthetic chemicals including DDT and other chlorinated hydrocarbons (such as Kepone) have been shown in experiments to disrupt the endocrine system and cause reproductive effects in many species (Davis and Bradlow 1995). Low levels of such compounds in water can be mislead-ing, because biomagnification up the food chain facilitates these impacts on wildlife, especially expressed in piscivorous fish and their avian preda-tors (Colborn and Thayer 2000). While much of the field information on EDCs came about follow-ing large and visible impacts to piscivorous bird and mammal populations following pesticide expo-sures decades ago, more subtle effects also occur. For instance, estrogenic activity has been detected in multiple sites in the lower Myakka River, Florida, primarily near areas of significant residential devel-opment. Runoff from urban areas and septic sys-tems were believed to be the likely sources of these compounds to the river water (Cox et al. 2006). While EDCs are particularly harmful to larval or otherwise immature stages of organisms, they can also impact adult organisms, and sex-related effects are seen in fish, reptiles, and amphibians (Czarny et al. 2017).

## 12.4.5 Pharmaceuticals and Personal Care Products

Humans consign all sorts of products such as unused prescription drugs, antibiotics, over-the-counter medications, birth control pills, illegal drugs, and various "natural" herbal medications (sometimes referred to as nutraceuticals) down the toilet and into the sewage stream (Cordy et al. 2004; Daughton 2007); these collectively are sometimes referred to as **pharmaceuticals and personal care products**, or **PPCPs** (Daughton and Ternes 1999). In

addition to outright dumping them into commodes and drains, some percentage of ingested PPCPs pass through human digestive tracts unabsorbed or as their breakdown products into the sewage stream. In contrast to the generally smaller amounts of pre-scription medications, much larger amounts of per-sonal care products enter the aquatic environment, including bath additives, shampoos, skin care prod-ucts, hair sprays and dyes, soaps, sunscreens, and perfumes (Daughton and Ternes 1999; Daughton 2007). Conventional wastewater treatment plants were not designed to remove such products; thus their removal varies from high for some products to virtually nil for other products (Drewes 2007). For example, the common artificial sweetener sucralose has been found in elevated concentrations in the river discharge of a large urban wastewater treat-ment plant, with dilute concentrations found in the lower estuary and offshore (Mead et al. 2009). In terms of septic systems, practically nothing is known of the fate of PPCPs entering such systems. While they may not be persistent in the sense that they cannot be broken down in a reasonable length of time, they have been termed pseudopersistent since they are being continually added to the envi-ronment (Richmond et al. 2017).

Some PPCPs mimic hormones and can serve as endocrine disruptors in fish and other organ-isms (Cox et al. 2006; Drewes 2007), and some have behavioral, growth, and other deleterious impacts on aquatic life (Jones et al. 2007; Rich-mond et al. 2017). Apparently most trophic levels can be adversely affected by certain PPCPs, includ-ing bacteria, algae, zooplankton, crustaceans, and aquatic insects, as well as fish (Richmond et al. 2017). The study of PPCPs is an ongoing one, with the breadth, magnitude, and severity of the prob-lem as it pertains to our waterways yet unclear. Available research suggests that at least some of these products are toxic to aquatic life (reviewed in Daughton and Ternes 1999; Jones et al. 2007). In addition to human PPCPs, pets and livestock are given antibiotics (in many cases when they are not needed), other food additives, and in the case of pets nutraceuticals as well. Besides the threat of endocrine-disrupting activity, there is a growing concern of increased antibiotic resistance as a result of overuse of veterinary pharmaceuticals and their

subsequent entry into the environment (Campagnolo et al. 2002; Gilchrist et al. 2007).

## 12.5 Acid Rain

Whereas the various mining operations discussed
in this chapter cause acid and other types of pollution to downstream waters, **acid precipitation**
caused by burning of fossil fuels has contributed
to widespread acidification of streams and lakes
even far from points of origin. Acid deposition is
a combination of dry and wet deposition of acids
and acid-forming compounds onto the Earth's surface. Acid deposition can also occur as acid fog
or snow. The normal pH of rain is 5.6; however,
owing to acid deposition the pH of rain in some
areas of the Northeastern USA was decreased to
4.0–4.2 during the latter twentieth century. Sources
of acid precipitation can be natural and stem from
major incidents such as volcanoes and forest fires.
However, anthropogenic acid precipitation is generated by electric power plants burning coal or oil;
the burning of wood fuel; and automobile exhaust.
A major component is $SO_2$, which is produced
when sulfur in fuel is oxidized during combustion
Such sulfur compounds are collectively termed SOx
and their major source is coal-fired power plants.
Use of tall smokestacks (see Fig. 12.1a) allowed
it to be spread widely downwind of generation
area; in the atmosphere $SO_2$ reacts with moisture
to form sulfuric acid. The second major source of
acid rain is nitrogen oxides (generally designated as
NOx) produced by nitrogen and oxygen combining
under combustion—especially internal combustion
engines of automobiles. The nitrogen interacts with
water and oxygen to form nitric acid.

Acid rain in the USA began in the mid-1950s.
The areas that were hardest hit by its impacts have
low buffering capacity such as the Adirondack and
Catskill lakes and the Canadian Shield lakes, which
lie upon granite overlain by little soil; other areas
in the USA vulnerable to acidification include parts
of Florida, the Midwest, and areas in the Western mountains (Greaves et al. 2012). In Europe the
most vulnerable areas are in Scandinavia, which
receives acid rain from industrial areas of Europe.
Buffering capacity is referred to as acid neutraliz-
ing capacity (ANC) and is controlled by the amount

of base cations ($Ca^{++}$, $Mg^{++}$, $Na^+$, and $K^+$) present.
The presence of the cations is largely controlled by
mineral weathering of the sediments and watershed
bedrock. In areas where the bedrock is resistant to
chemical weathering, ANC is low and acidification
can most easily occur. For instance, a survey showed
that, in the aforementioned Adirondack Mountains,
41% of the lakes were chronically acidic or subject to episodic acidification (Driscoll et al. 2001.
An EPA survey of 1180 lakes and 4670 streams in
acid-sensitive US areas in the late 1980s found that
atmospheric deposition was the principal cause of
acid anions in 75% of the acidic lakes and 47% of
the acidic streams, while acid mine drainage was
the principal acid source in 26% of the acid streams
(Baker et al. 1991). Not only are lakes and streams
acidified by acid deposition, but also soil waters and
shallow groundwaters become acidified (Driscoll
et al. 2001).

Acid deposition has a number of deleterious
effects on stream and lake life. It can cause the mobi-
lization of metals from sediments, including toxic
forms of copper and cadmium. The decrease in pH
can facilitate the conversion of inorganic mercury
to toxic methylmercury, which was discussed in
depth earlier. It especially can mobilize aluminum
into a toxic ionic form $Al^+$, which can kill fish and
plants. The $H^+$ ion itself can cause physiological
stress to organisms. Havas and Rosseland (1995)
concluded that the main acute impact of $H^+$ stress
is ionoregulatory failure, and aluminum toxicity
leads to respiratory and circulatory stress. Organisms that are particularly sensitive to acid water
include the salmonids among fish, along with gastropods, various crustaceans, and some insect larvae (Havas and Rosseland 1995). Acidification led
to complete or near-complete loss of sport fish populations in hundreds of lakes in the Adirondacks,
Canada, and elsewhere, both through direct toxicity and elimination of the minnow food base. Birds
that are strongly associated with streams are also
impacted by acidification; research in Scotland and
Wales found that dippers (see Chapter 7) strongly
avoided acidified streams or stream reaches, possibly due to scarcity of their preferred food items (caddisfly larvae, mayfly nymphs, molluscs, and crustaceans being scarce in acidified waters) (Ormerod
and Tyler 1993). Streams and lakes are indirectly

impacted as well. Over time, acid rain leaches base minerals from soils and thus reduces their buffering capacity (Driscoll et al. 2001). The losses in calcium and magnesium impact terrestrial vegetation and also result in longer recovery times of lakes that had been impacted by acid rain (Likens et al. 1996).

In 1980 the National Acid Precipitation Assessment Program (NAPAP) was begun in the USA; this was a huge research effort funded to investigate acid deposition, create models, and propose solutions. NAPAP produced much solid science, but not much guidance on reducing the problem was developed, in part due to changes in program heads and Administrations and political interference with the process (Roberts 1991; Schindler 1992). However, in 1990 amendments to the Clean Water Act led to significant reductions in sulfuric acid deposition (an approximate 50% reduction in the eastern USA, for instance). This was largely a result of the cap-and-trade system to cut sulfur emissions from power plants (Malakoff 2010). US SOx emissions peaked in 1973 at 28.8 million metric tons, from there declining to 17.8 million metric tons by 1998 (Driscoll et al. 2001). From 1995 to 2020, annual emissions of $SO_2$ from power plants fell by 93% and NOx by 87% (US EPA 2021). Along with the cap-and-trade system it is important to note that coal-fired power plants are continually going out of commission, replaced by natural-gas-powered plants and increasingly solar and wind generation. The decline in NOx and SOx has been an example of an environmental success story; however, recovery has not been even across affected areas. A large-scale survey of 205 lakes and streams conducted from 1980 to 1995 showed decreased lake and stream sulfate in most regions of North America and northern Europe (Stoddard et al. 1999). However, three North American areas, (south central Ontario, the Adirondack/Catskill Mountains, and the Midwest) did not show significant declines, nor did Great Britain. Also, increases in livestock produce ammonium emissions (see Chapter 13), some of which undergoes nitrification, and automobile driving and emissions increase along with human populations.

In 2014 the EPA under the Obama Administration promulgated a plan to reduce fossil-fuel-burning power plant emissions further, primarily with reduction of greenhouse gas production as a goal, but improvements should also be expected in reductions of SOx, mercury, selenium, and other power-plant-derived pollutants. This plan was subsequently stopped by the Trump Administration but later revived under President Biden; but note the US Supreme Court is having problems with its legality.

## 12.6 PFAS, the "Forever Chemicals"

Much of the material to this point in this chapter has been discussion of chemicals and metals with known toxicity to wildlife and humans. One would think that after the many aquatic chemical cautionary tales discussed earlier in Section 12.4 that knowing the environmental effects of heavily used chemicals would be a priority, but such is not the case. One of the major current issues in river pollution today concerns polyfluorinated compounds, emerging contaminants that have actually been in use for decades but only recently come under intense scientific and regulatory scrutiny. Many of these compounds have been discharged into rivers with virtually no research done on toxicity to wildlife and humans. Of particular interest are PFAS, which were produced for use in fire-fighting foams, nonstick coatings, food packaging, stain-repellent clothing and carpets, cleaners, electronics, and many other applications. Their carbon-to-fluorine bond is extremely strong and they are difficult to break down and remain in the environment for very long times (hence the nickname above); note, however, that research into microbial degradation of these products is at least somewhat promising (Wackett and Robinson 2020).

Production of long-chain (8–9 carbon) PFAS occurred from about 1950 to 2000, when alarms went up regarding their stability in the natural environment and bioaccumulation in organisms. Two of the key problematic compounds were long-chain PFAS, perfluorooctanesulfonate (PFOS) and perfluorooctanoic acid (PFOA), now called "legacy" PFAS, which have been banned or otherwise production ceased. In 2019 under the auspices of the Stockholm Convention on Persistent Organic Pollutants over 180 countries agreed to ban the production and use of PFOA and related compounds, although their use in current fire-fighting chemical stocks was permitted (Hogue 2019). However, industry continues to produce novel or emerging alternative PFAS with shorter chains that supposedly are less bioaccumulative and hopefully easier to degrade in the environment. One of these is hexafluoropropylene oxide dimer acid (HFPO-DA), commonly known as "GenX," a current subject of much regulatory and legal action and much ongoing research in the Cape Fear River basin (Cahoon 2019). The number of such compounds being produced or in the environment numbers in the hundreds. Current water pollution sources of PFAS are industrial and municipal wastewater outfalls, landfills, and military bases where they were used in fire-fighting foams; note that there are a lot of long-chain and short-chain PFAS in various environmental matrices.

For example, both legacy and novel PFAS have been found in river water and sediments, such as in the Cape Fear and Yadkin–Pee Dee Rivers in North and South Carolina, the Ohio River, and Alabama's Tennessee River (Nakayama et al. 2007; Strynar et al. 2015; Cahoon 2019; Penland et al. 2020; Saleeby et al. 2021). Significant atmospheric transport of both legacy and novel PFAS have been found in eastern North Carolina in both wet and dry deposition, with deposition occurring in all seasons and wet deposition as much as three times greater than dry deposition (Shimizu et al. 2021). Research in the Yadkin–Pee Dee Rivers has demonstrated PFAS throughout the food chain, in plants, microbial biofilms, insects, and fish, with some evidence demonstrating a maternal transfer in fish. Along the US East Coast striped bass (*Morone saxatilis*) are a much-prized game fish, although with unstable populations which regulatory organizations are trying to resuscitate. Researchers have discovered that striped bass in the Cape Fear River contained a number of PFAS (including GenX), and the fish contain some of highest concentrations of PFOS (mean 490 ng/ml) recorded in serum from any fish in North America (Guillette et al. 2020). PFOS and other PFAS concentrations in that study were correlated with increased serum biomarkers of liver and immune system function. Other research has shown legacy and novel PFAS in lower-estuary seabirds, far downstream from production source outfalls, in liver, kidney, lungs, and blood (Robuck et al. 2020, 2021).

Of course, given all this, there is great concern for human health effects from exposure to PFAS through drinking water or consuming fish. As an example, along the Cape Fear River a facility called the Chemours Fayetteville Works has manufactured fluorocarbons since the 1970s and discharged GenX as a byproduct for years, and is currently manufacturing GenX and until discharged its waste into the river. The city of Wilmington and other smaller municipalities draw their drinking water from this river downstream of Fayetteville Works (Cahoon 2019). As noted above, both legacy PFAS and novel compounds including GenX have been found in water, sediments, and wildlife throughout the river, but such compounds, especially GenX, have been found in high concentrations in drinking water in Wilmington (Sun et al. 2016; Cahoon 2019), where the author of this book lives. Typical water treatment facilities cannot remove PFAS compounds from incoming water, and activated carbon and/or reverse osmosis, both expensive, are required (see Chapter 13). This expensive treatment of drinking water has now become essential to protect human health in Wilmington and many other municipalities worldwide that have drinking water contaminated by legacy and novel PFAS.

Environmental and human health water standards for PFAS are few and in flux; the EPA avian wildlife value for PFOS is 43 ng/L, and in 2021 there was a much-argued EPA health advisory level for PFOA and PFOS of 70 ng/L, and there is a European environmental quality standard of 0.65 ng/L for PFOS and derivatives for surface waters; all of those advisories are regularly exceeded in rivers. In June 2022 the EPA tightened the lifetime health advisory for PFOA and PFOS down to 0.004 and 0.02 ng/L, respectively (Hogue 2023). Note that this only recommends that utilities notify customers when concentrations reach this limit; it doesn't require they reach this standard. Finally, what about humans? There have been a lot of epidemiological studies, mostly on legacy PFAS, that have associated human exposures (such as through drinking water and seafood) with immunotoxicity, thyroid issues, liver problems, and various cancers, among other problems (Sunderland et al. 2019; Fenton et al. 2020). This has led to numerous and ongoing lawsuits against dischargers by municipalities and citizens' groups, as well as petitions to the EPA to require extensive testing of these chemicals by manufacturers. As of October 2021, the EPA unveiled a set of actions, or strategic roadmap, concerning PFAS, for the period from fall 2021 to 2024. This plan consists of monitoring, identification of PFAS subgroups, toxicity studies, setting fish consumption limits, and setting maximum contaminant loads, among other actions. In January 2022 the EPA began requiring manufacturers to begin testing 30 PFAS chemicals for toxicity, disappointing citizens' groups who were requesting that 54 such compounds be tested. The toxicity, distribution, legalities, and overall impacts of PFAS on human health and wildlife health are a rapidly changing story that needs immediate enhanced regulatory emphasis, i.e. the regulatory system is constantly playing catch-up with the science.

## 12.7 Riverine Plastics Pollution

Last but certainly not least, plastics pollution, by both macro- and microplastics, is a whole field of aquatic pollution that really only came into its own since 2000. Such debris includes various polymers such as high-density polyethylene (HDPE), low-density polyethylene (LDPE), polyvinyl chloride (PVC), polystyrene (PS), polypropylene (PP), and polyethylene terephthalate (PET), and others (reviewed by Li et al. 2016). The visual impact of great mid-oceanic rafts of plastic debris coupled with marine beaches covered with plastic debris of all sizes has catapulted public interest, and funding, to the forefront. Macroplastics, such as packing debris of foods and other materials, has caused very visible impacts to seabirds, marine mammals, and fish through intentional or accidental swallowing and subsequent illness, and entanglement by fishing debris. Microplastics (materials < 5 mm diameter) can be sourced from decomposition of larger plastic debris or designed materials such as abrasives in cleaning agents; such particles are also accidently or deliberately ingested by invertebrates or fish (Li et al. 2016) and can physically injure wildlife or carry toxic chemicals into wildlife (Law and Thompson 2014).

A phenomenal amount of data is currently being collected on marine plastics pollution (Law et al. 2010; Li et al. 2016; and many others). However, studies specifically on riverine plastics pollution are currently less abundant than marine studies in the literature, even though rivers are a key source of such pollutants to the ocean and large lakes worldwide. Rivers carry macroplastic debris largely as industrial or food packing material (think of the ubiquitous plastic drink bottles), and a key source of microplastic debris is discharges from sewage treatment plants (McCormick et al. 2014). For instance, a study in a Chicago river found that concentrations of microplastics of all sorts were significantly greater downstream of a wastewater discharge outfall compared with upstream; the results also showed microbial biofilms on the downstream particles to be unique from the surrounding river water (McCormick et al. 2014), suggesting microbial transport capabilities of such particles. A large amount of riverine plastic debris ends up in sediments (Schwarz et al. 2019), where it is available to be re-entrained by storms. Storms are a major driver of plastic debris from watersheds to estuaries, with a study on the Cooks River in Australia (Hitchcock 2020) documenting a > 40-fold increase in microplastic abundance in the water column during storms compared with pre-storm conditions. Seasonally, the greatest loading of plastic to the world ocean occurs from rivers during May–October, coinciding with the Asian monsoons (Lebreton et al. 2017).Thus, stormwater runoff of plastic debris is a key pollution problem, as it is with so many other stormwater-driven pollutants (see Chapters 1, 9, 13, and 15). A modeling effort calibrated by actual river data from various areas (Lebreton et al. 2017) arrived at the following conclusions: approximately 1.15–2.41 million tons of plastic waste enters the ocean annually from rivers. Asian rivers contribute 86% of global inputs, with the Yangtze and Ganges the largest contributors.

## 12.8 Summary

- Power generation by fossil-fuel-burning plants leads to widespread aquatic pollution by emission of greenhouse gases, metals including mercury, selenium, and arsenic, and acid-rain causing sulfur oxide compounds; spills from waste ash ponds have also caused large-scale river pollution.

- Use of cooling water by fossil-fuel and nuclear power plants impacts river and reservoir fish and crustacean communities through impingement on screens and entrainment through cooling systems. Discharge of heated cooling water also impacts biota in adjacent reservoir and river-receiving waters.

- Mining for coal and metals pollutes streams with acid and waste metals, while mountaintop coal mining destroys headwaters and ephemeral streams, chemically pollutes areas well downstream, and reduces fish biomass and diversity.

- Oil extraction from tar sands causes local water-fowl kills in on-site waste ponds, produces PAH and metals pollution over a wide area, and causes deformation of fish downstream and potential human cancer clusters in downstream settlements.

- Mercury contamination even at low concentrations causes harmful, particularly neurological impacts to aquatic biota and human consumers. Selenium contamination injures fish and their predators, and other metals such as cadmium continue to cause pollution episodes.

- Pesticides, particularly chlorinated types, continue to pollute the aquatic environment even many years following their banning, as do the chlorinated organic compounds called PCBs. PAHs are urban pollutants that continue to be produced and impact stream life.

- Some of the organochlorine pollutants noted within this chapter are endocrine-disrupting compounds, impacting sex characteristics of fish and their consumers. Personal care products are an understudied group of aquatic contaminants with poorly known impacts on aquatic biota.

- Acid rain caused large and widespread damage to fish, amphibians, and other biota in the 1970s and 1980s; much progress has been made reducing acid rain, but subtle, long-term impacts to streams and lakes remain.

- The abundance, extent, and toxicity of PFASs are currently receiving much-needed attention by scientists and regulators worldwide. Many aquatic systems are polluted by both legacy and emerging PFASs, and a growing body of knowledge is demonstrating impacts to both aquatic wildlife and humans exposed to such substances through drinking water and seafood consumption.

- Aquatic pollution of macro- and microplastics is a rapidly growing field in the marine realm, with physical and chemical-caused damage to marine mammals, fish, and seabirds. It is understudied in riverine systems, although rivers are important vectors of such pollutants.

# References

Adams, A.P., M. Garbett, H.B. Rees and B.G. Lewis. 1978. Bacterial aerosols from cooling towers. *Journal of the Water Pollution Control Federation* 50: 2362–2369.

Baker, L.A., A.T. Herlihy, P.R. Kaufman and J.M. Eilers. 1991. Acid lakes and streams in the United States: the role of acidic deposition. *Science* 252:1151–1154.

Bamber, R.N. and R.M.H. Seaby. 2004. The effects of power station entrainment passage on three species of marine planktonic crustacean, *Acartia tonsa* (Copepoda), *Crangon crangon* (Decapoda) and *Homarus gammarus* (Decapoda). *Marine Environmental Research* 57: 281–294.

Beitinger, T.L., W.A. Bennett and R.W. McCauley. 2000. Temperature tolerances of North American freshwater fishes exposed to dynamic changes in temperature. *Environmental Biology of Fishes* 58:237–275.

Bernhardt, E.S. and M.A. Palmer. 2011. The environmental costs of mountaintop mining valley fill operations for aquatic ecosystems of the Central Appalachians. *Annals of the New York Academy of Sciences* 1223:39–57.

Bernhardt, E.S., B.D. Lutz, R.S. King, et al. 2012. How many mountains can we mine? Assessing the regional degradation of central Appalachian rivers by surface coal mining. *Environmental Science and Technology* 46: 8115–8122.

Billen, G., J. Garnier, C. Deligne and C. Billen. 1999. Estimates of early-industrial input of nutrients to river systems: implications for coastal eutrophication. *Science of the Total Environment* 243(244):43–52.

Bomgardner, M.M. 2012. Cleaner fracking. *Chemical and Engineering News* October 15:13–16.

Brown, M.P., R.J. Sloan and M.B. Werner. 1985. Polychlorinated biphenyls in the Hudson River. *Environmental Science and Technology* 19:656–659.

Cahoon, L.B. 2019. GenX contamination of the Cape Fear River, North Carolina: analytical environmental chemistry as a remedy for multiple system failures. Pages

341–354 in Ahuja, S. (ed.), *Evaluating Water Quality to Prevent Future Disasters*. Elsevier, Amsterdam.

Campagnolo, E.R., K.R. Johnson, A. Karpati, et al. 2002. Antimicrobial residues in animal waste and water resources proximal to large-scale swine and poultry feeding operations. *Science of the Total Environment* 299:89–95.

Carson, R. 1962. *Silent Spring*. Houghton Mifflin.

Chalmers, A.T., D.M. Argue, D.A. Gay, M.E. Brigham, C.J. Schmitt and D.L. Lorenz. 2011. Mercury trends in fish from rivers and lakes in the United States, 1969–2005. *Environmental Monitoring and Assessment* 175:175–191.

Claudio, L. 2002. The Hudson: a river runs through an environmental controversy. *Environmental Health Perspectives* 110:184–187.

Colborn, T. and K. Thayer. 2000. Aquatic ecosystems: harbingers of endocrine disruption. *Ecological Applications* 10:949–957.

Cordy, G.E., N.L. Duran, H. Bouwer, et al. 2004. Do pharmaceuticals, pathogens, and other organic waste water compounds persist when waste water is used for recharge? *Ground Water Monitoring & Remediation* 24:58–69.

Cox, H., S. Mouzi and J. Gelsleichter. 2006. Preliminary observations of estrogenic activity in surface waters of the Myakka River, Florida. *Florida Scientist* 69:92–99.

Cristol, D.A., R.L. Brasso, A.M. Condon, et al. 2008. The movement of aquatic mercury through terrestrial food webs. *Science* 320:335.

Cronin, J. and R.F. Kennedy Jr. 1997. *The Riverkeepers*. Scribner, New York.

Crutchfield, J.U. 2000. Recovery of a power plant cooling reservoir ecosystem from selenium bioaccumulation. *Environmental Science and Policy* 3:S145–S163.

Czarny, K., D. Szczukocki, B. Krawczyk, M. Zieliński, E. Miekoś and R. Gadzala-Kopciuch. 2017. The impact of estrogens on aquatic organisms and methods for their determination. *Critical Reviews in Environmental Science and Technology* 47:909–963.

Daughton, C.G. 2007. Pharmaceuticals in the environment: sources and their management. Pages 1–58 in Petrovic, M. and D. Barcelo (eds.), *Analysis, Fate and Removal of Pharmaceuticals in the Water Cycle*. Wilson & Wilson's Comprehensive Analytical Chemistry, Volume 50. Elsevier, Amsterdam.

Daughton, C.G. and T.A. Ternes. 1999. Pharmaceuticals and personal care products in the environment: agents or subtle change? *Environmental Health Perspectives* 107:907–938.

Davis, D.L. and H.L. Bradlow. 1995. Can environmental estrogens cause breast cancer? *Scientific American* 273:166–172.

De Francisco, N., J.D. Ruiz Troya and E.I. Agüera. 2003. Lead and lead toxicity in domestic and free living birds. *Avian Pathology* 32:3–13

De Jockheere, J. and H. van de Voorde. 1977. The distribution of *Naegleria fowleri* in man-made thermal waters. *American Journal of Tropical Medicine and Hygiene* 26:10–15.

DeLorenzo, M.E., G.I. Scott and P.E. Ross. 2001. Toxicity of pesticides to aquatic microorganisms: a review. *Environmental Toxicology and Chemistry* 20:84–98.

Drewes, J.E. 2007. Removal of pharmaceutical residues during wastewater treatment. Pages 427–449 in Petrovic, M. and D. Barcelo (eds.), *Analysis, Fate and Removal of Pharmaceuticals in the Water Cycle*. Wilson & Wilson's Comprehensive Analytical Chemistry, Volume 50. Elsevier, Amsterdam.

Driscoll, C.T., G.B. Lawrence, A.J. Bulger, et al. 2001. Acidic deposition in the Northeastern United States: sources and inputs, ecosystem effects, and management strategies. *BioScience* 51:180–198.

Entrekin, S., M. Evans-White, B. Johnson and E. Hagenbuch. 2011. Rapid expansion of natural gas development poses a threat to surface waters. *Frontiers in Ecology and the Environment* 9:503–511.

Erikson, B. 2013. Nations strike mercury deal. *Chemical and Engineering News* January 28:8.

Erikson, B. 2022a. Neonicotinoids likely to harm endangered species. *Chemical and Engineering News* June 27:14.

Erikson, B. 2022b. US EPA to reassess health risks of glyphosate. *Chemical and Engineering News* June 27:15.

Evers, D.C., Y. Han, C.T. Driscoll, et al. 2007. Biological mercury hotspots in the Northeastern United States and Southeastern Canada. *BioScience* 57:29–43.

Fenner, K., S. Canonica, L.P. Wackett and M. Elsner. 2013. Evaluating pesticide degradation in the environment: blind spots and emerging opportunities. *Science* 341:752–758.

Fenton, S.E., A. Ducatman, A. Boobis, et al. 2020. Per- and polyfluoroalkyl substance toxicity and human health review: current state of knowledge and strategies for informing future research. *Environmental Toxicology and Chemistry* 40:606–630.

Furness, R.W. 1993. Chapter 3 in Furness, R.W. and J.J.D. Greenwood (eds.), *Birds as Monitors of Environmental Change*. Chapman and Hall, London.

Galkovskaja, G.A. 1987. Planktonic rotifers and temperature. *Hydrobiologia* 147:307–317.

Gilchrist, M.J., C. Greko, D.B. Wallings, G.W. Beran, D.G. Riley and P.S. Thorne. 2007. The potential role of concentrated animal feeding operations in infectious disease epidemics and antibiotic resistance. *Environmental Health Perspectives* 115:313–316.

Gillespie, R.B. and P.C. Baumann. 1986. Effects of high tissue concentrations of selenium on reproduction by bluegills. *Transactions of the American Fisheries Society* 115:208–213.

Goss, L.B. and D.L. Bunting. 1976. Thermal tolerances of zooplankton. *Water Research* 10:387–398.

Greaves, T.L., T.J. Sullivan, J.D. Herrick, et al. 2012. Ecological effects of nitrogen and sulfur air pollution in the U.S.: what do we know? *Frontiers in Ecology and the Environment* 10:365–372.

Guillette, T.C., J. McCord, M. Guillette, et al. 2020. Elevated levels of legacy and novel per-and polyfluoroalkyl substances in Cape Fear River striped bass (*Morone saxatilis*) are associated with biomarkers of altered immune and liver function. *Environment International* 136:105358.

Hand, E. 2014. Injection wells blamed in Oklahoma earthquakes. *Science* 345:13–14.

Harkinson, J. 2008. Tar wars. Mother Jones May/June: 64–70.

Havas, M. and B.O. Rosseland. 1995. Response of zooplankton, benthos, and fish to acidification: an overview. *Water, Air and Soil Pollution* 85:51–62.

Hayes, T.B., A. Collins, M. Lee, et al. 2002. Hermaphroditic, demasculinized frogs after exposure to the herbicide atrazine at ecologically relevant doses. *Proceedings of the National Academy of Sciences of the United States of America* 99:5476–5480.

Hebert, C.E. 2019. The river runs through it: the Athabasca River delivers mercury to aquatic birds breeding far downstream. *PLoS ONE* 14(4):e0206192.

Heckle, D.G. 2012. Insecticide resistance after *Silent Spring*. *Science* 337:1612–1614.

Hitchcock, J.N. 2020. Storm events as key moments of microplastic contamination in aquatic ecosystems. *Science of the Total Environment* 734:139436.

Hitt, N.P. and D.B. Chambers. 2014. Temporal changes in taxonomic and functional diversity of fish assemblages downstream from mountaintop mining. *Freshwater Science* 33(3):915–926.

Hogue, C. 2012. Dioxins, assessed at last. *Chemical and Engineering News*, February 27.

Hogue, C. 2019. Governments endorse global PFOA ban, with some exceptions. *Chemical and Engineering News* May 13:5.

Hogue, C. 2023. Court dismisses industry suit on PFAS in drinking water. *Chemical and Engineering News*, pg. 15, January 30, 2023.

Hong, S., J.-P. Candelone, C.C. Patterson and C.F. Boutron. 1994. Greenland ice evidence of hemispheric lead pollution two millennia ago by Greek and Roman civilizations. *Science* 265:1841–1843.

Hong, S., J.-P. Candelone, C.C. Patterson and C.F. Boutron. 1996. History of ancient copper smelting pollution during Roman and medieval times recorded in Greenland ice. *Science* 272:246–248.

Huff, J. 2011. Primary prevention of cancer. *Science* 332:916–917.

Huggett, R.J. and M.E. Bender. 1980. Kepone in the James River. *Environmental Science and Technology* 14:818–823.

Irukayama, K. 1966. The Pollution of Minimata Bay and Minimata Disease. Pages 153–165 in *Proceedings of 3rd International Conference on Water Pollution Research, Volume 3, Munich*. Pergamon Press, Oxford.

Janssen, J. and J.P. Giesy. 1984. A thermal effluent as a sporadic cornucopia: effects on fish and zooplankton. *Environmental Biology of Fishes* 11:191–203.

Jones, O.A.H., N. Voulvoulis and J.N. Lester. 2007. Ecotoxicity of pharmaceuticals. Pages 388–424 in Petrovic, M. and D. Barcelo (eds.), *Analysis, Fate and Removal of Pharmaceuticals in the Water Cycle*. Wilson & Wilson's Comprehensive Analytical Chemistry, Volume 50. Elsevier, Amsterdam.

Kaji. M. 2012. Role of experts and public participation in pollution control: the case of itai-itai disease in Japan. *Ethics in Science and Environmental Politics* 12:99–111.

Kamman K, A.L. Blankenship, P.D. Jones and J.P. Giesy. 2000. Toxicity reference values for the toxic effects of polychlorinated biphenyls to aquatic mammals. *Human Ecological Risk Assessment* 6:181–201.

Kelly, E.N., D.W. Schindler, P.V. Hodson, J.W. Short, R. Radmanovish and C.C. Nielson. 2010. Oil sands development contributes elements toxic at low concentrations to the Athabasca River and its tributaries. *Proceedings of the National Academy of Sciences of the United States of America* 107:16178–16183.

Köhler, H.-R. and R. Triebskorn. 2013. Wildlife ecology of pesticides: can we track effects to the population level and beyond? *Science* 341:759–765.

Krabbenhoft, D.P. and E.M. Sunderland. 2013. Global change and mercury. *Science* 341:1457–1458.

Kurek, J., J.L. Kirk, D.C.G. Muir, X. Wang, M.S. Evans and J.P. Smol. 2013. Legacy of a half century of Athabasca oil sands development recorded by lake ecosystems. *Proceedings of the National Academy of Sciences of the United States of America* 110:1761–1766.

Larsen, M.C., P.A. Hamilton and W.H. Werkheiser. 2013. Water quality status and trends in the United States. Chapter 2 in Ahuja, S. (ed.), *Monitoring Water Quality: Pollution Assessment, Analysis and Remediation*. Elsevier, Amsterdam.

Law, K.L. and R.C. Thompson. 2014. Microplastics in the seas. *Science* 345:144–145.

Law, K.L., S. Morét-Ferguson, N.A. Maximenko, et al. 2010. Plastic accumulation in the North Atlantic Subtropical gyre. *Science* 329:1185–1188.

Lebel, J., D. Mergler, M. Lucotte, et al. 1996. Evidence of early nervous system dysfunction in Amazonian populations exposed to low levels of methyl mercury. *NeuroToxicology* 16:157–167.

Lebel, J., M. Roulet, D. Mergler, M. Lucotte and F. Larribe. 1997. Fish diet and mercury exposure in a riparian Amazonian population. *Water, Air and Soil Pollution* 97:31–44.

Lebreton, L.C.M., J. van der Zwet, J.-W. Damsteg, B. Slat, A. Andrady and J. Reisser. 2017. River plastic emissions to the world's oceans. *Nature Communications* 8:15611.

Lemly, A.D. 1985. Ecological basis for regulating aquatic emissions from the power industry: the case with selenium. *Regulatory Toxicology and Pharmacology* 5:465–486.

Lemly, A.D. 1993. Guidelines for evaluating selenium data from aquatic monitoring and assessment studies. *Environmental Monitoring and Assessment* 28:83–100.

Lemly. A.D. 2014. Teratogenic effects and monetary cost of selenium poisoning of fish in Lake Sutton, North Carolina. *Ecotoxicology and Environmental Safety* 104:160–167.

Li, W.C., H.F. Tse and L. Fok. 2016. Plastic waste in the environment: a review of sources, occurrence and effects. *Science of the Total Environment* 566–567:333–349.

Likens, G.E., C.T. Driscoll and D.C. Buso. 1996. Long-term effects of acid rain: response and recovery of a forest ecosystem. *Science* 272:244–246.

MacDonald, D.D., C.G. Ingersoll and T.A. Berger. 2000. Development and evaluation of consensus-based sediment quality guidelines for freshwater ecosystems. *Archives of Environmental Contamination and Toxicology* 39:20–31.

Malakoff, D. 2010. Taking the sting out of acid rain. *Science* 330:910–911.

Mallin, M.A. 1986. The feeding ecology of the blue tilapia (*T. aurea*) in a North Carolina reservoir. *Lake and Reservoir Management* 2:323–326.

Mallin, M.A. and W.E. Partin. 1989. Thermal tolerances of common Cladocera. *Journal of Freshwater Ecology* 5:45–51.

Martinelli, L.A., J. R. Ferreira, B. R. Forsberg and R. L. Victoria. 1988. Mercury contamination in the Amazon: a gold rush consequence. *AMBIO* 17:252–254.

Mascarelli, A. 2013. Growing up with pesticides. *Science* 341:740–741.

McCormick, A., T.J. Hoellein, S.A. Mason, J. Schluep and J.J. Kelly. 2014. Microplastic is an abundant and distinct microbial habitat in an urban river. *Environmental Science and Technology* 48:11863–11871.

Mead, R.N., J. Morgan, G.B. Avery Jr. and R.J. Kieber. 2009. Occurrence of the artificial sweetener sucralose in coastal and marine waters of the United States. *Marine Chemistry* 116:13–17.

Nakayama, S. M., J. Strynar, L. Helfant, P. Egeghy, X. Ye and A.B. Lindstrom. 2007. Perfluorinated compounds in the Cape Fear drainage basin in North Carolina. *Environmental Science and Technology* 41:5271–5276.

Normile, D. 2013. In Minimata, mercury still divides. *Science* 341:1446–1447.

Ormerod, S.J. and S.J. Tyler 1993. Birds as indicators of changes in water quality. Chapter 5 in Furness, R.W. and J.J.D. Greenwood (eds.), *Birds as Monitors of Environmental Change*. Chapman and Hall, London.

Otter, R.R., F.C. Bailey, A.M. Fortner and S.M. Adams. 2012. Trophic status and metal bioaccumulation differences in multiple fish species exposed to coal ash-associated metals. *Ecotoxicology and Environmental Safety* 85:30–36.

Palmer, M.A., E.S. Bernhardt, W.H. Schlesinger, et al. 2010. Mountaintop mining consequences. *Science* 327:148–149.

Papoulias, D.M. and A.L. Velasco. 2013. Histopathological analysis of fish from Acorn Fork Creek, Kentucky, exposed to hydraulic fracturing fluid releases. *Southeastern Naturalist* 12(Special Issue 4):92–111.

Penland, T.N., W.G. Cope, T.J. Kwak, et al. 2020. Trophodynamics of per- and polyfluoroalkyl substances in the food web of a large Atlantic slope river. *Environmental Science and Technology* 54:6800–6811.

Peterson, S.A., J. Van Sickle, A.T. Herlihy and R.M. Hughes. 2007. Mercury concentrations in fish from streams and rivers throughout the western United States. *Environmental Science and Technology* 41:58–65.

Pringle, H. 2009. Arsenic and old mummies: poison may have spurred first mummies. *Science* 324:1130.

Quinn, N.W.T. and P.T. Vorster. 1998. The role of science in resolution of environmental crises at Kesterson Reservoir and Mono Lake, California. *Lakes & Reservoirs: Research and Management* 3:187–191.

Richmond, E.K., M.R. Grace, J.J. Kelly, A.J. Reisinger, E.J. Rosi and D.M. Walters. 2017. Pharmaceuticals and personal care products (PPCPs) are ecological disrupting compounds (EcoDC). *Elementa: Science of the Anthropocene* 5:66.

Roberts, L. 1991. Learning from an acid rain program. *Science* 251:1302–1305.

Robuck, A.R., M.G. Cantwell, J.P. McCord, et al. 2020. Legacy and novel per- and polyfluoroalkyl substances in juvenile seabirds from the US Atlantic coast. *Environmental Science and Technology* 54:12938–12948.

Robuck, A.R., J.P. McCord, M.J. Strynar, M.G. Cantwell, D.N. Wiley and R. Lohmann. 2021. Tissue-specific distribution of legacy and novel per- and polyfluoroalkyl substances in juvenile seabirds from the US Atlantic coast. *Environmental Science and Technology Letters* 8:457–462.

Ruhl, L., A. Vengosh, G.S. Dwyer, et al. 2009. Survey of the potential environmental and health impacts in the immediate aftermath of the coal ash spill in Kingston, Tennessee. *Environmental Science and Technology* 43:6326–6333.

Sackett, D.L., D.D. Aday and J.A. Rice. 2009. A statewide assessment of mercury dynamics in North Carolina water bodies and fish. *Transactions of the American Fisheries Society* 138:1328–1341.

Saleeby, B., M.S. Shimizu, R.I.S. Garcia, et al. 2021. Isomers of emerging per- and polyfluoroalkyl substances in water and sediment of Cape Fear River, North Carolina, USA. *Chemosphere* 262:128359.

Scheuhammer, A.M., M.W. Meyer, M.B. Sand Heinrich and M.W. Murray. 2007. Effects of environmental methylmercury on the health of wild birds, mammals and fish. *AMBIO* 36:12–18.

Schindler, D.W. 1992. A view of NAPAP from north of the border. *Ecological Applications* 2:124–130.

Schmitt, C.J. 1998. Environmental contaminants. Pages 131–165 in *Status and Tends of the Nation's Living Resources*, Vol. 1. US Department of the Interior, National Biological Service, Washington, DC.

Schmidt, C.W. 2001. Of PCBs and the river. *Chemical Innovation* 31:48–52.

Schwarz, A.E., T.N. Ligthart, E. Boukras and T. van Harmelen. 2019. Sources, transport, and accumulation of different types of plastic litter in aquatic environments: a review study. *Marine Pollution Bulletin* 143:92–100.

Shimizu, M.S., R. Mott, A. Potter, et al. 2021. Atmospheric deposition and annual flux of legacy perfluoroalkyl substances and replacement perfluoroalkyl ether carboxylic acids in Wilmington, NC, USA. *Environmental Science and Technology Letters* 8:366–372.

Silva-Forsberg, M.C., B.R. Forsberg and V.K. Zeidermann. 1999. Mercury contamination in humans linked to river chemistry in the Amazon Basin. *AMBIO* 28:519–521.

Slabe, V.A., J.T. Anderson, B.A. Millsap, et al. 2022. Demographic implications of lead poisoning for eagles across North America. *Science* 375:779–782.

Souchon, Y. and L. Tissot. 2012. Synthesis of thermal tolerances of the common freshwater fish species in large Western Europe rivers. *Knowledge and Management of Aquatic Ecosystems* 405:19–26.

Stahl, L.L., B.D. Snyder, A.R. Olsen and J.L. Pitt. 2009. Contaminants in fish tissue from US lakes and reservoirs: a national probabilistic study. *Environmental Monitoring and Assessment* 150:3–19.

Stevens, A.R., R.L. Tyndall, C.C. Coutant and E. Willaert. 1977. Isolation of the etiological agent of primary amoebic meningoencephalitis from artificially heated waters. *Applied and Environmental Microbiology* 34:701–705.

Stoddard, J.L., D.S. Jeffries, A. StoLükewille, et al. 1999. Regional trends in aquatic recovery from acidification in North America and Europe. *Nature* 401:575–578.

Strynar, M., S. Dagnino, R. McMahen, et al. 2015. Identification of novel perfluoroalkyl ether carboxylic acids (PFECAs) and sulfonic acids (PFESAs) in natural waters using accurate mass time-of-flight mass spectrometry (TOFMS). *Environmental Science and Technology* 49:11622–11630.

Sun, M., E. Arevalo, M. Strynar, et al. 2016. Legacy and emerging perfluoroalkyl substances are important drinking water contaminants in the Cape Fear watershed of North Carolina. *Environmental Science and Technology Letters* 3:415–419.

Sunderland, E.M., X.C. Hu, C. Dassuncao, A.K. Tokraanov, C.C. Wagner and J.G. Allen. 2019. A review of the pathways of human exposure to poly- and perfluoroalkyl substances (PFASs) and present understanding of health effects. *Journal of Exposure Science & Epidemiology* 29:131–147.

Tillitt, D.E., D.M. Popoulias, S.S. Whyte and C.A. Richter. 2010. Atrazine reduces reproduction in fathead minnow (*Pimephales promelus*). *Aquatic Toxicology* 99:149–159.

Timoney, K.P. and R.A. Ronconi. 2010. Annual bird mortality in the bitumen tailings ponds in northeastern Alberta, Canada. *The Wilson Journal of Ornithology* 122:569–576.

US EPA. 2000a. Guidance for Assessing Chemical Contaminant Data for Use in Fish Advisories, Volume 1: Fish Sampling and Analysis. EPA-823-B-00-007. Office of Water, US EPA, Washington, DC.

US EPA. 2000b. Guidance for Assessing Chemical Contaminant Data for Use in Fish Advisories, Volume 2: Risk Assessment and Fish Consumption Limits. EPA-823-B-00-008. Office of Water, US EPA, Washington, DC.

US EPA. 2004. National Coastal Condition Report II. EPA-620/R-03/002. Office of Research and Development, Office of Water, US EPA, Washington, DC.

US EPA. 2021. Power Plant Emission Trends. https://www.epa.gov/airmarkets/power-plant-emission-trends. US EPA, Washington, DC.

Van der Elst, N.J., H.M. Savage, K.M. Keranen and G.A. Albers. 2013. Enhanced remote earthquake triggering at fluid-injection sites in the Midwestern United states. *Science* 341:164–167.

Ven Metre, P.C., B.J. Mahler and E.T. Furlong. 2000. Urban sprawl leaves its PAH signature. *Environmental Science and Technology* 34:4064–4070.

Wackett, L.P. and S.L. Robinson. 2020. The ever-expanding limits of enzyme catalysis and biodegradation: polyaromatic, polychlorinated, polyfluorinated, and polymeric compounds. *Biochemical Research* 477:2875–2891.

Walter, R.C. and D.J. Merritts. 2008. Natural streams and the legacy of water-powered mills. *Science* 319:299–304.

Wasserman, J.C., S. Hacon and M.A. Wasserman. 2003. Biogeochemistry of mercury in the Amazonian environment. *AMBIO* 32:336–342.

Weis, J.S. and P. Weis. 1989. Effects of environmental pollutants on early fish development. *Reviews in Aquatic Sciences* 1:45–73.

Woock S.E., W.R. Garrett, W.E. Partin and W.T. Bryson. 1987. Decreased survival and teratogenesis during laboratory selenium exposures to bluegill *Lepomis macrochirus. Bulletin of Environmental Contamination and Toxicology* 39:998–1005.

# CHAPTER 13

# Stream Pollution from Human Sewage and Animal Wastes

## 13.1 Human Sewage: Collection, Treatment, and Effects on Streams

Treatment of human sewage, or lack thereof, has been an integral part of stream ecology since the mid-twentieth century, when biologists and sanitary engineers (now called environmental engineers) seriously questioned the then-common practice of municipalities dumping raw or partially treated human waste into streams and rivers. In reality, human waste has been a stream pollution problem for millennia. Earliest treatments generally consisted of pit latrines—doing it in ditches and covering it up at some point. Exceptions to this were from a few advanced Bronze Age cities such as Mohenjo-daro in the Indus Valley and Knossus in Crete, which had indoor plumbing and toilets (Grey 1940). During this period, several civilizations discovered that human manure made good fertilizer, so at least some human waste was used to fertilize crops (Angelakis et al. 2018). However, all too frequently villages dumping it into the local stream or river became a convenient practice of choice—except for the village downstream of yours. If your town was small enough the waste could be absorbed by the stream's life (see stream zonation process below, Section 13.1.1.2), but as urban populations increased, human waste pollution led to severe human health problems. An example of this comes from medieval the Netherlands (see Hunt 2019). In fourteenth-century Leiden and other municipalities, most homes had underground brick cesspits in the backyards; periodically, men employed by the city came around in the evenings to clean out the pits and haul away the

accumulated human manure and whatever other trash found its way into the cesspits. The result was a relatively clean environment for such cities. In Leiden, however, the city fathers decided (in the seventeenth century) to construct a brick sewer system—unfortunately the sewers emptied directly into stagnant canals, polluting the waters and causing disease outbreaks (Hunt 2019). Until the mid-nineteenth century much of London's sewage was simply piped into the reeking Thames River; it was epidemic diseases, especially cholera that finally led to laws protecting the river (Grey 1940). While sewage treatment has come a long way in certain areas of the Earth, major problems remain today with untreated sewage (also called wastewater) that affects millions of people.

We have only to look at Haiti—the January 2010 earthquake destroyed the country's infrastructure and by September of that year a cholera epidemic was initiated that had led to over 820,000 infections and nearly 10,000 deaths; no further fatalities were reported since 2019. Humans can get cholera by drinking sewage-contaminate water, and reports state that many displaced people were drinking river water, polluted by human waste. In the Haitian incident it appears that the epidemic was initiated in a UN peacekeepers camp from Nepalese soldiers; contaminated human waste from their camp entered the Arbonite River, heavily used by locals for collecting drinking water, bathing, washing clothes, and irrigation and as a play area for children (Enserink 2011). Another example is Zimbabwe—a 2008–2009 epidemic led to 100,000 infections and over 4000 fatalities. This cholera epidemic was spread by water contaminated by human

*River Ecology.* Michael A. Mallin, Oxford University Press. © Michael A. Mallin (2023). DOI: 10.1093/oso/9780199549511.003.0014

excrement as public services shut down in the waning years of the Mugabe regime. According to news sources, in 2018 and 2019 cholera again occurred in Zimbabwe due to deteriorated water infrastructure and sewage contamination, infecting over 10,000 people and causing 69 fatalities. Worldwide, polluted water and poor sanitation has been blamed for some 10,000 human fatalities/day from microbial diseases (Vuorinen et al. 2007). India's Ganges River (Ganga) has been hit particularly badly: 50 cities release 6 billion L of untreated human waste into the river daily; note that a considerable amount of water upstream is diverted for irrigation and power production, concentrating the waste (Shah et al. 2018).

In the USA there are ongoing problems with sewage pollution. With the passage of the US Clean Water Act in 1972 and a subsequent large influx of funding into the construction and upgrades of wastewater treatment systems (i.e. point sources of pollution) there was an assumption that such raw sewage pollution would diminish to negligible levels. And clearly, things have greatly improved over the intervening decades. However, large problems still remain with sewage treatment, particularly with aging, undersized, or otherwise faulty infrastructure. The US EPA estimates that between 23,000 and 75,000 sanitary sewage overflows and spills occur annually in the USA, some of major proportions (US EPA 2023). In 1995 routine or accidental discharges from wastewater treatment systems were involved in 24% of the US public waters closed to shellfish harvest (according to the National Oceanic and Atmospheric Administration [NOAA 1998]). An analysis of US beach closings and advisories in 2009 showed that sewage spills, leaks, and overflows caused or contributed to 19% of the beach closing/advisory days where a cause of the pollution causing the closing could be determined (Dorfman and Rosselot 2010). Also, in the USA and Canada there remain some number of homes not served by either central sewer systems or septic systems but that rely on outhouses (i.e. discharge of untreated waste). With these issues in mind, it is time to examine what exactly our waste contains and how it affects human health and the stream environment.

### 13.1.1 What's In Our Waste?

#### 13.1.1.1 Human Health Issues

Human waste (feces) contains a variety of substances that can cause problems if they enter and pollute streams and rivers. These can be either human health or ecosystem problems, or in some cases both. People exposed to streams polluted by fecal matter can suffer health problems because our wastes contain copious amounts of bacteria, viruses, and protozoa, collectively called enteric microflora. Numerous pathogenic microbes can be found in human sewage (Table 13.1). There is always some percentage of the population that is sick, and these unfortunate individuals expel microbes that are pathogenic to other people who may contact them. By swimming, diving, surfing, or even wading in a polluted water body people can become infected by pathogenic fecal microbes entering the body through the mouth, nose, eyes, or open wounds. Symptoms and illnesses that people commonly display following an encounter in fecally polluted waters include diarrhea, cramping, gastroenteritis, skin rashes, and eye infections, and sometimes life-threatening illnesses such as cholera, hepatitis, typhoid fever, and Guillain–Barré syndrome.

Within the gut, enteric microbes live in a rather ideal environment, with constant body temperature, nutrients, and protection from elements. When ejected into the environment, however, most of the enteric microflora die off rapidly. The major factors that kill such microbes are UV radiation from sunlight and predation by the natural protozoan community in water and aquatic sediments (Burtchett et al. 2017). Additionally, elevated salinity accelerates death or deactivation of many enteric microbes, although some, such as *Enterococcus* bacteria, survive better than others in brackish and saltwater. Changes in pH or temperature are also factors that reduce survival of fecal bacteria. Some fecal bacteria will survive for extended periods (weeks to months) under good environmental conditions that may be found in some aquatic sediments that offer protection from UV radiation, constant temperature and pH, and abundant nutrients (C, N, and P).

Filter-feeding biota, such as bivalves, can concentrate fecal bacteria (such as *Vibrio cholera*) and

**Table 13.1** Pathogenic microbes found in human sewage effluent (West 1991; Smith and Perdek 2004).

**Bacteria**
*Campylobacter jejuni*
*Escherichia coli* 0157:H7
*Salmonella* spp.
*Shigella* spp.
*Yersinia enterocolitica*
*Vibrio cholerae*

**Protozoa**
*Balantidium coli*
*Cryptosporidium parvum*
*Entamoeba histolytica*
*Giardia lamblia*
*Toxoplasma gondii*

**Viruses**
Adenoviruses
Coxsackie virus
Echovirus
Hepatitis A virus
Human caliciviruses
Noroviruses
Reovirus
Rotovirus

viruses in their bodies. This may not harm the filter-feeder, but if humans consume raw shellfish from fecally polluted waters they can become ill or even die. Thus, governments have developed standards to ensure the safety of seafood, as well as what are termed recreational waters—those waters where people swim, surf, wade, snorkel, or otherwise contact the water. Since culturing pathogenic bacteria, viruses, and protozoans individually can be costly and time-consuming, fecal indicator bacteria are usually used to develop standards. These are bacteria found in the guts of warm-blooded animals that survive for a period outside of the body and are relatively easy to culture and quantify. Certainly there are problems with the use of such indicators, due to variable die-off, hardiness, and basically how well such bacterial indicators represent nonbacterial pathogens such as viruses and protozoans. In the USA individual states normally set their own recreational water standards, and many states utilize a human contact standard of 200 colony-forming units (CFU)/100 ml of fecal coliform bacteria for freshwater lakes, rivers, and streams. The EPA recommends using *Enterococcus* as a beach water standard, with 104 CFU/100 ml

the instantaneous standard and a geometric mean of 35 CFU/100 ml from a set of five samples within three weeks as a guide for closures. Other areas in the USA as well as other nations may utilize other indicator organisms such as total coliforms and *Escherichia coli*. In the USA shellfish water standards are set by the Public Health Service because shellfish are integral to interstate commerce (US FDA 1995). The current shellfish water standard is 14 CFU/100 ml for fecal coliform bacteria.

Tests for these indicators are found in *Standard Methods for the Examination of Water and Wastewater* (published by the American Public Health Association [APHA 2005]). In a nutshell, samples are collected from the field in sterile containers, placed on ice, and delivered to the laboratory within six hours. There they are filtered, utilizing various dilutions if high levels of pollution are suspected. The filters are placed on various media depending on the indicator microbe of interest and incubated, and after a prescribed period the number of colonies that were produced is counted, usually with the aid of a dissecting microscope. A tube fermentation method (APHA 2005) is also commonly used for some indicators. Thus, the indicator bacteria are quantified and reported as CFU/100 ml of sample. Other indicator microbes are constantly being tested in the academic world, but health agencies and municipalities are required to use government-approved indicators and methods.

### 13.1.1.2 Ecological Issues

Even a cursory look at raw sewage can tell the observer that this material is extremely turbid. Turbidity will serve to attenuate solar irradiance in a stream sufficient to block or reduce photosynthesis of rooted aquatic macrophytes, which are good habitat for biota as well as producing oxygen to the water column. Human (and animal) waste also contains large amounts of labile (i.e. easily broken down) organic matter, measured as BOD (Table 13.2). When such material enters a water body it provides a direct food source for naturally occurring bacteria. When this labile material is in large enough quantities, this causes the resident bacteria to rapidly consume the waste and multiply, exerting a BOD and using up DO through respiration. Where the BOD load is high, hypoxia or even

**Table 13.2** Average pollutant concentrations (in mg/L) in raw urban/suburban sewage. (Revised from Clark et al. 1977.)

| Pollutant | Concentration (mg/L) |
| --- | --- |
| TN | 35 |
| Ammonium | 15 |
| TP | 10 |
| Orthophosphate | 7 |
| BOD5 | 250 |
| TSS | 280 |

anoxia may result, causing fish and invertebrate kills and rendering areas of the stream uninhabitable to most fish and other organisms.

The term BOD has been used in several chapters in this book; here it is appropriate to describe its measurement. The strength of the BOD either in sewage or receiving water bodies is commonly measured using a five-day BOD test (BOD5), again found in *Standard Methods* (APHA 2005). A sample is collected from the field in a 250-ml specially designed BOD bottle and returned to the laboratory on ice, within six hours of collection. Samples are brought to constant temperature, any air bubbles are driven out, and DO and pH are measured; for wastewater samples, pH adjustment and bacterial seeding are usually required. Samples are incubated at 20°C for five days, after which the DO is remeasured and the oxygen "demand" by the bacteria is computed from the decrease in DO as mg/L. For samples of unpolluted (or reasonably unpolluted) streams and lakes BOD5 values ranging from 0.5 to 2.0 mg/L are common (Mallin et al. 2006). BOD5 concentrations exceeding 3.0 mg/L in streams or lakes may indicate organic inputs from sewage or stormwater runoff, or they may indicate organic matter produced as the result of an algal bloom (Mallin et al. 2006). The BOD5 test mainly measures labile organic material oxidation; for more recalcitrant organic waste, longer incubations are used such as BOD20, BOD60, or BOD180.

Another major ecosystem impact of human or animal waste comes from the concentrated nitrogen and phosphorus contained within (Table 13.2). Fecal matter is enriched in phosphorus compared to nitrogen, and human sewage generally has a molar N:P ratio < 5. The inorganic nitrogen component of sewage is dominated by ammonium due to the

strong reducing conditions. However, ammonium is rapidly utilized by phytoplankton and macroalgae. Thus, loads of human sewage (and animal waste) lead directly to algal blooms. These may be nuisance blooms that cause sight or odor problems, or blooms that create a BOD sufficient to lower DO to levels that kill or drive off fish and other aquatic organisms. In many nutrient-enriched situations BOD5 is significantly correlated with chlorophyll *a* concentration (Mallin et al. 2006; Miltner 2018). A number of algal taxa with known toxicity have been documented to bloom in response to sewage spills or inputs of poorly treated sewage. These include dinoflagellates (*Gymnodinium mikimotai, Pfiesteria piscicida* and *P. shumwayii*, and *Prorocentrum minimum*), chrysophytes (*Phaeocystis pouchetti* and *Chattonella antique*); diatoms (*Pseudo-nitzschia* spp.), and various blue-green algal taxa (Burkholder 1998, 2002; Burkholder et al. 2008). Toxic species in freshwater will tend to be mainly blue-green algae (cyanobacteria) and euglenophytes, while toxic species in brackish and marine systems will be mainly dinoflagellates, chrysophytes, and diatoms.

In addition to indirect formation of BOD from algal blooms, ammonium can contribute to the BOD load in a stream or river through nitrogenous BOD (see Chapter 2). In this situation ammonium (as a reduced compound) will undergo nitrification by bacteria (*Nitrosomonas* and *Nitrobacter*) to nitrite and then nitrate. This process consumes DO and thus exerts a BOD load on the water body. For that reason regulatory permits from point-source dischargers in the USA often have limits on the amounts of both BOD5 and ammonium they can discharge from a treatment facility into receiving waters. Additionally, phosphorus, particularly organic phosphorus, is readily utilized by ambient bacteria and directly contributes to BOD by stimulating growth of more bacteria in the stream or river. This has been demonstrated experimentally in several blackwater streams in the Southeastern USA (Mallin and Cahoon 2020). Phosphorus inputs are occasionally controlled in eutrophication reduction schemes in streams, rivers, and lakes, but not as a BOD stimulant.

Human sewage contains a variety of other potentially environmentally damaging substances in addition to the pathogens, BOD, and nutrients mentioned above. If sewage comes from industrial sources there may be excessive metals, acids, caustic substances, organic chemicals, and myriad lesser-known potential toxins entering streams and rivers. In addition, people dump all sorts of products such as unused prescription drugs including antibiotics, over-the-counter medications, birth control pills, and various "natural" herbal medications (sometimes referred to as nutraceuticals) into the sewage stream (Cordy et al. 2004). These collectively are sometimes referred to as pharmaceuticals and personal care products (PPCPs) (Daughton and Ternes 1999). In addition to dumping them, some quantity of these (or their breakdown products) passes through human digestive tracts unabsorbed into the sewage stream. In contrast to the generally smaller amounts of prescription medications, larger amounts of personal care products enter the aquatic environment including bath additives, shampoos, skin care products, hair sprays and dyes, soaps, sunscreens, and perfumes (Daughton and Ternes 1999). Some PPCPs mimic hormones and can serve as endocrine disruptors (Cox et al. 2006), with impacts to aquatic life. The study of PPCPs is an emerging one, with the breadth, magnitude, and severity of the problem as it pertains to our waterways yet unclear. Research suggests that at least some of these products are toxic to aquatic life (reviewed in Daughton and Ternes 1999). Impacts from one or combinations of many of the effluent pollutants described above will have trophic-level impacts on receiving stream food webs, including reduction of herbivores and sensitive top predators, and decrease in energy transfer efficiency between trophic levels (Mor et al. 2021).

So, from the above we can see that when raw sewage enters a stream multiple problems can occur. In sewered areas this occurs when lines leak or break or pump stations fail. In third-world countries (as well as some modernized nations) sewage frequently is dumped or drained into streams and rivers. As mentioned earlier, in the middle of the twentieth century such discharges were frequent enough that researchers compiled a large literature on the subject. Work by early public health researchers (Brinley 1942; Bartsch 1947; Bartsch and Ingram 1959) led to the following zonation scheme for sewage impacted streams:

- *Zone of clean water*: Upstream of the sewage slug, assuming that the stream is relatively unpolluted, the DO should be high, with the stream supporting a diverse assemblage of benthic invertebrates (including snails and mussels) and fish (including game fish). Rooted aquatic macrophytes are present and there is a diverse phytoplankton community present in low abundance (generally oligotrophic conditions are present).

- *Zone of degradation*: At and just downstream of the sewage slug is this relatively short zone where DO may fall to hypoxic or even anoxic conditions, fish and invertebrate kills occur, and other fish will flee downstream. High turbidity from the solid waste will block sunlight, retarding or stopping macrophyte photosynthesis, and sludge may appear on the bottom. Increases in chloride and organic N occur as well as very high BOD concentration. Bacterial activity is rampant from both the bacteria in the sewage itself and the stream bacteria multiplying to consume the organic BOD load. Fungi appear and may cover the bottom and protozoans bloom to feed on the bacteria. Only very tolerant invertebrates flourish (tubificid worms). In lighted areas blue-green algae and euglenoids occur.

- *Zone of active decomposition*: Downstream in this zone the waste is being processed. Here there is little light penetration and DO is low, generally no more than 40% DO saturation. Gases are being produced ($CO_2$, $H_2S$, methane [$CH_4$], and ammonia, with consequent offensive odors). Bacteria are abundant, as are their predators (protozoans). Fish are absent and surface blue-green and/or euglenoid algal blooms may be present. Macroalgae may bloom from the high nutrients.

- *Zone of recovery*: depending upon the strength of the sewage and the stream discharge, this may be a relatively long zone. Here there is gradually clearing water, gas production ceases, and DO increases above 40% saturation. N is present as ammonia, nitrite, and nitrate. Bacteria are decreased due to consumption; protozoans, rotifers, and crustacean zooplankton are abundant. Blue-green and green algal blooms may occur, with diatoms increasing at the lower end of the zone. Aquatic macrophytes appear with the clearing water. Chironomids and other insect larvae becoming reestablished; mussels, snails, and a few fish appear.

- *Zone of cleaner water*: The previous zone blends into this zone, where algal blooms decrease and there may be abundant macrophytes of various types. Microfauna are scarce and consist mainly of rotifers and some crustacean zooplankton. Insect larvae, larger crustaceans, mussels, and other benthos thrive and game fish are present.

To predict how much waste a stream could absorb, in 1925 sanitary engineers developed the Streeter–Phelps equation, commonly known as the "sag curve" as it predicted the DO sag that occurred in a stream following a waste input by accounting for the two opposing processes of deoxygenation and reaeration (Clark et al. 1977). The sag curve is essentially the grandfather of the predictive models that present-day agencies, university researchers, and consultants produce to assess point- and non-point-source waste inputs to streams and rivers impaired by low DO problems; more sophisticated models need to account for many forcing factors such as water releases from dams and tidal impacts in coastal rivers.

## 13.1.2 Wastewater Treatment in Today's Society

It wasn't until the mid-to-late nineteenth century that the role of polluted water in human disease was taken seriously (Vuorinen et al. 2007). Knowing the various health and ecosystem problems that human waste loads cause, and that they remain a problem to both third-world and industrialized countries, let's examine how human waste is commonly treated before it enters natural water bodies. First let's have a look at a modern municipal sewage treatment system; the typical system consists of several components.

### 13.1.2.1 The Collection System

In most homes in urbanized areas, the flush water from the toilets (sometimes termed black water) becomes mingled with used water from washing machines, dishwashers, and sinks (water that is termed gray water). From the house (or restaurant, bar, business, etc.) it enters the sewage collection system and is referred to as wastewater or sewage.

Sewer lines carry the sewage to a centralized treatment plant, sometimes very long distances. Wherever possible, engineers take advantage of topography to let gravity help carry the wastewater to the treatment plant. When this is not feasible the wastewater passes through pump stations (also called lift stations) to force it along the way. The pump stations require electrical power to run the pumps, so any interruption in power may cause sewage backups in the lines or leaks and spills. Pump stations are normally found on municipal property along road and stream right-of-ways and often need to pass across such streams and creeks. If a pump station suffers power failures then raw sewage may only need to travel a short distance to reach stream waters. The collection system is thus vulnerable chronically from leaks and acutely through environmental mishaps such as breaks in the lines and failures at pump stations. An example of the impacts of a 3,000,000-gallon human sewage spill on a tidal creek is presented in Box 13.1.

---

### Box 13.1 Anatomy of a sewage spill.

On Friday, July 1, 2005 the middle branch of Hewletts Creek, located in Wilmington, North Carolina was subjected to a raw sewage spill of 11,355,000 L (3,000,000 gallons). This occurred when a buried 60-cm force main coupling burst apart. This line carried sewage from a nearby beach community to a pump station on Hewletts Creek (near the beach), then on to the Wilmington South Side Wastewater Treatment plant located along the Cape Fear River. Some waste flowed into the creek or nearby swamp forest and some flowed into the nearest storm drains, which drain directly into Hewletts Creek. Local and state regulatory agencies were alerted that morning. As a result, the North Carolina Division of Marine Fisheries closed the creek and a large section of the Atlantic ICW and other nearby tributaries to shellfishing, and North Carolina Shellfish Sanitation closed the same area to swimming. Since the sewage leaked out over an 18-hour period, it contaminated the downstream areas of the creek and nearby areas of the ICW, and incoming tides pushed it back up into the tributary streams.

The BOD load drove the DO concentrations to daytime levels between 1 and 3 mg/L, and large numbers of fish were trapped in the upper and middle areas of the marsh by hypoxic water and died. Fecal coliform bacteria sampling of the surface waters found concentrations ranging from 3000 to 270,000 CFU/100 ml on day 2; on day 3 concentrations of 2000–21,000 were recorded, with decreases over the following days. However, a rain event 10 days after the spill pushed concentrations back up to 3000 in an upper creek station. Sediment fecal coliform counts before the spill were about 500 CFU/cm$^2$; five days after the spill they were 5000. Counts subsequently began to drop off but remained well above background levels for six weeks following the spill due to protection from the sun's UV radiation and abundant sediment nutrients. Several days after the spill the sampling crews photographed several ducks dead or dying on the banks, presumably killed by pathogenic microbes in the water (they had been feeding on the dead fish for a number of days). The elevated nutrients in the spilled sewage caused algal blooms ranging from 40 to 130 µg of chlorophyll *a*. Surfers and swimmers utilizing nearby beaches the day of the spill reported illnesses, and residents reported that dogs that visited the marshes in that period became ill. The ban on swimming in the ICW was lifted after two weeks, but Hewletts Creek was closed to swimming and shellfishing for the remainder of the summer.

(See Mallin et al. 2007 for details.)

Beyond outright acute accidents, the collection system is vulnerable to chronic damage referred to as I&I, or inflow and infiltration, especially in coastal areas (Flood and Cahoon 2011). Inflow occurs when surface flooding causes entry of floodwater to the collection system from the surface, while infiltration is when elevated groundwater enters through cracks and leaky pipes. Both of these issues have been demonstrated in coastal treatment systems (Flood and Cahoon 2011); this of course raises treatment cost while reducing effectiveness of treatment, and I&I occurring from high tides or sea level rise (see Chapter 17) can also introduce salinity to the wastewater, leading to enhanced corrosion.

### 3.1.2.2 Anatomy of a Typical Wastewater Treatment Plant

What happens when sewage reaches the treatment plant? Sewage enters the plant from the collection system (Fig. 13.1) and is screened (with certain larger objects sometimes manually removed from the waste stream with rakes) or ground up (by mechanisms called *comminuters*) to remove or break

down large particles that could interfere with subsequent processes. Removal of grit and other solid particulates that may damage equipment is also performed in this area. From here the sewage is piped into settling tanks (also called clarifiers) for **primary treatment**. Here the major settling of suspended solids occurs. The settled material is called sludge and is pumped to another area of the plant for further treatment.

The liquid supernatant (remaining waste) from the primary settling tanks is then pumped to other tanks for **secondary treatment**. This treatment involves the removal and disposal of dissolved and colloidal organic material. It is biologically converted to bacterial biomass and constituent chemical compounds through an *aeration* process.

$$C_{10}N_8O_6H_4P_2S \rightarrow (need\ O_2) \rightarrow$$

$$CO_2, H_2O, NO_3, SO_4, PO_4 + bacterial\ cells$$

Two forms of secondary treatment are commonly used: **activated sludge/aeration** or **trickling filter**. One or the other is used, not both. Secondary treatment is responsible for a large portion of the BOD removal as well as additional solids removal. This does not disinfect the sewage, however. Many polluting compounds such as organic contaminants and PPCPs can also pass through this treatment (Daughton and Ternes 1999; Cordy et al. 2004; Cox et al. 2006).

- *Activated sludge/aeration*: The supernatant from the primary system is piped into an aeration tank (aerated mechanically by air pumps). It is mixed with bacterial biomass (called **floc**), which in the presence of air converts the organic waste to more bacteria and the constituent chemical compounds shown above. Some of the resultant floc is recycled to the head of the aeration tank to reseed the process, while the remaining sludge is dewatered, as explained subsequently. Protozoans are in the latter portion of the tank and eat bacteria. The treated water is then sometimes pumped into a tertiary treatment system.
- *Trickling filters*: These are beds of gravel 1–3 m thick over which primary-treated wastewater is sprayed by rotating spraying arms. Bacterial matrices (along with protozoans and heterotrophic algae) form around the gravel and

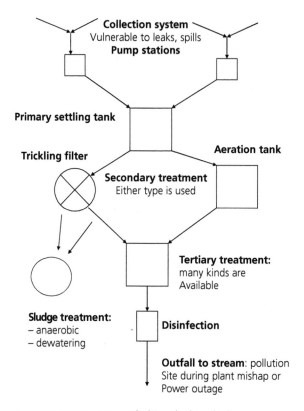

**Fig. 13.1** Idealized central wastewater treatment system, not to scale. (Figure by the author.)

consume waste, turning it into more bacterial biomass, and also mediate the production of $H_2O$, $CO_2$, $NO_3$, $PO_4$, and $SO_4$. Floc that dies and sloughs off is treated as sludge. The treated liquid that passes through the trickling filter then is pumped to a tertiary treatment process, or the liquid can be sent through the trickling filter again to achieve the desired level of treatment.

Several kinds of **tertiary treatment** exist to accomplish different functions. One or more may be used, depending on how sensitive the receiving waters are. By no means are these treatments standard, as they can add considerable cost to the process, which municipalities like to keep as low as possible.

1) Phosphorus removal—accomplished by chemical flocculation using alum or other compounds.
2) Nitrogen removal—biological nitrogen removal (BNR) is the conversion of N compounds to $NO_3$; then bacterial denitrification under anaerobic conditions converts it to $N_2$ gas (which is inert in

the atmosphere). Ammonia stripping is another process used in which N is converted to ammonia, which then becomes airborne—obviously this is not an environmentally sound technique as it can lead to eutrophication problems in water bodies downwind where the ammonia is deposited!

- Ultrafiltration to remove particulates—this can involve passing the secondarily treated sewage through screens, membranes, or sand filters.
- Removal of trace organic contaminants—these compounds can be stripped from the water by filtering it through activated carbon (essentially charcoal). Activated carbon chemically has many free sites to which ions become attached.
- Trace inorganic compounds (salts, metals)—can be removed by ion exchange, reverse osmosis, or electro dialysis.

Regardless of the degree of treatment, the treated sewage is required to be *disinfected* to kill pathogenic microbes. This can be accomplished by oxidation by chlorination (although this produces potentially harmful byproducts), ozonation, or iodine or bromine treatment; disinfection can also be accomplished using UV treatment. UV radiation kills bacteria, viruses, and protozoans by DNA disruption. A more recent addition to this arsenal is peracetic acid, which is a strong disinfectant and not known to produce harmful byproducts; it may be less effective against algae than chlorine, however (Bettenhausen 2020). Some pathogenic protozoans such as *Cryptosporidium* and *Giardia* have shown resistance to normal chlorination treatments, so additional chemical treatment or membrane filtration may be required to safely remove these pathogens (Hrudey et al. 2003).

The treated wastewater is then usually pumped or drained into the closest river or stream, or occasionally (as in San Diego and Boston) the ocean. Sometimes secondarily treated sewage is sprayed onto pine forests or other land areas, where the vegetation and soils either utilize or sequester nutrients and other remaining contaminants and denitrification can occur. This is a useful and environmentally sound technique, but obviously requires large amounts of available acreage to spray on. In some areas secondarily treated wastewater is chlorinated for disinfection and then pumped to golf courses or other public areas to be used in irrigation (a good recycling program that conserves water; Angelakis et al. 2018). The term "purple pipe" is used to describe this water, as that is the color of the pipes used to convey it (pipes are color coded). Another technique for the disposal of treated sewage is to pump it underground into an aquifer (the standards for the degree of treatment for this application are high). This allows for storage of water for later use during droughts, when it can be used for irrigation or other purposes. In some areas secondarily treated sewage is piped into recharge spreading basins to trickle downward through the soil for further treatment and to recharge the aquifer. However, experiments have shown that such passage through soils only removes a portion of the PPCPs and organic contaminants; thus some can reach the groundwater (Cordy et al. 2004).

Sludge can be anaerobically digested in tanks by microbes to achieve further *sludge treatment*.

This will break down the material further, producing $CO_2$, $H_2S$, and $CH_4$ as byproducts. The methane can be used to provide power if the facility is so equipped—other times it is simply burned off. Sludge also requires dewatering to reduce waste volume. This can be accomplished by spreading it and drying it on sand beds, vacuum filtration, or centrifugation. The dried sludge is typically trucked to a landfill. Municipal waste sludge, currently referred to as biosolids, has been used for decades as fertilizer for agriculture, landscaping, and remediation (Eliason 1952; Kinney et al. 2008) due to the elevated nutrient content. However, questions remain regarding other pollutants such as heavy metals or organic chemicals that make it through the treatment process and their effect on animal food chains and humans (Kinney et al. 2008). A new technology that is designed to remove and even recover N, P, and C from wastewater is called aerobic granular sludge treatment (Winkler and van Loosdrecht 2022). This uses a granular structure in which various nitrifying, phosphate-accumulating, and glycogen-accumulating microbes operate under alternating aerobic and anaerobic reactor conditions within a single reactor; this process is currently operational in some systems, but retrofitting current plants to accommodate this process remains challenging.

### 13.1.2.3 Package Plants

Smaller communities, subdivisions, and large private companies often utilize small wastewater treatment plants called package plants. These are prepared to scale by the manufacturer and generally consist of prescreening, aeration, and solids settling, with some solids returned to the aeration tank to seed the microbial breakdown process; more detail is presented in O'Driscoll et al. (2019). Some of these systems are designed to be computer controlled. However, the effectiveness of such systems is only as good as the operators. They can be effective when regularly maintained and inspected but serve as outlets for stream pollution when operation is lax. Such package plants may be owned and operated by a municipality, but operations of these systems are frequently contracted out to private companies.

One additional comment regarding sewage collection and treatment systems concerns combined sewer systems, where stormwater runoff is mixed with raw sewage during rain events. These systems are characteristic of older cities, and produce combined sewer overflow (CSO). In heavy rains the stormwater increases volume to the wastewater treatment plant so that the system can exceed treatment capacity.

## 13.2 On-Site Wastewater Treatment (Septic Systems)

Especially since the early 1970s centralized sewage collection and wastewater treatment plants have vastly improved the quality of surface waters in terms of both human health risks and ecological soundness. However, a centralized sewage system requires a large capital expenditure and a consistent tax base to support its construction and maintenance. In many rural or suburban areas (even resort areas) this may not be available. In these situations sewage is treated by on-site wastewater treatment systems (OWTS), commonly called septic systems. Septic tanks can be traced back to the mid-nineteenth century in England and Germany (Angelakis et al. 2018). Septic systems are used to treat human sewage from individual homes, multifamily structures, businesses, and even hotels in both urbanized and rural areas. In the USA approximately 23% of homes utilize septic systems and they are particularly abundant in North and South Carolina, Georgia, Alabama, Kentucky, West Virginia, Maine, Vermont, and New Hampshire (US EPA 2002). In the nutrient-sensitive Chesapeake Bay watershed septic systems are used by 25% of the population (Reay 2004). Wastewater entering septic systems contains elevated concentrations of fecal bacteria and other microbes (Table 13.1), nutrients, BOD, and potentially toxic chemicals and metals, depending on the source of the wastewater (US EPA 2002).

### 13.2.1 How Does a Septic System Function?

At its simplest, a septic system consists of a septic tank and an overflow pipe connecting to a set of perforated pipes that discharge liquid waste into a drainfield (Cogger 1988; US EPA 2002). Human wastewater from the toilet, sink, shower, and washing machine is piped into the septic tank (closed at the bottom), where solids are settled, grease retained, and some anaerobic microbial digestion occurs (Fig. 13.2). The supernatant passes through a filtered valve into a set of perforated pipes that discharge the liquid waste in a diffuse manner into a drainfield, also known as a distribution trench, leachfield, soil absorption field, or subsurface wastewater infiltration system (Fig. 13.2). The drainfield is generally located 1–2 ft (0.3–0.6 m) below the ground surface and lined with gravel; in cold climates it may be 3–4 ft (0.9–1.2 m) below surface to prevent freezing. The gravel and pipes are covered with a porous cloth which is covered with soil and planted with grass. The drainfield sits above a zone of aerated soil, called the vadose zone. The vadose zone is perched at some distance above the upper groundwater table, which is the eventual destination of the treated wastewater.

As the liquid waste is discharged over the gravel in the drainfield the pollutants within undergo removal or reduction. This occurs as a result of filtration, aerobic decomposition, chemical transformation, and microbial interactions. A biological mat (biomat) forms over the gravel in the drainfield of a regularly used septic system. Within the biomat, nitrogen and phosphorus (nutrients) are taken up by soil microbes and thus converted from inorganic nutrient to organic life. These bacteria in turn are fed upon by protozoans and other small invertebrates including some nematode species. Microbial transformations of elements also occur, such as nitrification, which is the oxidation of ammonium to nitrite and nitrate under aerobic conditions facilitated by the bacteria *Nitrosomonas* and *Nitrobacter*. Labile carbon, which creates the majority of the BOD in wastewater, is utilized as a substrate by bacteria as well. As the liquid percolates from the drainfield down through the vadose zone, other components of the wastewater, such as orthophosphate and ammonium, will be adsorbed by the soil particles. As mentioned, human sewage also contains abundant fecal microbes, including bacteria, viruses, and protozoans, some of which may be pathogenic (disease-causing). In terms of treatment, some fecal microbes are consumed by protozoans and other grazers, while some are filtered within

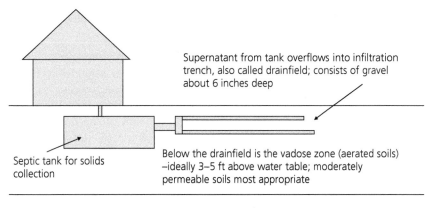

Supernatant from tank overflows into infiltration trench, also called drainfield; consists of gravel about 6 inches deep

Septic tank for solids collection

Below the drainfield is the vadose zone (aerated soils) –ideally 3–5 ft above water table; moderately permeable soils most appropriate

Upper or surface water table receives treated leachate, which will flow in the direction of the groundwater movement

Fecal microbes, BOD, and some nutrient species are treated in drainfield by a combination of filtration, microbial consumption, microbial transformation, and protozoan consumption; in underlying soil layer adsorption of phosphate and ammonium can occur as well. Nitrate is not significantly removed in the septic system process

**Fig. 13.2** Idealized residential septic system basic function, not to scale. (Figure by the author.)

the biomat, and others are adsorbed by soil particles beneath the drainfield. Death of enteric microbes also occurs over time because they are removed from their optimal (enteric) growth conditions into a hostile environment of different temperature, DO concentrations, and pH. As to treatment efficacy, the greater the contact between wastewater and the soil particles in this aerated zone the greater is the degree of treatment (Cogger 1988). Under ideal circumstances septic systems can achieve near-complete removal of fecal bacteria and BOD (US EPA 2002).

## 13.2.2 Geological Conditions Are Important for Proper Use and Functioning

The soils must permit moderate percolation; if they are too dense (like clay) the leachate can't percolate and surface ponding of septic leachate may occur and be carried by surface runoff to the nearest stream (Reneau et al. 1975). If soils are too porous (like sand) the material will pass through too readily without proper treatment (Cogger 1988; Cogger et al. 1988). The ground layer beneath the drainfield is called the vadose zone (zone of aerated soil). Aerated soil is required for the system to function properly: a minimum of 2 ft or more (preferably 3

ft or more) is needed for proper pollutant removal (Bicki and Brown 1990; US EPA 2002). A high water table saturates the soil so pathogens and nutrients flow easily through porous soils to enter streams, lakes, or drinking-water wells (Hagedorn et al. 1978; Lipp et al. 2001; US EPA 2002); such high water tables may be natural in a given area or occur sporadically for up to weeks at a time as a result of heavy rains (Humphrey et al. 2021). Steep slope of the land, cracks in the soil, and porous "karst" soils exacerbate movement of untreated pollutants away from the septic system toward water bodies (Chen 1988; Lapointe et al. 1990; Paul et al. 1997).

## 13.2.3 Microbial Pathogen Pollution from Septic Systems

Groundwater pathogen movement (bacteria, viruses) can occur as well through saturated soils, with infective viruses traveling several hundred meters (Yates 1985; Chen 1988; Cogger 1988; Cogger et al. 1988; Lipp et al. 1999, 2001; US EPA 2002). A number of documented disease outbreaks have been traced to drinking well contamination by fecal bacteria or viruses from septic system drainfields in the USA (Yates 1985; Borchardt et al. 2011). Sandy soils with minimal clay provide the most extensive off-site transport of potential viral and bacterial pathogens (Scandura and Sobsey 1997). Viruses have persisted as long as two months in groundwater (Scandura and Sobsey 1997). This type of contamination can impact recreational waters in rivers and lakes and contaminate shellfish beds in estuarine waters. Areas with sandy soils and high water tables are especially vulnerable to off-site movement of viruses and other pathogens (Scandura and Sobsey 1997). Karst topography includes sinkholes, fractures, and caves, and this sort of geological structure can be found in the USA from Florida to inland states such as Kentucky and Tennessee. In such situations pollutants can rapidly enter groundwaters from surface sources such as septic systems. In Tennessee a survey found enteric viruses and fecal bacteria indicators in community water supplies including both wells and springs (Johnson et al. 2011). Even when septic system design regulations are adhered to, unforeseen fracturing in karst soils can lead to disease outbreaks from septic system pollution (Borchardt et al. 2011). In south Florida such porous karst soils have been documented to transport fecal microbes from septic systems into nearshore coastal waters within hours of injection to the system (Paul et al. 1997). Thus, both sandy soils with high water tables and fractured karst soils are conduits for septic-system-sourced contamination of streams, lakes, estuaries, and drinking wells and springs.

Excessive septic system density in a given area, especially in areas of poor soils, is an important factor leading to the microbial contamination of surface and groundwater and nutrient loading to surface waters (Duda and Cromartie 1982; Yates 1985; Cahoon et al. 2006). When one considers that average water use in the USA ranges from 40 to 70 gallons (150 to 265 L)/average household water use, with some areas much higher, in areas of high septic system density this can account for considerable pollutant loading to ground- and surface waters if it is not effectively treated. Rainfall forcing applies downward pressure on the upper water table; if the water table is near the surface this will force lateral movement of the microbially contaminated groundwater into creeks and canals during wet periods (Lipp et al. 1999, 2001).

### 13.2.4 Nutrient Movement to Off-Site Waterways

Movement of nutrients from septic drainfields through the upper groundwater layer can stimulate the eutrophication process in nearby streams, rivers, lakes, and estuaries. The concentrated nutrients move as a plume downslope. Phosphate in the wastewater plume tends to bind readily to soils and is much less mobile than nitrate (Cogger 1988); thus, considerable phosphate sorption occurs in the vadose zone (Robertson et al. 1998; Reay 2004). Even so, under sandy soil conditions or conditions where long usage has led to saturation of phosphate sorption capacity in soils, phosphate concentrations exceeding 1.0 mg P/L in septic system plumes have been documented as far as 70 m from point of origin (Robertson et al. 1998), especially when drainage from several septic systems combine. If such plumes enter a P-limited system it can contribute to algal blooms and the eutrophication of freshwaters. Karst regions and coarse-textured soils low in aluminum, calcium, and iron present the biggest risk of phosphate movement and water contamination (US EPA 2002).

Nitrate moves much more readily through soils than phosphate, and plumes of nitrate from septic drainfields over 100 m long have been found with high concentrations (Chen 1988; US EPA 2002 and references within). The US EPA (and Canadian) nitrate standard for drinking water is 10 mg N/L. This has been set to protect the public from a disease called methemoglobinemia, also called *blue-baby syndrome*. This is a potentially fatal condition (mainly to infants) that is caused by ingestion of elevated nitrate concentrations in drinking water or food. The nitrate is reduced to nitrite by gut microflora, which then reacts with hemoglobin (which carries oxygen in the blood) and produces methemoglobin, which cannot transport oxygen and can lead to infant death (Johnson and Kross 1990). In the USA all documented cases of methemoglobinemia have been from consumption of water with nitrate concentrations in excess of this standard (Fan and Steinberg 1996). Studies have shown nitrate concentrations well in excess of 10 mg/L in groundwater plumes draining septic system drainfields (Cogger 1988; Cogger et al. 1988; Johnson and Kross 1990; Postma et al. 1992; Robertson et al. 1998). In Maryland elevated nitrate concentrations in drinking-well waters have been positively correlated with the number of septic systems in the area (Lichtenberg and Shapiro 1997). In the Chesapeake Bay region shoreline areas receiving septic system plumes have nitrate concentrations 50–100 times those of the open waters (Reay 2004). A study of 15 streams in Australia found stream nitrate and TN concentration was highly correlated with watershed septic system density (Hatt et al. 2004). Under reducing conditions where nitrification is suppressed, elevated ammonia concentrations will occur in septic plumes (Robertson et al. 1998). While not as mobile in the soil as nitrate, under sandy porous soil and waterlogged conditions groundwater ammonia plumes may also impact nearby surface waters. Septic system treatment efficacy of inorganic nitrogen has been inversely correlated with water table elevation ($r = -0.745$, $p = 0.031$; Humphrey et al. 2017), with greater ability to move off-site toward ambient waterways. If nearby streams, rivers, lakes, or estuaries receive plumes of nitrate or ammonium, this will stimulate the eutrophication process, especially in blackwater streams, estuarine systems, and other N-limited waters. In coastal areas it is worth considering what effect sea level rise will have on areas of abundant septic systems. Presumably, movement of nutrients off-site will be enhanced due to the now higher water table (see Chapter 17).

Again, high densities of septic systems in a given area will overcome the area's capacity to properly treat wastewater and allow off-site pollutant movement (Yates 1985; Lichtenberg and Shapiro 1997; Cahoon et al. 2006); land drainage by ditching will facilitate off-site movement of improperly treated wastewater. A study in the North Carolina Piedmont found that when septic system density in watersheds exceeded 0.4 systems/ha, nutrient, especially nitrate pollution, became problematic downstream (Iverson et al. 2018). That same study found that watersheds with high densities of septic systems exported four times the total dissolved N as low-density systems and two times the orthophosphate-P (Iverson et al. 2018).

### 13.2.5 Other Pollutants in Septic Systems

Various other less-studied contaminants enter residential and larger septic systems as well, including PPCPs and various chemicals people dump or excrete down their drains. Common septic systems do not have tertiary treatment; thus they are not designed to remove these contaminants from the waste stream. Such compounds may be removed if they settle with solids in the tank, adhere or bind to soil particles, or undergo chemical transformation through microbial activity. Those compounds that are not removed by those means may enter the water table and migrate to surface or well waters downslope. It is worth noting here that we give numerous PPCPs to our pets as well, in the form of prescription drugs, supplements, and even beauty products. Most of the excess from these pet-care PPCPs is likely to be deposited on the ground in manure and subject to downward leaching toward groundwater or laterally in the stormwater runoff flow, and thus not treated, except for deposition of some amount in landfills (Doughton and Ternes 1999). Septic systems can be outfitted with more advanced treatment technologies, however. Experiments have demonstrated that systems outfitted with sand filtration or aerobic treatment can significantly reduce the concentrations of certain PPCPs, as well as reducing the estrogen-impacting activities of such compounds (Wilcox et al. 2009).

### 13.2.6 Techniques for Tracking Sewage and Septic Spills and Leaks

Clearly, sewage and septic system contamination of surface and groundwaters can lead to severe human health and ecosystem problems. The detection of major spills is obvious, and public pressure will generally force an immediate response by agencies and municipalities to deal with the problem. However, the nonobvious sewage leaks and minor spills in streams also pose these problems in a chronic fashion and must be addressed as well. How are such problems detected if there are no obvious "red flags" such as dead fish and foul odors? There are several techniques that can be utilized in tracking leaks and minor spills:

- Utilize *bacterial source tracking methods*. Various molecular techniques such as PCR (polymerase chain reaction) and T-RFLP (terminal restriction fragment length polymorphism) have recently come into widespread usage that can pinpoint animal sources (i.e. human, canine, ruminant, avian, rat, etc.) of fecal bacteria in stream waters. Briefly, PCR involves heating a sample to separate (denature) the DNA double helix, then amplifying, or making numerous copies of, a target piece of DNA with the use of the enzyme polymerase and primers (short DNA strands that represent potential target sources, such as human, ruminant, canine). The anticipated/desired fragment is then size-separated using a technique called agarose gel electrophoresis to determine the presence of expected size fragments by comparing with known sizes of standard DNA fragments. T-RFLP is based on PCR amplification of a target gene, with primers having their end labeled with a fluorescent molecule (via a fluorescent dye). Restriction recognition sites are locations on a DNA molecule that are recognized by restriction enzymes. Upon completion of fingerprinting, each sample is represented by a profile. The profiles have representative peaks for each bacterial taxon present in the samples. The size of each fragment present is indicated in base pairs, and these fragments can be matched to a database of known fragments of DNA to identify what is present using a web-accessible genetic library. This is a continuously evolving field and these are technology intensive and generally expensive techniques, so there is as of yet no "standard method" widely accepted by regulatory agencies. Other techniques such as determining levels and types of bacterial antibiotic resistance present in environmental samples can perform a similar function. Such techniques require "libraries" of comparative material that often need to be site-specific, however.
- Measure nutrient concentrations in streams near suspected spill or leak sites. Ammonium is highly elevated in raw sewage, whereas nitrate concentrations are typically low. Phosphorus concentrations may also be elevated: the N:P ratio of raw sewage is generally quite low (Clark et al. 1977); thus unusually low ratios may indicate sewage or septic influence.
- Measure *optical brightener* concentrations in stream waters simultaneously with fecal bacterial indicator abundance (Hartel et al. 2007). Optical brighteners are used in nearly all detergents to produce "whiter whites" and brighter colors and enter the sewage collection system in gray water. As mentioned, gray water is a term used to describe water from sinks and washing machines—it is used in some recycling projects to fill toilets to flush human waste. If elevated optical brightener concentrations are statistically correlated with elevated fecal bacteria counts in a given location, it is good bet that they are both sourced from the same sewage/septic source (Hartel et al. 2007; Tavares et al. 2008).
- Measure *surfactants* in streams. Surfactants are a common component in soaps and detergents that break up grease globules. Thus, elevated concentrations of surfactants in a stream may be a signal of gray water or runoff from commercial areas where cleaning is done. However, surfactants are also used in the spraying of herbicides along roadways, etc., complicating their use as gray water signals.
- Use wintertime aerial *infra-red photography* to see where illegal or accidental wastewater plumes enter urban streams. The temperature of sewage is elevated above ambient surface waters in winter, so night-time flyovers using such technology can help pinpoint potential chronic areas of concern. Such areas can then be ground-truthed by sampling from boats or on foot for nutrients, fecal bacteria indicators, optical brighteners, etc.

## 13.3 Animal Waste

### 13.3.1 The Historical Farm

Humans first domesticated a number of animal species in several regions of the world ca. 6000 years ago (Diamond 1997). Early domestication of animals as food or draft animals allowed humans to exploit their abilities to convert large areas of forest or grassland into agriculture. Animal production was necessarily resource (i.e., feed or forage) limited and since production was tightly coupled to the productivity of the landscape, animal waste production would seldom have exceeded the assimilation capacity of the landscape. Even today one can still find tightly coupled agricultural systems, although they are becoming rarer. For example, following my MS degree (in Limnology) I worked in the environmental services section of a large power company. I lived in a trailer in what was then a rural (now heavily suburbanized) area. The farmer who owned the field next-door planted tobacco. In fall, after the harvest, he would plant a grass cover crop on the fields. When the cover crop attained sufficient growth he brought a herd of cows onto the fields, where they grazed (and defecated) for a number of months. After he removed the herd he disked the manure into the soil, then planted his cash crop. Thus, he prevented winter erosion by use of the cover crop, fed his livestock with the cover crop, and simultaneously fertilized his fields with manure (organic fertilizer to be sure!). Worldwide, animal waste has been used as a crop fertilizer for millennia.

## 13.3.2 The Modern CAFO

In recent decades livestock production, particularly that of swine, cattle, and poultry, has undergone a major change toward industrialization. The industrialization of the cattle and poultry industries began in the USA in the late 1950s, while industrialization of swine production began in the 1970s (Thu and Durrenberger 1998). Industrialization of livestock production basically consists of ceasing the millennia-old practice of pasturing livestock and instead moving them into large buildings (Fig. 13.3a), where they are confined and fed throughout

their lives until they are ready for market. Adoption of confined feeding techniques, together with the availability of large quantities of feedstuffs and efficient transportation systems, now allow animal producers to circumvent the ecological constraints otherwise imposed by the landscape. As a consequence, animal waste production often exceeds the assimilatory capacity of the landscape both locally and regionally.

Individual concentrated animal feeding operations (CAFOs) now house hundreds to thousands of animals in each confinement structure, and vast amounts of animal waste are generated by these facilities. The US EPA (2014) defines large CAFOs as containing $\geq$ 1000 head of beef cattle, 2500 swine > 25 kg, 10,000 swine < 25 kg, 125,000 chickens, 82,000 laying hens, or 55,000 turkeys. Swine waste is deposited on the floor of the structures by the animals, where it is periodically washed between slats in the floor into a system of trenches and pipes beneath the buildings. From there it is conveyed outside and into a cesspit called a "waste lagoon" (Fig. 13.3a). Some anaerobic treatment occurs in the lagoon and the liquid waste is periodically applied on surrounding fields by surface spraying (Fig. 13.3b), surface spreading or injection. The spraying of animal waste is usually limited by state or provincial governmental agencies to specific season. For instance, in North Carolina spraying is limited to March through September, which is considered the growing season where absorption by a cover crop would be optimal (Mallin and McIver 2018).

(a)                                                    (b)

**Fig. 13.3**  Fig. 13.3a (left) Swine CAFOs in eastern North Carolina, showing waste lagoons and sprayfields. Fig. 13.3b (right) "Big gun" liquid swine waste sprayer in action, Duplin County, North Carolina. (Photos by the author.)

Crops planted on the fields, such as Bermuda grass, cotton, corn, and soy, take up some of the plant nutrients in the waste material. Egg-laying poultry CAFOs utilize the lagoon system, but the majority of poultry CAFOs dispose of dry litter on the fields (Williams et al. 1999). In any case, concentrated waste material is spread onto fields, from where it can enter the environment through surface runoff or groundwater infiltration (Edwards and Daniel 1992; Mallin 2000). Cattle CAFOs may use lagoons, or spreading manure on fields, or disking it into the ground. Thus, individual CAFOs represent an ecologically anomalous concentration of animals whose waste production can easily exceed the assimilatory capacity of the local landscape.

Swine production in the USA is led by Iowa, North Carolina, Minnesota, Illinois, and Indiana; turkey production by Minnesota, North Carolina, Arkansas, South Carolina, and Virginia; broiler chicken production by Georgia, Alabama, Arkansas, North Carolina, and Mississippi; and cattle production by Texas, Nebraska, Kansas, Oklahoma, and Iowa (USDA 2021). Regional concentrations of CAFOs create circumstances in which very large imbalances of waste production versus waste assimilation capacity can arise (Barker and Zublena 1995; Cahoon et al. 1999; Jackson et al. 2000; Mallin and Cahoon 2003; Yang et al. 2016). The use of carefully formulated feeds, the need for large amounts of these feeds, and transportation cost considerations have led to the regional concentration of CAFOs around feed mills and meat-packing facilities. CAFOs have also had many acute pollution problems with their waste disposal systems, including lagoon ruptures and major leaks caused by mismanagement or weather (see below), and all too many of these facilities are located in river floodplains (Wing et al. 2002). The distance of some CAFOs to public waters can be uncomfortably short. Martin et al. (2018) used a GIS mapping approach to determine that 19% of CAFOs (primarily swine) were located within 100 m of a stream, with some less than 15 m from the stream's edge.

### 13.3.3 What's in Livestock Waste?

So what is contained in CAFO wastes that is a threat to streams, rivers, estuaries, and groundwater?

**Table 13.3** Average constituent concentrations (in mg/L) of five swine waste lagoons. (Revised from Westerman et al. 1990.)

| Pollutant | Concentration (mg/L) |
| --- | --- |
| TKN | 615 |
| Ammonium | 550 |
| TP | 100 |
| Orthophosphate | 89 |
| COD | 1830 |

Animal waste lagoons contain exceedingly concentrated amounts of nitrogen (mainly ammonium and organic nitrogen), phosphorus, BOD, chemical oxygen demand (COD), and fecal microbes. Some sampled swine lagoons contained the following concentrations (Table 13.3). It is worthwhile to compare the ammonium, TP, and orthophosphate concentrations in Table 13.3 with those of human sewage in Table 13.2.

North Carolina presents an excellent example of massive-scale industrialized livestock production, particularly that of swine. The lagoon waste disposal system was deployed with little foresight for the environmental consequences, and CAFOs were constructed with little regulation until lagoon construction standards, siting regulations, and waste management plans were legally required in 1993 (Burkholder et al. 1997). Following a series of major CAFO spills and river pollution incidents (see below) a moratorium on new CAFO production was begun in 1997; however, this did not take full effect until over 9,000,000 head of swine were present in eastern North Carolina, the vast majority in CAFOs (Burkholder et al. 1997; Mallin 2000). The number of poultry CAFOs in this state (and other US states) has continued to rise (Patt 2017; Glibert 2020; US DA 2021). This large number of swine, as well as poultry and cattle, requires vast amounts of animal feed, which contains N and P, nutrients that can lead to the eutrophication of water bodies (Carpenter et al. 1998; Correll 1998). The amount of nutrients generated annually by livestock waste is truly astonishing. Mallin and Cahoon (2003) estimated that on the North Carolina Coastal Plain, swine alone generated 101,000 metric tons of N and 22,700 tons of P and turkeys 12,600 metric tons of N and 3500 metric tons of P as manure. Again—since those data were published—poultry numbers, including turkeys, have continued to rise in this state (Patt 2017). For the USA as a whole, Glibert (2020) estimated total manure inputs to the environment as 4,000,000 metric tons of N and 1,400,000 metric tons of P.

As mentioned, livestock waste is loaded with fecal bacteria, viruses, and protozoans. Sobsey (1996) supplied estimates of fecal bacteria excreted from livestock. Applying Sobsey's (1996) conversion factors figures to livestock populations on North Carolina's CAFO-rich Coastal Plain yields estimated annual excretion of fecal coliform bacteria (as CFU) of $3.8 \times 10^{18}$ from swine, $1.7 \times 10^{18}$ from broilers, $1.8 \times 10^{17}$ from other chickens, and $3.3 \times 10^{17}$ from cattle (Mallin and Cahoon 2003).

In addition to high concentrations of N, P, BOD, and fecal microbes, confined swine, poultry, and cattle produce a variety of polluting compounds into the waste stream. Certain metals, such as copper and zinc, are added to the food supply of some confined livestock as micronutrients (Burkholder et al. 2007). Antimicrobial agents (antibiotics) are given to confined livestock both to prevent or reduce disease transmission and to enhance growth (Campagnolo et al. 2002; Burkholder et al. 2007). Natural hormones have often been used as additives to promote livestock growth (Cox et al. 2006; Burkholder et al. 2007).

## 13.3.4 Fate of Excreted Nutrients

Kellogg (2000) prioritized US watersheds in terms of vulnerability to manure nutrient contamination based on a number of factors, including soil percolation vulnerability in terms of nutrient leaching, soil runoff potential as an estimate of the amount of rainfall that runs off the surface, soil particle erosion potential, and the amount of animal nutrients applied to soils. The top four regions ranked by Kellogg (2000) were North Carolina's Cape Fear drainage, lower Arkansas, the Pee Dee River and coastal drainage, and the Susquehanna River system. North Carolina's Neuse and Pamlico River systems ranked 15th in the top 20 systems in vulnerability. Yang et al. (2016) mapped the distribution of livestock-generated N and P across the USA and found some of the densest concentrations in

riversheds draining into sensitive estuaries, including the Missouri and upper Mississippi Rivers (Gulf of Mexico), the North Carolina Coastal Plain (the Albemarle–Pamlico Sound system), and the eastern shore of Maryland and Delaware (Chesapeake Bay). Weldon and Hornbuckle (2006) worked in four CAFO-rich eastern Iowa watersheds and found that these systems made an outsized contribution of N loading to the Gulf of Mexico, relative to watershed size. Nutrients do indeed leave their source CAFOs and enter public waters, and do so via acute incidents and chronic leakage, leachate, and runoff.

### 13.3.4.1 Acute CAFO Pollution Events

As mentioned earlier, major storms and accidents are documented mechanisms by which large amounts of nutrients have been abruptly transported from CAFOs to receiving waters. In fact, several large and very public events in North Carolina in the mid-1990s brought the CAFO system to the headlines, while due to their presence in quite rural areas this system of production had avoided public attention. These events also marked the first time major spill events entered the peer reviewed literature (Box 13.2).

---

**Box 13.2  Acute CAFO incidents**

In July 1995, following heavy rains, 25 million gallons of liquid swine waste entered North Carolina's New River and its estuary following a waste lagoon rupture, polluting 22 miles of the river and much of the upper estuary, causing a large fish kill. The pollution load caused freshwater and estuarine fish kills and polluted the river and its sediments with fecal bacteria for months (Burkholder et al. 1997). In addition, the river was impacted by suspended sediment concentrations exceeding 70 mg/L and turbidity concentrations exceeding 70 NTU, and dramatic drops in DO. A leap in river nutrient levels, especially ammonium and orthophosphate, caused algal blooms in the lower river and estuary. The estuarine blooms included the toxic species *Pfiesteria piscicida* and *Phaeocystis pouchetti* (Burkholder et al. 1997). That same year a 9,000,000-gallon poultry lagoon breach in Limestone Creek and a 1,000,000-gallon swine waste lagoon leak into Harris Creek also caused algal blooms, hypoxia, fish kills, and microbial contamination in North Carolina's Cape Fear River basin (Mallin et al. 1997; Mallin 2000). In all of these cases the lagoons released large quantities of pollutants (nutrients, BOD, fecal microbes) that led to downstream ecosystem and animal health problems, as well as rendering the streams unfit for human recreation. For example, impacted areas of the New River and Harris Creek reached ammonium concentrations of 46 and 43 mg N/L, respectively, and phosphate concentrations in Limestone Creek and Harris Creek reached 6.9 and 11.5 mg P/L, respectively. Turbidity concentrations in the three spills ranged between 54 and 87 NTU, while fecal coliform bacteria counts were 3,400,000 CFU/100 ml in New River and 14,300 CFU/100 ml in Limestone Creek. Downstream hypoxia was severe in all cases, and in fact the Limestone Creek spill entered the Northeast Cape Fear River and produced a lengthy but classic DO "sag curve" that reached its nadir in the river at Castle Hayne, located nearly 100 km downstream from where the impacted stream entered the larger river (Mallin et al. 1997). Major CAFO accidents have also occurred in Iowa, Maryland, and Missouri (Thu and Durrenberger 1998; Mallin 2000). While the acute pollution caused by CAFOs is well documented, the sheer magnitude of their distribution and abundance merits an examination of the chronic effects that these facilities may have on our water resources.

### 13.3.4.2 Chronic CAFO Nutrient Pollution

CAFOs chronically export nutrients to water resources through several means during periods of normal operation. Normal rain events carry nutrients from swine sprayfields to nearby streams through surface and subsurface runoff and these inputs have caused stream nitrate-N to rise above 5 mg N/L and P above 1 mg P/L (Table 13.4). Nutrients, mainly nitrate and ammonium, also leach downwards into groundwater from animal waste lagoons, sprayfields, and litter fields. In a set of 11 North Carolina swine lagoons, Huffman

and Westerman (1995) found average inorganic (ammonium and nitrate) N concentrations of 143 mg/L in nearby groundwater and found that through leakage the lagoons exported on average 4.7 kg N/day to groundwater. Also, in North Carolina Westerman et al. (1995) found average concentrations of ammonium in downslope well fields that exceeded 50 mg N/L, compared with upslope wells that were less than 1 mg N/L. The nitrate form of N is especially mobile in soils and can pass readily through soils to contaminate groundwater. Liebhardt et al. (1979) found high levels of nitrate in soil groundwater beneath Delaware cornfields where poultry waste was applied as the sole fertilizer, with evidence that the nitrate moved laterally toward a nearby stream. In western Ohio, a study in a CAFO-rich area found that waste from livestock production, along with crop fertilizers, were the cause of very high nutrient concentrations in watershed streams and a nearby lake (Hoorman et al. 2008).

Harden (2015), in a large scale USGS study of Coastal Plain watersheds, compared 18 swine CAFO-influenced streams with 18 swine and poultry CAFO-influenced streams versus 18 control streams with no CAFOs. Statistical analyses demonstrated that the swine and swine + poultry influenced streams both had significantly higher concentrations of ammonium, TKN, TN, nitrate, and specific conductance compared to the control streams. The study also concluded that the effects of CAFO waste manures on stream water quality were most evident in watersheds with high swine barn densities and low wetlands coverage. Weldon and Hornbuckle (2006) working in row crop and CAFO-influenced eastern Iowa watersheds found frequent high nitrate (6 mg/L and above), which

**Table 13.4** Nutrient concentrations (in mg/L) in drainage waters from swine CAFOs. (Revised from Evans et al. 1984; Westerman et al. 1987; Stone et al. 1995; Gilliam et al. 1996.)

| Water body | Nitrate-N | Phosphorus-P | Ammonium-N |
|---|---|---|---|
| Surface runoff | 4.6 | 4.0 | |
| Subsurface | 21.0 | 0.6 | |
| Receiving stream | 5.4 | 1.3 | |
| Receiving stream | 7.7 | NA | |
| Drainage ditch | 2.1 | 3.1 | 7.1 |

was more strongly correlated with CAFO density than with row crops.

Season makes a difference when sampling CAFO-influenced areas where waste spraying is used. For example, as mentioned, in North Carolina swine waste spraying on Bermuda grass fields is limited to the March–September period. Mallin and McIver (2018) intensively sampled CAFO-dense Stocking Head Creek in early March and July; results found that the July results (mid-spray season) showed significantly higher specific conductance, nitrate, TN, TOC, and fecal coliform bacteria than the March samples. This has two major implications: one is that sampling CAFO-influenced streams must include spray season results, and two, the highest pollutant concentrations occurred in summer, when streams are already stressed for low DO by high water temperatures.

Using nitrogen isotopic techniques ($^{15}$N) Karr et al. (2001) traced nitrate generated from swine CAFOs through shallow groundwater into receiving stream waters, and at least 1.5 km downstream. A later and much broader $^{15}$N study in the Cape Fear River basin found highly enriched $^{15}$N signatures associated with streams draining CAFO-impacted areas (especially following the onset of spraying season), as well as two sites impacted by human wastewater discharges, whereas a near-pristine control stream and two sites near the ocean showed unenriched $^{15}$N signatures. In that study high nitrate concentrations were positively correlated with enriched $^{15}$N signatures, indicating that CAFO and sewage were the primary sources of high nitrate to the system rather than chemical fertilizers. On one occasion during the spraying season when elevated river discharge occurred, CAFO-generated $^{15}$N was detected scores of kilometers downstream to the upper estuary (Brown et al. 2020). The aforementioned Harden (2015) study also used $^{15}$N isotopic techniques and found that CAFO-influenced streams were significantly more enriched in $^{15}$N than the control streams, and the $^{15}$N values in the CAFO sites agreed with $^{15}$N signals common to manure signals.

Phosphorus is much less mobile and binds readily to soil particles. Cahoon and Ensign (2004) analyzed decades of North Carolina state soil P index data records and found that excessive soil P concentrations were most frequent on the Coastal Plain, and rose rapidly in the 1990s concurrent with the rapid increase in industrialized livestock production in that area. When the P content of soils is built up dramatically through excessive manure application, both surface export and subsurface loss of P occurs (Sharpley et al. 1999).

Mallin et al. (2015) investigated chronic pollutant loading to a second-order stream with a watershed containing 13 swine and 11 poultry CAFOs. At a stream site adjoining a sprayfield, ammonium ranged up to 38 mg/L and TP up to 11 mg/L, with fecal coliform counts up to 60,000 CFU/100 ml. Stream sites well downstream from sprayfields had average nitrate concentrations > 6 mg/L, ranging up to 13.6 mg/L (see also Table 8.3 for more detailed nutrient concentrations in this stream). Stream BOD5 concentrations ranged up to 26 mg/L and were most strongly correlated with TOC ($r$ = 0.833, $p$ < 0.0001), ammonium ($r$ = 0.666, $p$ < 0.0001), and TP ($r$ = 0.626, $p$ <0.0001). Thus, several studies have demonstrated that high nitrate concentrations occur in CAFO-influenced streams (Weldon and Hornbuckle 2006; Hoorman et al. 2008; Harden 2015; Mallin et al. 2015; Mallin and McIver 2018). Besides stimulating eutrophication, such elevated nitrate concentrations have been shown to be toxic to a number of wildlife species (see Chapter 2 for details on nitrate toxicity).

### 13.3.4.3 Airborne Ammonia Must Come Down

Anaerobic treatment of swine wastes with high concentrations of organic N promotes deamination, resulting in high concentrations of ammonium-N in lagoon liquid. Liming is used to maintain a pH above about 7, favoring ammonia formation. Ammonia volatilizes from sprayfields and waste lagoons and is transported downwind (McCulloch et al. 1998; Aneja et al. 2000; Walker et al. 2000). The North Carolina Department of Air Quality (NCDAQ) estimated that 70–80% of all swine waste N and a somewhat lesser percentage of poultry waste N is thus volatilized (NCDAQ 1997). Walker et al. (2000) documented a trend of increasing ammonium deposition in the coastal region of North Carolina, which they attribute to animal production sources. The lower Cape Fear River watershed, draining the largest concentration of CAFOS in the state, showed a 30% increase in airborne ammonium from the period 1988–2003 (Willey et al. 2006), which was attributed to increases in the swine industry airborne discharges. A GIS analysis (Costanza et al. 2008) indicated that the vast majority of airborne ammonia emissions from North Carolina CAFOs would be deposited in the nutrient-sensitive watersheds of lower rivers and estuaries including the Neuse, New, and Cape Fear systems.

On a nationwide scale Glibert (2020) estimated that 4.5 million metric tons of ammonia are emitted by livestock in the USA, with broilers and turkeys by far the biggest contributors, and swine, cattle, and dairy animals behind. On a regional scale across the USA, Glibert (2020) found that by far the highest ammonia emissions from livestock were in EPA Region VIII, which encompasses much of the Mississippi and Missouri River drainages, followed by Region IV, the Southeast USA, then Regions V (Upper Midwest), and Region VI (southern Mississippi River drainage and areas directly to the west of it).

What is the fate of such airborne ammonium emissions, as well as other nutrient losses from CAFOs? The Neuse River watershed contains approximately 25% of North Carolina's swine population and numerous poultry production facilities and is thus downwind of a large concentration of CAFOs. As mentioned, the Cape Fear River watershed contains approximately 50% of the state's swine production. The lower Neuse River showed a statistically significant, 500% increase in riverine ammonium from 1993 to 2003, while during 1995–2003 the Northeast Cape Fear River registered a significant 315% increase and the upper Cape Fear estuary a 100% increase in riverine ammonium (Burkholder et al. 2006), before later leveling off. The rise was concurrent with the rapid rise in the swine CAFO industry in the 1990s. The riverine ammonium increases were apparently responses to airborne loading of ammonium by CAFOs as well as overland runoff and groundwater lateral flow. While other anthropogenic sources of N may have contributed to this loading, the single major land use change in these areas has been the rapid proliferation of CAFOs during the late 1980s and the 1990s. In addition to airborne ammonia emissions,

livestock in the USA produce considerable volumes of greenhouse gases (Glibert 2020).

### 13.3.5 Animal Pathogens and Humans

Livestock are known to excrete many of the same pathogenic bacteria, viruses, and protozoans that can afflict humans (Tables 13.1, 13.5). The way animal waste is treated will affect pathogen survival and potential transmission to humans. Composting of manure raises temperatures high enough to kill most microbes, but animal waste slurries do not reach lethal temperatures (Mawdsley et al. 1995). Microbes in animal waste slurries such as lagoon liquid can survive for extended periods; *E. coli* has been known to survive up to 11 weeks in such an environment (Mawdsley et al. 1995). If CAFO-generated microbes enter the sediments of water bodies, organisms such as *E. coli* can find a favorable environment where they can remain viable for over two months (Davies et al. 1995). For example, following a large swine waste lagoon spill in the New River, North Carolina, Burkholder et al. (1997) found fecal coliform bacterial counts ranging from 1,000,000 to 3,000,000 CFU/100 ml of river water several kilometers downstream from the spill

**Table 13.5** Human pathogenic microbes that are found in animal waste (Hinton and Bale 1991; Cole et al. 1999; Berger and Oshiro 2002).

**Bacteria**
*Aeromonas* spp.
*Campylobacter jejuni*
*Clostridium* spp.
*Escherichia coli* 0157:H7
*Leptospira interrogans*
*Nocardia* spp.
*Salmonella* spp.
*Yersinia enterocolitica*

**Protozoa**
*Cryptosporidium parvum*
*Giardia lamblia*
*Balantidium coli*
*Encephalitozoon intestinalis*
*Enterocytozoon bieneusi*

**Viruses**
Hepatitis E virus
Reoviruses

site. These very high concentrations declined to the range of 1000–5000 CFU/100 ml after 14 days and to less than 1000 CFU/100 ml in 61 days. However, further sampling indicated that the river sediments maintained concentrations of fecal bacteria up to 5000 CFU/100 ml (sediment slurry) for 61 days.

When applied to fields in solid or liquid (Fig. 13.3b) manure most of these microbes are likely deactivated by UV radiation, microbial competition, and predation, or other means (Crane et al. 1983). However, because of the sheer volume of microbes deposited, there still remains a significant pollution potential from this material entering surface or groundwaters that humans will contact. The risk of large quantities of fecal microbes entering the lotic environment is high following acute CAFO mishaps, although the risk of human exposure to these microbes chronically through normal operations is yet undetermined. It is documented that normal rain events occurring shortly after animal waste is surface applied to fields cause vertical and horizontal movement of microbes to nearby water bodies (Crane et al. 1983; Mawdsley et al. 1995). Large-scale human microbial disease outbreaks have been traced to livestock vectors. In 1999 a disease outbreak occurred at a fair in Albany, New York attributed to pathogenic *E. coli* from water contaminated by a cattle husbandry area (MMWR 1999). During 2000 in Walkertown, Ontario, 2300 people became ill and several deaths occurred when a drinking-water well was contaminated by *E. coli* and *Campylobacter* entrained in runoff from a nearby cattle farm (Hrudey et al. 2003). It is critical to keep in mind that, unlike human waste, microbes generated by CAFOs are not exposed to secondary treatment or chlorination to disinfect the material.

As mentioned, various antibiotics are also administered to livestock in CAFOs. One problem of great concern is the possibility of enhancing antibiotic resistance in pathogenic microbes not only among confined livestock but also in humans (Gilchrist et al. 2007). There is clear evidence that many antibiotics excreted by CAFOs enter groundwater and nearby streams (Campagnolo et al. 2002). Repercussions to wildlife and humans downstream are presently poorly studied (Daughton and Ternes 1999; Burkholder et al. 2007).

## 13.4 Summary

- Inputs of human sewage have polluted streams and rivers for millennia, with serious consideration of the impacts to stream ecology first considered in a major way during the mid-twentieth century. Such wastes continue to be common in the developing world and all too frequently in fully modernized nations as well.
- Major pollutants such as nitrogen and phosphorus lead to stream eutrophication, BOD loads lead to hypoxia and anoxia, while fecal-derived microbes present a human health hazard. Various other compounds in the human waste stream such as pharmaceuticals, personal care products, and organic chemicals including endocrine-disrupting compounds are believed to impact stream life and potentially humans as well.
- Modern sewage treatment systems have the capability to remove major pollutants, but without special tertiary treatment (reverse osmosis, activated carbon, microfiltration) some of the latter contaminants pass through the system into waterways.
- Septic systems are still commonly used in many locations and have serious geological and hydrological limitations as to their effectiveness. Nutrient plumes and fecal microbes are known to travel hundreds of meters through sandy soils when water tables are high to pollute surface and groundwaters.
- In today's society, livestock are usually raised in CAFOs with minimal treatment of their concentrated wastes.
- Such facilities have been documented to pollute streams, rivers, estuaries, and groundwaters with high concentrations of nutrients, BOD, and fecal microbes, both acutely through major accidents and chronically through constant usage. CAFOs also produce other polluting compounds such as metals and antibiotic compounds that are known to enter nearby streams and groundwater but have impacts that are not understood or even quantified.
- While effective and safe treatment of CAFO wastes is technologically feasible, it is not widely practiced and pollution from these facilities remains a major threat to stream ecosystems.

# References

Aneja, V.P., J.P. Chauhan and J.P. Walker. (2000). Characterization of atmospheric ammonia emissions from swine waste storage and treatment lagoons. *Journal of Geophysical Research* 105(11):535–545.

Angelakis, A.N., T. Asano, A. Bahri, B.E. Jimenez and G. Tchobanoglous. 2018. Water reuse: from ancient to modern times and the future. *Frontiers in Environmental Science* 6:26.

APHA. 2005. *Standard Methods for the Examination of Water and Wastewater*. 21st Edition, American Public Health Association/American Water Works Association/Water Environment Federation, Washington DC.

Barker, J.C. and J. P. Zublena. 1995. Livestock manure nutrient assessment in North Carolina. Pages 98–106 in *Proceedings of the Seventh International Symposium on Agricultural and Food Processing Wastes (ISAFPW95)*. American Society of Agricultural Engineering, Chicago.

Bartsch, A.F. 1947. Biological aspects of stream pollution. *Sewage Works Journal* 20(2):202–302.

Bartsch, A.F. and W.M. Ingram. 1959. Stream life and the pollution environment. *Public Works* 90:104–110

Berger, P.S. and R.K. Oshiro. 2002. Source water protection: microbiology of source water. Pages 2967–2978 in Bitton, G. (ed.), *Encyclopedia of Environmental Biology*, Vol. 5. John Wiley & Sons, New York.

Bettenhausen, C.A. 2020. Treating wastewater with peracetic acid. *Chemical and Engineering News* April:16–18.

Bicki, T.J. and R.B. Brown. 1990. On-site sewage disposal: the importance of the wet season water table. *Journal of Environmental Health* 52:277–279.

Borchardt, M.A., K.R. Bradbury, E.C. Alexander Jr., et al. 2011. Norovirus outbreak caused by a new septic system in a dolomite aquifer. *Ground Water* 49:85–97.

Brinley, F.J. 1942. Biological studies, Ohio River pollution survey. I. Biological zones in a polluted stream. *Sewage Works Journal* 14(1):147–152.

Brown, C.N., M.A. Mallin and A.N. Loh. 2020. Tracing nutrient pollution from industrialized animal production in a large coastal watershed. *Environmental Monitoring and Assessment* 192:515.

Burkholder, J.M. 1998. Implications of harmful microalgal and heterotrophic dinoflagellates in management of sustainable marine fisheries. *Ecological Applications* 8:S37–S62.

Burkholder. J.M. 2002. Cyanobacteria. Pages 952–982 in Bitton, G. (ed.), *Encyclopedia of Environmental Microbiology*. Wiley, New York.

Burkholder, J.M., M.A. Mallin, H.B. Glasgow Jr., et al. 1997. Impacts to a coastal river and estuary from rupture of a swine waste holding lagoon. *Journal of Environmental Quality* 26:1451–1466.

Burkholder, J.M., D.A. Dickey, C. Kinder, et al. 2006. Comprehensive trend analysis of nutrients and related variables in a large eutrophic estuary: a decadal study of anthropogenic and climatic influences. *Limnology and Oceanography* 51:463–487.

Burkholder, J.M., B. Libra, P. Weyer, et al. 2007. Impacts of waste from concentrated animal feeding operations on water quality. *Environmental Health Perspectives* 115: 308–312.

Burkholder, J.M., P.M. Glibert and H.M. Skelton. 2008. Mixotrophy, a major mode of nutrition for harmful algal species in eutrophic waters. *Harmful Algae* 8:77–93.

Burtchett, J.M., M.A. Mallin and L.B. Cahoon. 2017. Micro-zooplankton grazing as a means of fecal bacteria removal in stormwater BMPs. *Water Science and Technology* 75:2702–2715.

Cahoon, L.B., J.A. Mickucki and M.A. Mallin. 1999. Nutrient imports to the Cape Fear and Neuse River basins to support animal production. *Environmental Science and Technology* 33:410–415.

Cahoon L.B, S.H. Ensign. 2004. Excessive soil phosphorus levels in eastern North Carolina: temporal and spatial distributions and relationships to land use. Nutrient Cycling in Agroecosystems 69: 111–125.

Cahoon, L.B., J.C. Hales, E.S. Carey, S. Loucaides, K.R. Rowland and J.E. Nearhoof. 2006. Shellfish closures in southwest Brunswick County, North Carolina: septic tanks vs. storm-water runoff as fecal coliform sources. *Journal of Coastal Research* 22:319–327.

Campagnolo, E.R., K.R. Johnson, A. Karpati, et al. 2002. Antimicrobial residues in animal waste and water resources proximal to large-scale swine and poultry feeding operations. *Science of the Total Environment* 299:89–95.

Carpenter, S.R., N.F. Caraco, D.L. Correll, R.W. Howarth, A.N. Sharpley and V.H. Smith. 1998. Nonpoint pollution of surface waters with phosphorus and nitrogen. *Ecological Applications* 8:559–568.

Chen, M. 1988. Pollution of ground water by nutrients and fecal coliforms from lakeshore septic tank systems. *Water, Air and Soil Pollution* 37:407–417.

Clark, J.W., W. Viessman Jr. and M.J. Hammer. 1977. *Water Supply and Pollution Control*, 3rd edition. IEP-A Dun-Donnelley Publisher, New York.

Cogger, C.T. 1988. On-site septic systems: the risk of groundwater contamination. *Journal of Environmental Health* 51:12–16.

Cogger, C.G., L.M. Hajjar, C.L Moe and M.D. Sobsey. 1988. Septic system performance on a coastal barrier island. *Journal of Environmental Quality* 17:401–408.

Cole, D.J., V.R. Hill, F.J. Humenik and M.D. Sobsey 1999. Health, safety and environmental concerns of farm animal waste. *Occupational Medicine: State of the Art Reviews* 14:423–448.

Cordy, G.E., N.L. Duran, H. Bouwer, et al. 2004. Do pharmaceuticals, pathogens, and other organic waste water compounds persist when waste water is used for recharge? *Ground Water Monitoring and Remediation* 24:58–69.

Correll, D.L. 1998. The role of phosphorus in the eutrophication of receiving waters: A review. *Journal of Environmental Quality* 27:261–266.

Costanza, J.K., S.E. Marcinko, A.E. Goewert and C.E. Mitchell. 2008. Potential geographic distribution of atmospheric nitrogen deposition from intensive livestock production in North Carolina, USA. *Science of the Total Environment* 398:76–86.

Cox, H., S. Mouzi and J. Gelsleichter. 2006. Preliminary observations of estrogenic activity in surface waters of the Myakka River, Florida. *Florida Scientist* 69:92–99.

Crane, S.R., J.A. Moore, M.E. Grismer and J.R. Miner. 1983. Bacterial pollution from agricultural sources: a review. *Transactions of the American Society of Agricultural Engineers* 26:858–872.

Daughton, C.G. and T.A. Ternes. 1999. Pharmaceuticals and personal care products in the environment: agents or subtle change? *Environmental Health Perspectives* 107:907–938.

Davies, C.M., J.A.H. Long, M. Donald and N.J. Ashbolt 1995. Survival of fecal microorganisms in marine and freshwater sediments. *Applied and Environmental Microbiology* 61:1888–1896.

Diamond, J. 1997. *Guns, Germs, and Steel*. W.W. Norton & Company, New York.

Dorfman, M. and K.S. Rosselot. 2010. *Testing the Waters: A Guide to Water Quality at Vacation Beaches*, 20th edition. Natural Resources Defense Council, New York.

Duda, A.M. and K.D. Cromartie. 1982. Coastal pollution from septic tank drainfields. *Journal of the Environmental Engineering Division, Proceedings of the American Society of Civil Engineers* 108:1265–1279.

Edwards, D.R. and, T.C. Daniel. 1992. Environmental impacts of on-farm poultry waste disposal—a review. *Bioresource Technology* 41:9–33.

Eliason, R. 1952. Stream pollution. *Scientific American* 186:17–21.

Enserink, M. 2011. Cholera linked to U.N. forces, but questions remain. *Science* 332:776–777.

Evans, R.O., P.W. Westerman and M.R. Overcash. 1984. Subsurface drainage water quality from land application of swine lagoon effluent. *Transactions of the American Society of Agricultural Engineers* 27:473–480.

Fan, A.M. and V.E. Steinberg. 1996. Health implications of nitrate and nitrite in drinking water: an update on methemoglobinemia occurrence and reproductive and developmental toxicity. *Regulatory Toxicology and Pharmacology* 23:35–43.

Flood, J.F. and L.B. Cahoon. 2011. Risks to coastal wastewater collection systems from sea-level rise and climate change. *Journal of Coastal Research* 27:652–660.

Gilchrist, M.J., C. Greko, D.B. Wallings, G.W. Beran, D.G. Riley and P.S. Thorne. 2007. The potential role of concentrated animal feeding operations in infectious disease epidemics and antibiotic resistance. *Environmental Health Perspectives* 115:313–316.

Gilliam, J.W., R.L. Huffman, R.B. Daniels, D.E. Buffington, A.E. Morey and S.A. Leclerc. 1996. *Contamination of surficial aquifers with nitrogen applied to agricultural land*. Report No. 306. Water Resources Research Institute of the University of North Carolina, Raleigh.

Glibert, P.M. 2020. From hogs to HABs: impacts of industrial farming in the US on nitrogen and phosphorus and greenhouse gas production. Biogeochemistry 150(2):139–180.

Grey, H.F. 1940. Sewerage in ancient and mediaeval times. *Sewage Works Journal* September:939–946.

Hagedorn, C., D.T. Hansen and G.H. Simonson. 1978. Survival and movement of fecal indicator bacteria in soil under conditions of saturated flow. *Journal of Environmental Quality* 7:55–59.

Harden, S.L. 2015. Surface-water quality in agricultural watersheds of the North Carolina Coastal Plain associated with concentrated animal feeding operations. Scientific Investigations Report 2015–5080, page 55. https://pubs.er.usgs.gov/publication/sir20155080.

Hartel, P.G., J.L. McDonald, L.C. Gentit, et al. 2007. Improving fluorometry as a source tracking method to detect human fecal contamination. *Estuaries and Coasts* 30:1–11.

Hatt, B.E., T.D. Fletcher, C.J. Walsh and S.L. Taylor. 2004. The influence of urban density and drainage infrastructure on the concentrations and loads of pollutants in small streams. *Environmental Management* 34:112–124.

Humphrey, C.P., G. Iverson and M. O'Driscoll. 2017. Nitrogen treatment efficiency of a large onsite wastewater system in relation to water table dynamics. *Clean: Soil, Air, Water* 45:1700551.

Humphrey, C.P., D. Dillane, G. Iverson and M. O'Driscoll 2021. Water table dynamics beneath onsite wastewater systems in eastern North Carolina in response to Hurricane Florence. *Journal of Water and Climate Change* 12(5):2136–2146.

Hinton, M. and M.J. Bale. 1991. Bacterial pathogens in domesticated animals and their environment. *Journal of Applied Bacteriology Symposium Supplement* 70:81S–90S.

Hoorman, J., T. Hone, T. Sudman Jr., T. Dirksen, J. Iles and K.R. Islam. 2008. Agricultural impacts on lake and stream water quality in Grand Lake St. Marys, western Ohio. *Water Air & Soil Pollution* 193:309–322.

Hrudey, S.E., P. Payment, P.M. Huck, R.W. Gillham and E.J. Hrudey. 2003. A fatal waterborne disease epidemic in Walkerton, Ontario: comparison with other waterborne outbreaks in the developed world. *Water Science and Technology* 7:7–14.

Huffman, R.L. and P.W. Westerman. 1995. Estimated seepage losses from established swine waste lagoons in the lower coastal plain of North Carolina. *Transactions of the American Society of Agricultural Engineers* 38:449–453.

Hunt, W. 2019. Of cesspits and sewers. *Archaeology* 72: 55–62.

Iverson, G., C.P. Humphrey Jr., M.A. O'Driscoll, C. Sanderford, J. Jernigan and B. Serozi. 2018. Nutrient exports from watersheds with varying septic system densities in the North Carolina Piedmont. *Journal of Environmental Management* 211:206–217.

Jackson, L.L., D.R. Keeney and E.M. Gilbert. 2000. Swine manure management plans in north-central Iowa: nutrient loading and policy implications. *Journal of Soil and Water Conservation* 55:205–212.

Johnson, C.J. and B.C. Kross. 1990. Continuing importance of nitrate contamination of groundwater and wells in rural areas. *American Journal of Industrial Medicine* 18:449–456.

Johnson, T.B., L.D. McKay, A.C. Layton, et al. 2011. Viruses and bacteria in karst and fractured rock aquifers in east Tennessee, USA. *Ground Water* 49:98–110.

Karr, J.D., W.J. Showers, J.W. Gilliam and A.S. Andres. 2001. Tracing nitrate transport and environmental impact from intensive swine farming using Delta nitrogen-15. *Journal of Environmental Quality* 30:1163–1175.

Kellogg, R.L. 2000. *Potential priority watersheds for protection of water quality from contamination by manure nutrients.* Paper presented at the Animal Residuals Management Conference 2000, November 12–14, Kansas City, MS.

Kinney, C.A., E.T. Furlong, D.W. Kolpin, et al. 2008 Bioaccumulation of pharmaceuticals and other anthropogenic waste indicators in earthworms from agricultural soil amended with biosolid or swine manure. *Environmental Science and Technology* 42:1863–1870.

Lapointe, B.E., J.D. O'Connell and G.S. Garrett. 1990. Nutrient couplings between on-site sewage disposal systems, groundwaters, and nearshore surface waters of the Florida Keys. *Biogeochemistry* 10:289–307.

Lichtenberg, E. and L.K. Shapiro. 1997. Agriculture and nitrate concentrations in Maryland community water system wells. *Journal of Environmental Quality* 26: 145–153.

Liebhardt, W.C., C. Golt and J. Tupin. 1979. Nitrate and ammonium concentrations of ground water resulting from poultry manure applications. *Journal of Environmental Quality* 8:211–215.

Lipp, E.K., J.B. Rose, R. Vincent, R.C. Kurz and C. Rodriquez-Palacios. 1999. Diel variability of microbial indicators of fecal pollution in a tidally influenced canal: Charlotte Harbor, Florida. Southwest Florida Water Management District, Technical Report.

Lipp, E.K., S.A. Farrah and J.B. Rose. 2001. Assessment and impact of fecal pollution and human enteric pathogens in a coastal community. *Marine Pollution Bulletin* 42: 286–293.

Mallin, M.A. 2000. Impacts of industrial-scale swine and poultry production on rivers and estuaries. *American Scientist* 88:26–37.

Mallin, M.A. and L.B. Cahoon. 2003. Industrialized animal production—a major source of nutrient and microbial pollution to aquatic ecosystems. *Population and Environment* 24:369–385.

Mallin, M.A. and L.B. Cahoon. 2020. The hidden impacts of phosphorus pollution to streams and rivers. *BioScience* 70:315–329.

Mallin, M.A. and M.R. McIver. 2018. Season matters when sampling streams for swine CAFO waste disposal impacts. *Journal of Water and Health* 16:78–86.

Mallin, M.A., J.M. Burkholder and M.R. McIver. 1997. Comparative effects of poultry and swine waste lagoon spills on the quality of receiving stream waters. *Journal of Environmental Quality* 26:1622–1631.

Mallin, M.A., V.L. Johnson, S.H. Ensign and T.A. MacPherson. 2006. Factors contributing to hypoxia in rivers, lakes and streams. *Limnology and Oceanography* 51: 690–701.

Mallin, M.A., L.B. Cahoon, B.R. Toothman, et al. 2007. Impacts of a raw sewage spill on water and sediment quality in an urbanized estuary. *Marine Pollution Bulletin* 54:81–88.

Mallin, M.A., M.R. McIver, A.R. Robuck and A.K. Dickens. 2015. Industrial swine and poultry production causes chronic nutrient and fecal microbial stream pollution. *Water, Air and Soil Pollution* 226:407.

Martin, K., R. Emanuel and J. Vose. 2018. Terra incognita: the unknown risks to environmental quality posed by the spatial distribution and abundance of concentrated animal feeding operations. *Science of the Total Environment* 642: 887–893.

Mawdsley, J.L., R.D. Bardgett, R.J. Merry, B.F. Pain and M.K. Theodorou. 1995. Pathogens in livestock waste, their potential for movement through soil and environmental pollution. *Applied Soil Ecology* 2:1–15.

McCulloch, R.B., G.S. Few, G.C. Murray Jr. and V.P. Aneja. 1998. Analysis of ammonia, ammonium aerosols and acid gases in the atmosphere at a commercial hog farm in eastern North Carolina, USA. *Environmental Pollution* 102:263–268.

Miltner, R.J. 2018. Eutrophication endpoints for large rivers in Ohio, USA. *Environmental Monitoring and Assessment* 190:55.

MMWR. 1999. Public health dispatch: outbreak of *Escherichia coli* 0157:H7 and *Campylobacter* among attendees of the Washington County fair—New York, 1999. *Morbidity and Mortality Weekly Report*. Centers for Disease Control and Prevention, Atlanta, GA.

Mor, J.-R., I. Munoz, S. Sabater, L. Zamora and A. Ruhi. 2021. Energy limitation or sensitive predators? Trophic and non-trophic impacts of wastewater pollution on stream food webs. *Ecology* 103:e03587.

NCDAQ. 1997. *Assessment Plan for Atmospheric Nitrogen Compounds: Emissions, Transport, Transformation, and Deposition.* North Carolina Department of Environment, Health and Natural Resources, Division of Air Quality, Raleigh.

NOAA. 1998. "Classified shellfish growing waters" by C.E. Alexander. NOAA's State of the Coast Report. NOAA, Silver Spring, MD. http://state_of_coast.noaa.gov/bulletins/html/sgw_04/sgw.html.

O'Driscoll, M., E. Bean, R.N. Mahoney and C.P. Humphrey Jr. 2019. Coastal tourism and its influence on wastewater nitrogen loading: a barrier island case study. *Environmental Management* 64:436–455.

Patt, H. 2017. *A Comparison of PAN and $P_2O_5$ produced from Poultry, Swine and Cattle Operations in North Carolina.* Division of Water Resources, North Carolina Department of Environmental Quality, Raleigh.

Paul, J.H., J.B. Rose, S.C. Jiang, et al. 1997. Evidence for groundwater and surface marine water contamination by waste disposal wells in the Florida Keys. *Water Research* 31:1448–1454.

Postma, F.B., A.J. Gold and G.W. Loomis. 1992. Nutrient and microbial movement from seasonally-used septic systems. *Journal of Environmental Health* 55:5–10.

Reay, W.G. 2004. Septic tank impacts on ground water quality and nearshore sediment nutrient flux. *Ground Water* 42:1079–1089.

Reneau, R.B. Jr., J.H. Elder Jr., D.E. Pettry and C.W. Weston. 1975. Influence of soils on bacterial contamination of a watershed from septic sources. *Journal of Environmental Quality* 4:249–252.

Robertson, W.D., S.L. Schiff and C.J. Ptacek. 1998. Review of phosphate mobility and persistence in 10 septic system plumes. *Ground Water* 36:1000–1010.

Scandura, J.E. and M.D. Sobsey. 1997. Viral and bacterial contamination of groundwater from on-site sewage treatment systems. *Water Science and Technology* 35:141–146.

Shah, T., C. Ray and U. Lele. 2018. How to clean up the Ganges? *Science* 362:503.

Sharpley, A.N., T. Daniel, T. Sims, J. Lemunyon, R. Stevens and R. Parry. 1999. *Agricultural Phosphorus and Eutrophication.* ARS–149. US Department of Agriculture, Agricultural Research Service, Washington, DC.

Smith, J.E. and J.M. Perdek. 2004. Assessment and management of watershed microbial contaminants. *Critical Reviews in Environmental Science and Technology* 34:109–139.

Sobsey, M.D. 1996. Pathogens and their indicators in North Carolina surface waters. Pages 6–10 in *Solutions: Proceedings of a Technical Conference on Water Quality.* North Carolina State University, Raleigh.

Stone, K.C., P.G. Hunt, S.W. Coffey and T.A. Matheny. 1995. Water quality status of a USDA water quality demonstration project in the Eastern Coastal Plain. *Journal of Soil and Water Conservation* 50:567–571.

Tavares, M.E., M.I.H. Spivey, M.R. McIver and M.A. Mallin. 2008. Testing for optical brighteners and fecal bacteria to detect sewage leaks in tidal creeks. *Journal of the North Carolina Academy of Science* 124:91–97.

Thu, K.M. and E.P. Durrenberger. 1998. *Pigs, Profits, and Rural Communities.* State University of New York Press, Albany.

USDA. 2021. *National Agricultural Statistics Service Census Data 2017.* USDA, Washington, DC.

US EPA. 2002. Onsite Wastewater Treatment Systems Manual. EPA/625/R-00/008. Office of Water, Office of Research and Development, US EPA, Washington, DC.

US EPA. 2014. National Pollutant Discharge Elimination System (NPDES). http://cfpub1.epa.gov/npdes/search.cfm. US EPA, Washington, DC.

US EPA. 2023. National Pollutant Discharge Elimination System (NPDES). http://cfpub.epa.gov/npdes/. Sanitary Sewer Overflows (SSOs) | US EPA, Washington, DC.

US FDA. 1995. Sanitation of shellfish growing areas. National Shellfish Sanitation Program Manual of Operations, Part I. 20204. Office of Seafood, United States Department of Health and Human Services, Food and Drug Administration, Washington, DC.

Vuorinen, H.S., P.S. Juuti and T.S. Katko. 2007. History of water and health from ancient civilizations to modern times. *Water Science and Technology: Water Supply* 7:49–57.

Walker, J.T., V.P. Aneja and D.A. Dickey. 2000. Atmospheric transport and wet deposition of ammonium in North Carolina. *Atmospheric Environment* 34:3407–3418.

Weldon, M.B. and K.C. Hornbuckle. 2006. Concentrated animal feeding operations, row crops, and their relationship to nitrate in eastern Iowa rivers. *Environmental Science and Technology* 40(10):3168–3173.

West, P.A. 1991. Human pathogenic viruses and parasites: emerging pathogens in the water cycle. *Journal of Applied Bacteriology Symposium Supplement* 70:107S–114S.

Westerman, P.W., L.D. King, J.C. Burns, G.A. Cummings and M.R. Overcash. 1987. Swine manure and lagoon effluent applied to a temperate forage mixture: II. Rainfall runoff and soil chemical properties. *Journal of Environmental Quality* 16:106–112.

Westerman, P.W., L.M. Safely Jr. and J.C. Barker. 1990. Lagoon liquid nutrient variation over four years for lagoons with recycle systems. *ASAE Publication* 5: 41–49.

Westerman, P.W., R.L. Huffman and J.S. Feng. 1995. Swine-lagoon seepage in sandy soil. *Transactions of the American Society of Agricultural Engineers* 38:1749–1760.

Wilcox. J.D., J.M. Nahr, C.J. Hedman, J.D.C. Hemming, M.A.E. Barman and K.R. Bradbury. 2009. Removal of organic wastewater contaminants in septic systems using advanced treatment technologies. *Journal of Environmental Quality* 38:149–156.

Willey, J.D., R.J. Kieber and G.B. Avery Jr. 2006. Changing chemical composition of precipitation in Wilmington, North Carolina, U.S.A.: implications for the continental U.S.A. Environmental Science and Technology 40:5675–5680.

Williams, C.M., J.C. Barker and T. Sims. 1999. Management and utilization of poultry wastes. *Reviews in Environmental Contamination and Toxicology* 162:105–157.

Wing, S., S. Freedman and L. Band. 2002. The potential impact of flooding on confined animal feeding operations in eastern North Carolina. *Environmental Health Perspectives* 110:387–391.

Winkler, M.-K.H. and M.C.M. van Loosdrecht. 2022. Intensifying existing urban wastewater. *Science* 373:377–378.

Yang, Q., H. Tian, X. Li, et al. 2016. Spatiotemporal patterns of livestock manure nutrient production in the conterminous United States from 1930 to 2012. *Science of the Total Environment* 541:1592–1602.

Yates, M.V. 1985. Septic tank density and ground-water contamination. *Ground Water* 23:586–591.

# CHAPTER 14

# Species Loss and Impacts of Invasive Species

Biodiversity refers to the number of different plant, animal, and microbial species within a given ecosystem and how well their populations are integrated to achieve efficient energy flow and system stability. A major problem in lotic ecosystems, both large and small, is loss of biodiversity and associated local, regional, and even continental loss of native species. In some cases these losses result in organisms being placed on state or federal threatened or endangered species lists (in the USA) or placed on the Red List of Threatened Species (the International Union for Conservation of Nature). Some of these species, especially large, easily identifiable fish or mammals, occasionally catch public attention and are considered "charismatic megafauna," e.g. manatees and whales often found injured by boats, pollutants, or aquatic debris. Such fauna fortunately draw attention to greater environmental problems (the famous panda logo for instance is widely identified with the nonprofit World Wildlife Fund). Regardless, on an everyday level, losses in biodiversity have real consequences for aquatic ecosystems. Reductions in biodiversity can be temporary, such as when a massive cyanobacterial bloom dominates the phytoplankton community of a reservoir, or permanent, such as when species are lost from a stream, river, or reservoir as a result of chronically changed conditions. The following discussion focuses on the latter.

## 14.1 Biodiversity Loss

Species losses from an aquatic ecosystem have both ecological and economic consequences. Ecologically, when an ecosystem has high plant and animal biodiversity there is greater variability of prey items from which to choose for predators as well as greater habitat availability for both predators and prey (many plant species make excellent animal habitat). Greater diversity, at both a species and genetic levels, has been positively associated with greater resistance to disturbances such as grazing or disease in plant communities (Hughes and Stachowicz 2004) as well as stronger resistance to exotic species invasions (Shurin 2000). For instance, greater diversity of native fish has been inversely correlated with numbers of introductions of exotic fish species in the USA (McKinney 2001). Such resistance to invasion is not universal, however, and exceptions occur such as when a biodiverse community is repeatedly invaded Ricciardi 2001); in fact, some authors (Jackson et al. 2017) conclude that there is no clear pattern regarding resistance to invasion by diverse communities, and results are very context-dependent.

### 14.1.1 Species Loss from Lotic Systems

Other chapters have demonstrated the wide variety of physical, chemical, and biological pressures placed on stream and river ecosystems. River fragmentation (see Chapter 10) is an important cause of species losses. Dam construction cuts off migration routes for aquatic fauna and changes lotic habitat to lentic habitat. Currently, for example, the rash of major dam construction in Southeast Asia and South America is threatening rare or otherwise important fish species. Pollution of all kinds (see Chapter 12) contributes to losses as well, with a

*River Ecology*. Michael A. Mallin, Oxford University Press. © Michael A. Mallin (2023). DOI: 10.1093/oso/9780199549511.003.0015

particularly stark example coming from the decimation of species in Adirondack streams and lakes from acid precipitation. Habitat loss (see Chapter 15) often results from clearcutting and construction activities that alter benthic areas through excessive sedimentation, drive down DO through BOD loading, and raise water temperatures. The harvesting of fish and game species has long contributed to species losses or decimations, such as with waterfowl in the nineteenth century. Physical contact with humans (and associated machinery) also plays a role in decimating certain species. In the abundant waterways of south Florida, impacts of motorboats on manatees are clearly evident in manatee scars and the signage warning boaters to be careful of manatees. The ever-increasing passage of container vessels in estuaries and rivers has an impact on endangered species, such as the federally endangered Atlantic sturgeon (*Acipenser oxyrhynchus*). As such, researchers working in the Delaware Estuary (Brown and Murphy 2010) showed that between 2005 and 2008 over 50% of the Atlantic sturgeon fatalities reported from that system were caused by large vessel strikes. Fish in other riverine systems that contain large ports are also subject to such fatal interactions (Fig. 14.1).

These various physical, chemical, and economic pressures have resulted in severe biodiversity losses, with some taxonomic groups much more threatened than others. A particularly impacted aquatic faunal group has been molluscs. Of the

freshwater mussels in North America, about 43% of the species are threatened or endangered, mainly by dam construction and degradation of water quality (Allan and Flecker 1993). The Southeast USA contains a particularly rich diversity of bivalves and gastropods; however, among the states in this region between 34 and 78% of the mussel species are considered endangered, threatened, or of special concern (Neves et al. 1997). Major threats to bivalve and snail populations come from agriculture (sedimentation and pesticide use) and mining, especially the mountaintop removal that devastates headwaters streams and extirpates the fauna within. Other threats to molluscan populations come from the aforementioned dam building, pollution from concentrated animal feeding operations, and dredging and channelization (Neves et al. 1997).

Stream and river fishes are valued for many reasons, including as a food source to subsistence fishermen, challenges to sport fishermen, an economic resource to commercial fishing operations, objects of art, and visible objects of enjoyment to nature lovers. In North American (Canada, the USA, and Mexico) it has been estimated that 39% of known fish species are imperiled, meaning vulnerable, threatened, or endangered (Jelks et al. 2008). Between 1900 and 2010 North America has suffered at least 57 fish species extinctions (Burkhead 2012) and Miller et al. (1989) determined that 15 North American species that went extinct in the twentieth century were

(a)  (b)

**Fig. 14.1** Fig. 14.1a (left) Researchers from the University of North Carolina Wilmington examine an Atlantic sturgeon cut in half (Fig. 14.1a; 14.1b) by a container vessel propeller in the lower Cape Fear River Estuary, 2001. Estimated age 26 years, estimated weight >79 kg, estimated length 2.3 m. (Photos courtesy of Dr. T.E. Lankford, UNC Wilmington.)

strictly from running water habitat. Areas where fish taxa are most imperiled include the Southeast USA from Virginia south through Georgia and west to the Mississippi River, north from there to the Great Lakes, and coastal California and Oregon as well as Nevada (Jelks et al. 2008). Areas with few imperiled fish include most of Canada and Alaska, the Great Plains states, northern Mexico, and southeastern Mexico (Jelks et al. 2008). The main threats to fish are considered to be habitat loss and introduction of non-native species (Jelks et al. 2008; Burkhead 2012); however, aquatic species extinctions are often due to a combination of factors. Miller et al. (1989) estimated that of fishes that went extinct during the twentieth century, 73% were associated with habitat loss, 68% were associated with exotic species introductions, 38% were associated with chemical pollution, and 15% were associated with overharvesting. The most imperiled fish species in the Southeast USA have been those closely associated with benthic habitat and thus subject to increases in sediment loading, accumulation of pollutants in the benthos, and direct physical alteration of habitat (Warren et al. 1997).

How do species invasions, either accidental or deliberate, contribute to native species losses? Ross (1991) found that in 31 case studies of fish introductions into streams, 77% reported a decline in native species. In some cases, exotic species were able to outcompete native species for food or habitat. The invasive species may be more naturally aggressive, or since it is newly arrived, it has not yet developed predators as have native species; however, Jackson et al. (2017) note that invasive species themselves can become a food resource to native consumers, creating new trophic pathways. Invasive species may bring with them diseases or parasites to which the natives have little or no resistance (Pyšek et al. 2020). Invasive species that can interbreed with natives can lead to a weakening of the native population, as shown by Miller et al. (1989) who found that species hybridization (interbreeding) was a factor in 38% of extinctions. Habitat loss and pollution are discussed in multiple chapters of this book, but the invasion of running water and reservoir habitats by non-native species requires significant discussion here.

### 14.1.2 Intercontinental Introduction of Non-Native Species

Besides the direct impact of invasive species on native species, such invasions lead to multiple ecological and economic problems. Non-native species have been introduced from one continent to another for centuries, even millennia (Allan and Flecker 1993; Erickson 2005), and are also variably referred to as exotic, introduced, invasive, non-indigenous, and even alien species. Approximately 50,000 exotics have been introduced into the USA (Pimentel et al. 2000). Most introduced species do not succeed for any number of reasons: they may be unable to tolerate the new environment physiologically, or are eliminated by competition with native species, or the initial population size was too low to establish itself (Ross 1991). Some of the ones that survive, and thrive, become immensely problematic.

## 14.2 Why and How Exotic Species Are Introduced

In recent decades there has been rapid expansion of international trade, and associated travel provides unprecedented means of increasing biological invasions worldwide (Ericson 2005). Almost 90% of international commerce is conducted by over 45,000 oceangoing merchant ships, which offers many opportunities for intercontinental transfer of non-native species (Tibbetts 2007). Oceangoing cargo vessels dock in US waters approximately 90,000 times a year (Strain 2012), thus creating a vast amount of exposure to potential invasive species. Intercontinental invasions of non-native species occur both accidently and deliberately; spread of locally invasive species within the borders of individual nations likewise occurs both accidently and deliberately (Carey et al. 2012). In the USA most invasive aquatic vertebrate animals and aquatic plants were brought in deliberately, while most aquatic invertebrate invasions were accidental (Pimentel et al. 2005). As human influence grows in a region, the reasons for introduction and the number of vectors involved in the influx of non-native species increases. Additionally, the hydrology of many regions has been altered, allowing for more interconnectivity of water bodies. In the USA human population size within a given region is a significant predictor of exotic plant and exotic fish establishment (McKinney 2001).

## 14.2.1 Accidental Species Introductions

Accidental introduction of non-native species occurs by a number of methods, but a few vectors are largely responsible for the vast majority of successful invasions. Oceangoing ships require large quantities of ballast water for stabilization on the high seas. Historically they have taken on such water in one port and released it in another (before loading with cargo), and numerous vertebrate and invertebrate organisms entered the ballast water tanks of ships and are discharged in distant ports. Many freshwater invasive species have been spread by ballast water taken on in European inland waterways and discharged into the Laurentian Great Lakes (see below). Within ship ballast water, species can travel as adults, larvae, eggs, cysts, spores, seeds, and fragments that can later reproduce in new waters. This means of introduction of non-native species has been ongoing for centuries and often follows established trade routes (Ericson 2005). In historic times stones were used as ballast (and can still be seen in the ruins of historic port settlements) and such ballast stones also served as vectors. Invasive attached organisms are also spread by attachment to ships' hulls and anchor chains,

In the USA estuarine areas have been severely impacted especially in popular trading ports such as San Francisco Bay, where some 250 invasive aquatic taxa have become established over time (Tibbetts 2007). Impacts from invasive species take on a particular significance where freshwater is concerned due to the interconnectivity of many inland water bodies, which offers a ready means of spreading. Channelization, canal construction, or other alteration of the environment enhances the ability of humans to move goods and services to distant areas. This of course also provides routes for invasive species to spread (species that are either brought in from another continent or "locally invasive" from within the continent), and essentially anything waterborne can take this route. Protection from

invasive species has been postulated as a reason for maintaining reduced hydrological activity (e.g. maintaining levees, reducing peak flows, retaining selected dams), at least under some circumstances (Jackson and Pringle 2010).

The Ponto-Caspian region in Eastern Europe refers to the Black, Caspian, and Azov Seas. This region has produced numerous salinity-tolerant freshwater species that spread throughout Europe following the construction of shipping canals, and some have since spread to the Laurentian Great Lakes in ballast of trans-Atlantic ships (Ricciardi and MacIsaac 2000). Interestingly, few Great Lakes species have become established in European waters (Ricciardi and MacIsaac 2000).

Once an invasive becomes established in a freshwater lake, river, or reservoir, a common accidental means of transfer is transport via small boats used by fishermen or pleasure boaters. Invasive plants adhere to propellers and boat trailers and can move from one water body to the next, and bivalves and planktonic organisms enter bilge water that is subsequently drained in another water body. Some inland waterways have been hit by multiple successful invasions of non-natives of several phyla (Fig. 14.2a), and invasives that enter one area of a continent will often spread to far-distant areas of that continent (Fig. 14.2b).

## 14.2.2 Deliberate Species Introductions

Deliberate introductions of non-native species to an area have been occurring for millennia, primarily for purposes of addition of a new freshwater food fish or game species. An example of this is from the American West, where fishermen moving from the east coast introduced some of their favorite game fish (walleye, bass, and various sunfish) into western water bodies (Rahel 2000), or introductions have been made by fish and game agencies (Carey et al. 2012). There have also been introductions for ornamental purposes which subsequently spread into the wild—mainly plants, such as from the aquarium trade which account for about 1/3 of the worst aquatic nuisance invaders (Padilla and Williams 2004). Some troublesome invasive species are readily available from nurseries and pet stores. There have been, and will continue to be, introductions for

(a)                                                              (b)

**Fig. 14.2**  Fig. 14.2a (left) Warning sign at a river boat ramp in Minnesota describing an invasive fish (Eurasian ruffe), bivalve (zebra mussel), and cladoceran (*Bythotrephes*). Fig. 14.2b (right) Zebra mussel warning sign on John Martin Reservoir, Colorado. (Photos by the author.)

purposes of biological control of another organism, mainly fish and insects, often by regulatory agencies. For all of the above, often there are no natural predators in the new area that can control the population of the introduced species. Importantly, both plant and animal invasives are generally free of their native parasites when introduced to a new region, a big aid in their success (Torchin and Mitchell 2004).

In the USA, south Florida has become the premier site of multiple invasions by exotic plant and animal species (Rodgers et al. 2022). In addition to the invasives arriving in association with the major international Port of Miami, the area is warm and sunny, with numerous interconnecting freshwater bodies. Unfortunately local residents are fond of acquiring exotic aquarium pets (fish, reptiles, and amphibians) and associated ornamental exotic plant species, many from South America. When the owners tire of their upkeep such exotics are frequently released into the Everglades, where species readily multiply and move north through the "river of grass," to the Lake Okeechobee environs, then northward into and through the newly restored Kissimmee River basin. Eventually most find their expansion limited by low winter water and air temperatures. As of 2022, over 130 invasive animal species have been documented in the Everglades and at least 62 invasive animal species in the Kissimmee River basin (Rodgers et al. 2022).

It has been theorized that when an area (such as the Laurentian Great Lakes or San Francisco Bay) has been successfully invaded many times,

these cumulative perturbations possibly allow for easier invasions the next time (there are also occasional facultative interactions between coevolved invaders). Simberloff and Von Holle (1999) coined the term "invasional meltdown" to describe what happens when a group of nonindigenous species facilitate each other's success through any number of means, including creating habitat, reducing predation on invasives, and facilitating dispersal. Ricciardi (2001) has pointed out a number of such facultative interactions between invasive species in the Great Lakes. On a worldwide basis some rivers are experiencing an actual increase in biodiversity as a result of multiple invasive species thriving (Su et al. 2021).

## 14.3 Exotic Animal Introductions

Exotic animal introductions into inland water bodies have been both accidental and deliberate (Table 14.1). The most visible exotics are fish and bivalve molluscs, but numerous invertebrates have also been introduced. In the USA invasives are particularly rampant in states with mild climates such as California, Florida, and Hawaii (Pimentel et al. 2000 and references therein). These are states with significant ports and some, such as Florida, contain numerous and often interconnecting waterways. Regarding very large watersheds, according to Pyšek et al. (2020) the Colorado and Mississippi river basins have been host to 100 and 73 invasive fish species, respectively.

**Table 14.1** Some invasive aquatic macrofaunal species that have significantly impacted the ecology of freshwater ecosystems in North America, including functional habitat group; when established; how they arrived (vector); principal area of infestation/colonization; and control (information from multiple sources).

Asian clam (*Corbicula fluminea*); filter-feeding bivalve; 1980; entered the Great Lakes via ballast water exchange; clogs water intakes; controlled mainly by biocides and manual cleaning.

Zebra (*Dreissena polymorpha*) and quagga (*D. bugensis*) mussels; filter-feeding bivalves; 1988; likely arrived in ship ballast in the Great Lakes; many Eastern USA river systems and now in the Colorado River system; also severely impacts other nations; clogs intakes; controlled mainly by biocides.

Snakehead (*Channa argus*); fish; native to East Asia; 2002; introduced through aquarium dumping; found in various Eastern USA states; voracious predation of native species; controlled by whole-lake poisoning in enclosed situations.

Bighead (*Hypophthalmichthys nobilis*) and silver (*H. molitrix*) carp; 1970s; introduced for aquaculture in Southern USA; now throughout Mississippi basin to Lake Michigan; outcompetes native fish for food resources; electric barriers used in Chicago ship canal; poisoning used.

Walking catfish (*Clarias batrachus*); 1960s; imported from Southeast Asia to Florida for aquarium trade; can impact aquaculture pond stock if they enter facilities; fenced out.

Nutria (*Myocaster coypus*); invasive rodent; 1930s; imported for fur production; throughout US South and other nations; extensive damage to agriculture and wetland plants from voracious feeding; outcompetes native muskrat in the USA; hunted for cash rewards.

Burmese python (*Python molurus bivittatus*); late 1900s; released from exotic pet industry; heavily colonized Florida Everglades; consumes vast numbers of native aquatic and terrestrial prey items; limited bounty hunting permitted as a control.

Sea lamprey (*Petromyzon marinus*); parasitic fish; 1800s to early 1900s; locally invasive in the Great Lakes through man-made canals; parasitic feeding impacts large native fish species. The most effective control is use of the lampricide TFM to kill larvae in tributary nursery areas.

## 14.3.1 Environmental Problems Resulting from Animal Introductions

### 14.3.1.1 Impacts on Native Species

Invasive animals can have major impacts on native species through direct predation (Carey et al. 2012; Pyšek et al. 2020). A well-known cautionary tale involves the Nile perch (*Lates niloticus*), deliberately introduced into Lake Victoria unofficially in 1954 and officially in 1962 to improve the fishery. This perch is a voracious predator that can grow to 200 kg and 2 m in length. It rapidly became a very valuable commercial species and is currently the base of a US$140 million industry among Kenya, Tanzania, and Uganda. This success has come at a major price as predation and competition by this species has caused the local extirpation of hundreds of native species, primarily cichlids (Goudswaard et al. 2008; McGee et al. 2015). Probably the most egregious recent example of overpredation by invasive species is the impact of introduced Burmese pythons (*Python bivittatus*) on the Florida Everglades, likely through deliberate releases by private owners of exotic pets. These are top predators that can reach > 6 m in length and weigh over 90 kg. Tens of

thousands reside in and around the Everglades National Park. They have successfully bred and are proliferating and spreading. A rather eye-opening study (Dorcas et al. 2012) found apparent massive predation by pythons severely reducing native organisms. Before 2000, mammals were encountered frequently during nocturnal road surveys within the Everglades. In contrast, road surveys totaling 56,971 km from 2003 to 2011 documented a 99.3% decrease in the frequency of raccoon observations and decreases of 98.9 and 87.5% for opossum and bobcat observations, respectively, and failed to detect rabbits. There are films and photos of battles between pythons and alligators in the park. University and National Park Service researchers study the pythons by capturing some and implanting tracking devices that provide data on movements and lead them to other pythons. Other captured specimens are euthanized and gut content analyses are performed. Some of the prey items found include marsh rabbits, bobcats, deer, alligators, woodrats, and waterbirds, some which are threatened or endangered species. Currently regional and state agencies conduct search and removal operations, and independent contractors are paid to remove and euthanize the pythons; as of mid-2021, 7293 Burmese pythons have been removed via these combined programs (Rodgers 2022).

An invasive animal can often outcompete a native species occupying the same or a similar niche. This can occur if the invasive is more prolific at breeding, or is a more voracious predator upon shared prey items, or can successfully utilize more of the habitat (i.e. deeper water, better tolerance of low DO, or more adapted to varying sediments). For example, the Asiatic clam (*Corbicula fluminea*) was introduced into the USA in the 1930s and has spread throughout every major drainage basin. It is highly prolific and directly competes for habitat with native bivalves (Neves et al. 1997). A successful invader can also make a new habitat more accommodating for a subsequent invasive. An example of an invasive invertebrate species outcompeting native species would be the amphipod *Echinogammarus ischnus*, a deposit feeder that lives in association with zebra mussels in Europe. It successfully invaded the Great Lakes and replaced local amphipods, potentially because biodeposits and shelter provided by the invasive zebra mussel facilitated its success (Ricciardi and MacIsaac 2000).

On rare occasions an invasive species can have a positive impact on native species. In the Florida Everglades, the native apple snail (*Pomacea paludosa*) is a principal prey item of snail kites (*Rostrhamus sociabilis*), whose nesting activity was low for the latter 1990s and early 2000s (Zhang et al. 2022). Around 2010 the exotic island apple snail (*P. maculato*), a larger species, became abundant around Lake Okeechobee and was believed to be associated with a resurgence in snail kite abundance and nesting, by being found in more environments and readily consumed by the kites (Zhang et al. 2022). However, questions remain as to the exotic apple snail's impact on native vegetation and potential disease transmission (Rodgers et al. 2022).

### 14.3.1.2 Economic Problems

Invasive animals can have severe economic impacts. Worldwide, since 1971 the economic cost of aquatic invasive animal and plant species (damages and controls) has been estimated at US$345 billion (Cuthbert 2021). Various invasive animal economic losses have been associated with recreation and tourism, reductions in native sport or commercial fisheries, biofouling (see below), and healthcare costs associated with disease-bearing mosquito invasions (Cuthbert et al. 2021). Invasive clams and mussels, in particular, have driven huge monetary costs in control attempts. These organisms form massive growths on the interior walls of water intakes of drinking water treatment plants, industrial users of water, and cooling water intakes for steam electric power generation plants. The aforementioned *Corbicula* is a filter-feeder; filter-feeders preferentially settle on hard substrata, with a constant flow of water carrying food, preferentially phytoplankton, to them. *Corbicula* is well known to clog intakes for power plants and municipal water systems. In the USA the nuisance overgrowths are controlled by chlorination or hand removal, at a cost of approximately $1 billion/year (Pimentel et al. 2000, 2005). The zebra mussel and quagga mussel (Box 14.1; Fig. 14.3) are later invaders (and filter-feeders) that also cause water intake clogging problems through overgrowths. Control costs of zebra mussels per facility can run from $30,000 to over $44,000/year, and removal techniques range from mechanical removal by divers to various chemical treatments (Connelly et al. 2007).

---

**Box 14.1  A major ecosystem disruption—zebra and quagga mussels and the changes they bring.**

An example of a major ecosystem disruption in a large freshwater body concerns zebra and quagga mussels (Fig. 14.3a). Zebra mussels, native to the Caspian Sea, entered the Great Lakes in 1986, probably via ballast water, and were established by 1988. They form extensive populations on hard substrates, causing problems to water intake systems to treatment plants and to cooling water intakes for electric power generation facilities. There is currently no good biological control agent, though some organisms do eat them. Zebra mussels graze phytoplankton and small zooplankton and have greatly altered the plankton communities of open waters (Caraco et al. 1997).

As noted (see Chapter 4), the filamentous green alga *Cladophora* formed massive blooms along shores in the Great Lakes, particularly Lake Erie, in the 1960s, due largely to phosphorus inputs from both point and non-point sources. These *Cladophora* balls, as they were called, fouled the beaches with huge decaying mats (this author recalls them from childhood). *Cladophora* needs rocky substrata and relatively clear water for its best growth and was largely confined to shallower waters. Nutrient controls in the 1970s following the Clean Water Act greatly reduced the problem. The later massive *Dreissena* growths have likely enhanced the recent reintroductions of *Cladophora* blooms in some areas of the Great Lakes through creation of additional substrata for attachment (the shells), with their intense filter-feeding clearing the water of phytoplankton so that solar irradiance transmission is increased, and these mussels rapidly recycle P back into the water. The greater irradiance penetration and increased P availability are believed to stimulate *Cladophora* growth (Higgins et al. 2008) and *Cladophora* infestations have become more widespread and extend to deeper waters. Heavy growths of *Cladophora* cause many problems including water intake clogging, displacement of other aquatic flora, reduction of local biodiversity, increases in BOD load, and subsequent DO decreases in areas of decaying material, and the decomposing mats along shorelines concentrate and protect potentially pathogenic fecal microbes deposited by shorebirds probing the wrack.

This is an example of introduced nuisance fauna creating improved habitat for a native nuisance organism. Zebra mussels also attach to the shells of native gastropods and contribute to local extirpations (Neves et al. 1997). Quagga mussels have moved west across the USA and have formed large infestations in Lake Mead on the Colorado River (Beaver et al. 2010) and threaten other western reservoirs (Fig. 14.2b). Apparently zebra mussels have also invaded US waters via the aquarium trade. Aquarists purchase "moss balls," which are live spheres of the green alga *Cladophora aegagropila*, from pet stores. Many of these were found to contain zebra mussels, which were subsequently traced back to their source, a zebra mussel-infested lake in Ukraine (Kobell 2021). Economic losses due to *Dreissena* in the USA have been estimated at $1 billion annually (Pimentel et al. 2005), while the worldwide cost of *Dreissena* damage and control (since 1971) has been estimated at $50 billion (Cuthbert et al. 2021).

### 14.3.1.3 Ecosystem Disruptions

Ecosystem disruptions can be highly variable depending upon the individual invader and the location of establishment. Such disruptions can involve alterations of the natural food webs,

changes in predator species, changes in game species, altered energy flow paths and nutrient cycling, and the creation of conditions that either greatly reduce phytoplankton and/or zooplankton or lead to algae bloom production. While much of the discussion in this chapter revolves around exotic species that have entered the USA, one of the major ecosystem disruptions is in regard to an invader going from the USA to Europe. *Mnemiopsis leidyi*, the American comb jelly, successfully invaded the brackish Black and Azov Seas in the early 1980s through ballast water transfer. It developed huge blooms and devastated the local zooplankton, leading to major crashes in the anchovy fishery. It also clogs fishermen's nets and eats eggs and larvae of commercial fish species (Ivanov et al. 2000). It is further successful because it has high tolerance for low DO and has a refuge within the hypolimnion. Invasives can not only directly alter local food webs but also create conditions for further disruption by other introduced species of native species (Box 14.1). Note that invasive fish have also been introduced to formerly fishless lakes, drastically altering their ecosystems (Strayer 2010).

While the majority of examples in this book are of North American invasions, an extremely concerning example of an Asian-sourced bivalve rapidly disrupting inland aquatic ecosystems is that of the golden mussel (*Limnoperna fortune*). A native of China, this small mollusc was first described in South America in 1991 (Moutinho 2021). It has since expanded at a rapid rate through Argentina, Brazil, Uruguay, and Paraguay at rates as high as 240 km/year. Although the expansion is upstream against the water flow, it has been aided by adhering to boat hulls and carried in fish stomachs. The golden mussels' filtration rates far exceed that of zebra mussels and thus they disrupt planktonic communities, clog water infrastructure, and change water flow patterns and sediment composition by their rapid accumulations (Moutinho 2021). A major worry is the ecosystem disruptions that may occur if they enter the Amazon watershed.

An invasive organism does not necessarily have to be aquatic to impact aquatic resources. Terrestrial animal invasions can alter local hydrological cycles through overgrazing, channel construction, or other local "ecological engineering." Some invaders can

(a)                                          (b)

**Fig. 14.3**  Fig. 14.3a (left) Size of quagga mussels *Dreissena bugensis*. Fig. 14.3b (right) Quagga mussels infesting the shoreline of Lake Mead in the Colorado River basin in Western USA. (Photos courtesy of Dr. John R. Beaver, BSA Associates.)

(a)                                          (b)

**Fig. 14.4**  Fig. 14.4a (left) Feral swine (*Sus scrofa*) in the USA are a mixture of swine imported for sporting and food purposes, mixed with escaped domestic swine. Fig. 14.3b (right) Feral swine readily invade new territories and increase their population rapidly; their rooting can cause localized water quality problems such as turbidity, as in the waterways of Congaree National Park, South Carolina. They also cause great amounts of crop damage; estimated crop and environmental economic damages in the USA have been estimated at $800 million (Pimentel et al. 2005). (Fig. 14.4a photo courtesy of Matthew R. McIver, UNC Wilmington; Fig. 14.4b photo by the author.)

alter water quality through defecation of nutrients or trampling and rooting (Fig. 14.4).

The phenomenon caused by the intersection of species loss and the introduction of new species to a region is known as **biotic homogenization**, which is exemplified by the homogenization of fish faunas (Rahel 2000). The phenomenon results in a reduction in regional differences among fauna and is associated with both natural waterways and created ecosystems (i.e. river impoundments). It is primarily caused by intentional or accidental introductions of new species to an area and secondarily caused by the loss of endemic or native species (which occurs in part due to the introductions). The main incidents

of homogenization have been introduction of eastern USA game fish into western USA waters (walleye, bass, sunfish). State agencies did much of this in the past, but that activity has declined more recently; however, such introductions apparently continue to occur illegally by anglers wishing to have their favorite game fish present to pursue. Natural lakes are scarce in the American West, and thus warmwater habitat should be likewise rare. The creation of large western impoundments to provide irrigation and municipal water (see Chapter 11) has, however, provided the necessary warmwater habitat to support eastern game fish species. For example, Arizona and Montana historically had no fish species in common but now share 33 species (Rahel 2000). Rahel (2000) further notes that in most of the western states (and some New England states) 25–50% of the fish species are introduced, and in Nevada, Utah, and Arizona more than half are introduced. Introductions of congeneric fish and other species can also lead to species hybridizations, which can contribute to a decline in native species (Allan and Flecker 1993). Biotic homogenization has occurred worldwide to many large river basins, largely as a result of introductions of nonindigenous species (Su et al. 2021).

## 14.4 Exotic Aquatic Macrophyte Introductions

### 14.4.1 How Are They introduced?

In a (relatively) pristine aquatic ecosystem overgrowths of macrophytes are normally not a problem because biological controls have evolved along with them (predators and parasites). Nuisance blooms may occur from time to time, but a balanced system will eventually right itself. When a successful introduction of an exotic macrophyte occurs, experience has shown that problems, both ecological and economic, can and do occur. The USA offers multiple examples of problematic introductions of exotic macrophytes in rivers, reservoirs, and lakes (Table 14.2). There are several reasons for this. First, many of these species are imported from other countries through the aquarium trade and/or as ornamental species (Fig. 14.5a). Dealers bring back the best specimens which are not infected with disease (Padilla and Williams 2004). Once they escape into the wild, they may have no natural consumers (grazers), and native fauna may take time to develop a "taste' for such species, or they might not graze them at all. Geographically, areas where such invasions are particularly successful include the southeastern USA, particularly Florida (Rodgers 2022), and Louisiana. These areas have long growing seasons, abundant sunlight, excessive nutrients from eutrophication, and numerous and interconnected waterways. Areas worldwide where such conditions occur likewise suffer from macrophyte invasions. Establishment from seeds, spores, and

**Table 14.2** Some well-known invasive macrophyte species that have significantly impacted the ecology of North American waterways including functional habitat group; when established; how they arrived (vector); and principal means of control (information from multiple sources).

Alligatorweed (*Alternanthera philoxeroides*); emergent; late 1800s from South America; arrived in ship ballast; biological control by alligatorweed beetle (*Agasicles hygrophila*) and stem borer ([*Vogtia*] *Arcola malloi*).

Australian melaleuca (punk tree) (*Melaleuca quenquenervia*); introduced 1907 for cultivation; South Florida, Louisiana; control thus far unsuccessful, but melaleuca snout beetle (*Oxyops vitiosa*) being tested.

Brazilian elodea (*Egeria densa*); submersed; late 1800s; aquarium trade; controlled by grass carp

Elodea (*Elodea canadensis*); submersed; native to the USA; locally invasive through aquarium trade and small boat spreading; invaded Europe 1836; confused with hydrilla; controlled through a variety of methods including herbicides.

Eurasian watermilfoil (*Myriophyllum spicatum*); submersed; 1940s; aquarium trade; herbicides; various biological agents being tested for control.

Giant salvinia (*Salvinia molesta*); free-floating fern; 1990s; accidental introduction; herbicides offer short-term control; biological control with weevil *Cyrtobaquos salviniae* is promising.

Hydrilla (*Hydrilla verticillata*) (Fig. 14.5b); submersed; 1960s from Indian subcontinent; aquarium trade; herbicides or grass carp control.

Hygrophila (*Hygrophila polysperma*); submersed; 1979; aquarium trade; poorly controlled; herbicide resistant, use of very high rates of sterile grass carp, hand pulling and raking.

Purple loosestrife (*Lithrum salicaria*); emergent; 1800s; deliberate introduction for ornamental and medicinal purposes; biological control by imported European beetles effective.

Water hyacinth (*Eichhornia crassipes*) (Fig. 14.5a); free-floating; 1800s; from Amazon watershed; deliberate introduction as an ornamental; readily sold in shops; partial control through mechanical, herbicide, and biological means.

Tamarisk (salt cedar) (*Tamarix* spp.); invasive trees in Southeast USA river riparian zones; early 1800s from Eurasia; introduced as ornamental; eradication plan currently suspended due to habitat value.

(a)                                                        (b)

**Fig. 14.5**  Fig. 14.5a (left) Water hyacinth *Eichhornia crassipes* on sale at a North Carolina nursery. Fig. 14.5b (right) *Hydrilla verticillata* is a major nuisance submersed macrophyte impacting waterways, here photographed in a regional wet detention pond. (Photos by the author.)

fragments carried in ballast water is another route of intercontinental invasion.

### 14.4.2  How Are They Spread?

Once invasive macrophytes become established in a new area, how are they spread? Many species can reproduce asexually through fragmentation (called clonal propagation; Barrett 1989). This allows for pieces to be carried through interconnecting waterways, both natural and canals. Exotic species can also be spread through movement of channel-dredging spoil materials (Erickson 2005). Fragments can be carried from lake to river to reservoir by small boat propellers and boat trailers, or within boat bilges that are subsequently drained in water bodies, or attached to fishing gear. This spread is exacerbated by construction of canals for residential, commercial, or industrial usage. Plant seeds are small, can adhere to surfaces, and are thus transferred to other waterways attached to humans and their vehicles, via birds and terrestrial animals, and can be spread by winds, including hurricanes and tornados. Plant fragments can also be spread this way (Barrett 1989).

### 14.4.3  What Problems Do They Cause?

Why are growths of invasive macrophytes so problematic to the environment and economy? There are numerous problems and the following list is not exclusive.

- The physical presence of macrophyte mats interferes with human use of waterways. Growths tangle in boat propellers, cause direct physical danger to swimmers and divers, and deny access to fishing grounds.
- Some invasives, such as *Hydrilla verticillata* (Fig. 14.5b), can outcompete and displace native macrophyte species after establishment (Langeland 1996).
- When reservoirs are afflicted by nuisance overgrowths this can cause major and costly issues with irrigation and power generation (both hydropower and cooling water). This is because water intakes, and protective screens, become coated and/or clogged with macrophyte growths. Operators then have to expend additional funds to control the growths by various means (herbicides, mechanical removal, physical removal by plant personnel, etc.). Such costs are then often passed on to the utility customer.
- Thick growths, particularly of submersed species, allow too much protection for small fish and may also stunt them (Nichols 1991). Too many survive predation and larger predators such as large-mouth bass (i.e. game fish) cannot forage in dense vegetation beds so they are denied important prey resources (Engel 1987).
- Heavy macrophyte growths increase water loss through transpiration (especially free-floating macrophytes)—more than three times normal water loss in the growing season. In reservoirs used for irrigation, municipal, or industrial water

supply, this can cause user conflicts and water rationing during drought periods.

- Nuisance overgrowths can be public health hazards for individuals with allergies to weeds; these growths can be home for snakes; some plants (water hyacinth and water lettuce [*Pistia stratiotes*]) are breeding areas for vectors of disease including malaria, encephalitis, and schistosomiasis (Barrett 1989); and as mentioned above, they are a danger for swimming children (and adults).
- Nuisance macrophyte growths lead to water quality and eutrophication problems. Excessive mats of macrophytes will die, either naturally or if they are killed by herbicide treatments. This will create a labile BOD load that can lead to hypoxia and anoxia and subsequent fish kills. Dying mats of vegetation also create odor problems in residential or recreational areas. Floating mats of macrophytes can reduce solar irradiance by shading to more desirable rooted macrophyte species and benthic microalgae, reducing fish habitat and microalgal food supplies. On a physical basis, extensive and repetitive macrophyte growths can cause more rapid filling in of shallow lakes with organic sediments and reduce the usable area of reservoirs.
- The USA spends \$100,000,000/year on control of non-native macrophyte growths (Pimentel et al. 2000).

Water hyacinth (*Eichhornia crassipes*) is native to the Amazon ecosystem and is a free-floating macrophyte. It was brought to New Orleans in the late 1800s as an ornamental plant for an international exhibition. It was then cultivated in ornamental ponds and from there entered the wild (Barrett 1989). Now it causes problems of obstruction of waterways and low DO; additionally, it makes good habitat for agents bearing diseases such as malaria, encephalitis, and schistosomiasis. This species was introduced to Africa a century ago and is still spreading: there, large growths of it have drawn down scarce water resources in some areas because of its high rate of evapotranspiration (Padilla and Williams 2004 and references therein); it is problematic in Asia as well. Reported worldwide costs of *Eichhornia* damages and control have been estimated at approximately US\$3 billion (Cuthbert

et al. 2021). An analysis of water hyacinth ecological impacts found that under some circumstances it can provide good habitat for fish and invertebrates (Villamagna and Murphy 2010). The author saw this species on sale at a greenhouse and also an ornamental on display at the North Carolina State Fair in fall 2012, demonstrating its apparent appeal to the public. No techniques offer complete eradication in lotic situations, but partial control is achieved through mechanical harvesting, herbicide treatments, and biological control with weevils of the genus *Neochetina* from its native range (Villamagna and Murphy 2010).

Hydrilla (*Hydrilla verticillata*) is a submersed macrophyte native to Sri Lanka, India, and Pakistan but now occurs on all major continents as well as Pacific islands. It was introduced into the USA via the aquarium trade in the 1960s, and dealers actually stored and raised the plants in Florida canals for sale (Fincham 2009). This species spreads readily through fragmentation and within the USA is a problem from Florida north through the Carolinas, even as far north as the Potomac River and as far west as California. It can also reproduce by seeds, turions, and tubers (subterranean turions), grows rapidly, can survive very low light conditions, and has been called "the perfect aquatic weed" (Langeland 1996). Florida spends \$14.5 million/year on control of hydrilla. Because of its ability to reproduce by fragmentation, mechanical control (i.e. cutting) can make infestations worse rather than better. Herbicides are used in control, and sterile grass carp have been used successfully to control it in enclosed situations (Fig. 14.5b).

The Kariba weed or giant salvinia (*Salvinia molesta*) is a fern native to Brazil. It has rapidly spread and caused massive nuisance overgrowths in Africa, Asia, Australia, and the Southern USA (Barrett 1989). It was at one time actually available commercially in the USA; it was first detected in nuisance growths in South Carolina in 1995. It accumulates in mats over 1 m thick, forces out other plants, and kills planktonic organisms. It can reproduce by fragmentation and has caused major economic and ecological problems by forming huge nuisance accumulations that interfere with water intake function, irrigation, and boating, and its thick growths can block solar irradiation to native macrophytes and benthic microalgae. It is usually treated by herbicide applications, but there has been success in biological control using a weevil, *Cyrtobagous salviniae*.

While most of the invasive species discussion in this chapter has concerned intercontinental or cross-continental movement of species into new territories, some species can be considered locally invasive. This occurs when a disturbance, generally human-caused, occurs in an area that allows or stimulates the rapid expansion of a species to new areas within its general range (Carey et al. 2012). The disturbance can be due to habitat change through development, timbering, canal construction, dam building or channelization, overhunting or overfishing of a top predator, aquarium dumping, or stocking of a desired native species elsewhere within its range. These local invasions can result in ecological changes such as alteration of food webs, changes in nutrient cycling, suppression of local native species, disease introduction and hybridization of species, and degradation of water quality through fecal microbial pollution, nutrient inputs, and increased turbidity.

Common reed (*Phragmites australis*) has both native and European genotypes present in the USA. It is locally invasive in that it readily colonizes disturbed areas (King et al. 2007) and outcompetes native emergent aquatic plants while doing so. Although found throughout the USA, road, bridge, and drainage ditch building and waterfront housing construction provide avenues for *Phragmites* to expand its presence within its overall range (Fig. 14.6a). This species can have deleterious impacts to natural ecosystems. For example, marshes that have been overtaken by invasive *Phragmites* support significantly less nekton abundance than natural *Spartina* marshes (Kimball and Able 2007), although other studies did not find such deleterious impacts. Additionally, anthropogenic nitrogen loading enhances *Phragmites* success as an invasive species (King et al. 2007) and in newly developed riparian areas the elevated nitrogen supplied by stormwater runoff or septic system leachate may facilitate *Phragmites* invasion. Common cattail (*Typha* spp.) can readily and rapidly move into new habitats within its native range as well and develop dense monocultures. If an organization constructs a wetland for stormwater runoff treatment, for instance, one can be assured that cattail will shortly appear even if not planted.

### 14.4.4 Exotic Plant Removal or Control Measures

It is far more difficult to remove a nuisance invasive macrophyte once it has become established in new waterways than to prevent its establishment in the first place. A broad array of tactics has been tried, with some showing success in removing or severely curtailing infestations within individual ponds, streams, or reservoirs. Even problematic macrophytes can have redeeming values (such as macroinvertebrate habitat or cover for young fish), so total eradication may not be necessary, just control (Nichols 1986). For control of nuisance macrophytes a combined campaign using two or more tactics is most appropriate in many cases. Several control measures are discussed below.

**Hand-pulling** of nuisance invasive macrophytes can be at least temporarily successful in shoreline areas. There are success stories of lake-wide control if a diver campaign is funded by a municipality or local homeowners association (Gallagher 2015). The benefit is that harmful herbicides are not needed and fragmentation is reduced. If hand-cutting is used, however, spread of the infestation by fragmentation remains a problem.

**Mechanical control** of nuisance macrophyte masses most often is done by large mechanical harvesters. These machines are designed to cut the vegetation and scoop up the cut fragments. When dealing with species that reproduce readily through fragmentation there is an obvious issue of spread of the infestation around the targeted area. Further, if equipment is moved from system to system, fragments can be introduced elsewhere if proper cleaning is not accomplished. In heavily infested reservoirs, such mechanical harvesting or screening can clear "lanes" to allow piscivorous fish access to prey using the vegetation for hiding and at the same time open areas for human recreation (Engel 1987).

**Chemical control** by the use of herbicides has been and remains a common approach to controlling noxious macrophyte invasions. It can be highly effective at killing the targeted plant and can be used

(a)                                          (b)

**Fig. 14.6** Fig. 14.6a (left) Locally invasive *Phragmites* near a bridge over a waterway in Cape Hatteras National Seashore. Fig. 14.6b (right) Decaying mass of aquatic vegetation creating a BOD load in Sutton Reservoir, North Carolina, following herbicide treatment which led to a subsequent fish kill through hypoxia. (Photos by the author.)

in some situations to restructure littoral zones to create a more balanced habitat (Nichols 1986). However, over the years many herbicides have turned out to have unintended consequences to nontarget biota (see Chapter 12). Another problem that may occur concerns the invasive species dying in place—this leads to a labile source of BOD that can drive DO down to dangerous levels (Fig. 14.6b). Also, dying macrophytes can release considerable loads of nutrients (Dierberg 1993; O'Dell et al. 1995) that can cause noxious phytoplankton blooms that can also lead to hypoxia or be toxic. Herbicide use is optimal when there are small infestations that can be readily accessed and targeted, which reduces the chance of system-wide disruptions. In Lake Okeechobee and Kissimmee River watersheds in south Florida, shoreline invasive plant infestations have been successfully controlled by **controlled burns**, sometimes in combination with herbicide use (Rodgers 2022).

There is a rich and occasionally successful history of **biological control** of invasive macrophyte infestations. Addition of grass carp is a commonly used method for lake and reservoir nuisance vegetation control (Leslie et al. 1987; Ross and Lembi 1998), and tilapia have been used on occasion. Unfortunately, such herbivores may consume not only the target macrophyte but also others (Crutchfield et al. 1992); thus such control attempts are best suited to unconnected lakes and ponds. Further,

introductions of grass carp to control infestations is usually permitted only if the fish are sterile (i.e. triploid), to minimize the introduction of reproducing populations to open waters. Biological control of alligatorweed (*Alternanthera philoxeroides*) has been successful primarily by the alligatorweed beetle (*Agasicles hygrophila*) and the stem borer ([*Vogtia*] *Arcola malloi*). Outbreaks of purple loosestrife (*Lithrum salicaria*) in the USA have been controlled by additions of the black-margined loosestrife beetle, imported from Germany. The beetle will defoliate 80% of the loosestrife in an area and allows regrowth of native vegetation. Of course, any biological treatment, especially using a deliberately imported grazer, must be undertaken with care to avoid unanticipated impacts to the food web or fish community, and thus involves thorough planning (Nichols 1991). Currently in south Florida biological control agents are continually being developed and used against some of the many invasives in that region (Rodgers 2022).

**Physical control** of the environment has taken a variety of forms in the battle against invasive macrophytes. Some tactics have included environmental manipulations such as reservoir water level control to dry out infested shoreline areas, and solar irradiance reductions through dyeing the water to increase light attenuation, or simply shading small local patch infestations by cloth coverings (Nichols 1991).

A unique solution: Since 2000, South Africa has built up a program called Ukuvuka, wherein thousands of workers are employed to dig up, rip up, cut down, and remove numerous alien plant species manually, both aquatic and terrestrial. It is cost effective when placed in perspective of mitigating the many problems non-natives create (such as massive water loss) and good for employment of unskilled laborers, employing up to 29,000 workers/year full- or part-time (Koenig 2009).

## 14.5 Invasive Plankton and Microbes

As noted above, transport in ship ballast water is a major vector for movement of invasive aquatic species. Carlton and Geller (1993) brought this to the forefront of invasive species research when they sampled the ballast of 159 cargo ships in Coos Bay, a marine port in Oregon, and found 367 identifiable taxa representing 47 different orders. A major increase in harmful estuarine algal blooms (mainly dinoflagellates) that was linked strongly to intercontinental transport via ballast water occurred in the 1980s (summarized by Hallegraeff 1993). Not only commercial vessels but also military vessels can carry a broad variety of invasive species from port to port. Burkholder et al. (2007) performed a study of 28 US military (nonwarship) vessels from which a total of 60 ballast water tanks were sampled. In general these vessels maintained excellent records of ballasting and over 90% of tanks examined had undergone some sort of ballast water exchange before port. Nonetheless, 100 phytoplankton species were found, with the largest contributions from chain-forming diatoms and dinoflagellates; however, 23 potentially harmful species, mostly dinoflagellates and diatoms, were found and most species were culturable. With greater age of the ballast water, fewer species were culturable, and after 33 days at sea culturability of phytoplankton was nil. Many of the organisms were common coastal species that can survive low salinities in riverine estuaries.

The transfer of zooplankton species intercontinentally into new freshwater habitats has occurred on a number of occasions in the Laurentian Great Lakes; from there these organisms have spread locally through canals, bilges, bait buckets, and other means to many other river systems and reservoirs (Figs. 14.2 and 14.3). Examples of such organisms include the predaceous cladoceran *Bythotrephes longimanus*, known as the spiny water flea. This organism was first found in Lake Huron in 1984, fouls fishing lines, has altered local zooplankton communities, and competes with local planktivorous fish for prey. A similar invasive cladoceran is the fishhook water flea (*Cercopagus pengoi*), which was first detected in Lake Ontario in 1998. It also competes for planktonic food with local planktivores, it is a fouling organism, and its long spiny tail discourages planktivores from feeding on it. An example of a cladoceran introduced to southern USA waters and spreading north from there is the cladoceran *Daphnia lumholtzi*, which may have been introduced to Texas reservoirs in the 1980s along with Nile perch. This organism also competes with native zooplankton and has predation-resistant properties by virtue of its long helmet and tail spine.

Besides algae and zooplankton, other nonobvious organisms that can be transferred to new areas include microbes, some of which are pathogenic. Pathogenic microbes can be spread by some of the same vectors as invasive plants and animals are spread. When plants or animals are infected and transported to a new area there is the potential for disease transmission to local hosts (Pimentel et al. 2000). In the study of military vessel ballast water (Burkholder et al. 2007), species-specific assays detected *Escherichia coli* in 23 ballast tanks and *Pseudomonas aeruginosa* in four tanks, with minor detections of *Aeromonas* sp., *Mycobacterium* sp., and *Listeria monocytogenes*. Again, these vessels generally had good ballast water treatment procedures in place. While no *Vibrio cholerae* were detected in the military vessels, another study of three commercial ships entering Chesapeake Bay and two ships entering the Great Lakes detected *V. cholerae* in half the water samples (Drake et al. 2005). This study also detected the presence of the toxic dinoflagellate *Pfiesteria piscicida* in 25% of the samples taken. The authors of that study noted that biofilms that form along the walls of ballast tanks harbor bacteria and cysts, and common methods used to reduce risks (chemical or physical treatments or ocean exchanges) may not impact such embedded organisms.

Other important aquatic disease organisms (reviewed by Strayer 2010) spread widely include the chitrid fungus (*Batrachochytrium dendrobatidis*), which has caused major mortalities among amphibian populations; the crayfish plague fungus (*Aphanomyces astaci*), introduced to Europe from North America and heavily impacting European crayfish; and the parasite *Myxobolas cerebralis*, which causes whirling disease mainly among farmed salmonids.

## 14.6 Reducing Aquatic Invasive Species Introductions: General Tactics

Due to the many successful exotic species invasions that have entered the USA from ballast water exchanges, this country now requires overseas ships to exchange their ballast water with open ocean water at least 200 miles offshore (Tibbetts 2007). Dumping ballast waters sourced from freshwater and estuarine waters will kill such organisms in high salinities, and the open ocean replacement ballast water has far fewer planktonic organisms than estuarine water or freshwater. Replacement of ballast water (fresh) by marine water before entering freshwater ports in places like the Great Lakes has been an improvement, but some non-native species are still introduced to new waters (Ricciardi and MacIsaac 2008). Some reasons have to do with specific policy issues and regulations, but a major reason is even with ballast water exchanges there remains tons of residual bottom water and sediment in large commercial vessels, and organisms can be resuspended after reballasting and subsequently released at the next freshwater port (Holeck et al. 2004). Aside from ballast water exchange, what are some of the strategies to reduce the transport of non-native species?

- Ballast water filtration, followed by chemical treatment to kill organisms and their larvae has been tested in oceangoing vessels. The use of chlorination and ozonation is generally successful; UV treatment is also being tested (note that these are common means to treat sewage). Many of these techniques are proprietary, as finding a "silver bullet" can result in a lucrative business opportunity, given the size of the shipping industry. Heat treatment of ballast water has also been investigated. The US Coast Guard now requires that vessels built after 2013 and planning to enter US waters must have a ballast water treatment system installed (Strain 2012).

- Implementing strict controls on the aquarium trade—what plant and animal species they may import—seems to be a logical means to reduce exotic species invasions. However, a major hurdle is that the aquarium trade is well organized and strict controls are seen as interfering with commerce; thus implementing meaningful regulations is not an easy task (Padilla and Williams 2004). Also, as mentioned, trading on the internet provides ways around these bans.

- Many aquatic plants and invertebrates are carried from water body to water body in small boats. Resource authorities often provide extensive signage at boat launch facilities describing invasives and instructing boat owners to empty bilges and clean propellers and fishing tackle before leaving the facility so they do not spread invasive species to other waterways (Fig. 14.2).

- Offering rewards to kill nuisance invasive species is a tactic that has been used in a number of settings. For instance, in the Columbia River basin the introduced northern pikeminnow (*Ptychocheilus oregonensis*) preys on endangered salmon. As a control means, anglers can receive $4–8/fish for pikeminnows they take, resulting in millions being harvested over the years, and this has had some success in relieving predation pressure on salmon (Carey et al. 2012). Other examples include rewards offered for invasive nutria in Louisiana and for the deliberately imported toxin-producing cane toad (*Bufo marinus*) in Australia.

- Before any management program, especially one that involves outright removal of an invasive organism, a thorough understanding of the invasive's current role in the ecosystem is required. In some circumstances an entrenched invasive may be providing a needed ecosystem service, such as maintaining a needed predator role, providing habitat for key organisms, or providing food resources (Lampert et al. 2014). Thus, a set of goals should be first established that include desired restoration without further system disruption.

## 14.7 Summary

- Maintaining high natural biodiversity in waterways encourages ecological stability and complex food webs and yields economic benefits through ecotourism and useful products derived from aquatic species.
- Threats to a waterway's biodiversity include pollution, habitat loss, overexploitation of key species, invasions of non-native species, and homogenization of species.
- Invasive animal problems include clogging of water intakes by filter-feeding clams and mussels, severe impacts on plankton by the same organisms as well as introduced jellyfish, decimation of native fishes by introduced fish taxa, homogenization of regional fisheries by introductions, and major food-web disruptions by introduced zooplankton, molluscs, crustaceans, and fish.
- Invasive plant problems include massive overgrowths that clog waterways to boat traffic and recreation, block water intakes, harbor microbes dangerous to public health, crowd habitat enough to stunt or change native fauna, outcompete or shade out native macrophytes, and create large BOD loads through overgrowths.
- The principal vectors for overseas transport of invasive plants and animals are ballast water, adherence to ship hulls and anchor chains, and attachment to oceangoing debris. The legal and illegal overseas aquarium trade is also a major vector.
- The principal vectors for within-continent spreading of invasive species are transport in small boat bilges, attachment to anchor lines, engines, and fishing tackle, passage through man-made canals, aquarium dumping, wind and flood transport during major storms, and deliberate introductions of fish and game species to new areas, both legal and illegal.
- Major invasive animal control measures include removal of fouling organisms by hand or toxic chemicals and selective hunting or fishing targeting invasive species.
- Invasive plant control includes hand-pulling or -cutting, chemical control by herbicides, mechanical control by harvesters, shading by dyes or cloth, import of herbivorous fish or insects to graze the invasive, and lake and reservoir drawdown.
- Prevention of the spread of invasive species includes ballast water dumping at sea, treatment of ballast water by filtration, chemicals, or UV radiation, local signage warning to clean boat propellers and bilges, laws against import of invasive species, and electric barriers to prevent fish passage.

## References

Allan, J.D. and A.S. Flecker. 1993. Biodiversity conservation in running waters. *BioScience* 43:32–43.

Barrett, S.C.H. 1989. Waterweed invasions. *Scientific American* October:92–97.

Beaver, J.R., T.E. Tietjen, B.J. Blasius-Wert, et al. 2010. Persistence of *Daphnia* in the epilimnion of Lake Mead, Arizona–Nevada, during extreme drought and expansion of invasive quagga mussels (2000–2009). *Lake and Reservoir Management* 26:273–282.

Brown, J.J. and G.W. Murphy. 2010. Atlantic sturgeon vessel-strike mortalities in the Delaware estuary. *Fisheries* 35:72–83.

Burkhead, N.M. 2012. Extinction rates in North American freshwater fishes, 1900–2010. *BioScience* 62:798–808.

Burkholder, J.M., G.M. Hallegraeff, G. Melia, et al. 2007. Phytoplankton and bacterial assemblages in ballast water of U.S. military ships as a function of port of origin, voyage time, and ocean exchange practices. *Harmful Algae* 6:486–518.

Caraco, N., J.J. Cole, P.A. Raymond, et al. 1997. Zebra mussel invasion in a large, turbid river: phytoplankton response to increased grazing. *Ecology* 78:588–602.

Carey, M.P., B.L. Sanderson, K.A. Barnas and J.D. Olden. 2012. Native invaders—challenges for science, management, policy and society. *Frontiers in Ecology and the Environment* 10:373–381.

Carlton, J.T. and J.B. Geller. 1993. Ecological roulette: the global transport of nonindigenous marine organisms. *Science* 261:78–82.

Connelly, N.A., C.R. O'Neill Jr., B.A. Knuth and T.L. Brown. 2007. Economic impacts of zebra mussels on drinking water treatment and electric power generation facilities. *Environmental Management* 40:105–112.

Crutchfield, J.U., D.H. Schiller, D.D. Herlong and M.A. Mallin. 1992. Establishment and impact of two *Tilapia* species in a macrophyte-infested reservoir. *Journal of Aquatic Plant Management* 30:28–35.

Cuthbert, R.N., Z. Pattison, N.G. Taylor, et al. 2021. Global economic costs of aquatic invasive alien species. *Science of the Total Environment* 775:145238.

Dierberg, F.E. 1993. Decomposition of desiccated submersed aquatic vegetation and bioavailability of released phosphorus. *Lake and Reservoir Management* 8:31–36.

Dorcas, M.E., J.D. Willson, R.N. Reed, et al. 2012. Severe mammal declines coincide with proliferation of invasive Burmese pythons in Everglades National Park. *Proceedings of the National Academy of Sciences of the United States of America* 109:2418–2422.

Drake, L.A., A.E. Meyer, R.L. Forsberg, et al. 2005. Potential invasion of microorganisms and pathogens via 'interior hull fouling': biofilms inside ballast water tanks. *Biological Invasions* 7:969–982.

Engel, S. 1987. The impact of submersed macrophytes on largemouth bass and bluegills. *Lake and Reservoir Management* 3:227–234.

Ericson, J.A. 2005. The economic roots of aquatic species invasions. *Fisheries* 30:30–32.

Fincham, M.W. 2009. Travels with *Hydrilla*: the unnatural history of an accidental invader. *Chesapeake Quarterly* 8(2):14–16.

Gallagher, M. 2015. Hand-pulling aquatic invasives. *Lake-Line* 35:40–43.

Goudswaard, K., F. Witte and E.F.B. Katunzi. 2008. The invasion of an introduced predator, Nile perch (*Lates niloticus*, L.) in Lake Victoria (East Africa): chronology and causes. *Environmental Biology of Fishes* 81:127–139.

Hallegraeff, G.M. 1993. A review of harmful algal blooms and their apparent global increase. *Phycologia* 32:79–99.

Higgins, S.N., S.Y. Malkin, E.T. Howell, et al. 2008. An ecological review of *Cladophora glomerata* (Chlorophyta) in the Laurentian Great Lakes. *Journal of Phycology* 44:839–854.

Holeck, K.T., E.L. Mills, H.J. MacIsaac, M.R. Dochoda, R.I. Colautti and A. Ricciardi. 2004. Bridging troubled waters: biological invasions, transoceanic shipping, and the Laurentian Great Lakes. *BioScience* 54:919–929.

Hughes, A.R. and J.J. Stachowicz. 2004. Genetic diversity enhances the resistance of a seagrass ecosystem to disturbance. *Proceedings of the National Academy of Sciences of the United States of America* 101:8998–9002.

Ivanov, V.P., A.M. Kamakin, V. B. Ushivtzev, et al. 2000. Invasion of the Caspian Sea by the comb jellyfish *Mnemiopsis leidyi* (Ctenophora). *Biological Invasions* 2:255–258.

Jackson, C.R. and C.M. Pringle. 2010. Ecological benefits of reduced hydrologic connectivity in intensely developed landscapes. *BioScience* 60:37–46.

Jackson, M.C., R.J. Wasserman, J. Grey, A. Ricciardi, J.T.A. Dick and M.E. Alexander. 2017. Novel and disrupted trophic links following invasion in freshwater systems. *Advances in Ecological Research* 57:55–97.

Jelks, H.L., S.J. Walsh, N.M. Birkhead and 13 others. 2008. Conservation status of imperiled North American freshwater and diadromous fishes. *Ecological Applications* 33:372–407.

Kimball, M.E. and K.W. Able. 2007. Nekton utilization of intertidal salt marsh creeks: tidal influences in natural *Spartina*, invasive *Phragmites*, and marsh treated for *Phragmites* removal. *Journal of Experimental Marine Biology and Ecology* 346:87–101.

King, R.S., W.V. Deluca, D.F. Whigham and P.P. Marra. 2007. Threshold effects of coastal urbanization on *Phragmites australis* (common reed) abundance and foliar nitrogen in Chesapeake Bay. *Estuaries and Coasts* 30:469–481.

Kobell, R. 2021. Moss balls, whelks and snakeheads. *Chesapeake Quarterly* 20:9–11.

Koenig, R. 2009. Unleashing an army to repair alien-ravaged ecosystems. *Science* 325:562–563.

Lampert, A., A. Hastings, E.D. Grosholz, S.I. Jardine and J.N. Sanchirico. 2014. Optimal approaches for balancing invasive species eradication and endangered species management. *Science* 344:1028–1031.

Langeland, K.A. 1996. *Hydrilla verticillata* (L.F.) Royle (Hydrocharitaceae), "the perfect aquatic weed". *Castanea* 61:293–304.

Leslie, A.J. Jr., J.M. Van Dyke, R.S. Hestand III and B.Z. Thompson, 1987. Management of aquatic plants in multi-use lakes with grass carp (*Ctenopharyngodon idella*). *Lake and Reservoir Management* 3:266–276.

McGee, M.D., S.R. Borstein, R.Y. Neches, H.H. Buescher, O. Seehausen and P.C. Wainwright. 2015. A pharyngeal jaw evolutionary innovation facilitated extinction in Lake Victoria cichlids. *Science* 350:1077–1079.

McKinney, M.L. 2001. Effects of human population, area and time on non-native plant and fish diversity in the United States. *Biological Conservation* 100:243–252.

Miller, R.R., J.D. Williams and J.E. Williams. 1989. Extinctions of North American fishes during the past century. *Fisheries* 14:22–38.

Moutinho, S. 2021. A golden menace: an invasive mussel is devastating ecosystems as it spreads through South American rivers, threatening the Amazon basin. *Science* 374:390–393.

Neves, R.J., A.E. Bogan, J.D. Williams, S.A. Ahlstedt and P.W. Hartfield. 1997. Status of aquatic mollusks in the southeastern United States: a downward spiral of diversity. Chapter 3 in Benz, G.W. and D.E. Collins (eds.), *Aquatic Fauna in Peril: The Southeastern Perspective*. Special Publication 1. Southeast Aquatic Research Institute, Lenz Design and Communications, Decatur, GA.

Nichols, S.A. 1986. Community manipulation for macrophyte management. *Lake and Reservoir Management* 2:245–249.

Nichols, S.A. 1991. The interaction between biology and management of aquatic macrophytes. *Aquatic Botany* 41:225–252.

O'Dell, K.M., J. Van Arman, B.H. Welch and S.D. Hill. 1995. Changes in water chemistry in a macrophyte dominated lake before and after herbicide treatment. *Lake and Reservoir Management* 11:311–316.

Padilla, D.K. and S.L. Williams. 2004. Beyond ballast water: aquariums and ornamental trades as sources of invasive species in aquatic systems. *Frontiers in Ecology and the Environment* 2:131–138.

Pimentel, D., L. Lach, R. Zuniga and D. Morrison. 2000. Environmental and economic costs of nonindigenous species in the United States. *BioScience* 50:53–64.

Pimentel, D., R. Zuniga and D. Morrison. 2005. Update on the environmental and economic costs associated with alien-invasive species in the United States. *Ecological Economics* 52:273–288.

Pyšek, P., P.E. Hulkme, D. Simberloff, et al. 2020. Scientists' warning on invasive alien species. *Biological Reviews* 95:1511–1543.

Rahel, F.J. 2000. Homogenization of fish faunas across the United States. *Science* 288:854–856.

Ricciardi, A. 2001. Facultative interactions among aquatic invaders: is an "invasional meltdown" occurring in the Great Lakes? *Canadian Journal of Fisheries and Aquatic Sciences* 58:2513–2525.

Ricciardi, A. and H.J. MacIsaac. 2000. Recent mass invasion of the North American Great Lakes by Ponto-Caspian species. *TREE* 15:62–65.

Ricciardi, A. and H.J. MacIsaac. 2008. Evaluating the effectiveness of ballast water exchange policy in the Great lakes. *Ecological Applications* 8:1321–1323.

Rodgers, L., C. Mason, E. Metzger, et al. 2022. Chapter 7: Status of Invasive Species. 2022 South Florida Environmental Report, Volume 1. South Florida Water Management District, West Palm Beach.

Ross, M.A. and C.A. Lembi.1998. *Applied Weed Science*, 2nd edition. Prentice Hall, Hoboken, NJ.

Ross, S.T. 1991. Mechanisms structuring stream fish assemblages: are there lessons from introduced species? *Environmental Biology of Fishes* 30:359–668.

Shurin, J. 2000. Dispersal limitation, invasion resistance and the structure of pond zooplankton communities. *Ecology* 81:3074–3086.

Simberloff, D. and B. Von Holle. 1999. Positive interactions of nonindigenous species: invasional meltdown? *Biological Invasions* 1:21–32.

Strain, D. 2012. Researchers set course to blockade ballast invaders. *Science* 336:664–665.

Strayer, D.L. 2010. Alien species in fresh waters: ecological effects, interactions with other stressors, and prospects for the future. *Freshwater Biology* 55(S1):152–174.

Su, G., M. Logez, J. Xu, S. Tao, S. Villéger and S. Brosse. 2021. Human impacts on global freshwater fish biodiversity. *Science* 371:835–838.

Tibbetts, J.H. 2007 Knocking back biological invaders. *Coastal Heritage* 21(4):3–11.

Torchin, M.E. and C.E. Mitchell. 2004. Parasites, pathogens and invasions by plants and animals. *Frontiers in Ecology and the Environment* 2:183–190.

Villamagna, A.M. and B.R. Murphy. 2010. Ecological and socio-economic impacts of invasive water hyacinth (*Eichornia crassipes*): a review. *Freshwater Biology* 55:282–298.

Warren, M.L., P.L. Angermeier, B.M. Burr and W.R. Haag. 1997. Decline of a diverse fish fauna: patterns of imperilment and protection in the southeast United States. Chapter 5 in Benz, G.W. and D.E. Collins (eds.), *Aquatic Fauna in Peril: The Southeastern Perspective*. Special Publication 1. Southeast Aquatic Research Institute, Lenz Design and Communications, Decatur, GA.

Zhang, J., P. Jones and A. Betts. 2022. Chapter 8B: Lake Okeechobee Watershed Protection Plan Annual Progress Report. 2022 South Florida Environmental Report, Volume 1. South Florida Water Management District, West Palm Beach.

# CHAPTER 15

# Ecology and Pollution of Urban Streams

Streams come in many types. They may be sinuous streams in foothills, or in mountain or formerly glaciated regions with riffles, pools, and runs; they may be lowland multibranching systems with copious nearby wetlands such as on the Coastal Plain or in Sandhills regions; they may be tidal creeks lined with marshes in northern regions or creeks lined by mangrove forests in tropical regions. However, few streams and rivers in either developed or developing countries can be considered pristine. Unless they lie in remote or otherwise inaccessible regions most lotic systems become impacted by agriculture, urbanization, and industrial usage. The conversion of pristine streams to urban streams occurred centuries ago in Europe and much of North America. Historically, people moving into an area built near streams for access to drinking water, fishing, irrigation water, water to turn mills, etc. Even centuries ago human use led to significant alterations in stream hydrology through channelization and construction of water intakes for mills and other facilities (Groffman et al. 2003) and further led to massive in-stream sedimentation from damming for millpond creation (Walter and Merritts 2008). As urbanization intensified to modern times, often these original human uses of the stream became lost and many urban steams became little more than waste depositories. Modern times has brought with it urban sprawl, the outward movement of development from an urban center into rural areas that envelopes and alters streams, a process that is ongoing.

In the USA the urban sprawl process is very evident in the southeastern states, especially along tidal creeks of the Coastal Plain but also in the Piedmont and mountain regions. Urbanization and consequent stream impacts are also commonly seen in tropical areas of Florida where mangrove swamp forests are being lost, and in western mountain states, which have seen an influx of people from California, among other states. In these areas relatively unimpacted streams are primarily being affected by new housing developments as well as recreational facilities such as golf courses and marinas. Stream change due to land alteration continues to occur to a major degree in South America, largely through road building, agriculture, ranching, and mining, where much valuable rainforest (along with its carbon sequestration function) is also being lost (Wasserman et al. 2003; Escobar 2020). The following discussion will focus on the urbanization of unimpacted streams.

## 15.1 What Happens to a Stream During and After Watershed Urbanization?

### 15.1.1 Clearcutting, Hydrology, and Erosion

When people move into an unimpacted watershed for any purpose, be it farming, housing, or industry, invariably the first thing to occur is that the forest is **clearcut**. This leads to several immediate impacts to nearby surface waters, as well as a cascading series of impacts to the stream and watershed (Table 15.1). When trees are removed their water removal function of transpiration is lost, so more rainfall becomes runoff once the soil infiltration capacity is reached. A denuded area of the watershed will cause rapid runoff of stormwater (at a much greater volume) into the stream. This

*River Ecology*. Michael A. Mallin, Oxford University Press. © Michael A. Mallin (2023). DOI: 10.1093/oso/9780199549511.003.0016

**Table 15.1** Physical changes expected to stream watersheds during and following the clearcutting and urbanization process.

**On-land changes**

Loss of riparian forest, generally as a result of clearcutting

Increased stormwater runoff

Increased soil erosion resulting in downslope movement of suspended sediments

Road construction with hydrological change and runoff pollution

Increased cover by impervious surfaces (roads, roofs, sidewalks, parking lots, compacted soils)

Decreased infiltration of rainwater through watershed soils

Installation of sewage and septic infrastructure (subject to spills and leaks)

**In-stream changes**

Changed hydrograph: faster, larger, and more frequent peak flows

Increased stream suspended sediment and turbidity concentrations

Formation of mid-channel bars from excessive sediment inputs

Incision and widening of stream channel through bank erosion and undercutting

Animal habitat loss from scouring by floods and sediment loading

Consequent loss of woody debris supply to stream

Increased water temperature from riparian shade loss and urban heat island effect

Lowered stream base flow from less infiltration of rainwater and groundwater recharge

Decreased stream interaction with the floodplain

changes the stream hydrograph by causing faster and steeper peak flows (see Chapter 1). The rapid changes in the hydrograph are often termed "flashiness" (Arnold and Gibbons 1996).

One of the ecological functions of trees is to protect against erosion through a variety of means. Their roots anchor the soil, but following clearcutting stumps are removed to make room for future structures. Normally, trees, especially mature ones, intercept rainfall on their leaves, branches, trunks, and epiphytes, greatly reducing the amount reaching the ground (which is called throughput) during rainfall (Asadian and Weiler 2009). However, following a clearcut the tree canopy no longer exists so rainfall strikes the soil unimpeded rather than buffered by the canopy, thus having a greater impact on loosening the soil and further contributing to erosion. The increased runoff combined with the looser soils increases surface runoff and

subsequent sedimentation of the stream. This is readily visible to anyone who visits a watercourse downstream of a newly clearcut area—along with visible turbidity in the water sediment can be seen a covering of the natural stream bed and its associated vegetation (Fig. 15.1a), whereas in an unimpacted nearby stream a variety of boulders and cobbles can be seen with associated clear water and periphyton (Fig. 15.1b).

Additionally, following a clearcut the N and P assimilation function by the trees no longer occurs and elevated nutrients enter the stream. Presumably the disruption of the soil leads to the remineralization of organic nutrients into usable forms as well. For example, following a clearcut along a Coastal Plain stream, TN, TP, and orthophosphate increased significantly. This combined with increased sunlight reaching the stream from canopy loss led to a significant, seven-fold increase in average stream chlorophyll $a$, and the stream suffered unprecedented cyanobacterial blooms for two summers following the logging (Ensign and Mallin 2001). The decomposing algal blooms in combination with organic inputs from clearcut runoff led to subsequent significantly lower stream DO over a two-year period, including periods of anoxia.

Tree removal itself has severe consequences for aquatic habitats. Trees provide shade and cool stream waters in summer, and during autumn leaf fall they provide small particulate organic material that serves as food to aquatic organisms (see Chapter 6). Trees also provide stream habitat in the form of a regular supply of larger woody parts (snags) as substrata for microbes, periphyton, invertebrates, and fish (Forman et al. 2003), especially in soft-bottomed streams. The presence of large woody debris can also buffer the stream biota from high flood velocities generated downstream of clearcuts because it provides eddies where fish can rest (Finkenbine et al. 2000). Whereas there may be an immediate input of woody debris following the clearcut, over time the loss of forest, particularly riparian forest, has many negative impacts on stream biota. The presence of the streamside forest thus imparts multiple benefits to a stream. In a comparison of deforested versus forested reaches on a number of northeastern stream Sweeney et al. (2004)

(a)                                                                (b)

**Fig. 15.1**  Fig. 15.1a (left) North Carolina Piedmont stream covered by clay sedimentation from upstream construction project. Fig. 15.1b (right) Stream draining adjoining forested watershed 1 km from impacted stream. (Photos taken on same trail on same day by the author in Lake Johnson Park, Raleigh, North Carolina.)

found that the forested reaches had more invertebrates, greater habitat per unit reach, greater N processing rates, and higher metabolism than the unforested reaches.

## 15.1.2 Road Building and Stream Impacts

Following the clearcutting, road construction commences, which itself contributes to stream degradation in numerous ways. Carving the landscape to proper grade and bringing in fill dirt to support road beds exposes more loose soil to rainfall, with consequent runoff and increases in stream TSS and turbidity. Road construction can create major changes in an area's **hydrology**. As mentioned above, cutting and removing of trees will eliminate their water removal service of transpiration of water vapor to the atmosphere, leading to more flooding and stormwater runoff. Roads that are then constructed intercept surface streamflow and subsurface (groundwater) flows and reroute the water into roadside ditches. The ditches carry the water into streams that did not originally receive the water. This can physically fragment natural habitats, reduce groundwater recharge in a given area by channeling rainwater elsewhere as stormwater runoff, and alter existing surface water bodies hydrologically or ecologically through increased stormwater inputs. Stormwater inputs from roadside ditches often carry elevated concentrations of suspended sediments, organic debris, N and P, fecal bacteria, heavy metals, and organic chemicals into

receiving surface waters (Wu et al. 1998; Forman et al. 2003; Davidson et al. 2010).

Many stream organisms as well as organisms depending upon streams for some part of their life cycle or daily regimen require connectivity between the aquatic areas they use and nearby terrestrial areas. Roadways, especially large roads, can divide wildlife areas and create habitat fragmentation in the now-smaller wildland portions (Forman et al. 2003). This has caused documented losses in local biodiversity through an impact that has been termed "habitat split" (Becker et al. 2007). This is a human-induced disconnect between habitats that are used by different life history stages of a species. For instance, amphibians that live in the forest but breed in streams or rivers have to pass through denuded areas (clearcuts), cross roads and polluted areas, etc. to reach the stream. Field work has shown that amphibian species richness in Brazil varies inversely with the degree of habitat split (Becker et al. 2007). This is not only true of wildlife but of plant communities and birds as well. Connectivity is important as it allows for continued reproductive success and genetic diversity of the various native species that populate an ecosystem.

Stream crossings are areas where some of the most severe impacts of road construction are seen (Waters 1995). Heavy equipment will cause direct physical damage to stream beds and banks during bridge construction and bank shaping. The in-stream and streamside construction will cause elevated turbidity and increased sedimentation of

**Fig. 15.2** Heavy sedimentation into stream caused by construction. (Photo by the author.)

the channel (Fig. 15.2). Earlier (see Chapter 1) the impacts of suspended sediments and turbidity to lotic systems were detailed. In regards to clearcutting and road building these are briefly summarized below. Suspended sediments will interfere with shellfish filter-feeding by clogging gills and causing them to reject large amounts of particulates (called pseudofeces) and spend way too much energy in feeding to survive (Rothschild et al. 1994). It has been demonstrated experimentally that increasing turbidity will lower the feeding rates for sight-feeding fish such as bluegill (Gardner 1981) as well as feeding rates of zooplankton (Kirk 1991) which are the prey items for most species of larval and juvenile fish. Elevated turbidity will reduce light penetration through the water and lead to reduction of rooted aquatic plant photosynthesis and loss of SAV beds (Dennison et al. 1993). Large inputs of suspended sediments will change the natural bottom habitat by coating gravel or vegetated sediments with dirt and organic particles (Fig. 15.1a) and smothering the benthic microalgae (Cahoon et al. 1999), which are the major food resource for a myriad number of aquatic organisms. In coldwater

fish habitats suspended sediment from land construction can smother the gravel beds needed for salmonid spawning (Waters 1995) and in tidal creek areas such sediment loads can smother the available oyster reef habitat. Suspended sediment particles accumulate fecal bacteria, phosphate, and other pollutants and transport them downstream into shellfishing waters and recreational waters; strong correlations have been found between pollutants such as fecal bacteria, phosphate, and BOD with turbidity and/or suspended sediments (Sayler et al. 1975; Baudart et al. 2000; Mallin et al. 2009).

Paved roads are **impervious surfaces**, which are surfaces that do not permit rainwater to infiltrate into the underlying soils. Rainfall striking a roadway will wash accumulated materials, including pollutants, into roadside ditches as stormwater runoff. What are some of the most common pollutants that come from road use and what are they generated by? Chemicals on roads originate from roadway construction material, application of chemicals to roads and roadsides, and vehicle-sourced material. Included in the latter (Fig. 15.3) are oil, grease, hydraulic fluids, tire

**Fig. 15.3** Petrochemicals and trash entering storm drain leading to urban stream in Wilmington, North Carolina. (Photo by the author.)

abrasion particles, brake lining abrasion particles, metals plating and rust, and metals including zinc, lead, copper, chromium, and nickel (Muschak 1990; Pitt et al. 1995; Forman et al. 2003). Some breakdown products include monocyclic aromatic hydrocarbons and PAHs; the latter are known to be toxic to aquatic life (MacDonald et al. 2000). High concentrations of suspended solids are also produced from highways and urban environments (Wahl et al. 1997; Wu et al. 1998; Gobel et al. 2007) and can impact nearby aquatic communities, as outlined above. The various pollutants leave a chronic legacy; for instance, strong gradients of heavy metals are found outward up to 100 m from roadways in soil, air, and plants (Forman et al. 2003). Additionally, chemical spills occur on highways. While such spills are rare for a given area, when they occur they can cause severe pollution damage to terrestrial and aquatic communities. All of these pollutants can accumulate in receiving streams, lakes, marshes, and ponds and become toxic to aquatic life (MacDonald et al. 2000; US EPA 2000).

Pertinent to stream eutrophication is the high amount of N and P that comes from road runoff (Wu et al. 1998). N in vehicle exhaust is emitted as nitric

oxide, nitrogen dioxide, nitrous acid, and ammonium. A study of roads in Cape Cod estimated that 10 kg of dissolved N were produced annually per hectare of roadway in stormwater runoff (Davidson et al. 2010). Along with these vehicle-produced pollutants are any airborne pollutants that happen to become deposited on the road. In a natural area, nutrients and other airborne pollutants are adsorbed or absorbed and processed by vegetation, soil, and the resident bacteria within soils. However, they become concentrated on roadways and right-of-ways, are then carried off the surfaces in a rainstorm, enter streamside ditches, and are carried into creeks and streams. There these nutrients and other pollutants disrupt the natural stream communities, as outlined above.

### 15.1.3 Stream Impacts from Area-Wide Development

Following or even simultaneous with road construction is the development of the adjoining area into residential, commercial, or industrial uses. During construction there is continuous earth-moving, leading to more bare earth as well as disturbed

**Fig. 15.4** Soil compaction following a clearcut and heavy equipment usage, three days after a rain event. (Photo by the author.)

earth and subsequent stream damage (Figs. 15.1a; 15.2). Construction activity leads to large increases in suspended sediments to streams (Waters 1995; Arnold and Gibbons 1996; Line et al. 2002). During the development process much of the watershed becomes increasingly covered by impervious surfaces. Homes and businesses are built and roofed, and parking lots and sidewalks are constructed and surfaced in concrete or blacktop. In general, watersheds with less than 10% impervious cover are considered lightly developed. Residential watersheds are rather heavily developed by the time the watershed contains more than 20% impervious coverage, and industrial areas may have impervious coverage exceeding 50% coverage.

Some unpaved areas also become impervious over time. During construction the earth-moving process at first disturbs and loosens the soil, making it amenable for runoff of suspended solids. However, where heavy equipment continually rolls over bare areas such activity eventually leads to severe **soil compaction**, with the ground becoming

essentially impervious (Fig. 15.4). All of this development activity greatly **decreases** infiltration of rainwater and **increases** surface runoff. Total impervious surface coverage, or total impervious area (TIA), is considered to be an excellent indicator of the degree of urbanization within a watershed (Griffin et al. 1980; Schueler 1994; Arnold and Gibbons 1996).

With the large increases in impervious coverage the stream hydrograph is further altered by the establishment of the **stormwater drainage system**. In urbanized areas the common practice for many decades has been to usher rainwater (stormwater) off the streets, yards, and parking lots as rapidly as possible and dispose of it downslope into a surface water body. As such, a system of concrete storm drains along roadways and parking lots was devised as the most common practice—these are visible everywhere in urban and suburban areas. In many areas there are municipal codes dictating the standard design of such storm drain systems. The vast amount of impervious surface area in urban areas ensures the rapid transit of stormwater into the drain system. In terms of the stream hydrograph, peak discharge increases and occurs much more rapidly that when the watershed was forested; i.e. discharge is much higher in streams draining developed areas than undeveloped areas (Gold et al. 2019). A naturally vegetated stream should have runoff of approximately 10% of rainfall; an increase of 10–20% TIA causes runoff to double to 20%; a 35–50% increase causes it to triple to 30%; and a 75–100% increase leads to a runoff increase of 55% (Arnold and Gibbons 1996).

Flooding from developed areas makes the habitat for stream biota much more unstable; note that the clearcut itself already caused higher discharge and flashiness and the land development intensifies this. The increased stream water velocity and volume caused by the greater stormwater runoff inputs leads to greater channel widening and deepening (incision) because the stream is forced to handle much greater stormwater discharge (Fig. 15.5), and more downstream flooding occurs (Schueler 1994). The degree of stream incision has been strongly significantly associated ($R^2 = 0.79$, $p < 0.05$) with degree of watershed impervious surface coverage (Hardison et al. 2009). Flashiness (the rapidity of change in the hydrograph) increases notably (Arnold and Gibbons 1996; Holland et al. 2004; Cuffney et al. 2010). During this process there is severe erosion of banks and bottom sediments well downstream of the construction area; thus in urbanized streams

suspended sediment loads can be high (Lenat and Crawford 1994). Riparian trees are undercut by the downstream erosion and fall in (Fig. 15.5a); there is loss of riparian vegetation and habitat, and loss of shade leading to stream water temperature increases. The increased erosion also destabilizes habitat for stream biota.

An unarmored "natural" stream will physically adjust itself to compensate for the increased flow brought on by urbanization, given sufficient room to expand. It will reduce sinuosity and enlarge itself through bank erosion and deepen through incision until an eventual "urban equilibrium" is reached (Fig. 15.5a, 15.5b). This will occur after the watershed is developed as much as it is going to be (Finkenbine et al. 2000). The impervious surface coverage coupled with stable landscaping will cause the inputs of fine sediments to diminish. However, if the floodplain is limited by excessive near-stream urbanization then such lateral expansion is of course limited; such streams become culverts with little animal habitat other than the sediments. On occasion urban streams are converted into outright culverts (Fig. 15.6); this of course removes fish and wildlife habitat (Finkenbine et al. 2000). Engineers, with the blessing of town boards, may physically alter urban streams by use of heavy equipment to redirect flows, straighten channels, or even armor banks with concrete (Fig. 15.6). This may be done for "aesthetic" purposes, flood control, to enhance drainage, or simply to move a stream

(a)

(b)

**Fig. 15.5**  Fig. 15.5a (left) Bank undercutting from severe erosion caused by upstream watershed urbanization and increased runoff. Fig. 15.5b (right) Bank collapse caused by stream widening from upstream watershed urbanization in Wake County, North Carolina. (Photos by the author.)

**Fig. 15.6** Armored, engineered urban stream in New Mexico. (Photo by the author.)

that is in the way of a planned structure. This clearly leads to further disconnect with the natural floodplain and loss of riparian habitat.

Another impact of the stormwater runoff system in urbanized areas concerns water table changes. Studies have shown that area-wide stream base flow can decrease because the increased impervious surface area intercepts rainfall that would normally percolate downward and recharge the groundwater aquifer (Klein 1979; Simmons and Reynolds 1982; Finkenbine et al. 2000). The intercepted rainfall is then directed into storm drains (Fig. 15.3) and carried to surface water bodies including streams, urban lakes, and estuaries (Klein 1979). This has been termed "hydraulic drought" and can change soil chemistry and the type of riparian vegetation toward more upland species (Groffman et al. 2003). The decrease in base flow can concentrate pollutants or reduce the flushing and dilution effect created by a significant base flow. Reduced base flow will degrade fish habitat by leading to shallower water, reduced pool volume, and less physical space

(Finkenbine et al. 2000). In some areas this reduction of base flow may be offset to a degree if there is significant leakage in sewage infrastructure or water distribution systems, which often import water into an area from another watershed (Walsh et al. 2005a). A review by O'Driscoll et al. (2010) also notes that under developed conditions groundwater recharge can be "redistributed" through leaking pavements and leaking infrastructure, and in "greener" developed areas groundwater recharge can occur through best management practices (BMPs).

Another urbanization impact that can impact stream fish habitat is the urban "heat island" effect; this is caused by the loss of shade from riparian forest loss coupled with heat absorption by concrete and other construction materials. Lowered stream base flows from urbanization can also enhance stream heating (O'Driscoll et al. 2010). The net effect will raise both day and night-time water temperatures, impacting micro- and macrobiota physiology, and alter moisture flux from the surface (Grimm et al. 2008). Stream water temperature has been positively correlated with degree of watershed urbanization (Hatt et al. 2004).

Increased urbanization leads to increased **water pollution** for several reasons. First, with urbanization the communities will need to treat their sewage by 1) constructing a central sewage system, or 2) utilizing smaller package treatment plants, or 3) relying on septic systems (detailed in Chapter 13). Invariably there will be wastewater treatment system or septic system malfunctions, with associated fecal microbial, BOD, and nutrient loading into the nearest stream (US EPA 2002; Mallin et al. 2007; Tavares et al. 2008). The second urbanization process leading to increased water pollution will occur if ditching and draining is performed within the watershed to permit more development in areas with a high water table (often ecologically valuable wetlands are impacted). This can lead to increased pollution of downslope streams especially in areas where septic system usage is common (Cahoon et al. 2006). The third reason, which is exceedingly widespread, concerns pollution by urban stormwater runoff. Unfortunately, the imperviousness character of urban areas also ensures that most of the stormwater will flow untreated into the stormwater collection system (Fig. 15.3) and from there into receiving waters. Some of these waters may be detention ponds or other stormwater control devices (see Chapter 16), whereas others will be

public streams, lakes, and rivers. During dry periods between rain events, polluting materials will collect on impervious surfaces. When rain arrives, the runoff thus contains a concentrated dose of pollutants that are carried into the storm drain system (as discussed above; Fig. 15.3) and in many cases from there into urban streams and lakes and estuaries.

Thus, hydrological changes, loss of habitat, increased water temperatures, and water pollution significantly degrade the biotic communities of urban streams relative to forested streams or even agricultural streams (Karr 1981; Garie and McIntosh 1986; Lenat and Crawford 1994; Cuffney et al. 2010). The amount of urbanization is well reflected by the amount of total impervious surface coverage within the stream's watershed. Early work on the impacts of impervious surface coverage in stream watersheds provided a simple and useful model showing expected impacts of increasing impervious surface areal coverage on freshwater streams (Schueler 1994; Arnold and Gibbons 1996). This model described stream health "breakpoints" for TIA at approximately 10% (between unimpacted and impacted biotic communities) and approximately 30% (between impacted and severely impacted biotic communities) based primarily on fish and benthic macroinvertebrate community indices (Table 15.2). An estimator that

**Table 15.2** Biological and chemical impacts on stream ecosystems determined to occur at various levels of watershed impervious surface.

| Impervious surface Coverage percent | Impact on the ecosystem |
|---|---|
| 0–8% | Highly variable fish community health (Wang et al. 2001) |
| 0–10% | Some deterioration in benthic community (Booth et al. 2004) |
| 5–10% | Deterioration of benthic community (Cuffney et al. 2010) |
| 10% | Closures of tidal creek shellfish waters due to fecal bacteria counts (Mallin et al. 2000) |
| 6–14% | Deterioration of benthic community (Walsh et al. 2005a) |
| 8–12% | Fish community health threshold (Wang et al. 2001) |
| 14% | Fish community health decline (Miltner et al. 2004) |
| 12% | Significant decrease in fish diversity (Klein 1979) |
| 12% | Poor quality fish community (Wang et al. 2001) |
| 27% | Severe decrease in fish community health (Miltner et al. 2004) |
| 20–30% | Deterioration in tidal creek benthic community (Lerberg et al. 2000; Holland et al. 2004) |
| 20–30% | Severe fecal microbial pollution in tidal creeks (Mallin et al. 2000; Holland et al. 2004; Sanger et al. 2015) |
| 40% | Significant chemical pollution of tidal creeks (Holland et al. 2004) |
| 40–50% | Major increase in stream TN and TP (Sanger et al. 2015) |
| 40–50% | Major COD increase |

is even more precise in explaining the negative impacts on the stream environment has been termed effective impervious area (EIA) (Hatt et al. 2004; Walsh et al. 2005b) or connected imperviousness (Wang et al. 2001). This refers to the amount of watershed impervious cover that is connected by pipes or solid surfaces to a receiving stream. Even at relatively low levels of impervious coverage (< 10%) significant impacts can be had on streams from runoff if the impervious coverage is highly connected to receiving streams by pipes and lined channels (Hatt et al. 2004). Other stream health impacts not measured by percentage TIA can include leachate of chemical pollutants into stream waters, proximity of wastewater treatment plant outfalls (Karr and Chu 2000), leachate from septic systems, width of vegetated buffers, or other such factors that can lead toward stream degradation at < 10% TIA. Regardless, TIA is by and large a strong indicator of stream ecosystem health in most situations. Total impervious coverage can be estimated by several means (Capiella and Brown 2001); however, the watershed coverage by effective or connected impervious coverage is more difficult to determine than total impervious coverage in a catchment (according to the National Research Council [NRC 2009]).

The health of stream biotic communities, principally those of benthic invertebrates and fish, is frequently used to assess overall impacts of urbanization on streams. Since benthic organisms are at least relatively restricted to small areas of streams, they make good candidates for such analyses, and researchers commonly use measures such as overall abundance, presence or absence of pollutant-sensitive species, number and abundance of stress-tolerant species, species richness, species diversity, and other indices such as EPT taxa richness (see Chapter 5). Benthic invertebrate identification to species level requires a good deal of skill and experience to master. Fish identifications, however, generally require less training. Fish generally have a lot of life history information available, represent a range of trophic levels, have toxicity information available, readily display disease or other damage, and are well visualized by the general public; thus their community health is an excellent assessment tool (Karr 1981; Karr and Chu 2000). An IBI was

developed that utilizes fish community attributes including number of species, presence of tolerant and intolerant species, proportion of hybrid species, omnivore versus insectivore proportions, populations of top carnivores present, and number that are diseased or damaged otherwise, as well as several other parameters (Karr 1981). The IBI has been adapted to many regional systems and is commonly used in urbanization impact studies, as in Table 15.2 (see also Karr and Chu 2000; Miltner et al. 2004).

Table 15.2 indicates that in terms of biotic communities a considerable amount of data indicates significant degradation in the range of 5–12% impervious surface cover, for both benthic invertebrates and fish communities. Such changes may be due to a combination of factors including sedimentation, flashiness, temperature, DO changes, and chemical pollution. Increased impervious coverage has also been directly linked to animal habitat factors including lowered base flow in streams (about 12% EIA; Wang et al. 2001) and major disappearance of large woody debris (about 20% EIA; Finkenbine et al. 2000). Significant deterioration in tidal creek benthic communities in terms of sensitive species lost and diversity decreases occurs at a somewhat higher impervious cover level (20–30%), possibly due to tidal action enhancing pollutant flushing. In terms of human health a breakpoint occurs centered in the neighborhood of 10% impervious coverage. Above this percent coverage shellfishing beds in tidal creek areas become closed to harvest from elevated fecal bacteria counts. When a watershed reaches > 20% impervious coverage major fecal microbial pollution is present in tidal creeks. Above 40% significant chemical pollution is present in fresh and tidal streams, including elevated nutrients and COD, and metals and chemical contaminants. There is some evidence that the amount of TIA in close proximity to the stream has an especially strong impact on stream health. Wang et al. (2001) found that impervious surface coverage within 100 m of the stream was more important in impacting fish community health than urbanization farther away from the stream corridor. Thus, the extended riparian zone is critically important to urban streams as littoral habitat, a continuous supply of large woody debris, and an immediate source of runoff pollution.

What are some of the problematic pollutants originating from different types of urban areas? In terms of direct human health threats, fecal bacterial pollution has been documented as especially high in residential areas compared with commercial or industrial areas, especially lawn and feeder street runoff (Bannerman et al. 1993). The fecal bacteria (and presumably viruses and protozoans) most likely originate in pet manure and manure from urban and suburban wildlife (deer, raccoons, rats, gulls, etc.). Additionally, urban storm drain systems have been determined to serve as reservoirs for fecal bacterial survival and even serve as areas where such microbes can multiply (Marino and Gannon 1991); thus these pollutants can be flushed out into nearby streams in the next rainstorm.

Suspended solids have been found in high concentrations in road and parking lot runoff (Bannerman et al. 1993; Pitt et al. 1995; Wu et al. 1998; Gobel et al. 2007). Suspended solids of course will carry other pollutants with these particles by absorption or adsorption; as such, TSS and turbidity concentrations in urban streams have been strongly correlated with fecal coliform bacteria, BOD, and TP (Mallin et al. 2009). Residential lawns have also been cited as important sources of P in runoff (Bannerman et al. 1993), likely due to fertilizer use. An issue very relevant to areas receiving snow is stream salinization due to washoff of road salt applied to de-ice roads; in northern streams and rivers road salt deposition in winter will lead to increases in sodium and chloride following melting and runoff (Interlandi and Crockett 2003). Researchers in the US Northeast (Kaushal et al. 2005) discovered that in winter urban streams contained up to 100 times the chloride concentrations of forested streams, and in watersheds with impervious surface coverage exceeding 40%, mean annual chloride concentrations exceeded tolerance levels for freshwater biota.

Scrap and recycling yards and vehicle maintenance yards have been found to be sources of elevated BOD and COD in runoff; this has been attributed to the considerable heavy equipment use at such facilities and spillage of oil and grease (Line et al. 1996). BOD and COD can also be found in high concentrations in runoff from heavily trafficked roads (Gobel et al. 2007). Urban stream BOD5 and BOD20 have been statistically correlated with rainfall, indicating its susceptibility to stormwater runoff (Mallin et al. 2009). Both DOC and POC are sources of BOD in stormwater runoff, with more particulates from urbanized watersheds (McCabe et al. 2021). The BOD load presumably comes from petroleum products and organic yard waste; such waste can be large, but note also that grass clippings (which are ubiquitous) decompose after mowing and break into smaller pieces, as does much human-sourced street trash. A rain pulse will send these organic particles into urban streams where they exert a biochemical demand as bacteria and fungi decompose such organic waste in the aquatic environment. Interestingly, the correlation between rainfall and BOD20 was stronger than the correlation with BOD5, suggesting that much of the material was not immediately labile—consistent with yard waste source rather than sewage or septic leachate. In contrast, the urbanization process leads to lower CDOM, which is largely refractory, in favor of more labile DOM (Wahl et al. 1997; Gold et al. 2020). CDOM is largely a product of natural vegetation leachate, which is lost with deforestation.

Metals pollution in urban areas can be from industrial sources (Line et al. 1996). Relevant to that is that roofs in commercial and industrial areas can be significant sources of zinc and copper in runoff (Bannerman et al. 1993; Pitt et al. 1995; Gobel et al. 2007). Collectively, zinc and copper are the metals found in greatest abundance on roads, with lead less so following introduction of unleaded fuels. Urban or industrial sources have been associated with elevated tidal creek sediment metals concentrations (Sanger et al. 1999a).

Organic pollutants are also common in urban areas. PAHs (see Chapter 12) are sourced from autos, asphalt, solvents, spills, etc. (US EPA 2000) and have been found in high concentrations in runoff from parking lots (Fig. 15.3) and roads (Gobel et al. 2007). Many airborne pollutants settle on impervious surfaces and are flushed into storm drains and urban streams by runoff (Klein 1979; Garie and McIntosh 1986). PCBs and other airborne pollutants are subject to airborne deposition on impervious surfaces; PCBs in estuarine sediments and fish have been strongly linked to various measures of anthropogenic development in watersheds (Sanger et al. 1999b; King et al. 2004). Other potential and less-investigated urban pollutants that may accumulate in urban areas and potentially enter streams include pesticide and herbicide loading from yards, gardens, landscaping, and golf courses.

Runoff from several different source areas was subject to bioassay testing to determine toxicity to bacteria (Pitt et al. 1995); these tests found the highest toxicity in runoff from industrial storage and parking areas; elevated toxicity was also found in runoff from landscaped areas—potentially indicating pesticide runoff. The impacts of environmental estrogens on aquatic communities are not fully known, but evidence points toward potentially significant deleterious effect (Colborn and Thayer 2000). Thus, it is important to point out that a study in the Myakka River, Florida attributed estrogenic activity to inputs of stormwater runoff, possibly in combination with septic leachate (Cox et al. 2006). Urbanization is likely to cause elevated stream concentrations of contaminants not yet fully understood or anticipated.

Thus, during and following the urbanization process the native stream biota are subject to stress from increased sedimentation, flooding, in-stream erosion and scouring, loss of complex habitat, increased water temperatures, hypoxia, and chemical pollution (Table 15.3). How do these pressures

**Table 15.3** Stream water quality and ecosystem responses to watershed urbanization. (Expanded and modified from Walsh et al. 2005a.)

Increased N and P loading to stream

Increased phytoplankton biomass from nutrient loading and loss of shade

Increased BOD loading to stream from stormwater inputs and algal bloom decay

Lower DO minima

Increased fecal microbial loading from runoff and leaking sewage/septic infrastructure

Increased surfactant concentrations

Increased PAH and other toxicant concentrations in sediments

Increased metals concentrations in sediments

Loss of pollution-sensitive fish and benthic invertebrate taxa

Increase in pollution-tolerant benthic invertebrate taxa

Decrease in stream fish and benthic invertebrate diversity

Increase in non-native species through human additions to stream biota

affect stream biota? There will be an overall reduction in stream benthic invertebrate diversity and abundance, with a reduction especially of sensitive taxa such as Ephemeroptera, Trichoptera, Plecoptera, and many mussels; concurrent with this loss there will be an increase in tolerant taxa such as oligochaetes (worms) and chironomids (midge larvae) (see Chapter 5). With the habitat changes and water quality degradation sensitive fish species will be lost. Fish communities in urbanizing streams suffer reductions in diversity, shifts in trophic structure, changes in variability, and increases in herbivorous fish due to increased habitat instability, loss of spatial heterogeneity, and losses of refugia from harsh conditions as a result of urbanization (Schlosser 1995). The natural stream biota is likely to be even further stressed by a flux of **invasive species** (fish, macroinvertebrates and macrophytes) stemming from human dumping of aquaria as well as deliberate dumping into the stream of fish caught elsewhere (see Chapter 14).

The nutrient loading into urban streams from fertilizers, road runoff, animal manure, airborne nitrogen, autos, sewage, and/or septic leaks and spills makes urban streams subject to algal blooms (Mallin et al. 2009; Gold et al. 2019). These blooms are exacerbated by loss of streamside shade, and cyanobacterial blooms can thrive in warm waters (Burkholder 2002) provided flow is low. Bloom formation typically also leads to reduction of algal species diversity in favor of bloom monocultures. With the increase in impervious surface coverage there will be lower DO minima as well as wider DO fluctuations (Holland et al. 2004). This will occur because urban stormwater runoff is a direct source of BOD to streams and algal blooms will turn into BOD upon death and decay (Heiskary and Markus 2001; Mallin et al. 2006, 2009).

**Streamside recreation** in urban areas typically consists of city parks, which may or may not strongly impact the stream ecosystem. A passive "nature park" can preserve riparian areas and buffers and be ecologically informative to the public with proper signage and trails. However, an excessively manicured park can have fertilized grass and ornamental vegetation, impervious areas such as tennis courts that drain runoff into the stream, lenient attitudes toward dog manure pick-up, and overzealous management that mows down to the water's edge. All of these practices can impact the steam by nutrient and fecal microbial runoff and loss of streamside habitat. Thus, it is a task for local biologists and urban planners to convince city fathers (and mothers) to utilize as light a human touch as possible to make urban parks an asset to stream ecosystem health rather than a detriment. A park that is located downstream of a significantly urbanized area may unfortunately have to deal with the consequences of the "urban syndrome" (Fig. 15.7), with severe erosion and polluted water.

One of the most popular forms of recreation is golfing, and golf courses are designed for aesthetics as well as providing a challenging course to play. Numerous golf courses have been built along the shores of (and sometimes incorporating) streams and tidal creeks, rivers, and estuaries, i.e. very scenic areas. Runoff from golf courses, especially greens and fairways, can produce copious amounts of nitrogen and phosphorus as fertilizers that enter on-course or near-course streams (Kunimatsu et al. 1999; Line et al. 2002), and the amount that reaches the stream or estuary and contributes to algal blooms depends largely on the type of runoff controls present on the course (Mallin and Wheeler 2000). In South Carolina, nutrient-rich golf course ponds have hosted a variety of toxic algal species and had fish kills (Lewitus et al. 2003). Golf courses also produce runoff of pesticides and herbicides that have the potential to adversely impact stream and estuary aquatic life; this avenue of research needs expansion.

## 15.2 Riverside Development Issues

Larger rivers are subject to the same environmental insults from urbanization that are experienced in the smaller streams discussed above (Tables 15.2 and 15.3; Fig. 15.8). Additionally, along the shores of rivers (stream order 4+) are large human structures that can strongly impact river ecosystems and hydrology, including dams and canals (discussed in Chapter 10) and industrial water withdrawals (such as power plants) and industrial pollution outfalls (discussed in Chapter 12). Many large rivers worldwide host significant barge traffic of industrial and agricultural goods; in the USA this is an important economic enterprise within rivers such as the Mississippi and its major tributaries. Such aquatic commerce requires river dredging and construction of numerous lock structures that alter flow patterns (White et al. 2005) and can impact river biota (Wehr and Thorp 1997).

Rivers of course also host numerous ports (sometimes massive ones) as well as shopping, dining, and residential riverside structures. Ports are found in estuaries and rivers and represent major shoreline alterations (Fig. 15.9). Some ports have been in existence for centuries, such as ones serving the great cities of Europe; some in North America are relatively old as well. Major ports in developed

(a)

(b)

**Fig. 15.7** Fig. 15.7a (left) Deeply incised urban stream in Macon, Georgia. Fig. 15.7b (right) Cautionary signage on the same stream after it flows into Ocmulgee National Park, Macon. (Photos by the author.)

**Fig. 15.8** Untreated stormwater runoff draining directly into Cape Fear River, Wilmington, North Carolina. (Photo courtesy of Matthew R. McIver, UNC Wilmington.)

**Fig. 15.9** The Port of Wilmington, North Carolina, showing containers and warehouses sitting on a vast impervious surface area. (Photo by the author.)

countries undergo periodic upgrades, and in some developing (and developed) nations new ports have recently been built or are planned. Large, international ports can be located in essentially marine waters (such as the Ports of Miami, Boston, and New York), in estuarine systems (such as the Ports of Charleston, Savanna, and Baltimore), in riverine areas (such as the Port of Philadelphia), and in freshwaters (such as in the North American Great Lakes and along major European rivers such as the Danube, Seine, and Rhine). Container vessels are major users of international ports. These ships carry numerous 20-ft-long metal containers stacked aboard; many pass through the Panama Canal. The sizes of these vessels are continually increasing; thus ports have to expand to handle the size and draft of the vessels or else new ports must be constructed. Container ships that traverse oceans are ever-increasing in size. The newest breed are called "Post-Panamax" vessels, which are up to 1260 ft long and have a 185-ft beam; their 50-ft draft requires channel depths of approximately 54 ft (CH2M Hill 2008). Thus, for most ports to be able to handle these new vessels the present ship channels will need deepening and widening; when a port is located many kilometers upriver from the ocean such dredging can be a massive undertaking. The next section will discuss how construction of ports impacts river (or estuary) ecology. This will be followed by a discussion of day-to-day impacts of shipping and port operations on aquatic ecosystems.

## 15.2.1 Impacts to Rivers of Port Construction and Operation

Ports contain administrative facilities, warehouses and other storage structures, rail yards, access roads, the dockage facilities, and vast areas of impervious surfaces upon which vehicles are parked and driven and containers are stored (Fig. 15.9). Thus, constructing a port involves causing major changes to the riparian landscape. Pertinent to riverside construction, riparian areas host associated wetlands such as intertidal marshes in estuarine areas and riparian forests in freshwater areas. As will be elaborated in Chapter 16, wetlands are key habitats for a huge variety of organisms.

Wetlands serve as roosting spots for birds and nesting areas and foraging areas for many species including endangered species such as alligators; if fresh they provide water for drinking to numerous animals both resident and visiting; and all types of wetlands serve as spawning or nursery areas for numerous species of fish and shellfish. Port construction removes these riparian wetlands and their ecological functions permanently and covers these ecosystems with impervious pavement and bulkheading.

The construction of a large port complex will displace large numbers of animals, including birds, mammals, reptiles, and amphibians. Bird habitat on the site will be converted from supporting a natural community of seed-eaters, wading birds, aerial-searching birds, and probing shorebirds to a community of scavenging birds, primarily gulls. Resident mammals in undeveloped riparian areas are normally able to utilize several habitats; i.e. they have access to the river, wooded areas, farmed fields, and old fields. This allows for high diversity and ready access to food resources as well as cover. The conversion of a riparian area to a port removes all such habitat and river access, although there are likely to be rats and other scavengers utilizing the site for human leavings.

A natural riparian shoreline will support a plant community that includes benthic microalgae, periphyton on living and nonliving surfaces, and macroalgae, in addition to vascular plants. All of these primary producers are food organisms for grazers such as zooplankton, snails, aquatic worms, insect larvae, crabs and other crustaceans, and fish. Rocks, tree roots, and tree debris (snags) in the shallows provide substrata for attached organisms and places of concealment for fish and other organisms. Birds utilize shoreline areas for hunting and scavenging, as do mammals, reptiles, and amphibians. A port requires a large, sometimes kilometers-long area to be bulkheaded with either concrete or steel. When a shoreline area is bulkheaded for a port facility (or to support any large riverside structure), most of the natural functions of the shoreline are lost for good. A complex habitat becomes a one-dimensional surface that may support some fouling organisms but little else in the way of a community. Research has demonstrated that the nearshore waters of bulkheaded areas contain significantly lower abundance and species richness of fishes compared with natural shoreline areas or riprap shorelines (Waters and Thomas 2001; Street et al. 2004). Additionally, the wharf structure will extend out over the water, shading out any SAV or benthic microalgae in the area and losing that food and habitat resource.

Bulkheading is considered armoring a shoreline. Whether it is done for the purpose of converting a shoreline for a structure (like a port) or to protect human residential areas from erosion it deflects the energy of waves and tidal and downstream currents. However, it diverts that energy to unarmored shoreline areas down-current from the structure, causing erosion (Pilkey 2003). Extreme cases of this down-current erosion are seen especially in Maryland along Assateague Island (Pilkey 2003).

Once a port is constructed there are numerous shipping-related activities that impact the river's hydrology, biotic communities, and water quality. One of the most important and damaging of the environmental impacts of international shipping is **introduction of non-native species** including algae, bivalves, zooplankton, and bacteria through in-port ballast water exchanges (Pimentel et al. 2000; Holck et al. 2004; Burkholder et al. 2007). As of now, in the USA it is states with major ports and abundant water resources that are most frequently and severely impacted by invasive species transferred by ballast water. As noted in Chapter 14, the USA spends $100,000,000/year on control of non-native aquatic plants that impact waters, and much more on control of invasive bivalves that impact water treatment and power plant cooling water intakes and other habitats (Pimentel et al. 2000). Ballast water exchanges in-harbor bring in non-native species to the area, including dinoflagellates and other algae, zooplankton, molluscs, fish, and bacteria. Some of these introduced species create major food-web alterations and large financial problems; the Great Lakes and San Francisco Bay are examples (Pimentel et al. 2000; Ricciardi and MacIsaac 2000).

Shipping activities cause chemical pollution to the aquatic environment. Ships at anchor in ports leak petroleum products into the water. In fact, small-scale oiling in total amounts to eight times that of the Exxon Valdez disaster (Ng and Song

2010). The author of this book has personally witnessed and photographed, while on sampling cruises, container ships at the Port of Wilmington leaking fuel and leaving plumes kilometers long (reported by us to the Coast Guard). Besides petroleum spills and leaks, routine shipping often releases other pollutants as well, including antifouling agents, garbage, gray water, and floating trash (Ng and Song 2010). Of course, metals leach from ships, paints, and ground vehicles and are thus subject to stormwater runoff into the river or estuary.

In addition to water pollution, large oceangoing vessels produce copious amounts of **air pollution** that can contribute to degrading human health. Some of the pollutants released include particulate matter, sulfur oxides, nitrogen oxides, hydrocarbons, and carbon monoxide (Corbett and Fischbeck 2000; Corbett et al. 2007). It has been estimated that shipping-related particulate matter emissions are responsible for approximately 66,000 cardiopulmonary and lung cancer deaths annually worldwide (Corbett et al. 2007). In addition, such emissions contribute to ocean acidification and global warming. While such air pollution is of immediate human health interest (i.e. breathing) it is important to keep in mind that airborne pollutants from ships often return to Earth on local water surfaces or on impervious surfaces where they are subject to stormwater runoff.

Ship channels as well as ship berthing areas require maintenance dredging. With any large river draining developed inlands, large amounts of suspended sediments come downriver (Benedetti et al. 2006) as well as entering ship channels from local sources. As this material will continue to settle in the channel, maintenance dredging is commonly performed. This periodic dredging will spawn periodic river and estuary problems, including increased turbidity and suspended sediments and their effects on benthic invertebrates, benthic microalgae, fish, oysters, clams, and blue crabs, fecal bacteria, and pollution of a fisheries primary nursery area.

Fecal bacteria from stormwater runoff, leaking septic systems, and wastewater discharges that enter water bodies are often carried to the bottom sediments along with sedimenting suspended solids. Channel dredging has long been known to increase the abundance of fecal bacteria in the water column by displacing them from the bottom and bringing them to the surface, and separating them from the inorganic particles that they have been adhering to (Grimes 1975). If shellfishing waters are located downstream of a dredging area the increase in fecal bacteria is likely to cause these areas to be closed to harvest, creating an economic burden on the commercial fishermen who utilize the area.

Blasting is a common technique used to deepen channels if hardpan is present. Blasting can be hazardous to the health of residential fishes. This issue is even more significant if endangered species are present in the area. For example, in the Cape Fear River during 1998 and 1999 there were a number of studies performed to examine the impact of blasting on native species, including striped bass, mullet, and shortnose sturgeon (Moser 1999). Caged fish were located 35, 70, 140, and 280 ft from the test blast areas to study the impact on the fish. In some tests an experimental air bubble curtain was used to mitigate in the hope of lessening blast effects outside of the area (according to the US Army Corps of Engineers [USACOE 2010]). The test results (Moser 1999) showed that mortality of the caged fish was highest in the 35-ft location (26% with no curtain and 28% with the air curtain), next highest at the 70-ft location (24% with no curtain and 7% with air curtain), and less at 140 ft (2% with no curtain and 6% with the air curtain), and negligible beyond 280 ft. Thus, the air curtain did not successfully reduce fish death from blasting. Mortalities were highest among striped bass, second among mullet, and least among sturgeon, probably due to the larger size of the sturgeon. The types of injuries commonly seen on the fish were loss of equilibrium, distended swim bladder, and hemorrhaging (Moser 1999). The air bubble curtains were deemed not effective in those previous analyses and physical barriers were deemed impractical (USACOE 2010). Thus, blasting to increase channel depth brings with it significant physical danger to resident species.

Some endangered species utilize shipping channels leading to ports in rivers and estuaries. Such animals presumably could be impacted by channel maintenance dredging operations or physical encounters with large vessels. An example of this is the federally endangered Atlantic sturgeon (*Acipenser oxyrhynchus*). This fish can grow to

spectacular lengths (up to 15 ft) and weigh up to 800 lb, and adults typically reach 6–8 ft in length and 300 lb. However, a study in the Delaware Estuary by Brown and Murphy (2010) found that between 2005 and 2008 over 50% of the fatalities reported from that system were caused by large-vessel strikes (see Chapter 14). Vessel strikes are also likely to take a toll on other listed species such as sea turtles, manatees, and alligators. At present there is no recommended mitigation for loss of these large endangered species by large-vessel encounters. Use of ultrasound, lights, and odors have been suggested but not tested on sturgeon. Reducing vessel speed is believed to be one strategy that may be effective, as it has worked in the case of marine mammals (Brown and Murphy 2010).

Of course, as indicated above, port areas are prime places for stormwater runoff pollution to enter adjacent waters. When flying over a large port area one thing that is stunningly visible is the vast sea of impervious concrete (or blacktop), upon which containers are stacked and cars and trucks and various heavy equipment are parked and operate, and in many cases tanks and drums of petroleum products, and sometimes piles or stores of organic material (grains, wood chips, etc.), fertilizers, chemicals, and other bulk products waiting for loading onto ships or trucks (Fig. 15.9). These large areas of impervious surface (such as readily seen in the Ports of Philadelphia, Wilmington, Morehead City, and Charleston, for example) concentrate pollutants and provide a source of stormwater runoff to rivers and estuaries. Such runoff may contain PAHs (from cars and trucks and asphalt), fecal microbes, N and P from feces of gulls, rats, and other wildlife, surfactants, P and other chemicals from cleaning compounds, and herbicides and pesticides used on-site. Especially in older ports, runoff from these surfaces is often not treated and flows directly into the receiving waters.

### 15.2.2 Impacts to Rivers of Small- to Medium-Scale Structures

Ports, power plants, wastewater plants, and major industries can be massive structures that may impact a river in many ways. However, populated areas along river shores also contain numerous medium-scale shoreline structures and alterations that change the natural riparian zone and often will continue to have adverse impacts on river biota long after actual construction is completed. Marinas are found on streams from 2nd order on up to major rivers. The boats anchored within produce fecal bacterial pollution from houseboats and other craft that practice legal or illegal boat head dumping (Sobsey et al. 2003; Mallin et al. 2010). Boat repair and painting are standard at many marinas; thus sediments are often contaminated with metals (copper, chromium, arsenic, and zinc) and other chemicals (Riggs et al. 1991, 1993) from paints and other boat repair operations. In addition to marinas, many commercial districts take advantage of river views, and some cities showcase their riversides as tourism destinations. Thus, such areas may contain restaurants and bars, shopping areas, roads and parking lots, and boardwalks, all of which serve the public but alter natural shoreline habitat. These areas can contribute significant stormwater runoff to the adjacent river if not controlled (Fig. 15.8). Trash accumulates in the water below, while restaurants attract gulls and other pests.

Other structures can be found along the shores of smaller rivers, streams, and tidal creeks. Some of the most common shoreline structures are boat ramp areas—boat ramp locations, especially in areas protected from strong tides and currents, maintain higher fecal bacteria counts than non-ramp control areas (Harrington 2007). The sediments contain elevated fecal bacteria counts which release fecal microbes into the overlying water column when sediments are stirred up by wading or boat launching (Harrington 2007). The microbes likely come from defecation in the ramp area by dogs, raccoons, gulls, and other wildlife attracted to trash left by humans. The boat ramp is of course usually impervious and the slope of the ramp is ideal to send stormwater runoff directly into the water. Ducks also often congregate in boat ramp areas and defecate nutrients and fecal microbes. Small boats at marinas are vectors for bringing in invasive species from other systems (on props and in bilges), mainly aquatic weeds and molluscs (see Chapter 14). As mentioned earlier, other shoreline alterations include urban streamside parks; unfortunately too many people allow their pets to defecate

without cleaning it up, leaving the manure (and associated fecal microbes and nutrients) subject to runoff.

Streamside communities often have local or community boat dock facilities (i.e. small docking facilities with less than 10 slips); these can provide some structural habitat for fish. Lumber treated with creosote can leach pollutants into the water, however, and nearby seagrass or marsh vegetation gets some shading. Also, some of the same problems as marinas may occur, but this issue is understudied (Street et al. 2004). Additionally, cumulative effects of several such structures in a given area may be problematic. Individual docks, as well as community docks, can alter natural flow patterns, creating eddies and allowing for the sedimentation of suspended matter including pollutants such as fecal bacteria (Mallin et al. 2010). However, analysis of the direct pollutant impact of individual docks in small tidal creeks found their impact to be generally limited in scope (Sanger et al. 2004).

The aforementioned bulkheads are built to protect shores from eroding in front of waterfront homes of well-heeled waterfront residents. Bulkheading causes currents to wrap around them and increase erosion on the down-current shoreline, making it somebody else's problem. Bulkheads remove the natural sloping and vegetated habitat, and the scouring and turbulence created at the foot of the bulkhead prevents vegetation from reestablishing (Street et al. 2004). Several studies have documented lower relative abundance of fish and invertebrates adjacent to bulkheads compared with unaltered shorelines (summarized in Street et al. 2004). If a shoreline area needs to be protected, a much better alternative to bulkheads is riprap (Street et al. 2004), which provides good habitat for periphyton, grazers, and fish—also vegetation can take root and grow (i.e. mangrove propagules; see Fig. 15.10).

Within the water proper are various structures. Bridges that cross rivers or estuaries can provide habitat for larger fish, and attract fishermen as well. However, their surfaces get polluted by airborne pollutants, metals, and petrochemicals, including PAHs from cars and trucks, which are subject to concentration on the bridge surface and eventual deposition into the water during stormwater runoff

**Fig. 15.10** Mangrove propagule taking root in rock riprap along the shore of a tidal creek in West Florida. (Photo by the author.)

events (Marsalek et al. 1997; Wu et al. 1998). These pollutants can concentrate in sediments near the bridge at levels problematic to the health of benthic organisms (Marsalek et al. 1997). Additionally, birds are attracted to bridges and defecate into the local waters. Buoys—channel markers mainly—are habitats for periphyton, macroalgae, and fouling animals and also attract larger fish. These are well known to sport fishermen as good sites (as a completely nonstatistical example with no replication, this author caught a 5-ft hammerhead shark at the channel marker at the entrance to the Lockwood Folly River, North Carolina).

Relegated mainly to estuaries and offshore areas but potentially impacting riverine communities are oil derricks—on the good side these structures provide habitat for numerous fouling organisms and the larger fish that feed on them. They are often the only hard "bottom" habitat in a region characterized by soft sediments. Their use by recreational fishermen and divers has been well studied in the Gulf of

Mexico. Via surveys, catch per unit effort by sportsmen utilizing bottom fishing has been estimated as 2.32 fish/angler hour and 0.94 for fishermen trolling near oil platforms and derricks (Stanley and Wilson 1989). Presumably offshore windmills, a growing form of energy production, will likewise benefit certain fisheries. Oil derricks also are toppled by storms, or deliberately toppled by explosives to form artificial reefs. Toppled structures serve as a pioneering site for fouling communities and a recruitment site for young fish; the community develops larger and older fishes as the structure ages (Bull and Kendall 1994).

However, active and even retired oil structures can leak oil—sometimes major quantities in hurricanes (i.e. the Gulf of Mexico). Also, the pipes that service these structures will leak and breach, causing some massive pollution incidents. During Hurricane Ike (September 2008, Category II storm) at least 52 oil platforms were destroyed and 32 severely damaged. Approximately 500,000 gallons of crude oil spilled into the Gulf of Mexico, mainly from pipelines that snapped or were punctured. Of course, the estimated 4,400,000-barrel oil spillage (Crone and Tolstoy 2010) from infrastructure below the Deepwater Horizon platform during July 2010 in the Gulf of Mexico was unprecedented, ecologically damaging, and extremely expensive to taxpayers and various oil companies. Oil infrastructure leaks and spills are not confined to estuarine waters. For example, in July 2010 an estimated 877,000 gallons of oil spilled from a ruptured pipe into Talmadge Creek and from there into the Kalamazoo River, Michigan. In July 2011 an estimated 50,000 gallons of oil spilled from a ruptured pipe into the Yellowstone River near Billings, Montana, fouling at least 30 miles (48 km) of river. The pipe had been buried in some places only 5 ft below the river bed, and scouring during a storm and associated flooding led to the rupture. During peak cleanup over a thousand contractors were involved in cleaning up the oil and removing fouled vegetation, with preliminary cleanup cost estimates of at least $42 million by ExxonMobil. In June 2013 Exxon was fined $1.7 million by the US Department of Transportation for the incident. Months later the pipe was reburied 60 ft below the river bed and pumping was restarted. On the other side of the world, in Dalian, China, a July 2010 oil pipeline explosion polluted 165 miles$^2$ (420 km$^2$) of harbor with at least 400,000 gallons of oil. On March 22, 2014 an oil barge collided with a ship in the Houston Ship Channel, releasing up to 170,000 gallons of crude. The port, which is the busiest petroleum port in the world, had to be closed for several days while cleanup activities were ongoing. In June 2018 the Rock River in Iowa was polluted by 230,000 gallons of crude oil when a train derailed and 14 cars leaked oil, 150 miles upstream of Omaha. Sadly, well away from urbanization even remote freshwater rivers are not safe from human development oil pollution. In 2014 a pipeline running 525 miles (845 km) from Amazon oil fields to the coast leaked about 2000 barrels into a canal feeding the Cuninico River in Peru. The interconnecting local river system became polluted for a year, and the local villages, which heavily rely on fish for their diet, were forced to avoid much of the contaminated fishery (Fraser 2016). The point being that oil production mishaps impact marine, estuarine, and freshwater resources, and will continue to do so.

## 15.3 Summary

- Development of a natural area leads to a cascading series of impacts to nearby receiving streams. First the area to be developed is generally clearcut. This causes greatly increased stormwater runoff and erosion of the landscape.
- The hydrograph of the nearest stream downslope will be altered, with more rapid and steeper flood peaks. The stream will receive loads of suspended sediments, and in-stream erosion will occur as well, damaging animal habitat.
- Clearcutting is followed by road building, which causes hydrological disconnects, terrestrial and aquatic habitat loss, and pollution inputs to local water bodies. This is followed by construction of residential, commercial, or industrial areas. All of this leads to a large increase in impervious surface coverage of the watershed.
- Impervious surfaces prevent infiltration of rain and runoff, collect pollutants on the surfaces, and during rain events cause rapid and direct runoff of polluted water into storm drain systems and receiving streams. Local steam hydrographs are further altered, with larger and more rapid flooding leading to stream incision and widening, and more habitat loss.
- Increased urbanization (which is well represented by impervious surface coverage) causes degradation of fish and invertebrate communities such as loss of sensitive species, decreased diversity, and increases in pollution-tolerant species. This is collectively due to habitat loss, chemical pollution, decreased DO, eutrophication, and stream warming—collectively known as the "urban syndrome."
- Human development along rivers brings with it ecological impacts and water pollution from wastewater treatment plants, industrial outfalls, industrial and municipal water intakes, and non-point-source runoff from riverside commercial districts. Smaller-scale riverside development includes marinas, which are a source of fecal microbial and metals pollution.

- Bulkheading, or shoreline armoring, is done to protect ports, riverside industries, and residential areas from erosion. However, bulkheading destroys riparian habitat, disconnects the river from its floodplain, and causes increased erosion to shorelines down-current from the bulkheading.
- Ports and the shipping industry are major commercial enterprises along rivers, great lakes, and estuaries. These endeavors impact river ecology by channel dredging, which increases fecal microbial pollution, impact commercial fishing, and change river hydrology. Shipping increases water and air pollution, imports non-native species to new areas, and poses threats to endangered species.

## References

Arnold, C.L. Jr. and C.J. Gibbons. 1996. Impervious surface coverage: the emergence of a key environmental indicator. *Journal of the American Planning Association* 62:243–258.

Asadian, Y. and M. Weiler. 2009. A new approach in measuring rainfall interception by urban trees in coastal British Columbia. *Water Quality Research Journal of Canada* 44:16–25.

Bannerman, R.T., D.W. Owens, R.B. Dodds and N.J. Hornewer. 1993. Sources of pollutants in Wisconsin stormwater. Water Science and Technology 28: 241–259.

Baudardt, J., J. Grabulos, J.-P. Barussean and P. Lebaron. 2000. *Salmonella* spp. and fecal coliform loads in coastal waters from a point vs. non-point source of pollutants. *Journal of Environmental Quality* 29:241–250.

Becker, C.G., C.R. Fonseca, C.F.B. Haddad, R.F. Batista and P.I. Prado. 2007. Habitat split and the global decline of amphibians. *Science* 318:1775–1777.

Benedetti, M.M., M.J. Raber, M.S. Smith and L.A. Leonard. 2006. Mineralogical indicators of alluvial sediment sources in the Cape Fear River basin, North Carolina. *Physical Geography* 27:258–281.

Booth, D.B., J.B. Karr, S. Schauman, et al. 2004. Reviving urban streams: land use, hydrology, biology and human behavior. *Journal of the American Water Resources Association* 40:1351–1364.

Brown, J.J. and G.W. Murphy. 2010. Atlantic sturgeon vessel-strike mortalities in the Delaware estuary. *Fisheries* 35:72–83.

Bull, A.S. and J.J. Kendall Jr. 1994. An indication of the process: offshore platforms as artificial reefs in the Gulf of Mexico. *Bulletin of Marine Science* 55:1086–1098.

Burkholder, J.M. 2002. Cyanobacteria. Pages 952–982 in Bitton, G. (ed.), *Encyclopedia of Environmental Microbiology*. Wiley, New York.

Burkholder, J.M., G.M. Hallegraeff, G. Melia, et al. 2007. Phytoplankton and bacterial assemblages in ballast water of U.S. military ships as a function of point of origin, voyage time, and ocean exchange practices. *Harmful Algae* 6:486–518.

Cahoon, L.B., J.E. Nearhoof and C.L. Tilton. 1999. Sediment grain size effect on benthic microalgal biomass in shallow aquatic ecosystems. *Estuaries* 38:735–741.

Cahoon, L.B., J.C. Hales, E.S. Carey, S. Loucaides, K.R. Rowland and J.E. Nearhoof. 2006. Shellfish closures in southwest Brunswick County, North Carolina: septic tanks vs. storm-water runoff as fecal coliform sources. *Journal of Coastal Research* 22:319–327.

Capiella, K. and K. Brown. 2001. Impervious Cover and Land Use in the Chesapeake Bay Watershed. Report Prepared for US EPA Chesapeake Bay Program. Center for Watershed Protection, Ellicott City, MD.

CH2M Hill. 2008. North Carolina International Terminal Infrastructure Report. Prepared for North Carolina State Ports Authority, September 2008. CH2M Hill, Englewood, CO.

Colborn, T. and K. Thayer. 2000. Aquatic ecosystems: harbingers of endocrine disruption. *Ecological Applications* 10:949–957.

Corbett, J.J. and P.S. Fischbeck. 2000. Emissions from waterborne commerce vessels in United States continental and inland waterways. *Environmental Science and Technology* 34:3254–3260.

Corbett, J.J., J.J. Winnebrake, E.H. Green, P. Kasibhatla, V. Eyering and A. Lauer. 2007. Mortality from ship emissions: a global assessment. *Environmental Science and Technology* 41:8512–8518.

Cox, H., S. Mouzi and J. Gelsleichter. 2006. Preliminary observations of estrogenic activity in surface waters of the Myakka River, Florida. *Florida Scientist* 69: 92–99.

Crone, T.J. and M. Tolstoy. 2010. Magnitude of the 2010 Gulf of Mexico oil leak. *Science* 330:634.

Cuffney, T.F., R.A. Brightbill, J.T. May and I.R. Waite. 2010. Responses of benthic macroinvertebrates to environmental changes associated with urbanization in nine metropolitan areas. *Ecological Applications* 20: 1384–1401.

Davidson, E.A., K.E. Savage, N.D. Bettez, R. Marino and R.W. Howarth. 2010. Nitrogen in runoff from residential roads in a coastal area. *Water, Air and Soil Pollution* 210: 3–13.

Dennison, W.C., R.J. Orth, K.A. Moore, et al. 1993. Assessing water quality with submersed aquatic vegetation. *BioScience* 43:86–93.

Ensign, S.E. and M.A. Mallin. 2001. Stream water quality following timber harvest in a Coastal Plain swamp forest. *Water Research* 35:3381–3390.

Escobar, H. 2020. Deforestation in the Brazilian Amazon is stil rising sharply. *Science* 369:613.

Finkenbine, J.K., J.W. Atwater and D.S. Mavinic. 2000. Stream health after urbanization. *Journal of the American Water Resources Association* 36:1149–1160.

Forman, R.T.T., D. Sperling, J.A. Bissonette, et al. 2003. *Road Ecology: Science and Solutions*. Island Press, Washington, DC.

Fraser, B. 2016. Oil in the forest. *Science* 353:641–643.

Gardner, M.B. 1981. Effects of turbidity on feeding rates and selectivity of bluegills. *Transactions of the American Fisheries Society* 110:446–450.

Garie, H.L. and A. McIntosh. 1986. Distribution of macroinvertebrates in a stream exposed to urban runoff. *Water Resources Bulletin* 22:447–455.

Gobel, P., C. Dierkes and W.G. Coldeway. 2007. Stormwater runoff contamination matrix for urban areas. *Journal of Contaminant Hydrology* 91:26–42.

Gold, A.C., S.P. Thompson and M.F. Piehler. 2019. The effects of urbanization and retention-based stormwater management on Coastal Plain stream nutrient export. *Water Resources Research* 55:7072–7046.

Gold, A.C., S.P. Thompson, C.L. Magel and M.F. Piehler. 2020. Urbanization alters Coastal Plain carbon export and dissolved oxygen dynamics. *Science of the Total Environment* 747:141132.

Griffin, D.M. Jr., T.J. Grizzard, C.W. Randall, D.R. Helsel and J.P. Hartigan. 1980. Analysis of non-point pollution export from small catchments. *Journal of Water Pollution Control Federation* 52:780–789.

Grimes, D.J. 1975. Release of sediment-bound fecal coliforms by dredging. *Applied Microbiology* 29:109–111.

Grimm, N.B., S.H. Faeth, N.E. Golubiewski, et al. 2008. Global change and the ecology of cities. *Science* 319: 756–760.

Groffman, P.M., D.J. Bain, L.E. Band, et al. 2003. Down by the riverside: urban riparian ecology. *Frontiers in Ecology and the Environment* 1:315–321.

Hardison, E.C., M.A. O'Driscoll, J.P. DeLoach, R.J. Howard and M.M. Brinson. 2009. Urban land use, channel incision, and water table decline along Coastal Plain streams, North Carolina. *Journal of the American Water Resources Association* 45:1032–1046.

Harrington, R. 2007. Investigation of microbial contamination of sediment and water at boat ramps. MS Thesis, University of North Carolina Wilmington, Program in Marine Science, Wilmington.

Hatt, B.E., T.D. Fletcher, C.J. Walsh and S.L. Taylor. 2004. The influence of urban density and drainage infrastructure on the concentrations and loads of pollutants in small streams. *Environmental Management* 34: 112–124.

Heiskary, S. and H, Markus. 2001. Establishing relationships among nutrient concentrations, phytoplankton abundance and biochemical oxygen demand in Minnesota, USA, rivers. *Journal of Lake and Reservoir Management* 17:251–267.

Holck, K.T., E.L. Mills, H.J. MacIsaac, M.R. Dochoda, R.I. Colautti and A. Ricciardi. 2004. Bridging troubled waters: biological invasions, transoceanic shipping, and the Laurentian Great Lakes. *BioScience* 54: 919–929.

Holland, A.F., D. M. Sanger, C.P. Gawle, et al. 2004. Linkages between tidal creek ecosystems and the landscape and demographic attributes of their watersheds. *Journal of Experimental Marine Biology and Ecology* 298:151–178.

Interlandi, S.J. and C.S. Crockett. 2003. Recent water quality trends in the Schuylkill River, Pennsylvania, USA: a preliminary assessment of the relative influences of climate, river discharge and suburban development. *Water Research* 37:1737–1748.

Karr, J.R. 1981. Assessment of biotic integrity using fish communities. *Fisheries* 6:21–27.

Karr, J.R. and E.W. Chu. 2000. Sustaining living rivers. *Hydrobiologia* 422/423:1–114.

Kaushal, S.S., P.M. Groffman, G.E. Likens, et al. 2005. Increased salinization of fresh water in the Northeast United States. *Proceedings of the National Academy of Sciences of the United States* 102:13517–13520.

King R.S., J.R. Beamon, D.F. Whigham, A.F. Hines, M.E. Baker and D.E. Weller. 2004. Watershed land use is strongly linked to PCBs in white perch in Chesapeake Bay subestuaries. *Environmental Science and Technology* 38:6546–6552.

Kirk, K.L. 1991. Suspended clay reduces *Daphnia* feeding rate: behavioural mechanisms. *Freshwater Biology* 25:357–365.

Klein, R.D. 1979. Urbanization and stream quality impairment. *Water Resources Bulletin* 15:948–963.

Kunimatsu, T., M. Sudo and T. Kawachi. 1999. Loading rates of nutrients discharging from a golf course and a neighboring forested basin. *Water Science and Technology* 39:99–107.

Lenat, D.R. and J.K. Crawford. 1994. Effects of land use on water quality and aquatic biota of three North Carolina Piedmont streams. *Hydrobiologia* 294:185–199.

Lerberg, S.B., A.F. Holland and D.M. Sanger. 2000. Responses of tidal creek macrobenthic communities to the effects of watershed development. *Estuaries* 23: 838–853.

Lewitus, A.J., L.B. Schmidt, L.J. Mason, et al. 2003. Harmful algal blooms in South Carolina residential and golf course ponds. *Population and Environment* 24:387–413.

Line, D.E., J.A. Arnold, G.D. Jennings and J. Wu. 1996. Water quality of stormwater runoff from ten industrial sites. *Water Resources Bulletin* 32:807–816.

Line, D.E., N.M. White, D.L. Osmond, G.D. Jennings and C.B. Mojonnier. 2002. Pollutant export from various land uses in the upper Neuse River basin. *Water Environment Research* 74:100–108.

MacDonald, D.D., C.G. Ingersoll and T.A. Berger. 2000. Development and evaluation of consensus-based sediment quality guidelines for freshwater ecosystems. *Archives of Environmental Contamination and Toxicology* 39:20–31.

Mallin, M.A. and T.L. Wheeler. 2000. Nutrient and fecal coliform discharge from coastal North Carolina golf courses. *Journal of Environmental Quality* 29:979–986.

Mallin, M.A., K.E. Williams, E.C. Esham and R.P. Lowe. 2000. Effect of human development on bacteriological water quality in coastal watersheds. *Ecological Application* 10:1047–1056.

Mallin, M.A., V.L. Johnson, S.H. Ensign and T.A. MacPherson. 2006. Factors contributing to hypoxia in rivers, lakes and streams. *Limnology and Oceanography* 51: 690–701.

Mallin, M.A., L.B. Cahoon, B.R. Toothman, et al. 2007. Impacts of a raw sewage spill on water and sediment quality in an urbanized estuary. *Marine Pollution Bulletin* 54:81–88.

Mallin, M.A., V.L. Johnson and S.H. Ensign. 2009. Comparative impacts of stormwater runoff on water quality of an urban, a suburban, and a rural stream. *Environmental Monitoring and Assessment* 159:475–491.

Mallin, M.A., M.I. Haltom, B. Song, M.E. Tavares and S.P. Dellies. 2010. Bacterial source tracking guides management of boat head waste in a coastal resort area. *Journal of Environmental Management* 91:2748–2753.

Marino, R.P. and J.J. Gannon. 1991. Survival of fecal coliforms and fecal streptococci in storm drain sediment. *Water Research* 25:1089–1098.

Marsalek, J., B. Brownlee, T. Mayer, S. Lawal and C.A. Larkin. 1997. Heavy metals and PAHs in stormwater runoff from the Skyway Bridge, Burlington, Ontario. *Water Quality Research Journal of Canada* 32:815–827.

McCabe, K.M., E.M. Smith, S.Q. Lang, C.L. Osburn and C.R. Benitez-Nelson. 2001. Particulate and dissolved organic matter in stormwater runoff influences oxygen demand in urbanized headwater catchments. *Environmental Science and Technology* 55:952–961.

Miltner, R.J., D. White and C. Yoder. 2004. The biotic integrity of streams in urban and suburbanizing landscapes. *Landscape and Urban Planning* 69:87–100.

Moser, M.L. 1999. Wilmington Harbor Blast Effect Mitigation Tests: Results of Sturgeon Monitoring and Fish Caging Experiments. Final Report. CZR, Wilmington, NC.

Muschack, W. 1990. Pollution of street run-off by traffic and local conditions. *Science of the Total Environment* 93:419–431.

Ng, A.K.Y. and S. Song. 2010. The environmental impacts of pollutants generated by routine shipping operations on ports. *Ocean and Coastal Management* 53:301–311.

NRC. 2009. *Urban Stormwater Management in the United States*. National Research Council, The National Academies Press, Washington, DC.

O'Driscoll. M., S. Clinton, A. Jefferson, A. Manda and S. McMillan. 2010. Urbanization effects on watershed hydrology and in-stream processes in the southeastern United States. *Water* 2:605–648.

Pilkey, O.H. 2003. *A Celebration of the World's Barrier Islands*. Columbia University Press, New York.

Pimentel, D., L. Lach, R. Zuniga and D. Morrison. 2000. Environmental and economic costs of nonindigenous species in the United States. *BioScience* 50:53–64.

Pitt, R., R. Field, M. Lalor and M. Brown. 1995. Urban stormwater toxic pollutants: assessment, sources and traceability. *Water Environmental Research* 67:260–275.

Ricciardi, A. and H.J. MacIsaac. 2000. Recent mass invasion of the North American Great Lakes by Ponto-Caspian species. *TREE* 15:62–65.

Riggs, S.R., J.T. Bray, E.R. Powers, et al. 1991. Heavy Metals in Organic-Rich Muds of the Neuse River Estuarine System. Report No. 90-07, Albemarle–Pamlico Estuarine Study. North Carolina Department of Environment, Health and Natural Resources, Raleigh.

Riggs, S.R., J.T. Bray, R.A. Wyrick, et al. 1993. Heavy Metals in Organic-Rich Muds of the Albemarle Sound Estuarine System. Report No. 93-02, Albemarle–Pamlico Estuarine Study. North Carolina Department of Environment, Health and Natural Resources, Raleigh.

Rothschild, B.J., J.S. Ault, P. Goulletquer and M. Heral. 1994. Decline of the Chesapeake Bay oyster population: a century of habitat destruction and overfishing. *Marine Ecology Progress Series* 111:29–39.

Sanger, D., A. Blair, G. DiDonato, et al. 2015. Impacts of coastal development on the ecology and human well-being of tidal creek ecosystems of the US Southeast. *Estuaries and Coasts* 38:S49–S66.

Sanger, D.M., A.F. Holland and G.I. Scott. 1999a. Tidal creek and salt marsh sediments in South Carolina coastal estuaries: I. Distribution of trace metals. *Archives of Environmental Contamination and Toxicology* 37: 445–457.

Sanger, D.M., A.F. Holland and G.I. Scott. 1999b. Tidal creek and salt marsh sediments in South Carolina coastal estuaries: I. Distribution of organic contaminants. *Archives of Environmental Contamination and Toxicology* 37:458–471.

Sanger, D.M., A.F. Holland and D.L. Hernandez. 2004. Evaluation of the impacts of dock structures and land use on tidal creek ecosystems in South Carolina estuarine environments. *Environmental Management* 33: 385–400.

Sayler, G.S., J.D. Nelson, A. Justice and R.R. Colwell. 1975. Distribution and significance of fecal indicator organisms in the upper Chesapeake Bay. *Applied Microbiology* 30:625–638.

Schlosser, I.J. 1995. Stream fish ecology: a landscape perspective. *BioScience* 41:704–712.

Schueler, T.R. 1994. The importance of imperviousness. *Watershed Protection Techniques* 1:100–111.

Simmons, D.L. and R.J. Reynolds. 1982. Effects of urbanization on base flow of selected south-shore streams, Long Island, New York. *Water Resources Bulletin* 18: 797–806.

Sobsey, M.D., R. Perdue, M. Overton and J. Fisher. 2003. Factors influencing faecal contamination in coastal marinas. *Water Science and Technology* 47:199–204.

Stanley, D.R. and C.A. Wilson. 1989. Utilization of offshore platforms by recreational fishermen and scuba divers off the Louisiana coast. *Bulletin of Marine Science* 44:767–775.

Street, M.W., A.S. Deaton, W.S. Chappell and P.D. Mooreside. 2004. North Carolina Coastal Habitat Protection Plan. North Carolina Department of Environment and Natural Resources, Division of Marine Fisheries, Morehead City.

Sweeney, B.W., T.L. Bott, J.K. Jackson, et al. 2004. Riparian deforestation, stream narrowing, and loss of ecosystem services. *Proceedings of the National Academy of Sciences of the United States of America* 101:14132–14137.

Tavares, M.E., M.I.H. Spivey, M.R. McIver and M.A. Mallin. 2008. Testing for optical brighteners and fecal bacteria to detect sewage leaks in tidal creeks. *Journal of the North Carolina Academy of Science* 124:91–97.

USACOE. 2010. Section 905(b) Analysis, Wilmington Harbor Navigation Improvement Project, North Carolina International Terminal. US Army Corps of Engineers, Wilmington, NC.

US EPA. 2000. Guidance for Assessing Chemical Contaminant Data for Use in Fish Advisories, Volume 2: Risk Assessment and Fish Consumption Limits. EPA-823-B-00-008. Office of Water, US EPA, Washington, DC.

US EPA. 2002. Onsite Wastewater Treatment Systems Manual. EPA/625/R-00/008. Office of Water, Office of Research and Development, US EPA, Washington, DC.

Wahl, M.H., H.N. McKellar and T.M. Williams. 1997. Patterns of nutrient loading in forested and urbanized coastal streams. *Journal of Experimental Marine Biology and Ecology* 213:111–131.

Walsh, C.J., A.H. Roy, J.W. Feminella, P.D. Cottingham, P.M. Groffman and R.P. Morgan III. 2005a. The urban stream syndrome: current knowledge and the search for a cure. *Journal of the North American Benthological Society* 24:706–723.

Walsh, C.J., T.D. Fletcher and A.R. Ladson. 2005b. Stream restoration in urban catchments through redesigning stormwater systems: looking to the catchment to save the stream. *Journal of the North American Benthological Society* 24:690–705.

Walter, R.C. and D.J. Merritts. 2008. Natural streams and the legacy of water-powered mills. *Science* 319: 299–304.

Wang, L., J. Lyons, P. Kanehl and R. Bannerman. 2001. Impacts of urbanization on stream habitat and fish across multiple spatial scales. *Environmental Management* 28:255–266.

Wasserman, J.C., S. Hacon and M.A. Wasserman. 2003. Biogeochemistry of mercury in the Amazonian environment. *AMBIO* 32:336–342.

Waters, C.T. and C.D. Thomas. 2001. Shoreline Hardening Effects on Associated Fish Assemblages in Five North Carolina Coastal Rivers. Final Report. Coastal Fisheries Investigations, Federal Aid in Fish Restoration Project F-22. North Carolina Wildlife Resources Commission, Division of Inland Fisheries, Raleigh.

Waters, T.F. 1995. *Sediment in Streams: Sources, Biological Effects and Control.* American Fisheries Society Monograph 7. American Fisheries Society, Bethesda, MD.

Wehr, J.D. and J.H. Thorp. 1997. Effects of navigation dams, tributaries, and littoral zones on phytoplankton communities in the Ohio River. *Canadian Journal of Fisheries and Aquatic Sciences* 54:378–395.

White, D., K. Johnston and M. Miller. 2005. Ohio River basin. Pages 375–424 in Benke, A.C. and C.E. Cushing (eds.), *Rivers of North America.* Elsevier, Amsterdam.

Wu, J.S., C.J. Allan, W.L. Saunders and J.B. Evett. 1998. Characterization of pollutant loading estimation for highway runoff. *Journal of Environmental Engineering* July:584–591.

# CHAPTER 16

# Protecting and Restoring River Ecosystems

Human-induced impacts both pollute streams and change their physical habitat, often severely. In order to restore the natural values of impacted streams, such as repositories of biotic diversity, functional fisheries, places for human recreation, etc., at least one and sometimes two major goals must be accomplished. The first is to remove or reduce the pollutant load or physical impact affecting the stream and degrading it. In some cases that is sufficient to return some of its ecological functions or at least prevent further degradation. In severely impacted systems, stream or river restoration may be necessary, which requires major capital and construction efforts well beyond pollutant reduction. The first part of this chapter will focus on degradation source reduction or removal and the various techniques that have been demonstrated to be effective. The second part will focus on stream restoration and its more complicated and costly counterpart, river restoration.

## 16.1 Protection of Streams and Rivers from Pollution

A critically important driver of stream pollution is non-point-source pollution, of which a major component is stormwater runoff. Stormwater runoff, the rainfall-driven washing of polluting substances from the landscape into the nearest body of water, is considered to be one the most important sources of impairment to freshwater and estuarine water bodies in the USA (NRC 2009; see also Chapter 15). Stormwater runoff occurs when rainfall exceeds the capacity of the local soils to infiltrate water; the remaining water that is not evaporated from the landscape or transpired by vegetation flows downslope. As discussed in Chapter 15, runoff alone becomes a major hydrological driver of in-stream erosion and habitat loss and should be controlled by interception and decoupling its input from reaching the stream (Booth et al. 2004). Pollution carried by runoff occurs in rural situations as nutrient and pesticide runoff from farmlands as well as nutrient and bacterial runoff from CAFO areas and feedlots (see Chapter 13); in urban and suburban areas runoff contains numerous pollutants, as detailed in Chapter 15.

It is important to note that stormwater runoff pollution has two pollutant measures. One is the concentration of any pollutant in question, such as fecal bacteria (CFU/100 ml), suspended solids (mg/L), turbidity (NTU), nutrients (µg/L), etc. Water quality regulations are usually based on pollutant concentrations; clearly, stormwater inputs drive up stream water pollutant concentrations in some cases far above ambient stream standards. The second component is pollutant load, which is the pollutant mass carried within a volume of stormwater over a given time period. This might be presented as milligrams of N entering a stream from an agricultural area of given size per hour during the duration of a storm, or milligrams of TSS draining a construction site per hour of a storm. This is an important measure because a receiving stream with a gauged discharge carries a pollutant load and when one knows the approximate stream discharge and the approximate stormwater pollutant load one can estimate (model) the stormwater impact on the stream's pollutant concentration. When one plans on-land schemes to reduce stream pollution (such as for

*River Ecology*. Michael A. Mallin, Oxford University Press. © Michael A. Mallin (2023). DOI: 10.1093/oso/9780199549511.003.0017

TMDL requirements in the USA), load reductions from various techniques can be estimated to help plan the most ecological and cost-effective ways of stream pollutant reduction. These techniques, which may be behavioral or structural, are frequently termed best management practices (BMPs). Note that physical-/construction-based practices to reduce stormwater inputs are often called stormwater control measures (SCMs). This section on stream protection will begin with an explanation of natural stream protections provided by wetlands (nature's BMPs) and will later address various BMPs, both construction-based and behavior-based.

## 16.1.1 Function of Natural Wetlands

Wetlands perform numerous key functions in supporting and protecting the water quality, ecosystem stability, and health of streams and rivers. There are many different types of wetlands (classified in detail by Cowardin et al. 1979, among others). Some of these ecosystems are isolated from running water systems, including pocosins (swamp-on-a-hill) and lacustrine (wetlands associated with lakes). All types of wetlands will contribute to the water quality protection of area water resources, however, by collecting and processing rainfall-driven runoff. The amount of natural wetland coverage in a watershed is positively related to water quality. For example, a set of 11 North Carolina blackwater streams was analyzed in terms of rainfall/runoff impacts on stream turbidity and fecal bacteria pollution (Mallin et al. 2001). In most of the streams both turbidity and fecal coliform bacterial counts were significantly correlated with rainfall within the previous 48 hours. However, in the stream watersheds where natural wetlands areal coverage exceeded about 17%, those runoff relationships were nonsignificant, indicating a pollution prevention effect of large wetland coverage. Among numerous regulatory authorities and environmentally oriented NGOs, a major tenet is that to protect streams from pollution (as well as deleterious physical changes) a key approach is the conservation of natural wetlands. In terms of protecting and enhancing the environmental quality of streams and rivers there are several wetland types that provide immediate streamside protection. Riparian swamp forests are streamside hardwood forests located in nontidal areas and are periodically flooded during wet periods. Freshwater marsh areas consist of fleshy, succulent emergent, free-floating and floating leaved plants, along with some woody shrubs. In tidal freshwater areas both forests and marshes may be present as well. Farther downstream toward the estuary where tidal brackish waters dominate, shorelines contain oligohaline marshes and salt marshes. Between approximately 25°N and 25°S latitude (about Mid-Florida and southward in the USA) mangrove forests dominate the riparian areas of brackish and salt marsh tidal creeks and rivers.

Riparian and other types of wetlands serve numerous functions besides water quality protection. Wetlands are areas of water storage and water removal from the system (Ewel 1990). They serve as key habitats and feeding areas for all sorts of biota associated with streams and rivers, including invertebrates, amphibians, reptiles, fish, birds, and mammals (Clark 1978; Thayer et al. 1978). Wetland edges serve as ecotones where animal life can pass between habitats, and wetlands serve as corridors for animal movement. Isolated wetlands, including temporary ponds, serve as spawning areas for amphibians and reptiles (Clark 1978; Becker et al. 2007), while riparian wetlands, especially along estuaries, serve as fisheries primary nursery areas where young fish find both abundant food and shelter from larger predators. Along freshwater rivers the seasonal flood pulse (at least in the rare undammed system) opens the food resources of the riparian swamp forest to river fishes (Bayley 1995; see also Chapter 6). Riparian swamp forests provide woody habitat to the stream as deadfall or storm blow-overs that become stream and river snags, vastly important habitat for fish and other biota especially in soft-bottomed streams. Likewise, riparian swamps provide allochthonous food to the stream as POM and DOM.

In terms of stream water quality protection, wetlands, especially riparian wetlands, function in several ways to reduce pollutant loads to lotic ecosystems. These functions reduce pollution by physical, chemical, and biological means. As mentioned above, stormwater carries a pollutant load. Pollutants entrained in stormwater runoff include suspended sediments and other particulates;

fecal microbes including bacteria, viruses, and protozoans; organic and inorganic nitrogen and phosphorus; BOD materials; metals; and various chemical pollutants. When wetlands absorb stormwater the pollutant load to downstream water bodies is consequently reduced. Wetlands by nature are usually heavily vegetated with woody and succulent species. The basic plant function of **transpiration** draws water into a plant from saturated soils and releases some amount of that water back to the atmosphere. Transpiration from a large well-vegetated wetland can have a considerable effect on reducing stormwater volume itself. This will of course vary with length of the growing season; wetlands in warmer climates will have a greater transpiration impact on an annual basis. Coupled with transpiration is evaporation. Stormwater retained in a wetland will be subject to evaporative losses as well as transpiration losses. Season, or ambient temperature, plays a big role here. For instance, in North Carolina the amount of evaporation

increases by a factor of four times from January to July (Robinson 2005). Thus, as with transpiration, evaporative losses from wetlands will be greater in warmer seasons as well as warmer climates.

In addition to enhancing water losses from stormwater and causing subsequent load reductions, wetland vegetation physically causes the reduction of entrained pollutants from stormwater entering a wetland. This occurs through **baffling and settling**. As discussed in Chapter 1, when currents slow in flowing waters (be it within streams or stormwater flows), the load of suspended solids settles out, with the heavier particles settling first and the lighter particles settling as the current slows even more. In marsh wetlands the stems of the plants serve as a baffle to slow down water flow, causing suspended sediments and other particulate matter to settle out of the water column (Fig. 16.1). This effect has been measured in lotic waters and natural marsh wetlands for emergent macrophyte vegetation (Leonard and Luther 1995;

**Fig. 16.1** Stormwater inflow baffling in a constructed wetland; species present in this forebay include Eastern bur-reed *Sparganium americanum*, alligatorweed *Alternanthera phileroxoides*, soft rush *Juncus effuses*, and giant cutgrass *Zizaniopsis miliacea*. (Photo by the author.)

Gurbisz et al. 2016). Submersed macrophytes also reduce current velocity, particularly within the first 5 cm of a bed of macrophytes (Madsen and Warncke 1983); the amount of plant surface area available for interacting with the incoming suspended matter was found to account for 70% of the variability in flow reduction in one study (Petticrew and Kalff 1992). The interaction with macrophytes allows for sedimentation of inorganic and organic particles to occur, reducing turbidity and providing food for benthic organisms. Many types of pollutants become attached to sediment particles either physically or chemically, including fecal bacteria, ammonium, phosphate, and chemical pollutants. As stormwater enters a wetland and is slowed by current baffling by macrophytes, these attached pollutants settle to the bottom along with the suspended sediments. There they are subject to burial, consumption, and/or microbial and chemical degradation and transformation. Note that by clarifying the overlying waters macrophytes permit their own beds to expand due to enhanced light for photosynthesis (Gurbisz et al. 2016).

Natural wetlands are known to retain and remove nutrients from inflowing stormwater runoff by a variety of chemical and biological means (Johnston 1991; Weisner et al. 1994; Knox et al. 2008). Organic material in the sediments as well as plant uptake will serve to adsorb nutrients, metals, and other pollutants (Crumpton 1995; Pond 1995). P is removed from the water by assimilation by wetland bacteria, algae, and the resident aquatic macrophytes and woody vegetation. Other important removal mechanisms are adsorption and absorption of inorganic P by the wetland sediments; the precipitation of P-associated suspended sediment particles to the wetland sediment; and the incorporation of organic P into soil peat (Richardson 1985; Woltemade 2000). Richardson (1985) determined that increased P removal in wetlands is dependent upon higher concentrations of Al and Fe in the wetland sediments and subsequent adsorption to those sediments.

Nitrogen removal occurs through settling of organic particles and assimilation by bacteria, algae, macrophytes, and woody vegetation, and especially by microbial denitrification (Weisner et al. 1994; Crumpton 1995; Woltemade 2000). Denitrification is most effective when soils are anaerobic, organic carbon is available for use as a substrate, and there are multiple habitats provided by a mixture of aquatic plant species and their roots (Weisner et al. 1994). Denitrification is more efficient in association with macrophyte rhizomes than in bare soil, and some macrophyte species (pickerelweed [*Pontederia cordata*], for instance) are more effective than other macrophytes in enhancing denitrification and anammox (Song et al. 2014; see also Chapter 2). Additionally, ammonium can be nitrified to nitrate and then denitrified in wetlands (Woltemade 2000); and oxygen release by macrophyte roots into the rhizosphere may enhance this nitrification–denitrification process (Wiesner et al. 1994).

Stormwater inputs contain both labile and recalcitrant BOD (see Chapter 13) as particulates and dissolved inputs (McCabe et al. 2021). Particulate BOD materials can be removed in wetlands by sedimentation, whereas dissolved BOD materials can become attached to surfaces by adsorption, and both particulate and dissolved BOD materials are subject to microbial metabolism within the wetland (Karathanasis et al. 2003; Knox et al. 2008), thus reducing the BOD load to adjacent streams.

Fecal bacteria and other microbes can be removed from incoming water by a variety of methods (Gopal 1999; Knox et al. 2008). Since many fecal microbes are associated with particulate matter a major removal mechanism is sedimentation. Fecal microbes, especially those unassociated with particles, are killed or inactivated by exposure to UV radiation from the sun. As in any ecosystem there is an active microbial food web in wetlands and fecal bacteria are subject to consumption by protozoans and other microbial predators (Stenström and Carlander 2001; Vymazal 2005; Burtchett et al. 2017). Wetland vegetation has been demonstrated to provide much more efficient fecal microbe removal than bare sediments in ponds (Davies and Bavor 2000), likely by enhancing settling of fine particles and associated bacteria and by providing increased physical contact between the bacteria and plant structure and the protozoan predators associated with the plants (Gerba et al. 1999; Burtchett et al. 2017). Exposure of pathogens to antimicrobial vegetation-produced exudates and exposure of anaerobes to the oxic layer produced by aquatic plants have also been proposed as removal mechanisms (Vymazal 2005). Note that for heavy microbial pollutant loads a good functioning wetland can remove 99% of the fecal bacteria and still there may be concentrations exceeding standards (Mallin et al. 2012).

Metals are adsorbed most readily by organic as opposed to inorganic sediments, and some species, particularly cattail (*Typha latifolia*) are particularly effective in contributing to such removal (Pond 1995). The efficacy of other chemical pollutant removal in wetlands is understudied, likely because of the increased costs of analyzing the numerous organic pollutants water resources are subjected to.

Wetlands thus provide a set of environmental services including habitat, water storage and removal, and pollution control that some researchers have quantified in dollar figures (Costanza et al. 1998). For instance, the wetland functions of wastewater and stormwater runoff treatment can be computed with some degree of accuracy based on comparisons with engineered chemically based systems. In fact, for several decades natural wetland areas have been utilized and demonstrated to remove high percentages of pollutants from raw or partially treated inflowing sewage effluents (Ewel 1990; Gopal 1999). However, raw sewage can severely degrade the water quality and animal life of natural wetlands (Ewel 1990) and regulatory agencies frown on natural wetlands to be used as repositories for human wastewater, although engineered wetland systems are commonly utilized as tertiary treatment in municipal and industrial wastewater treatment (Gerba et al. 1999; Gopal 1999; Karanthanasis et al. 2003).

## 16.1.2 Constructed Wetlands as Stream Protection BMPs

Since natural wetlands provide a variety of processes that reduce the pollution and degradation of streams, environmental and ecological engineers have in recent years concentrated much effort on designing and constructing wetlands to treat stormwater runoff before it enters lotic systems. Such systems are most effective for controlling runoff from large drainage basins rather than small, highly urbanized ones. As suggested above,

a variety of mechanisms within wetlands contribute to pollutant removal, depending upon the pollutant in question, summarized below.

Since suspended solids are removed primarily by settling, constructed wetlands should be engineered to promote decreased flow caused by wetland basin construction (deeper areas enhance such settling) and baffling by emergent or submersed aquatic macrophytes (Petticrew and Kalff 1992; Gopal 1999; Davies and Bavor 2000; Knox et al. 2008) (Fig. 16.1).

A hugely important factor in the ultimate efficacy of constructed wetlands in pollutant removal is having a large enough wetland to process the runoff from the targeted drainage basin. If it is too small it may export pollutants rather than retain them. The wetland should be constructed to maintain a base flow, encourage sheet flow across the wetland surface with a depth of less than 1 m, and have an outlet control structure that can regulate release during wet and dry periods (Field et al. 2006; England and Stuart 2007). A forebay that settles suspended sediments is an important factor, especially if planted (Fig. 16.1) with emergent and submersed vegetation to enhance settling (Mallin et al. 2012). High plant diversity in constructed wetlands appears to increase wetland function especially in terms of increasing nutrient removal/retention (Englehardt and Ritchie 2001). Data from a limited number of SCM projects found relatively high TSS and nitrate removal but much lower P removal in constructed wetlands (Table 16.1). Regardless, constructed stormwater treatment wetlands have become a popular means of treating stormwater runoff. Their pollutant removal efficacies, however, are highly variable and thus scientists, engineers, and managers continually seek new data on wetland design and pollutant removal capacity. A constructed wetland project that has provided excellent pollutant load and concentration reductions is described in Box 16.1.

A more recent technique for reducing nutrients is the use of **floating treatment wetlands** (FTLs), which are emergent wetland plants fixed to a rack that can be floated in a detention pond or a natural body of water (Fig. 16.2). The roots and rhizomes are hanging in the water but do not touch the sediments—they are essentially hydroponic systems. Nutrients can be taken up by assimilation

**Table 16.1** Median pollutant removal efficacy (percentage removal) of four SCMs analyzed from studies in the US Southeast and Mid-Atlantic (including number of sites analyzed). A negative value signifies export of parameter. (Revised from Wossink and Hunt 2003.)

| Parameter | Wet ponds | Constructed wetlands | Sand filters | Bioretention |
|---|---|---|---|---|
| TSS | 65 (27) | 61 (14) | 79 (12) | NA |
| Nitrate | 42.5 (16) | 55 (8) | −56.5 (11) | 16 (4) |
| TN | 28 (27) | 22 (14) | 41 (12) | 45 (4) |
| TP | 46 (28) | 32.5 (14) | 59 (11) | 71 (5) |

N/A, no data available.

---

**Box 16.1  A high-performance constructed wetland.**

Hewletts Creek in Wilmington, North Carolina drains a large 3001- ha (7436-acre) suburban watershed and as such is impacted by high fecal bacteria loads and periodic algal blooms from nutrient loading. During 2007 the J.E.L. Wade stormwater wetland, a 3.1-ha (7.6-acre) facility, was constructed to treat stormwater runoff from a 238-ha (589-acre) watershed within the Hewletts Creek drainage. The wetland contains two inflows with deep forebays for TSS settling, has a large stormwater storage capacity, and was designed to treat up to 2.54 cm (1 inch) of rainfall runoff. The forebays and wetland proper developed a diverse array of emergent and submersed vegetation, including soft rush, cattail, pickerel weed, Eastern burr-reed, giant cutgrass, alligator-weed, and parrot feather, some through plantings and some through invasion (Fig. 16.1); additionally woody shoreline wetland vegetation became abundant.

A rain event sampling program was carried out in 2009–2010 to evaluate the efficacy of the wetland in reducing pollutant loads (fecal bacteria, nutrients, suspended solids, and metals) from the stormwater runoff passing through the wetland. During the eight storms sampled, the wetland served to greatly moderate the hydrograph, retaining and/or removing (through groundwater recharge, transpiration, and evaporation) 50–75% of the inflowing stormwater volume within the wetland. High removal rates of fecal coliform bacteria were achieved (based on "first flush"), with

an average load reduction of 99% and overall concentration reduction of > 90%. Particularly high (> 90%) load reductions of ammonium and orthophosphate loads also occurred and lesser but still substantial reductions of total phosphorus (89%) and TSS loads (88%) were achieved. Removal of nitrate was seasonally dependent, with lower removal occurring in cold weather and high percentage (90%+) nitrate load removal occurring in the growing season when water temperatures exceeded 15°C. Both the planted and invasive macrophytes enhanced denitrification, some more so than others (Song et al. 2014). Most metals tested had concentrations too low to be measured in inflowing and outflowing waters, except for zinc, for which an average load reduction of 87% was achieved. A sampling program in Hewletts Creek downstream of where discharge from the wetland entered the creek was in operation both before and after wetland construction. Creek sampling showed statistically significant reductions in both nitrate and ammonium and a 50% reduction in median fecal coliform bacteria counts as well, following wetland construction. Since the principal source of impairment in Hewletts Creek is fecal bacteria contamination and a secondary source is algal blooms (limited by nitrogen in this system), this constructed wetland was successful in reducing both concentrations and loads of polluting substances to the receiving waters

(See Mallin et al. 2012 for details.)

---

or N can be removed via denitrification, microbial cycling and transformation is locally enhanced, and sedimentation of organic and inorganic solids occurs below the FTL. In a New Zealand experiment N removal between a detention pond and a second pond that included a floating wetland was compared. The pond with the floating wetland did experience enhanced N removal, although enhanced

removal only occurred in the warm months (Borne et al. 2013). Recall from Chapter 2 that denitrification requires local anoxia. This is difficult to achieve if there is a constant current where the FTL is located. However, in still waters or very slow currents anoxia can occur below an FTL due to several processes (Borne et al. 2015). First, the plants and rack above impede light to the phytoplankton,

**Fig. 16.2** Installed floating treatment wetland stretching across nutrient-polluted Squash Branch, a tributary to eutrophic Greenfield Lake, Wilmington, North Carolina. A duckweed *Lemna* infestation was covering the branch at the time. (Photo by the author.)

benthic microalgae, and macrophyte below, reducing photosynthesis. The plant roots will excrete DOC, increasing the localized BOD; sedimentation of organic POC from the FTL will likewise increase local BOD. Finally, with a larger FTL size compared to the wet pond or stream width, anoxia is more likely to occur to enhance denitrification. Of course, if local fish and aquatic invertebrate enhancement is desired, anoxia works against that. Borne et al. (2015) noted that pollutant removal is enhanced in general with larger FTL size compared to water surface area (as above), slower water current, and shallower depth—although the roots from the FTL need to be above the sediment surface.

### 16.1.3 Other Constructed Stream Pollutant Protection Mechanisms

One of the most widespread means of treating stormwater runoff is the use of **wet detention ponds**, sometimes called wet retention ponds (Schueler and Galli 1995; US EPA 1999; England and Stein 2007). These range widely in size and shape and treat runoff from apartment complexes, housing subdivisions, shopping areas, businesses, academic institutions, and even large suburban and urban areas (regional wet detention ponds). In these systems detention of stormwater for 36–72 hours is desirable (England and Stein 2007), and they are designed so the central pond areas are in contact with the water table and base flow is maintained. While these systems can be effective in treating a variety of pollutants (Table 16.1) (Pennington et al. 2003) they are primarily designed to remove suspended solids (generally 80–85% of incoming; according to the South Carolina Department of Health and Environmental Control [SCDHEC 2007]) rather than targeting fecal bacteria or nutrients. In a North Carolina study Cahoon (1996) found that most ponds surveyed were eutrophic based on chlorophyll *a* and TP concentrations, had low (average 3.3) N:P ratios, and supported considerable cyanobacterial growth. Consequently, some ponds likely support conditions that enhance N fixation. In South Carolina a variety of ponds that receive nutrient loading (located in housing developments and on golf courses) have been afflicted by toxic algal blooms and fish kills (Lewitus et al. 2003). Additionally, some ponds export nutrients in storm events (SCDHEC 2007) and of course planktonic algal blooms are subject to export with their organic nutrient and BOD load (McCabe et al. 2021). Some stormwater ponds, through summer heating and discharge of their waters, can also amplify the heating of receiving urban streams (Schueler and Galli 1995).

There have been efforts to improve the functioning of wet detention ponds. One technique that improves retention of solids and associated pollutants is the use of a forebay at the inflow to settle most suspended solids where the runoff enters. As noted in the wetland discussion in Section 16.1.2, the use of rooted aquatic macrophytes (native species) greatly improves nutrient removal in ponds via assimilation and denitrification. Thus, a shallow shelf around the edge of the pond is recommended to support macrophyte growth (according to the North Carolina Division of Water Quality [NCDWQ 2007]). Planted macrophytes of course die and decompose, which increases the organic content of the sediments. Increased organic content of sediments increases their capacity to absorb metals and nutrients (Pond 1995). Use of wetland areas in conjunction with the pond can increase pollutant removal efficacy (Mallin et al. 2002). The pond should have a deep middle (> 2 m) so macrophytes don't fill the pond and overly restrict flow. However, pond depth should not be deep enough to be subject to thermal stratification, creating bottom water anoxia (US EPA 1999). Anoxic waters will reduce biological activity and serve to mobilize phosphate and some metals from the sediments back into the water column.

Shape of the pond is also important. Designing a pond with inflows near the outfall short-circuits the pollutant removal capacity, especially for nutrients. Optimally the pond should be designed with an increased length/width ratio of 3/1 to 7/1, with stormwater inputs directed into one end with the discharge at the other end so the entire pond can be used for treatment (Mallin et al. 2002; England and Stein 2007). Wet ponds are also sometimes designed with islands or baffles to increase detention time and treatment efficacy.

Sediments accumulate over time in such ponds and infilling will reduce the pond's storage capacity and effectiveness. Bottom sediments are subject to accumulation of elevated loads of phosphorus, metals, and chemical pollutants, and the biota are exposed as well in such ponds (SCDHEC 2007). Thus, bottom sediments, especially in the forebay, should be removed every two to five years (US EPA 1999). Pond construction usually requires a permit by state or municipal authorities and there is often a checklist associated with their upkeep. Inspections are technically required periodically but, in reality, depend on regulatory agency manpower availability and priorities. For example, a North Carolina study showed that of 249 stormwater pond projects investigated, only 26.9% were in full compliance with construction and maintenance regulations (NCDWQ 2005).

**Dry detention ponds** are sometimes used for stormwater treatment as well. These are constructed so their bottoms are above the water table and they have control structures so that the pond volume can be released to downstream waters over a 36- to 72-hour period (England and Stein 2007). Primarily particulate matter is removed from stormwater with the use of dry ponds; dissolved nutrient removal is limited due to lack of biological processing (England and Stein 2007).

**Bioretention** is a process in which stormwater runoff from roofs, lots, roads, and parking lots is directed by gravity flow into a depression consisting of layers of mulch, sand, other soils, and organic materials that is planted with shrubs and trees (Lloyd et al. 2002). Pollutant removal is achieved by a variety of physical, chemical, and biological processes (Field et al. 2006). The designs of such systems vary considerably and are often proprietary, and sizes of these systems vary as well. The stormwater may infiltrate downward into the water table if soils permit, or if fine soils predominate below the bioretention cell then the system may be underlain by an underdrain and outlet to carry away filtered water into the stormwater system. The plants in the bioretention area utilize water and assimilate nutrients, and the mulch on top of the system absorbs and retains water for later percolation (Wossink and Hunt 2003). Pollutant removal data for bioretention cells are not abundant in the literature; those that are indicate good to moderate pollutant removal, with the exception of nitrate, unless anaerobic zones develop or are engineered to enhance denitrification (Table 16.1) (Hunt 2003; Hunt et al. 2006). Additionally, higher pollutant removal rates appear to occur in warmer seasons as opposed to winter (Hunt 2003).

On the smallest scale are **rain gardens**, which are small, low-vegetated areas built adjacent to parking lots of private and public facilities or within a homeowner's yard to collect, retain, and treat runoff (Dunnett and Claydon 2007). The rain garden is sized to approximately 1/20th of the area to be drained and planted with native species that can handle flooded soils for a period of up to two days while the stormwater percolates into the soil. They can be attractively landscaped for aesthetic purposes (Fig. 16.3) (Dunnett and Claydon 2007) and will require some care.

Stormwater runoff from parking lots and roadways can be treated by a variety of engineered devices based on **sand filtration**, the principle of

which has been used since the early 1800s (England and Stein 2007). Most of these devices (the designs of which some are proprietary) consist of dual-chambered devices, where settling of solids occurs in one chamber followed by filtration through sand or other porous media in an adjacent chamber (Wossink and Hunt 2003). Upon and within the upper sand layer a biomat forms (similar to the effect in a septic system; see Chapter 13) where bacteria and BOD can be removed by biological processing (England and Stein 2007). Such devices are highly effective for removal of suspended sediments, fecal bacteria, and some other pollutants, but other approaches may be more effective if nitrate reduction is of primary interest (Table 16.1). In pervious soils, treated runoff from the filters enters the surficial aquifer, or if the device is underlain by drains the treated runoff can enter the stormwater conveyance system. Sand can be taken advantage of as a natural filtration system in coastal areas or inland areas where sand is abundant (such as certain locations near the Laurentian Great Lakes). Such

**Fig. 16.3** Aesthetically designed rain garden at a public park. (Photo by the author.)

infiltration to remove pollutants can be achieved by directing stormwater flows into sand-filled chambers (Bright et al. 2011; Price et al. 2013) or into engineered, perforated metal or PVC chambers that are buried above sand (Mallin et al. 2016; Grogan and Mallin 2021). High removal efficacies of suspended solids, fecal bacteria, and TN have been achieved using sand in this way. Use of these types of devices is targeted to small, urban drainages; maintenance will be required since clogging can become an issue.

Standard curb and gutter systems, seen everywhere in urban and suburban areas, do not treat stormwater runoff. They are built to remove road runoff as quickly as possible and send it somewhere else, sometimes into a wet detention pond and sometimes directly into an urban stream, tidal creek, or urban lake. As mentioned in Chapter 15, excessive removal of stormwater from an area by these systems can actually lower the groundwater and base flow of streams in a given area (Simmons and Reynolds 1982). If **curbside treatment** is designed into the road construction plans it can result in cleaner runoff entering the nearby streams while improving groundwater recharge. Such engineered projects can direct road runoff through gaps in the curb into a median where it flows over rock riprap to settle solids, then flows sideways through the grassed median, before it finally encounters a storm drain. Thus, it allows for percolation through the grass to reduce runoff (reducing effective impervious coverage) and settling of suspended solids in the riprap and grass, and filtration of pollutants through the grass and through the soil (Fig. 16.4).

There are many (again mainly proprietary) small pollution reduction units that are designed to fit into grated road storm drains or curbside storm drains (some are referred to as **drain traps**). Most of these are based on filtration, absorption, or adsorption of pollutants onto some type of material, sedimentation, and trapping, or hydrodynamically enhanced separation, and thus require regular maintenance. A number of these systems are detailed in England and Stein (2007).

A technique to control stormwater on a smaller scale is the use of **infiltration trenches** (Field et al. 2006; England and Stein 2007). These are trenches that have been excavated, lined with porous material, and refilled with coarse gravel. Stormwater

is routed into these systems (which can be either at the ground surface or buried) and the water collects within and slowly infiltrates into the surrounding soils. This process is also called "exfiltration." Provided the system is not overwhelmed by excessive inflow, pollutant removal efficacy can be high (England and Stein 2007). Related techniques are infiltration chambers of various designs which are constructed of metal or PVC and perforated; stormwater is directed into such buried structures and slowly filters out into the surrounding gravel, sand, or earth.

### 16.1.4 Reducing Runoff at the Source

Reducing runoff at the source is an effective way of reducing the stormwater pollutant load to streams. Regardless of how, this tactic reduces pollutant loading to and physical degradation of the receiving stream and reduces treatment costs of the runoff. This can be accomplished by both conservation practices and technological means.

Directly reducing water use through **water conservation** as well as **reusing runoff water** have the dual benefits of decreasing pollutant loading to streams and other receiving waters and reducing monetary costs (water purchase) to the consumer. This involves the **decoupling** of local runoff generation from standard stormwater removal systems and the direction of it either to beneficial uses or into passive treatment areas. This decoupling reduces the hydrological impacts on receiving streams as well as pollutant impacts on the stream biota (Hatt et al. 2004; Walsh et al. 2005). On an individual homeowner basis, rooftop runoff, which typically is directed into rain gutters, then the street, and then enters the stormwater system, can instead by directed into rain barrels (Fig. 10.7b) so the roof runoff can be used for irrigation of flowers, shrubs, and vegetable gardens. On a larger scale, buildings with extensive roof areas can collect and store rainwater in cisterns (Fig. 10.7a) that can be later used for irrigation of landscaping. These practices disconnect the discharge of runoff from the municipal curb and gutter systems described above, with the additional benefit of recharging the aquifer and contributing to stream base flow. Homeowners can also perform some simple behavioral changes

**Fig. 16.4** Innovative curbside stormwater treatment design directing road runoff into rock riprap, then a grassy swale, before reaching a drain to the stormwater conveyance system. (Photo by the author.)

to carefully utilize water and avoid its loss. For instance, by washing cars on the lawn rather than on an impervious driveway one avoids runoff issues. By watering lawns and gardens during cooler hours of the morning or evening the homeowner reduces runoff and saves money by avoiding evaporation losses that occur during mid-day hours.

Major reductions in stormwater generation and losses to base flow can be achieved for housing and commercial and institutional developments by planning to **maximize greenspace** and minimize impervious surface coverage. Trees, especially large mature individuals, should be retained on-site rather than removed whenever possible. Large trees utilize large quantities of water and expel excess amounts to the atmosphere via evapotranspiration. This reduces stormwater volume (and the subsequent pollutant load to nearby streams) and aids in reducing the impacts of flooding during storms. Trees intercept rainfall with their leaves, branches, and trunks and buffer it from hitting the ground immediately; this both reduces ground erosion by

diminishing the force of the strike and smooths the hydrograph by delaying the throughfall of the rain. Research has demonstrated that individual trees in urban areas have a stronger impact in intercepting throughfall of rain than individual trees in forested areas (Asadian and Weiler 2009). Thus, urban trees, especially mature ones left in place during the building process, have a direct positive impact on reducing stormwater runoff and stream pollution. In addition to these ecosystem services, trees are habitat for many species of wildlife, reduce noise pollution, provide cooling with shade, are aesthetically pleasing, and can add to home values for nature-appreciating buyers. If trees need to be moved for a building footprint, some can be dug up and replanted on-site elsewhere (a process called **tree spading**), which maximizes the services provided by a larger tree rather than growing a new one from a seedling. In terms of other vegetation used in the landscaping of residential and commercial areas, it is important to utilize native species—non-native imports usually require more watering and more fertilizer usage, adding to the runoff pollution load.

Stormwater generation can also be reduced by conservative construction processes designed to minimize impervious surface area. Some of these tactics include reducing the width of roads within developments and placing sidewalks on only one side of the street instead of both. These types of practices both reduce the amount of impervious surface coverage and subsequent stormwater runoff and reduce the cost of materials and labor to the developer. This type of development is often termed **low-impact development (LID)**. This is a general term for a set of building practices that includes treating rainfall or runoff at or near where it falls rather than diverting it away by standard runoff conveyances (Field et al. 2006). Other related terms sometimes used are water-sensitive urban design (WSUD) and sustainable urban drainage (SUD), both of which concern utilizing various practices to minimize the impact of urban development on the water cycle (Lloyd et al. 2002; Gurnell et al. 2007). While it is most effective to incorporate LID techniques into plans for new developments, some older developments can be successfully retrofitted to improve water quality in stream watersheds. As these techniques have achieved popularity only in recent years, some zoning ordinances (i.e. wide streets, sidewalk ordinances, standard curb and gutter) are not conducive to LID practices and new regulations may be required for broad use. The following management practices fall under the LID umbrella, which includes a mixture of engineered as well as human behavioral solutions. However, some of the engineered practices are still generally considerably cheaper than standard conveyance systems, again saving money as well as providing water quality protection.

Technological advances in construction materials have led to various types of **pervious pavement**, also known as **permeable pavement** (Fig. 16.5). Such pavements provide support for vehicles while allowing rainwater to drain through pores in the pavement to enter groundwater rather than the

**Fig. 16.5** Pervious and impervious pavement together on Ocracoke Island, North Carolina. (Photo by the author.)

stormwater conveyance system. There are several types of pervious pavement currently available including porous concrete and porous asphalt, which are produced with a high percentage of void space (17–22%), thus allowing water to be stored in the pavement and percolate through to the underlying soils (Field et al. 2006). Concrete grid pavers, variously known as turfstone or "grassy" pavers, have large voids within the pavers as well as along the sides to permit percolation and to support grass as well. Permeable interlocking concrete pavers have voids at the corners which permit drainage. Cost can be a deterrent to wider use; pervious concrete is currently about 20% more expensive than standard concrete and some permeable pavement is more expensive. Such pavements are most effective in areas where soils are sandy; there, rainfall filtering downward can move unimpeded toward the upper aquifer. In areas dominated by finer soils like clay, percolation through the soil will be retarded and underdrains may be required to move the filtered runoff off-site. Pervious pavement does require a regular (at least annual) program of vacuuming. The use of pervious pavements can be required legally and encouraged financially by government agencies. In some areas, when a given impervious parking lot threshold is reached the municipality may require any further parking spaces be fitted with pervious pavement. Municipalities can offer water quality or pollution control credits to developers who utilize such pavements by lowering built-upon area and subsequent fees (NCDWQ 2007).

Passive treatment of stormwater results from use of the techniques above. Passive means non-labor intensive, with only periodic site maintenance to consider and of course no costs for chemicals other than those used in landscaping efforts. When stormwater is directed through greenspace the pollutants entrained in the water are treated (removed) by physical filtration through soil particles, adsorption to soil particles, uptake by plants, and soil bacterial transformations. Some organic pollutants may pass through such passive treatment relatively intact, however.

Another alternative to standard curb and gutter systems is the use of roadside **grassy swales**, in which road stormwater runoff is directed into grassed or otherwise vegetated low areas along the roadside. This allows for some pollutant treatment by settling, filtration, and/or plant uptake, if present. The bulk of pollutant removal in swales is through infiltration rather than biological processes or adsorption (Schueler 1995a). If underlain by a gravel-filled trench (French drain) this will move the water along rather than allowing it to plond up in the swale for extended periods. Some studies have shown swales can be highly effective for suspended solids removal but only marginally effective for nutrient removal, and fecal bacteria removal may be ineffective (Schueler 1995a; Pennington et al. 2003; England and Stein 2007). However, in a sandy coastal municipal area the use of grassy swales provided very effective reduction of suspended solids and fecal bacteria from stormwater runoff (Mallin et al. 2016).

### 16.1.5 Streamside Vegetated Buffers

Most of the above discussion concerned reducing pollutant loads and runoff volume as close to the source areas as possible. However, it is also imperative to reduce pollution at the stream, since streams (as well as ditches) are highly subject to sheet flow stormwater runoff generated by agriculture, suburban areas, and urban situations. A common tactic is to employ streamside treatment of runoff by use of **vegetated buffer zones**, sometimes referred to as filter strips. Buffer strips have been researched and used extensively in urban and rural situations (Welsch 1991; Landry and Thurow 1997; Lowrance et al. 1997). The wider the buffer the better it is for pollution control (until a point of diminishing returns is reached), although in practice the width will depend on how wide a strip streamside property owners will permit for such use and how much room there is available for such systems. According to Schueler (1995b) urban streamside buffers have multiple benefits, including effective pollutant removal, protection from streambank erosion, reducing impediments to migrating fish, reducing the percentage watershed impervious cover by up to 5%, mitigating stream warming, providing habitat and food for wildlife, and providing corridors for animal migration and other movement. While urban buffers range from 20 to 200 ft in width, a 100-ft buffer per shoreline is recommended to provide adequate stream protection (Schueler 1995b). Buffers may be ineffective with slopes > 15%, should be vegetated throughout, and with soils that permit infiltration (Field et al. 2006). Buffers with grasses alone can help reduce pollutant runoff, but mixed vegetation buffers are considerably more effective (Lee et al. 2000).

How does a streamside vegetated buffer function to reduce stream pollution? It is based on reduction of sheet flow pollutant loading; thus every effort should be made in design and maintenance to avoid formation of channels through the buffer (Lowrance et al. 1997). Physically, the vegetation in the buffer (a mixture of native grasses, shrubs, and trees) slows overland flow so pollutants such as suspended solids and associated pollutants settle out and become part of the soil layer. As stormwater flows over the buffer some amount will infiltrate into the soil where inorganic nutrients can be assimilated by plants. Porous soils enhance the removal of pollutants by permitting infiltration of more stormwater as well as infiltration of fine particles and dissolved nutrients (Lee et al. 2000). Soil particles adsorb fecal bacteria, ammonium, orthophosphate, and metals. However, some organic contaminants can pass through buffers and soils (Cordy et al. 2004). Trees both assimilate nutrients and take up vast quantities of water and transpire it away, reducing flooding. Again, buffers should be planted with mixed species; the different species have mixed root depths so groundwater-borne nitrate can be intercepted before reaching the stream. The nitrate can be effectively denitrified by microbes associated with plant roots in a well-vegetated buffer, with denitrification effectiveness enhanced in waterlogged soils and where groundwater is high in DOC (Spruill 2003). Forested buffers appear to be very effective in enhancing nitrate removal from shallow groundwater, especially in the root zone, through assimilation and denitrification (Lowrance et al. 1997). The hydraulic drought in riparian areas caused by lowering of the water table and stream base flow by urbanization (see Chapter 15) may lead to reduction of denitrification from riparian soil drying (Groffman et al. 2003). The amount of in-stream habitat structure and organic matter is primarily a function of vegetation adjacent to a stream (Wang et al. 2001). Thus, establishment of forested buffers also ensures a future supply of in-channel wood to the stream or river to serve as habitat (Palmer et al. 2005). Additionally, shading from the buffer will reduce stream water temperature (Finkenbine et al. 2000). When considering the various services that large, vegetated buffer systems provide, especially in areas under heavy agriculture (erosion control, channel dredging relief, wetlands protection, fisheries protection, flood control, drinking water quality protection, etc.), the installation of vegetated buffer systems can have extensive quantifiable economic benefits to society (Rein 1999).

In some agricultural situations artificial drainage, either surface or subsurface, is used to maintain proper soil aeration for planting and growth. This leads to increased annual outflow from fields; coupled with fertilizer applications this can increase the eutrophication of receiving streams. However, when drainage is controlled (such as by use of flashboard risers in ditches) the off-site movement of total nitrogen and phosphorus can be significantly reduced (Evans et al. 1995). Some of this removal is due to enhanced denitrification of nitrate when ditch soils are waterlogged for extended periods in the process.

One final note on the physical protection of streams from pollution is that many urban streams receive inputs of human sewage from leaks and spills or else from septic system leachate (see Chapter 13). These sewage-based stream pollution problems can be acute or chronic. Many of the non-point-source stream protection techniques discussed above and the stream rehabilitation techniques discussed in the following section can be less than effective unless such basic sewage infrastructural pollution problems are dealt with as well.

## 16.1.6 Stream Protection by Conservation Easement

An alternative means of stream protection that does not involve construction of active or passive stormwater treatment is based on landowner goodwill coupled with capitalism and tax breaks. This involves a riparian buffer zone, of any size, that is purchased or leased by a third party—this is the

**conservation easement approach**. By this approach a private environmental conservancy or some government entity can purchase streamside property to keep it from being developed or buy a long-term lease agreement (a conservation easement) in which the landowner retains the ownership of the land but his uses are limited by the terms of the agreement. Owners can get tax breaks in many locations. The entity can manage the easement in some cases (such as when government money is involved) to permit passive recreation (i.e. riding, walking, canoeing) for the public.

### 16.1.7 Stream Protection Through Incentives to Developers

During the discussion on pervious pavement above, it was brought out that developers who use that technology can receive financial incentives to reduce or minimize pollution. Hopefully such financial incentives can be expanded to the use of other LID techniques. This is a function of government regulatory organizations keeping abreast of evolving LID techniques.

Finally, another incentive that local and regional municipalities can use to encourage developers to build in an environmentally sound manner is to offer environmental stewardship awards, perhaps on an annual basis. These can offer positive public relations for developers and can serve as positive (nonregulatory) feedback. There will be a certain segment of the home-buying population that will gravitate toward such an environmentally sound development or home—and even pay a premium for it. Such awards can be offered to both individual homeowners and to developers of commercial or housing developments, with winners determined by a panel of judges with various types of expertise including planning, green building, engineering, and pollution science. These awards can consider water quality protection, wildlife protection and enhancement, and energy-efficient building conservation practices.

## 16.2 Stream Restoration

The previous sections of this chapter demonstrated many of the ways that streams and rivers can be protected from the damaging effects of pollution and physical changes such as erosion. What about restoring streams to their original function? In many cases a stream may be so degraded, or at least so changed, that restoration to its original state is impractical if not impossible. Work by Walter and Merritts (2008) indicated that the sinuous gravel-bedded stream with regular riffles and pools that was believed by many to be the "natural" state for Mid-Atlantic streams was in fact an incorrect model. Before European colonization and the consequent construction of thousands of mill dams, the streams in such areas were actually braided wetland streams with forested meadows and organic sediments. Restoration to such conditions in many locations is obviously impractical if not impossible. However, measures can be taken to restore or at least rehabilitate some of the many ecosystem services that stream formerly supplied, before urbanization or heavy agricultural land use degraded the stream or river (Booth et al. 2004). Some of the specific reasons that stream restoration is desirable include saving endangered species in the stream, protecting a trout (or other) fishery, improving biodiversity, enhancing in-stream habitat, protecting a community water supply, enhancing water quality to improve human recreation, achieving or maintaining an "Outstanding Waters" designation, recovering a stream from pollution such as acid mine drainage or metals contamination, providing erosion control and streambank stabilization, and improving aesthetics.

Stream restoration can be a complicated procedure, but streams are often located in only one or two political districts or geographic provinces. However, river restoration may involve working across numerous political borders (even international), dealing with some combination of mountain, foothill, prairie, sandhill, and coastal ecoregions, and so on. This greatly complicates river restoration; thus, let us first look at restoration for streams of third order or less.

### 16.2.1 Urban Stream Restoration

If a government or user group is pushing for restoration of a given stream for any one or a combination of the above factors, they will find they cannot go it alone and will need allies. Thus, stream restoration needs solid **planning** and the involvement of a diverse group of users (stakeholders) and skills: landowners, environmental groups, state and local agencies, local governments, scientists, engineers, sport fishermen groups, etc. It is always productive to conduct an on-site walk-through for interested parties to show everyone what needs protecting or restoring and how to do it (Izaak Walton League 2006). As a matter of practicality, when politicians and managers are involved, it is easier to enlist their support if a stream is visibly degraded with erosion and bank slumping, algal blooms, obviously fouled water, and other gross signs of abuse and neglect. Whereas chemical pollution and fecal microbial contamination may be very hazardous to humans as well as stream life, it is not always visible to the untrained eye. If the pollution or degradation has affected the fish community or rendered the stream or tidal creek closed to human contact or shellfishing from fecal pollution, then community support (especially if the local news media is involved) can help stimulate local government action. When an urban stream receives direct lateral drainage from streamside private homes, this presents an additional and often considerable challenge. What a stream needs in terms of riparian protection and habitat is often vastly different from what landowners, even concerned landowners, think is ecologically sound or aesthetically pleasing (Booth et al. 2004). Thus, a strong public education effort is needed as well.

How is funding obtained for stream rehabilitation projects? When the economy is in (relatively) good shape, funding may come from federal, state, provincial or local government, or agency sources. Bear in mind that when government budgets are tight, spending for the environment, in terms of both protection and restoration, is often the first affected. However, restoration funds can often be obtained as grants from foundations and private individuals, and if a project is not too expensive groups such as sport fishing clubs will contribute.

Stream restoration projects usually involve up to several hundred linear meters of shoreline. Depending on the amount of work involved, stream restoration costs US$50–200/linear foot (Southeast Watersheds Forum 2003). In extensive projects it is best to perform restoration work in the upstream reaches first, followed by downstream—if done in the opposite order the downstream sites can be compromised by impacts from unrestored areas upstream (Izaak Walton League 2006). Ideally, restoration sites should be ranked and chosen by need and potential effectiveness; however, on many occasions reaches have been targeted simply because land was available for the restoration work (Sudduth et al. 2007).

In planning a stream restoration project collection of all sorts of data will help provide restoration goals, provide useful engineering data for projects involving construction, and help determine what is feasible. These data include historical information on the condition of the stream before it was degraded (if available), reference reaches (if available), and any type of current information on physical, chemical, and biological stream water quality, fish and invertebrate communities, and vegetation communities, as well as sources of impairment. There are many types of useful hydrological and morphological data including base flow and storm flow discharge, condition and slope of the banks at various points, stability, sediment composition, and numerous other measures (as detailed in Doll et al. 2003 and Izaak Walton League 2006). Maps and photos of current conditions and drawings of planned goals are essential.

Heavy **erosion** is a common symptom of degraded streams and leads to cascading problems (see Chapters 1 and 15). This is a result of increased runoff generated upstream by clearcutting and urbanization. As detailed earlier, suspended sediments carry associated pollutants downstream, bury habitat, increase turbidity that will shade and degrade bottom vegetation, and interfere with filter-feeders and sight-feeding aquatic life. Sand bars along the inner curve of a meander are normal; however, sand bars in the stream center (mid-channel bars) mean elevated erosion upstream (see Chapter 15). Erosion-forced infilling will lead to shallowing of the stream, in turn leading to stream widening by increased erosion along the banks. Banks will be undercut and slumping will occur, and streamside trees may collapse, removing their shade and leading to elevated water temperatures. In such a situation the first tactic would be to attempt to decrease the source of erosion upstream (if possible) by decoupling runoff from curbs and gutters and impervious surfaces, increasing infiltration, and installing vegetated buffers, as detailed in the previous section. This part of the stream restoration process can of course involve the installation of BMPs, including in the watershed but away from the stream bed proper.

In-stream tactics have been developed to relieve and reverse the impacts of erosion on a stream. In locations where storm flow discharge is destabilizing the bank, natural materials have been successfully used to divert the current energy toward the stream middle. **Root wads**, also called root balls, are the root mass of a tree trunk placed so the trunk is buried and the root mass faces into the stream flow (Fig. 16.6). This deflects the current and protects the bank, and can take many years to decompose (Doll et al. 2003). It has the added benefit of providing substrata for periphyton, and fish and invertebrate habitat as well.

Another tactic is to utilize **rock vanes**, which are boulders covered by heavy fabric faced 30° into the stream flow (Fig. 16.7). This deflects the erosive

energy of the current and a zone of deposition behind the rock will accumulate gravel. Combinations of tree trunks and boulders are also used. Other stream-bank protection devices such as tree revetments (logs placed longitudinally and anchored to the bank) and various other deflectors are also used (Izaak Walton League 2006). Root wads and rock vanes create habitat and can develop over time into ecologically functional and visually attractive stream subunits (Figs 16.8 and 16.9).

Degraded stream banks can be **revegetated** with native species to develop "natural" bank stabilizers, once the erosive currents are reduced or deflected away. Grading will likely be required as vegetation will not anchor on banks with slopes greater than 3:1 or 4:1 horizontal to vertical ratio (Izaak Walton League 2006). Once the bank is built to the desired height and slope, brush mattresses of live, dormant tree branches (called *live fascines*) can be laid down and covered with a foot of soil and seeded with native grasses. While roots of grasses are shallow, they can provide a temporary protection from rainfall-driven local erosion until the larger, planted vegetation can take root. Restoration

**Fig. 16.6** (left) root wads, also called root balls (left). (Photo by the author.)

**Fig. 16.7** Newly installed rock vanes in Asheville, N.C. –root wads and root vanes both are used to deflect current and prevent erosion. (Photo by the author.)

crews can further stabilize the bank by staking down coir (Fig. 16.10), a coconut fiber mat that will last two to three years to prevent erosion while plants grow; *jute mesh* is a plant fiber that performs a similar function. The stream bank slopes can be staked with **live stakes** (Fig. 16.11), which are tree shoots that take root—willow, elderberry, and dogwood are commonly used (Doll et al. 2003). This stabilizes the bank and eventually provides shade to cool the stream temperatures to permit greater fish diversity. It is best to finish the project just before the growing season begins (Doll et al. 2003; Izaak Walton League 2006). In the case of urban streams in commercial or residential districts, traditional plantings of ornamentals can be utilized if aesthetics are desired; however, the plantings need to be of native species rather than exotics. Exotics tend to require additional pesticide and fertilizer, exactly what should be avoided for streams undergoing rehabilitation. A restored stream planted with native vegetation can be very aesthetically pleasing (Fig. 16.9).

*Channel realignment* can be used to perform several functions, provided floodplain space (and sufficient funds) are available. One goal to help reduce erosion is to spread the stream's energy out over the floodplain instead of against the bank. This is an attempt to restore a natural floodplain and its function to a stream that has been widened and straightened by increased upstream runoff (Doll et al. 2003). An obstacle to this is that when restoring an urbanized stream, it will require creating a larger floodplain to handle the increased runoff from the upstream impervious surface coverage. If space is available a goal can be to restore stream sinuosity—this increases base flow retention, creates pools for fish habitat, increases pollutant uptake and removal, reduces erosion and flooding, and improves habitat. There is a growing amount of literature from agencies and academics on the engineering aspects of stream restoration that is beyond the scope of this book.

Other stream restoration objectives may require less construction work. If restoration

**Fig. 16.8** Rock vane 3.5 years after placement, showing value as habitat. (Photo by the author.)

or improvement of a local fishery is desired, techniques are used that increase fish structural habitat, increase shade for promoting cooler areas, and create woody habitat in a sand-bottom stream. Rehabilitating a stream from infestation of invasive macrophytes is of course a very challenging goal, as indicated by the discussion in Chapter 9. Removal of terrestrial invasive plant species along shorelines in areas undergoing rehabilitation is essential to promote plant diversity and avoid monocultures.

## 16.2.2 Rural Stream Restoration

The above discussion has focused primarily on protecting and restoring streams from urban and suburban pollution and impacts. However, rural streams are also subject to different, site-specific stressors. Streams draining agricultural areas host benthic and fish communities that can be as degraded as those in urban streams (Lenat and Crawford 1994; Wang et al. 2001).

Fertilizers are heavily applied in agricultural areas (see Chapter 2) as well as pesticides (see Chapter 14). Buffer strips with grasses and diverse vegetation species can be an effective means of reducing nutrients and suspended sediments (see Section 16.1.5); buffers can trap pesticides when associated with particulates, but when in the dissolved form stream protection from pesticide pollution is more problematic (NRCS 2000). Nutrient and fecal microbial pollution to rural streams caused by swine, poultry, and cattle CAFOs is discussed in Chapter 13. Additionally, in rural areas stream pollution and physical degradation can be caused by free-ranging grazing cattle using the stream directly for drinking purposes. The deposition of manure into or along the stream leads to pollution by fecal microbes and nutrient loading. In addition, cattle are heavy beasts with hard hooves, and their movements along streams lead to bank crumbling and erosion. The simple act of *fencing* along a stream can lead to significant decreases in stream nutrient

**Fig. 16.9** Creek on the grounds of Asheville, North Carolina Arboretum 3.5 years after restoration with rock vanes and streamside plantings, showing improved aesthetics as well as habitat function. (Photo by the author.)

(TKN, TP) and TSS concentrations (Line et al. 2000). Creation of *alternative watering systems* for the cattle away from the stream is also an effective measure and installing animal waste storage facilities will reduce nutrient and fecal microbial runoff into the stream. Once the cattle are prevented from entering the stream bed area then the erosion recovery techniques discussed above, followed by replanting riparian buffers can serve to rehabilitate the stream. Detailed information on rangeland stream protection and restoration are found elsewhere (Izaak Walton League 2006; note there are many other handbooks currently available from agencies and NGOs).

A technique that combines the natural rural stream ecology and in-stream construction is called a *beaver dam analog* (BDA) (summarized in Goldfarb 2018). Beavers were discussed in detail in Chapter 6, including their ecological engineering feats producing small to massive dams in rural areas, along with

the ecological benefits. However, in many areas beavers were essentially wiped out by trapping over a century ago. Since 2009 researchers began experiments in which they chose degraded streams in areas with few beavers, pounded logs upright into the sediments, and wove willow sticks through the logs. The beavers then took over, improving the dams. The BDAs caused a rise in beaver populations, spread water over the floodplain, caused the growth of side channels, and improved the fisheries, including endangered species. Since these early experiments BDAs have been used by public authorities, nonprofit groups, and private citizens including ranchers for fish and wildlife improvements and cattle irrigation. However, the technique is not widely accepted and has run into permitting issues in various states based largely on concerns about altering the natural stream function to unknown conditions, or else based on funding concerns.

**Fig. 16.10**  Restored bank being revegetated and secured with coir mats Asheville, North Carolina. (Photo by the author.)

### 16.2.3 In-Stream Nutrient Removal

While many of the stream protection techniques can reduce nutrient loading to a eutrophic stream, it is also important to try to reduce nutrient loads within streams, particularly nitrogen (summarized in Craig et al. 2008). Most of these techniques concern enhancing denitrification by additions of artificial debris dams to slow waters and adding carbon substrates where needed, channel widening to increase contact with the benthos, meander building, adding more geomorphic features, floodplain reconnection, and flow path alteration through adjacent wetlands, side channels, etc. A relatively new in-stream engineering technique is called regenerative stormwater conveyance (RSC), which is designed to slow flow, force subsurface seepage through media, and enhance nutrient removal (Duan et al. 2019; Cizek et al. 2019). This has shown some success at N removal/retention, but the associated physical and chemical factors enhancing N and P removal are as yet understudied. Some of these proposed in-stream nutrient removal methods are clearly engineering-heavy and costly, so it is always best to reduce/remove nutrients at the source.

In any stream restoration program it is important to follow the progress both photographically and with the gathering of physical, chemical, and biological stream monitoring data for several years at least. This is essential because this information will provide evidence that the techniques did or did not work to achieve the goals—important for planning future efforts. Additionally, this information is critical to obtaining public and private monies for future stream restoration projects. If it can be demonstrated (by monitoring) that given techniques work, then granting agencies and municipal governing councils will be more inclined to provide funding for additional efforts.

## 16.3 River Restoration

River restoration is obviously much more technically difficult than stream restoration, as well as a

**Fig. 16.11** Live staking to restore riparian vegetation in urbanizing creek, Asheville, North Carolina. (Photo by the author.)

political nightmare in some cases because you need to cross municipal, state, provincial, or even international boundaries! Rivers may be simultaneously affected by numerous plights including industrial pollution, nutrient loading, floodplain decoupling, and channel modification and erosion, as was the case of the Rhine River (Farmer and Braun 2002; Jungwirth et al. 2002). Goals for river restoration include many of the ones listed above for stream restoration; additional goals may include modifying river flows for various reasons (including possible dam removal), improving recreation or aesthetics, reconnecting floodplains, and channel restoration. In a large-scale analysis of stream and river projects the most commonly stated goals were, in order, enhancing water quality, managing riparian zones, improving in-stream habitat, improving fish passage, stream bank stabilization, flow modification, aesthetics and recreation enhancement, and channel modifications (Bernhardt et al. 2005). Despite such needs and goals as above, in many rivers it is also

critical to maintain channels with sufficient depth for shipping needs.

### 16.3.1 Planning

River restoration involves extensive planning and it is essential to involve multiple stakeholders in setting the goals and deciding on the approaches to take. For stream restoration, as mentioned earlier, this involves various interested local parties and regulators. In international situations this usually means having different countries represented, as was the case with the International Commission for the Protection of the Rhine, involving Switzerland, France, Germany, Luxembourg, and the Netherlands (Farmer and Braun 2002; Jungwirth et al. 2002). A major charge to the stakeholders is deciding on what the goals of the project or series of projects are to be. In almost all cases returning a river to its natural predevelopment state is impossible. However, there should be a condition that can be obtained in which many of the lost ecosystem functions can be returned, at least to a degree, with the technology and political willpower available (Gore and Shields 1995). To help set these goals the stakeholders can utilize historical maps, photos, water quality, and ecosystem information, and relatively undisturbed reference sites in the same watershed or one of similar qualities (Palmer et al. 2005). As part of goal setting it is important to set measures, or standards, by which to judge the degree of ecological success (Palmer et al. 2005).

## 16.3.2 Strategies

Rivers can become disconnected from their floodplains through levee construction, dam construction, and flow regulation, or decoupled from their floodplains through riparian zone modifications such as deforestation, urban development, excessive cattle grazing, draining for agricultural activity, and even beaver trapping. In extreme cases some urban rivers were engineered to become little more than concrete tubes draining water, such as the Los Angeles River (although local officials there have recently undertaken extensive rehabilitation efforts). In Japan the Itachi River was an armored, straight channel with no recreational value, which in the 1990s underwent significant restoration including addition of sediments, creation of riffle and pool habitats, and provision of vegetated streamside riparian zones (Nakamura et al. 2006). *Reconnected floodplains* should lead to increased biodiversity, improved water quality through the many wetland-based physical, chemical, and biological runoff treatment means detailed earlier in this chapter, increased carbon sequestration by the plant life, and improved fish production, as described in the flood-pulse concept (Junk et al. 1989; Bayley 1995—detailed in Chapter 6). As such Bayley (1995) noted that river restoration (in addition to whatever other goals are hoped for) should include a return of the natural flooding to the floodplain, at least at a reduced scale, and regular access of water and biota to previously isolated floodplain to increase the productivity of multiple fish species. Such floodplain reconnection can also help the completion of needed life-cycle migrations for animal life, especially for endangered amphibians,

that suffer population declines due to "habitat split" when urbanization disconnects breeding areas from other needed habitats (Becker et al. 2007). Reconnecting parts of the original floodplain has been a key element of the restoration of the Kissimmee River in Florida, a project showing solid fish and wildlife benefits (Box 16.2). Floodplain restoration efforts in sections of the Rhine River have also shown success (Jungwirth et al. 2002).

---

### Box 16.2 Restoration of the Kissimmee River, Florida.

A great example of the multiple benefits of floodplain reconnection has been the 30+-year process of restoration of the Kissimmee River in central Florida. This river was once a 167-km-long meandering river draining a headwaters chain of lakes that supported abundant and diverse wildlife. Its floodplain was of course subject to flooding, which became the primary reason for its conversion (straightening and channeling) between 1962 and 1971 into a 95-km canal (named C-38) that consists of a series of compartments complete with locks, water control structures, and levees. This controlled the flooding and the subsequent draining of floodplain areas allowed for grazing, dairy operations, and crop agriculture on the former floodplain. The various agricultural efforts on the former floodplain led to increased eutrophication of downstream water bodies. The river meanders were either disconnected or only loosely connected to C-38 and became subjected to low water level and flow rates, and stagnation resulting in low DO. Fish and wildlife populations were severely impacted. The loss in aesthetics combined with the wildlife and ecosystem quality degradation led to citizen action to restore the river, at least partially.

The planning process began rather haphazardly in the early 1970s but was focused after a 1988 symposium to adapt a goal of reestablishment of ecological integrity. Fortunately there were decades of studies on the ecosystem to guide the process. The first phase of the Kissimmee River Restoration Project (KRRP) was completed in 2001 and consisted of demolishing a lock and a water control structure, building 2.4 km of new river channel, reestablishing 24 contiguous kilometers of river channel, and backfilling 12 km of C-38. River metabolism studies in an old channel that was restored showed that after restoration average DO levels increased from < 2.0 to 4.7 mg/L, while P/R ratios increased from 0.29 to 0.51. GPP increased by nearly an order of magnitude and R rates increased by a factor of nearly six. Metabolism rates after restoration became similar to other blackwater rivers in the US Southeast. A 2009 analysis by the ACOE found that in the restored areas bottom organic deposits had decreased, and sand bars became reestablished. The percentage of the fish community comprised by largemouth bass and sunfish increased from 38 to 63%, long-legged wading bird populations increased significantly, ducks returned to the river, and eight shorebird species that were absent before restoration have returned to the river and floodplain. Since those early successes, work has continued, and as of 2021 the final construction phases of the KRRP were completed, restoring > 100 km$^2$ of floodplain, > 8000 ha of wetlands, and 70 km of the historic river channel. It also involved backfilling 35 km of canal and removing two water control structures. There are still periodic problems that occur in some newly restored areas, with incidents of hypoxia and fish kills, as reengineering a more natural flow is currently a work in progress.

(Information from Whalen et al. 2002; Colangelo 2007; US ACOE 2009; Ollis et al. 2022; https://www.sfwmd.gov/our-work/kissimmee-river.)

Large-scale floodplain reconnection will serve to reduce flood damage risk in two ways (Opperman et al. 2009). First, if the floodplain is reconnected, land use on it will of necessity be limited to compatible uses such as periodic grazing, sustainable timber production, and flood-tolerant or seasonal agriculture. When expensive human infrastructure such as housing is excluded from floodplains, flood damage and subsequent insurance costs to riparian landowners will be far less. Second, reconnected floodplains will store and convey floodwaters, protecting upland and downstream areas from flood damage. An example of an engineered floodplain that has successfully prevented numerous floods in populated areas is the Yolo Bypass in California's Sacramento Valley (Sommer et al. 2001). Designed and completed in the 1930s, this system supports seasonal agriculture in low flow periods, then provides a fish and wildlife cornucopia during winter when it receives flood flows from several sources so that the heavily engineered Sacramento River and its associated populated areas do not flood. The Yolo Bypass has a mean depth of 3 m, supports 42 species of fish and vast numbers of waterfowl as part of the Pacific flyway, and is used by many shorebird species. Researchers (Sommer et al. 2001) have

shown that the Bypass provides diverse habitat and abundant invertebrate food as drift for species such as Chinook salmon and is an important source of phytoplankton and carbon to help support the food web in San Francisco Bay.

The ACOE has spent vast amounts of taxpayer dollars over the past century in projects designed to control flooding or increase river barge traffic. In recent decades it has been involved in river restoration, including reversing some of its own previous work (see Box 16.2). The construction of levees of course decouples the river from its floodplain and there is evidence that the presence of levees leads to increased flooding (see Chapter 17) in unprotected areas, while floodplains protect communities from flooding (Opperman et al. 2009). Thus, *levee removal* is a viable tactic to modify flows, reduce downstream flooding, and improve fish and wildlife production. During extreme floods, the presence of levees can serve to have a "bottling" effect on the river, causing it to rise rapidly, where it will overtop, or even blowout, levee sections. To ameliorate this, as well as reconnect with the floodplain, a tactic being used in areas such as along the Missouri River is to create "setback levees," more distant from the river proper in place of the more nearshore levees (Gore and Shields 1995). As with smaller streams, the physical restoration of river sinuosity across a floodplain is also a viable tactic, sometimes enhanced with erosion-control devices as outlined above (Whalen et al. 2002; Southeast Watershed Forum 2003). Bear in mind that obtaining the land to do so can be challenging.

**Dam modification** is another means of flow and channel modification. In many locations dams cannot be removed due to their flood control, water supply, or power production functions. Thus, dams in many locations are equipped with fish ladders, modified locks, and rock arch rapids to promote migratory fish passage (see Chapter 10). Where some rivers downstream of dams are sediment starved due to accumulation behind the dam, sediment flushing gates or sediment bypass facilities can be created to alleviate that condition (Nakamura et al. 2006). Dams are not built for the benefit of ecosystems; they are built for the benefit of society's various needs. Wildlife and vegetation require flood conditions during key periods in their life cycles to thrive, however. Unfortunately these key periods are usually at odds to human uses. A thorough understanding of the downstream ecosystem's needs coupled with the willingness to revise water release times accordingly can restore downstream riparian zones. Rood et al. (2005) describe how such "systemic restoration" has been successfully accomplished in rivers in the Western USA, especially regarding the cottonwood and willow riparian forests. As mentioned in Chapter 10, there have been experimental controlled releases (floods) from Glen Canyon Dam on the Colorado River to work toward providing a more natural ecosystem downstream of Lake Powell in Grand Canyon (Patten et al. 2001; Valdez et al. 2001; Cornwall 2017). Bear in mind that climate change has induced recent droughts in the Colorado River watershed that has precipitously dropped water levels in its great reservoirs, so such experimental floods may become rare, if done at all in the future. Some fishes are dependent upon certain natural water movement regimes to migrate or reproduce and dam operators can regulate water releases during critical seasons to enhance the survival and reproduction of historically important resident fishes (Rulifson and Manooch 1990). As the water behind a dam concerns many stakeholders (power companies, farmers, ranchers, cities, industries, fisheries biologists and other scientists, sport fishermen, kayakers) restoration plans will need to accommodate many needs when devising flow modifications. Richter et al. (2006) recommend a science-based collaborative process called adaptive management to develop and subsequently fine-tune below-dam flow regulation to satisfy ecological as well as anthropogenic needs. This process includes orientation meetings of stakeholders, scientific literature reviews that feed into a flow recommendation workshop, and implementation of flow management, all followed by monitoring, research, and data collection to subsequently fine-tune the flow recommendations if necessary.

Rehabilitation of a river and its estuary may involve **removal of pollutants** that have accumulated in the bottom sediments. The levels of pollutant may be directly harmful to the organisms or their animal predators (see Chapter 14) or they may be at levels that do not permit human consumption and fishing (or fish sale bans have been applied to the system). In the case of Kepone pesticide pollution in the James River, Virginia, fishing was banned for 13 years until the polluted sediments were buried sufficiently that the aquatic life was at safe levels. The classic case of polluted sediment removal is the PCB contamination of Hudson River, New York, which required a lengthy and expensive court battle to force the polluting entity (GE) to dredge and dispose of vast amounts of sediment (see Chapter 14). It is notable that microbial activities do biodegrade pollutants in natural settings, although it can take a long time (Wackett and Robinson 2020).

Stream and river degradation can cause local extirpation of threatened or endangered species (mussels or fish, for instance), decreases in biodiversity, and/or loss of game fish (such as striped bass) or one or more commercially important species. These losses can be a result of sedimentation from erosion covering needed habitat, organic pollution leading to hypoxia or anoxia, overfishing of selected species, exotic species introductions competing with or preying upon native species, dam construction and subsequent migratory movement impediments, chemical pollution, or any combinations of the above. **Restoration to restore native organisms** generally will involve large-scale river restoration (or a significant portion thereof), i.e. habitat more extensive than a single stream is likely required. Of course if pollution or sedimentation is involved, removing the sources of these insults through the stream protection methods listed above is needed first. If dams are involved, releases to alter flow patterns may be necessary to approximate or restore original habitat conditions. Dam-released water may require mechanical reoxygenation to provide appropriate living conditions for the target species (Southeast Watersheds Forum 2004). In terms of human behavior, catch limits or bans on critical species, as occurred with striped bass in the Southeastern USA, will need to be included (Southeast Watersheds Forum 2004).

However, for highly impacted organisms, reintroduction to their native habitat will be needed to revive the local population. For species that are common elsewhere but locally extirpated or reduced, such species can be obtained from commercial or government-run aquaculture facilities. If rare or endangered species are involved, rearing quantities of them under laboratory conditions is likely required, until significant quantities are produced to reintroduce into the river. Developing successful rearing methods and producing sufficient quantities of the organism may be a multi-year process, so patience is required. As examples that this process does work; the endangered interrupted rock-snail (*Leptoxis foremani*) was successfully reintroduced to the Coosa River, Alabama, and the smoky madtom (*Noturus baileyi*) and yellowfin madtom (*N. flavipinnis*) were successfully reintroduced to Abrams Creek, Tennessee (Southeast Watersheds Forum 2004).

In order to maintain or reintroduce aquatic biota to a degraded system, appropriate habitat (providing food and shelter) must be available for the organisms. In some systems large artificial riffle areas have been introduced, with success depending upon the specific locations of the enhancement (Gore and Shields 1995). Habitat can also be created by additions of large woody debris (which can be anchored if need be). Such debris will interact with stream flow to create pools, alter flow patterns, and hide organisms from predators (Finkenbine et al. 2000). In blackwater rivers (see Chapter 8) it is essential to have woody substrata available due to the soft-bottomed streams and rivers. In general, reforestation of floodplains and streamside buffers will provide a constant source of in-stream woody habitat and organic matter (Wang et al. 2001; Palmer et al. 2005). As mentioned above, current-deflecting rock vanes and root wads also serve as habitat (Fig. 16.7). In the lower, tidal portions of rivers and their estuaries major loss of habitat as SAV, especially seagrasses, has occurred in many locations. While natural diseases impact SAV, anthropogenic reasons for SAV losses include direct toxicity to eelgrass by elevated nitrate pollution (Burkholder et al. 2007), shading by phytoplankton blooms (Dennison et al. 1993; Greening et al. 2014), periphyton or macrophyte overgrowths from eutrophication (Tomasko and Lapointe 1991; Hauxwell et al. 2001), or high turbidity from erosion (Dennison et al. 1993). Revegetation of SAV in lower rivers and estuaries to restore habitat is a viable tactic (Thorhaug 1986), especially when it is coupled with nutrient

and TSS runoff reduction on-land. It is best accomplished in former beds; revegetation in areas of high current and resuspension may not be successful (Thorhaug 1986), and planting depth is critical to avoid problems of light limitation, heat exposure, or scouring in shallows (Aoki et al. 2020). In some cases use of a constructed sill made from native materials that offers protection to the newly planted shoreline bed will encourage growth.

Restoration to achieve improved **aesthetics** and **human recreation** is a sought-after goal in many situations (Bernhardt et al. 2005). In most restoration activities aesthetics will improve along with successful implementation of many of the tactics discussed earlier in this chapter. Stream bank stabilization coupled with revegetation of mixed native species leads to a visually attractive stream corridor, along with improved ecological function (Fig. 16.8). Removal of upstream or upslope sources of sedimentation and other pollution will likewise improve a stream or river's looks (and smell, in some cases). Establishment of streamside forests offers shade and a source of wood to improve a stream or river fishery, a boon to the angler. Vegetated buffers and restored floodplains can be used for passive recreation such as hiking and canoe and kayak launches, and even primitive camping can be established to offer improved recreation. Where river depths are sufficient to support tour boats, a diverse and visible wildlife assemblage is a major plus. Indeed, creating trails in such riparian areas by volunteer or nonprofit groups integrates biological sciences with social sciences (Groffman et al. 2003).

While restoration is usually a very positive thing for rivers, Jackson and Pringle (2010) provide some cautionary statements, especially regarding reconnection of river areas. They note how in some cases barriers need to be maintained when invasive species are seeking to move into a new area (using the example of bighead and silver carp in the Chicago ship canal moving toward Lake Michigan). Another example is connecting heavily polluted waters (and their sediments) with nearby clean waters. Also they note that in some cases off-channel habitats have become refuges for threatened species and connection to larger water bodies may have unpredictable consequences to their health. Thus, the potential negatives must be considered along with the positives in the restoration planning process.

To summarize, large-scale river restoration will likely involve combinations of techniques (including erosion prevention techniques as discussed above, runoff controls, revegetating buffer zones, enhanced point-source pollutant controls, dam removal, reestablishment of connectivity of water bodies and wetlands, etc., as well as organism reintroductions. Unfortunately few river restoration projects have had concurrent extensive monitoring projects to determine the efficacy of the project. A synthesis of information on thousands of stream and river restoration projects in the USA is reported online in the National River Restoration Science Synthesis (NRRSS) database (Bernhardt et al. 2005).

## 16.4 Summary

- Whether from urbanization or intense agricultural usage, stream watershed land runoff can severely degrade stream ecological and economic values. Thus, stream protection is a very active applied scientific field.
- The most important tactic is to reduce stormwater runoff at or near the source. Such approaches include water conservation, stormwater reuse, purchase of conservation easements, collecting runoff in rain barrels or cisterns, and breaking up contiguous impervious surfaces with natural areas; much of this is included in pre-planning for low-impact development.,
- Some tactics successfully treat stormwater pollution as well as reduce its volume. These approaches can include use of pervious pavement instead of impervious, creation of treatment wetlands and enhanced wet detention ponds, use of vegetated streamside buffers, installation of bioretention systems and rain gardens, and use of infiltration devices fitted to defined areas.
- Stream and river restorations are currently areas of focused research, with the latter a much more difficult task due to high cost, scale, complexity, and cross-border issues. For either it is critical to realize that you can never return a system to its pristine state, but instead you can use committees of stakeholders to set achievable goals. In

almost all circumstances financial and legal assistance from government organizations is likely to be critical to success.

- Stream restoration techniques include removal or reduction of pollution sources, restoration of some degree of sinuosity, revegetation of restored banks, in-stream use of root balls or rock vanes to reduce in-stream erosion, use of other installed devices to provide shade and cover to fish, and reintroduction of degraded native species.
- River restoration can involve some of the same techniques at scale, as well as use of dam flow modifications, removal or setback of levees in selected areas, reconnection with the floodplain, and reintroduction of sport or even commercial fisheries; again, note that multiple stakeholders have sometimes competing uses in mind.

## References

Aoki, L.R., K.J. McGlathery, P.L. Wiberg and A. Ai-Haj. 2020. Depth affects seagrass restoration success and resilience to marine heat wave disturbance. *Estuaries and Coasts* 43:316–328.

Asadian, Y. and M. Weiler. 2009. A new method in measuring rainfall interception by urban trees in coastal British Columbia. *Water Quality Research Journal of Canada* 44:16–25.

Bayley, P.B. 1995. Understanding large river-floodplain ecosystems. *BioScience* 45:153–158.

Becker, C.G., C.R. Fonseca, C.F. B. Haddad, R.F. Batista and P.I. Prado. 2007. Habitat split and the global decline of amphibians. *Science* 318:1775–1777.

Bernhardt, E.S., M.A. Palmer, J.D. Allan, et al. 2005. Synthesizing U.S. river restoration efforts. *Science* 308: 636–637.

Booth, D.B., J.B. Karr, S. Schauman, et al. 2004. Reviving urban streams: land use, hydrology, biology and human behavior. *Journal of the American Water Resources Association* 40:1351–1364.

Borne, K.E., C.C. Tanner and E.A. Fassman-Beck. 2013. Stormwater nitrogen removal performance of a floating treatment wetland. *Water Science and Technology* 68(7):1657–1664.

Borne, K.E., E.A. Fassman-Beck, R.J. Winston, W.F. Hunt and C.C. Tanner. 2015. Implementation and maintenance of floating treatment wetlands for urban stormwater management. *Journal of Environmental Engineering* 14(11):04015030.

Bright, T.M., M.R. Burchell, W.F. Hunt and W. Price. 2011. Feasibility of a dune infiltration system to protect

North Carolina beaches from fecal bacteria contaminated storm water. *Journal of Environmental Engineering* 137(10):968–979.

Burkholder, J.M., D.A. Tomasko and B.W. Touchette. 2007. Seagrasses and eutrophication. *Journal of Experimental Marine Biology and Ecology* 350:46–72.

Burtchett, J.M., M.A. Mallin and L.B. Cahoon. 2017. Micro-zooplankton grazing as a means of fecal bacteria removal in stormwater BMPs. *Water Science and Technology* 75:2702–2715.

Cahoon, L.B. 1996. Water quality variability in stormwater detention ponds in New Hanover County, North Carolina. Pages 75–78 in Solutions: A Technical Conference on Water Quality (Proceedings), March 19–21. North Carolina State University, Raleigh.

Cizek, A.R., J.P. Johnson, F. Birgand, W.F. Hunt and R.A. McLaughlin. 2019. Insights from using in-situ ultraviolet-visible spectroscopy to assess nitrogen treatment and subsurface dynamics in a regenerative stormwater conveyance (RSC) system. *Journal of Environmental Management* 252:109656.

Clark, J. 1978. Freshwater wetlands: habitats for aquatic invertebrates, amphibians, reptiles and fish. Pages 330–343 in Greeson, P.E., J.R. Clark and J.E. Clark (eds.), *Wetland Functions and Values: The State of Our Understanding*. American Water Resources Association, Minneapolis, MN.

Colangelo, D.J. 2007. Response of river metabolism to restoration of flow in the Kissimmee River, Florida, U.S.A. *Freshwater Biology* 52:459–470.

Cordy, G.E., N.L. Duran, H. Bouwer, et al. 2004. Do pharmaceuticals, pathogens, and other organic wastewater compounds persist when wastewater is used for recharge? *Ground Water Monitoring and Remediation* 24:58–69.

Cornwall, W. 2017. U.S.–Mexico water pact aims for a greener Colorado delta. *Science* 357:635.

Costanza, R., R. d'Arge, R. de Groot, et al. 1998. The value of the world's ecosystem services and natural capital. *Ecological Economics* 25:3–15.

Cowardin, L.M., V. Carter, F.C. Golet and E.T. LaRoe. 1979. Classification of Wetlands and Deepwater Habitats of the United States. FWS/OBS-79/31. Office of Biological Services, Fish and Wildlife Service, US Department of the Interior, Washington, DC.

Craig, L.S., M.A. Palmer, D.C. Richardson, et al. 2008. Stream restoration strategies for reducing river nitrogen loads. *Frontiers in Ecology and the Environment* 6:529–538.

Crumpton, W.G. 1995. Pollution dynamics within stormwater wetlands II. Mesocosms illuminate role of organic matter in nutrient removal. Technical Note 54. *Watershed Protection Techniques* 1:214–216.

Davies, C.M. and H.J. Bavor. 2000. The fate of stormwater-associated bacteria in constructed wetland and water pollution control pond systems. *Journal of Applied Microbiology* 89:349–360.

Dennison, W.C., R.J. Orth, K.A. Moore, et al. 1993. Assessing water quality with submersed aquatic vegetation. *BioScience* 43:86–93.

Doll, B.A., G.L. Grabow, K.R. Hall, et al. 2003. *Stream Restoration: A Natural Channel Design Handbook*. North Carolina Stream Restoration Institute, North Carolina State University, Raleigh.

Duan, S., P.M. Mayer, S.S. Kaushal, B.M. Wessel and T. Johnson. 2019. Regenerative stormwater conveyance (RSC) for reducing nutrients in urban stormwater runoff depends upon carbon quantity and quality. *Science of the Total Environment* 652:134–146.

Dunnett, N. and A. Claydon. 2007. *Rain Gardens: Managing Water Sustainably in the Garden and Designed Landscape*. Timber Press, Portland, OR.

Engelhardt, K.A. and M.E. Ritchie. 2001. Effects of macrophyte species richness on wetland ecosystem functioning and services. *Nature* 411:687–689.

England, G. and S. Stein. 2007. *Stormwater BMPs: Selection, Maintenance and Monitoring*. Forester Press, Santa Barbara, CA.

Evans, R.O., R.W. Skaggs and J.W. Gilliam. 1995. Controlled versus conventional drainage effects on water quality. *Journal of Irrigation and Drainage Engineering* 121:271–276.

Ewel, K.C. 1990. Multiple demands on wetlands: Florida cypress swamps can serve as a case study. *BioScience* 40:660–666.

Farmer, A. and M. Braun. 2002. Fifty Years of the Rhine Commission: A Success Story in Nutrient Reduction. SCOPE Newsletter, No. 47. Institute for European Environmental Policy, Brussels.

Field, R., A.N. Tafuri, S. Muthukrishnan, B.A. Acquisto and A. Selvakumar. 2006. *The Use of Best Management Practices (BMPs) in Urban Watersheds*. DEStech Publications, Lancaster, PA.

Finkenbine, J.K., J.W. Atwater and D.S. Mavinic. 2000. Stream health after urbanization. *Journal of the American Water Resources Association* 36:1149–1160.

Gerba, C.P., J.A. Thurston, J.A. Falabi, P.M. Watt and M.M. Karpiscak. 1999. Optimization of artificial wetland design for removal of indicator microorganisms anad pathogenic protozoa. *Water Science and Technology* 40:363–368.

Goldfarb, B. 2018. Beavers, rebooted: artificial beaver dams are a hot restoration strategy, but the projects aren't always welcome. *Science* 360:1058–1061.

Gopal, B. 1999. Natural and constructed wetlands for wastewater treatment: potentials and problems. *Water Science and Technology* 40:27–35.

Gore, J.A. and F.D. Shields Jr. 1995. Can large rivers be restored? *BioScience* 45:142–152.

Greening, H.A., A. Janicki, E.T. Sherwood, R. Pribble and J.O.R. Johansson. 2014. Ecosystem responses to long-term nutrient management in an urban estuary: Tampa Bay, Florida, USA. Estuarine, Coastal and Shelf Science 151:A1–A16.

Groffman, P.M., D.J. Bain, L.E. Band, et al. 2003. Down by the riverside: urban riparian ecology. *Frontiers in Ecology and the Environment* 1:315–321.

Grogan, A.E. and M.A. Mallin. 2021. Successful mitigation of stormwater-driven nutrients, fecal bacteria and suspended solids loading in a recreational beach community. *Journal of Environmental Management* 281:111853.

Gurbisz, C., W.M. Kemp, L.P. Sanford and R.J. Orth. 2016. Mechanisms of storm related loss and resilience in a large submersed plant bed. *Estuaries and Coasts* 39: 951–966.

Gurnell, A., M. Lee and C. Souch. 2007. Urban rivers: hydrology, geomorphology, ecology and opportunities for change. *Geography Compass* 1(5):1118–1137.

Hatt, B.E., T.D. Fletcher, C.J. Walsh and S.L. Taylor. 2004. The influence of urban density and drainage infrastructure on the concentrations and loads of pollutants in small streams. *Environmental Management* 34:112–124.

Hauxwell, J., J. Cebrian, C. Furlong and I. Valiela. 2001. Macroalgal canopies contribute to eelgrass (*Zostera marina*) decline in temperate estuarine ecosystems. *Ecology* 82:1007–1022.

Hunt, B. 2003. Bioretention Use and Research in North Carolina and Other Mid-Atlantic States. NWQEP Notes, Number 109. North Carolina State University Cooperative Extension Service, Raleigh.

Hunt, W.F., A.R. Jarrett, J.T. Smith and L.J. Sharkey. 2006. Evaluating bioretention hydrology and nutrient removal at three field sites in North Carolina. *Journal of Irrigation and Drainage Engineering* 132:600–608.

Izaak Walton League. 2006. *A Handbook for Stream Enhancement & Stewardship.* The MacDonald and Woodward Publishing Company, Blacksburg, VA.

Jackson, C.R. and C.M. Pringle. 2010. Ecological benefits of reduced hydrological connectivity in intensely developed landscapes. *BioScience* 60:37–46.

Johnston, C.A. 1991. Sediment and nutrient retention by freshwater wetlands: effects on surface water quality. *Critical Reviews in Environmental Control* 21:491–565.

Jungwirth, M., S. Muhar and S. Shmutz. 2002. Reestablishing and assessing ecological integrity in riverine landscapes. *Freshwater Biology* 47:867–887.

Junk, W.J., P.B. Bayley and R.E. Sparks. 1989. The flood-pulse concept in river-floodplain systems. *Canadian Journal of Fisheries and Aquatic Sciences* 106:110–127.

Karathaniasis, A.D., C.L. Potter and M.S. Coyne. 2003. Vegetation effects on fecal bacteria, BOD, and suspended solid removal in constructed wetlands treating domestic wastewater. *Ecological Engineering* 20:157–169.

Knox, A.K., R.A. Dahlgren, K.W. Tate and E.R. Atwill. 2008. Efficacy of natural wetlands to retain nutrient, sediment and microbial pollutants. *Journal of Environmental Quality* 37:1837–1846.

Landry, M.S. and T.L. Thurow. 1997. *Function and Design of Vegetated Filter Strips: An Annotated Bibliography.* Texas State Soil and Water Conservation Board Bulletin No. 97–1. Texas State Soil and Water Conservation Board, Temple.

Lee, K.-H., T.M. Isenhart, R.C. Schultz and S.K. Mickelson. 2000. Multispecies riparian buffers trap sediment and nutrients during rainfall simulations. *Journal of Environmental Quality* 29:1200–1205.

Lenat, D.R. and J.K. Crawford. 1994. Effects of land use on water quality and aquatic biota of three North Carolina Piedmont streams. *Hydrobiologia* 294:185–199.

Leonard, L.A. and M.E. Luther. 1995. Flow hydrodynamics in tidal marsh canopies. *Limnology and Oceanography* 40:1474–1484.

Lewitus, A.J., L.B. Schmidt, L.J. Mason, et al. 2003. Harmful algal blooms in South Carolina residential and golf course ponds. *Population and Environment* 24: 387–413.

Line, D.E., W.A. Harman, G.D. Jennings, E.J. Thompson and D.L. Osmond. 2000. Nonpoint source pollutant load reductions associated with livestock exclusion. *Journal of Environmental Quality* 29:1882–1890.

Lloyd, S.D., T.H.F. Wong and B. Porter. 2002. The planning and construction of an urban stormwater management scheme. *Water Science and Technology* 45:1–10.

Lowrance, R., L.S. Altier, J.D. Newbold, et al. 1997. Water quality functions of riparian forest buffers in Chesapeake Bay watersheds. *Environmental Management* 21:687–712.

Madsen, T.V. and E. Warncke. 1983. Velocities of currents around and within submerged aquatic vegetation. *Archiv für Hydrobiologie* 97:389–394.

Mallin, M.A., S.H. Ensign, M.R. McIver, G.C. Shank and P.K. Fowler. 2001. Demographic, landscape, and meteorological factors controlling the microbial pollution of coastal waters. *Hydrobiologia* 460:185–193.

Mallin, M.A., S.H. Ensign, T.L. Wheeler and D.B. Mayes. 2002. Pollutant removal efficacy of three wet detention ponds. *Journal of Environmental Quality* 31:654–660.

Mallin, M.A., J. McAuliffe, M.R. McIver, D. Mayes and M.R. Hanson. 2012. High pollutant removal efficacy of a large constructed wetland leads to receiving stream improvements. *Journal of Environmental Quality* 41: 2046–2055.

Mallin, M.A., M.I.H. Turner, M.R. McIver, B.R. Toothman and H.C. Freeman. 2016. Significant reduction of fecal bacteria and suspended solids loading by coastal Best Management Practices. *Journal of Coastal Research* 32:923–931.

McCabe, K.M., E.M. Smith, S.Q. Lang, C.L. Osburn and C.R. Benitez-Nelson. 2021. Particulate and dissolved organic matter in stormwater runoff influences oxygen demand in urbanized headwater catchments. *Environmental Science and Technology* 55:952–961.

Nakamura, K., K. Tockner and K. Amano. 2006. River and wetland restoration: lessons from Japan. *BioScience* 56:419–429.

NCDWQ. 2005. *State Stormwater Management (15A NCAC 2H.1000): Project Characteristics and Compliance Account for Five Selected Coastal Counties in Southeastern North Carolina.* North Carolina Division of Water Quality, North Carolina Department of Environment and Natural Resources, Raleigh.

NCDWQ. 2007. *Stormwater Best Management Practices Manual.* North Carolina Division of Water Quality, North Carolina Department of Environment and Natural Resources, Raleigh.

NRC. 2009. *Urban Stormwater Management in the United States.* National Research Council, Water Science and Technology Board. The National Academies Press, Washington, DC.

NRCS. 2000. *Conservation Buffers to Reduce Pesticide Losses.* Natural Resources Conservation Service, United States Department of Agriculture, Washington, DC.

Ollis, S., S. Davis and A. Chelette. 2022. Northern Everglades and Estuaries Protection Program—Annual Progress Report. Chapter 8A in 2022 South Florida Environmental Report, Volume 1. South Florida Management District, West Palm Beach.

Opperman, J.J., G.E. Galloway, J. Fargione, J.F. Mount, B.D. Richter and S. Secchi. 2009. Sustainable floodplains through large-scale reconnection to rivers. *Science* 326:1487–1488.

Palmer, M.A., E.S. Bernhardt, J.D. Allan, et al. 2005. Standards for ecologically successful river restoration. *Journal of Applied Ecology* 42:208–217.

Patten, D.T., D.A. Hartman, M.I. Vaita and T.J. Randle. 2001. A managed flood on the Colorado River: background, objectives, design and implementation. *Ecological Applications* 11:635–643.

Pennington, S.R., M.D. Kaplowitz and S.G. Witter. 2003. Reexamining best management practices for improving water quality in urban watersheds. *Journal of the American Water Resources Association* 39:1027–1041.

Petticrew, E.L. and J. Kalff. 1992. Water flow and clay retention in submerged macrophyte beds. *Canadian Journal of fisheries and Aquatic Sciences* 49:2483–2489.

Pond, R. 1995. Pollution dynamics within stormwater wetlands I. Plant uptake. Technical Note 53. *Watershed Protection Techniques* 1:210–213.

Price, W.D., M.R. Burchell II, W.F. Hunt and G.M. Chescheir. 2013. Long-term study of dune infiltration

systems to treat coastal stormwater runoff for fecal bacteria. *Ecological Engineering* 52:1–11.

Rein, F.A. 1999. An economic analysis of vegetative buffer strip implementation case study: Elkhorn Slough, Monterey Bay, California. *Coastal Management* 27: 377–390.

Richardson, C.J. 1985. Mechanisms controlling phosphorus retention capacity in freshwater wetlands. *Science* 228:1424–1427.

Richter, B.D., A.T. Warner, J.L. Meyer and K. Lutz. 2006. A collaborative and adaptive process for developing environmental flow recommendations. *River Research and Applications* 22:297–318.

Robinson, P.J. 2005. *North Carolina Weather and Climate.* The University of North Carolina Press, Chapel Hill.

Rood, S.B., G.M. Samuelson, J.H. Braatne, C.R. Gourley, F.M.R. Hughes and J.M. Mahoney. 2005. Managing river flows to restore floodplain forest. *Frontiers in Ecology and the Environment* 3:193–201.

Rulifson, R.A. and C.S. Manooch III. 1990. Recruitment of juvenile striped bass in the Roanoke River, North Carolina, as related to reservoir discharge. *North American Journal of Fisheries Management* 10:397–407.

SCDHEC. 2007. State of Knowledge Report: Stormwater Ponds in the Coastal Zone. October. South Carolina Department of Health and Environmental Control, Office of Ocean and Coastal Resource Management, North Charleston.

Schueler, T.R. 1995a. Pollutant removal pathways in Florida swales. Technical Note 65. *Water Protection Techniques* 2:299–301.

Schueler, T.R. 1995b. Architecture of urban stream buffers. *Water Protection Techniques* 1:155–163.

Simmons, D.L. and R.J. Reynolds. 1982. Effects of urbanization on base flow of selected south-shore streams, Long Island, New York. *Water Resources Bulletin* 18: 797–806.

Sommer, T., B. Harrell, M. Nobriga, et al. 2001. California's Yolo Bypass: evidence that flood control can be compatible with fisheries, wetlands, wildlife and agriculture. *Fisheries* 26:6–16.

Song, B., M.A. Mallin, A. Long and M.R. McIver. 2014. Factors Controlling Microbial Nitrogen Removal Efficacy in Constructed Stormwater Wetlands. Report No. 443. Water Resources Research Institute, University of North Carolina, Raleigh.

Southeast Watersheds Forum. 2003. *A Guide Book for Local River Restoration.* Southeast Watersheds Forum, vol. 6, issue 1, Spring/Summer. Southeast Watersheds Forum, Nashville, TN.

Southeast Watersheds Forum. 2004. *A Community Guide for Restoration of Fish and Aquatic Species.* 37229. Southeast Watersheds Forum, Nashville, TN.

Spruill, T.B. 2003. Effectiveness of riparian buffers in controlling ground-water discharge of nitrate to streams in selected hydrogeologic settings of the North Carolina Coastal Plain. *Water Science and Technology* 49:63–70.

Stenström, T.A. and A. Carlander. 2001. Occurrence and die-off of indicator organisms in the sediment in two constructed wetlands. *Water Science and Technology* 44:223–230.

Sudduth, E.B., J.L. Meyer and E.S. Bernhardt. 2007. Stream restoration practices in the southeastern United States. *Restoration Ecology* 15:573–583.

Thayer, G.W., H. Stuart, W.J. Kenworthy, J.F. Ustach and A.B. Hall. 1978. Habitat values of salt marshes, mangroves and seagrasses for aquatic organisms. Pages 235–247 in Greeson, P.E., J.R. Clark and J.E. Clark (eds.), *Wetland Functions and Values: the State of Our Understanding*. American Water Resources Association, Minneapolis, MN.

Thorhaug, A. 1986. Review of seagrass restoration. *AMBIO* 15:110–117.

Tomasko, D.A. and B.E. Lapointe. 1991. Productivity and biomass of *Thalassia testudinum* as related to water column nutrient availability and epiphyte levels: field observations and experimental studies. *Marine Ecology Progress Series* 75:9–17.

US ACOE. 2009. *Kissimmee River Restoration Progress Facts*. November. US Army Corps of Engineers, Jacksonville, FL.

US EPA. 1999. Storm Water Technology Fact Sheet: Wet Detention Ponds. EPA 832-F-99-048. Office of Water, US EPA, Washington, DC.

Valdez, R.A., T.L. Huffnagle, C.C. McIvor, T. McKinney and W.C. Liebfried. 2001. Effects of a test flood on fishes of the Colorado River in Grand Canyon, Arizona. *Ecological Applications* 11:686–700.

Vymazal, J. 2005. Removal of enteric bacteria in constructed treatment wetlands with emergent macrophytes: a review. *Journal of Environmental Science and Health* 40:1355–1367.

Wackett, L.P. and S.L. Robinson. 2020. The ever-expanding limits of enzyme catalysis and biodegradation: polyaromatic, polychlorinated, polyfluorinated and polymeric compounds. *Biochemical Journal* 477:2875–2891.

Walsh, C.J., T.D. Fletcher and A.R. Ladson. 2005. Stream restoration in urban catchments through redesigning stormwater systems: looking to the catchment to save the stream. *Journal of the North American Benthological Society* 24:690–705.

Walter, R.C. and D.J. Merritts. 2008. Natural streams and the legacy of water-powered mills. *Science* 319:299–304.

Wang, L., J. Lyons, P. Kanehl and R. Bannerman. 2001. Impacts of urbanization on stream habitat and fish across multiple spatial scales. *Environmental Management* 28:255–266.

Weisner, S.E.B., P.G. Eriksson, W. Graneli and L. Leonardson. 1994. Influence of macrophytes on nitrate removal in wetlands. *AMBIO* 23:363–366.

Welsch, D.J. 1991. Riparian Forest Buffers: Function and Design for Protection and Enhancement of Water Resources. NA-PR-07-91. United States Department of Agriculture, Forest Service, Northeastern Area, Forest Resources Management, Radnor, PA.

Whalen, P.J., L.A. Toth, J.W. Koebel and P.K. Strayer. 2002. Kissimmee River restoration: a case study. *Water Science and Technology* 45:55–62.

Woltemade, C.J. 2000. Ability of restored wetlands to reduce nitrogen and phosphorus concentrations in agricultural drinking water. *Journal of Soil and Water Conservation* 55:303–309.

Wossink, A. and B. Hunt. 2003. The Economics of Structural Stormwater BMPs in North Carolina. Report No. 344. Water Resources Research Institute, University of North Carolina, Raleigh.

# CHAPTER 17

# Floods, Hurricanes, and Climate Change

Meteorological events cause disturbances to riverine ecosystems. Such disturbances are a normal part of the life of a river. Storm-induced floods rearrange channel units and subunits, destroy and create habitat (Parsons et al. 2005), displace organisms and bring in new organisms, and can reduce or enhance river productivity. It has been argued that flooding should not be considered a disturbance at all (Sparks et al. 1990). In general, after natural disturbances (flooding or droughts) there is recovery of the system, and in some cases rapid recovery, or system resets (Minshall et al. 1985). However, extreme weather events can cause extreme disturbances to occur. Periods of excessive rain, tropical storms, hurricanes, and winter storms (called nor'easters in the Eastern USA) result in massive flooding and severe winds that cause disasters to human life and property and the natural environment. The unnatural human development along rivers and estuaries exacerbates environmental impacts (Sparks et al. 1990). This chapter looks at some key impacts of floods and major storms to river ecosystems. Also, the climate is rapidly changing due to excessive greenhouse gas production, and some realized and potential impacts are discussed below. Note that at this point some of the riverine impacts from climate change remain largely speculative but well worth planning for.

## 17.1 Floods

Floods (defined here as stream waters overtopping the banks) are natural events that are caused by excessive rainfall occurring over an extended period, which saturates the soils and forces excess water to flow over the surface. Flooding can happen in any season, but in the northern hemisphere flooding is more likely to occur in colder weather when evapotranspiration is minimal.

### 17.1.1 Seasonal Floods as Disturbances

On a broad scale seasonal flooding is an expected and even desired phenomenon. In pre-Aswan Dam times seasonal flooding of the Nile brought nutrient-rich silt to farmers' fields to serve as an annual fertilizer (see Box 10.2). The seasonal change in winds on the Indian subcontinent, called the monsoon, brings annual rains needed for crop irrigation; but note that some monsoons bring massive rains and destructive flooding. In unaltered rivers the excess water flows over the floodplain, allowing fish access to the resources of the palustrine forest and enhancing biotic productivity. This was discussed in the flood-pulse concept in Chapter 6. Such flooding also allows for connectivity between the main channel and floodplain lakes such as oxbows, which as noted previously can serve as incubators of fish and invertebrate life (Penczak et al. 2003).

Floods can occur far inland; an excellent example is the Upper Midwest of the USA. In recent decades this area has seen a number of powerful floods, such as in 1993, 2008, 2009, and 2011, with some having significant impacts on the Mississippi River. Such flooding can arise in spring following a winter of heavy snowfall, which then proceeds to melt. When this is coupled with a rainy spring the problems begin, with devastation to flooded farm fields as well as residential areas, weakening or failure of dams, and compromised levees.

*River Ecology.* Michael A. Mallin, Oxford University Press. © Michael A. Mallin (2023). DOI: 10.1093/oso/9780199549511.003.0018

Dams and levees, of course, force river fragmentation (see Chapter 10) and exacerbate flooding damage and extent in adjacent nonleveed reaches. In summer 2021 unprecedented and unexpected large-scale flooding occurred in Europe following massive rains (as much as 15 cm in 24 hours), with huge property losses especially in Germany and Belgium (Cornwall 2021) and with news reports of > 180 human deaths. Many organizations attributed the flooding to climate change, with warmer air holding more moisture, leading to heavier rains. The Dutch appeared to be better prepared and suffered less damage, allowing for larger channels and larger amounts of "natural" floodplain to absorb the impact (Cornwall 2021).

## 17.1.2 Impacts to Streams

Streams in general are resilient and adapted to flooding, droughts, and other disturbances such as pollution incidents (Yount and Niemi 1990); note that if a stream is physically altered in a major way it might never recover to a similar state. Research on US Great Plains prairies (Dodds et al. 2004) found that following a scouring flood, the stream microbial community begins recovery within a couple days, the algal community within two weeks, with invertebrates and fish following soon thereafter. In general, waterborne microbes, algae, and invertebrates recolonize from upstream areas, and organisms of various taxa can recolonize from refuges in the hyporheic zone (Wallace 1990; Dodds et al. 2004). Some invertebrates may also repopulate from downstream areas if they are moving upstream to exploit new spaces for food or reproductive purposes (Wallace 1990). Fish and highly mobile invertebrates can recolonize from side channels, as well as upstream or downstream reaches. The same colonization pattern can follow droughts. Even in desert streams floods occur, and recovery is rapid, with much of the biota recolonizing from two weeks to two months after the event (Fisher et al. 1982). The first algal recolonizers in desert streams tend to be diatoms, followed by filamentous green algae and cyanobacteria. Grazers follow and their predators. Stream reaches can be repopulated from the air as well; adult insects fly in, while microbes and algae can be airborne and drift in on the wind.

## 17.1.3 Impacts to the River Ecosystem

Well-documented riverine impacts from flooding have occurred in the Missouri and Mississippi River watersheds, the largest lotic systems in the USA. For centuries the Mississippi River has been a major commercial conduit for goods; thus, maintaining navigable waters within is of interest and this is carried out by the US Army Corps of Engineers. The construction of dams, dikes, and levees (river armoring) over many decades was done to enhance navigation as well as protect adjacent municipalities and farmland from the natural flooding that occurs. Armoring alters the access of the channel waters to the floodplain, forcing them into scarce unarmored stretches and enhancing destructive flooding in such locations (Sparks et al. 1990; Myers and White 1993). Thus, despite the Mississippi being heavily altered with dams and levees, destructive floods still occur, with property damage from the 1993 flood approximately $12 billion (Myers and White 1993). Regarding river–estuarine impacts, the 1993 flood had major impacts on the Mississippi's receiving water, the Gulf of Mexico (Rabalais et al. 1998), which can serve as harbingers of future climate change on large rivers. The massive discharge of freshwater considerably lowered Gulf surface salinities, contributing to stronger and longer-lasting coastal stratification, and it greatly increased the summer estuarine load of nitrate and silicate, leading to much higher surface chlorophyll a concentrations. Surface cyanobacterial counts greatly increased, and the Si load allowed diatoms to thrive much longer into summer than normal. The net result was that the much higher surface summer primary production, upon death, sedimented to the bottom waters where it became a labile BOD load. This led to much more severe and extensive bottom-water hypoxia in the Gulf, with dead and moribund fish and shellfish seen in bottom waters by divers and submersibles (Rabalais et al. 1998).

Thus, while extensive reengineering of natural river systems has undoubtedly saved lives and property, very large floods will remain uncontrollable and cause extensive damage to structures and crops, especially on floodplains, with some writers viewing levee failures as rivers reclaiming their natural terrain (Myers and White 1993). Heavily armored rivers have no "escape valves;" such constrained rivers can burst levees and dikes and even dams. Researchers note that designing human areas on floodplains to allow periodic flooding will greatly reduce the extensive damages. This might include greatly limiting building on floodplains (limiting floodplain insurance would help there), allowing more river access to riparian wetlands, and accounting for occasional crop losses from extensive floods.

Hurricanes also cause major floods. Such floods can result from as little as a one-day downpour to a three-day drenching, depending on the rain generated by the hurricane, where the hurricane passes, and how long it takes to pass (translational speed; see Section 17.3.3). Flooding, especially coastal flooding, is expected to increase with global warming, as is rainfall (Sobel et al. 2016). Flooding also increases with land subsidence, which in many areas is caused by groundwater depletion, i.e. where groundwater recharge is outpaced by natural loss and human-caused withdrawals (Herrera-Garcia et al. 2021). Coastal areas are particularly at risk where there are large groundwater withdrawals coupled with exposure to sea-level rise from climate change (Sacatelli et al. 2020).

## 17.2 Hurricanes

**Hurricanes** are storms generated off the African coast and carried across the Atlantic Ocean; a given storm intensifies according to contact with warm waters (hurricanes are most abundant June through October); the US East Coast is also afflicted by nor'easters in winter. Hurricanes impact the US Mid-Atlantic to the US Southeast, all around Florida, the Gulf of Mexico, and south through Central America and the Eastern Pacific. In the Western Pacific such storms are called **typhoons** and in the Indian Ocean they are **cyclones**. The warmer the ocean waters are, the stronger is the storm when it makes landfall. Hurricane strength is rated by the Saffir–Simpson scale (Table 17.1); as the storm passes over land it loses wind strength. The areas that are most vulnerable to such storms are coastal lowlands, but major hurricane impacts, including

**Table 17.1** Hurricane categories according to the Saffir–Simpson scale.

| Hurricane category | Wind speed |
|---|---|
| 1 | 120–152 km/h (75–95 mph) |
| 2 | 153–176 km/h (96–110 mph) |
| 3 | 177–208 km/h (111–130 mph) |
| 4 | 209–248 km/h (131–155 mph) |
| 5 | >249 km/h (>155 mph) |

river flooding, can also occur hundreds of kilometers inland along the course of the weakening storm. For example, in 2004 category 2 Hurricane Francis passed through Florida, Georgia, and South Carolina and had weakened to a tropical depression by the time it passed through central North Carolina, but heavy rains caused by Francis caused very damaging flooding in the Appalachian Mountains, well to the west of the storm track. In the mountain city of Asheville the waters of the Swannanoa River breached their banks and flooded Biltmore Village, a popular shopping mecca. In 2021 category 4 Hurricane Ida came ashore in Louisiana, causing great damage, then passed across the center of the USA to also do great flood damage to New York and New Jersey on the East Coast.

Such storms lead to destruction by wind and wave action, rainfall, excessive runoff and flooding, erosion, and anthropogenic environmental impacts of many types. Wind speed and flooding are not always correlated; if the hurricane arrives at low tide less flood damage occurs near the coast, but if it arrives near high tide the storm surge can be devastating and felt well upriver. Climate change (see below) is likely to exacerbate the already formidable destruction and ecological damage caused by these major storms.

Damage from hurricanes increases exponentially with wind speed; the higher the wind speed is on arrival, the more damage is likely to power grids. Major hurricanes of category 5, 4, or 3 will lead to power failures, and hurricanes of category 2 and 1 may do so, depending on the hurricane path and speed. We will concentrate on riverine and riverine estuary impacts here: note that coastal and general estuarine impacts are detailed in Estuaries and Coasts (2006).

## 17.2.1 Impacts

### 17.2.1.1 Wind

Wind speeds exceed 155 mph for a category 5 storm on the Saffir–Simpson scale; as of 2023, there is no category 6 yet. Wind impacts to biota (Table 17.2) include habitat damage such as stripping of leaves and branches of upland forests and uprooting trees, creating a BOD load when washed into streams and rivers (Tabb and Jones 1962; Van Dolah and Anderson 1991; Tomasko et al. 2006). When trees and large forest debris fall or are blown into waterways they cause debris dams that are disruptive to boat traffic and stream flow. Note that powerful winds can also cause the spread of non-native species, including plants that can reproduce via fragmentation or microbes that cause plant diseases (Mallin and Corbett 2006). Wind may also cause the airborne distribution of chemical and biological pollutants, toxicants, or microbiological agents, particularly those associated with dust particles.

Most visible to humans is the destruction of human infrastructure in or near the storm track. Powerful winds cause the collapse of buildings and creation of debris that covers natural habitats, and severe damage to marinas and other littoral structures. Such winds cause the loss of electrical power through downed lines and substation damage.

Power losses to sewage treatment plants and the pump station associated with them will cause systems without backup power to go offline and reroute untreated or partially treated sewage into receiving waterways. This loads excessive BOD, nutrients, and fecal bacteria into those systems (Table 17.2).

### 17.2.1.2 Wave and Tidal Action

Wave and tidal action are generally thought of as marine phenomena, but a hurricane or nor'easter can bring reverberations of these actions well up rivers and into coastal lakes; note that wave and tidal action can have a multiplier effect (storm surge) if the storm hits on a high tide. A modeling effort in Rhode Island indicated that flooding due to such hurricane-driven storm surges is likely to adversely affect septic systems, with the number of threatened systems rising from about 2000 in a category 1 storm to approximately 4600 in a category 4 storm (Cox et al. 2020), causing thousands of dollars in repairs to homeowners. Another headache to humans is the sinking of boats in marinas within or near the storm track, and the release of multiple chemicals from flooded households, storage sheds, and businesses into the environment such as cleaning agents, propane tanks, auto fluids, and

**Table 17.2** Known environmental impacts of hurricanes to river ecosystems.

*Riparian forest damage*—will deposit large amounts of leaf litter and other forest debris into streams and river, creating a BOD load.

*Increases in snags and debris*—floodplain wash-in will greatly increase large woody debris that is hazardous to boating.

*Inputs of industrial chemicals to rivers*—flooded riverside chemical storage lagoons and industrial waste treatment facilities will release metals and organic chemicals of numerous types into the river and estuaries (Fig. 17.1a). Infrastructure involved in oil production and transport will fail and lead to both small and large spills into receiving waters.

*Floodplain flooding*—can allow fish access to floodplain (flood-pulse effect is positive for fish), but flooding can also release stored animal waste from CAFO lagoons (Figs. 17.1b–d) and flood human sewage infrastructure and septic systems.

*Release of human sewage to rivers*—flooding of human sewage and septic systems, coupled with power outages that cause pumps and treatment systems to go offline, releases vast amounts of human sewage to streams, rivers, estuaries, and lakes.

*Increase of river BOD, nutrients, and fecal microbe concentrations*—from human sewage inputs and CAFO damage and flooding.

*Decreases in DO in streams, rivers, estuaries, and lakes*—BOD loads from the sources listed above often cause anoxia and fish kills, sometimes on a massive scale (Figs. 17.1e, f).

*Freshening of estuarine waters*—the rainfall-driven increase in river discharge will decrease estuarine salinities, displacing estuarine fish and invertebrates and allowing freshwater organisms to move downstream in normally brackish water.

*Suppression of primary productivity*—in rivers and riverine estuaries increased turbidity from erosion and increase in water color from swamp water inputs will severely attenuate solar irradiance to phytoplankton, periphyton, and macrophytes.

*Algal blooms*—in contrast to the above, lakes, bays, and coastal lagoons will receive storm-driven nutrients, but their open waters will allow sufficient light transmission to lead to algal blooms.

industrial chemicals (Tilmant et al. 1994). Wave and tidal action also cause the release of large amounts of pollutants stored near or transported on water. A frequent example of this is in nearshore marine waters, where oil pumping and transport infrastructure is frequently damaged by hurricanes, such as in the Gulf (rigs, pipelines, tanks, etc.). An example of pollutant release by riverine erosive action occurred during Hurricane Florence in 2018 when coal ash (see Chapter 12) stored in a pit adjacent to a cooling pond entered the pond upon wall failure in the storm; the dike separating the pond from the Cape Fear River also failed and allowed elevated Se and As into the river (Fig. 17.1a). Another example in the Cape Fear River occurred during Hurricane Floyd in 1999 when an industrial storage lagoon containing waste chromium was flooded, draining into the river. As a result of Hurricane Harvey in 2017 local Houston watersheds rich in industrial sources loaded considerable amounts of metals and other pollutants into the floodwaters from damaged facilities (Kiaghadi and Rifai 2019).

### 17.2.1.3 Rainfall

In Section 17.1.1 we discussed flooding, primarily from a more natural, seasonal viewpoint, but also noted that human landscape changes greatly increase the pollution impacts of excessive rainfall. This has been clearly established in post-hurricane studies. During hurricanes rainfall can be massive. When hurricanes follow a river upstream to the headwaters regions, the flooding of lower-order streams combines to greatly raise river levels and subsequent river discharge downstream to the riverine estuary and ocean proper. As noted, flooding causes great damage to human residences and other structures in low-lying areas through outright physical damage, but it also causes massive mold infestation. Mold infestation ruins water-damaged furnishings, and many humans are allergic to mold (including this author); injuries and disease incidents among humans spike following hurricanes (Mallin and Corbett 2006).

Rural areas can be important source areas of pollution following hurricanes. While small rural sewage package plants are subject to flooding, the major sources are the animal waste lagoons that serve swine, some poultry, and some cattle CAFOs

(Fig. 17.1b–d). Numerous incidents of this have been detailed in Mallin 2000 and Mallin et al. 2002. As with human sewage, animal waste contains concentrated BOD, nutrients, and fecal microbes (see Chapter 13). This waste, along with human sewage, drives down river DO to near-anoxic levels. These post-hurricane decreases in DO are common and often severe. Massive fish and invertebrate kills have resulted from hypoxia and anoxia caused by the anthropogenic (and natural) BOD loads that enter rivers during and following hurricanes (Fig. 17.1e, f).

The flooding of residential human landscapes (and widespread impervious surface areas) in river watersheds leads to large and rapid increases in urban stormwater runoff and non-point-source pollution. Flooding of sewer lines and septic systems occurs, adding to the pollutant loads mentioned above regarding sewage rerouting from power losses. For instance, in the riverside city of Fayetteville, North Carolina, flooding from Hurricane Florence in 2018 caused approximately 7.6 million gallons (28.8 million L) of untreated sewage to enter surface waters from sewage pump stations that went offline or sewer lines that flooded or were otherwise compromised (North Carolina Division of Environmental Quality records (NCDEQ), 2018). Flooding following Hurricane Florence in eastern North Carolina caused flooding of septic systems at three sites to stay elevated for > 3.5 weeks (Humphrey et al. 2021). Thus, hurricane-caused urban flooding and runoff load huge amounts of pollutants into the major tributaries and main stem of affected rivers, including suspended solids, BOD, N, P, fecal microbes, metals, chemicals, and trash of all sorts, the latter which is often left piled on shorelines after water recession (Tilmant et al. 1994). Flooding of forested coastal lowlands can also release mercury (Hg), including toxic MeHg, into rivers and coastal ocean. Tsui et al. (2020) quantified such release of Hg from a forested blackwater watershed in South Carolina during Hurricane Joaquin in 2015 and Hurricane Matthew in 2016 and found large releases of MeHg as floodwaters subsided, becoming available to be taken up and biomagnified within the riverine food web.

Compromised upstream sewage treatment and animal waste sources load large amounts of N and

**Figure 17.1** Fig. 17.1a (top left) Flooded power plant ash pond and cooling pond breaching into Cape Fear River, North Carolina, following Hurricane Florence in 2018. (Photo courtesy of North Carolina Department of Environmental Quality.) Fig. 17.1b (top right) Flooded CAFO in Northeast Cape Fear River basin following Hurricane Florence. (Photo courtesy of Walker Golder, Audubon.) Fig. 17.1c (middle left) Flooded swine CAFO lagoon in Neuse River basin, North Carolina, following Hurricane Floyd in 1999. (Photo courtesy of Rick Dove, Waterkeeper Alliance.) Fig. 17.1d (middle right) Flooded CAFO and waste lagoon following Hurricane Florence in 2018. (Photo courtesy of Rick Dove, Waterkeeper Alliance.) Fig. 17.1e (bottom left) Variety of fish killed by hypoxia in Northeast Cape Fear River following Hurricane Bonnie in 1998. (Photo by the author.) Fig. 17.1f (bottom right) Federally endangered Atlantic sturgeon *Acipenser* sp. killed by hypoxia along Cape Fear River following Hurricane Florence in 2018. (Photo courtesy of Fritz Rohde, National Oceanic and Atmospheric Administration.)

P into streams and rivers. Well downstream, algal blooms can form in open bays and sounds from hurricane-driven nutrient loading (Tilmant et al. 1994; Peierls et al. 2003), but blooms are depressed in rivers or riverine estuaries from light limitation by increased water color and/or turbidity (Burkholder et al. 2004; Mallin and Corbett 2006). Note that hurricanes can release large amounts of DOC from lowland swamps, much of it refractory, that can flow well out into oceanic waters (Avery et al. 2004). Algal species can change too; in 2004 massive inland flooding caused by several hurricanes led to authorities pumping excess water from Florida's Lake Okeechobee into the Caloosahatchee and St. Lucie Rivers, freshening the estuaries and causing blooms of the toxic cyanobacteria *Microcystis aeruginosa* in those downstream estuaries (Lapointe et al. 2015). Also in Florida, the hurricane-caused massive drainage from the Caloosahatchee and other west-flowing rivers, along with increased groundwater pumping into the Gulf, caused elevated nutrients that fed blooms of the toxic dinoflagellate *Karenia brevis* for nearly the entire following year (Hu et al. 2006).

Beyond pollutant loading, hurricane-induced flooding can have many effects. Freshwater inputs lead to alteration of estuarine fish and benthic communities by displacing marine fauna with freshwater or upper estuarine fauna (Knott and Martore 1991; Mallin et al. 2002; Walker et al. 2020). Terrestrial animals are displaced by floodwaters and enter human residential areas—this can have unpleasant consequences when some of these organisms (poisonous snakes, poisonous insects, bees, etc.) are encountered by humans. Flooding can spread non-native aquatic species: for example, exotic fish such as *Tilapia* or grass carp that have been introduced to ponds for weed control can become spread into rivers by floodwaters entering the ponds (Mallin et al. 2002).

Well downstream in the estuary and along the coast, floodwaters from inland cause beach and shellfish bed closures to occur from the loading of fecal bacteria, protozoans, and viruses from compromised upstream sewage, septic, and CAFO sources; lakes can be similarly polluted (Mallin et al. 2002). Whereas hypoxia and anoxia causes massive fish kills up in the rivers, what about estuarine fish

populations well downstream? Post hurricane commercial fishing may be good or bad, depending upon the target item, mobility of the organism to escape hypoxic waters, and concentration of organisms in areas outside of impacted waters, which can lead to overharvesting (Burkholder et al. 2004).

Recent research (Wang and Toumi 2021) demonstrates that the frequency of hurricanes entering the coastal zone (defined as < 200 km) from the coast has been increasing for several decades. While the data do not indicate that actual land strikes have also increased, major storms in offshore waters certainly bring major increased rains and wind events; thus such nearshore hurricanes also have strong impacts on river watersheds.

## 17.2.2 Recovery

River ecosystems do recover from hurricanes, with recovery times dependent upon several variables (Table 17.3). The amount of rainfall determines the degree of local and downstream flooding, as well as the magnitude of river discharge and downstream freshening of riverine estuaries or receiving lagoons. When, as in which month, the storm arrives is highly important to the degree of hypoxia inflicted upon waterways and the time to recovery. A mid-summer, warmwater hurricane will severely drive down DO, while a cool-weather, late-fall hurricane may have minimal impact on hypoxia in the cooler receiving waters (Mallin and Corbett 2006). However, flood recovery during mid-summer when evapotranspiration is maximal may be faster than a late-fall hurricane, when cooler temperatures may extend duration of the flooding. The path of the storm is critical; if the storm merely crosses a river the recovery can be rapid; but if its path takes it upstream into the headwaters the recovery can be long. If municipal, industrial, or agricultural pollutants are concentrated in the storm's path this will impact severity and recovery time, increasing hypoxia, fecal microbial pollution, metals and organic chemical pollutants, and severity of fish and invertebrate kills in the rivers and estuaries. In general, most biotic communities are reasonably resilient and recovery occurs in weeks to months after the hurricane for most taxa (Wallace 1990; Knott and Martore 1991).

Storm impacts on the benthic community of lower rivers and riverine estuaries have been well studied (Knott and Martore 1991; Mallin et al. 2002; Walker et al. 2020); invertebrates, whether displaced by freshwater pulses or killed by anoxia, will recolonize as water quality conditions improve, aided by opportunistic taxa. Rivers are often restocked by wildlife agencies with sunfish, catfish, and bass following hurricanes (Table 17.3). What has been little reported on are hurricane impacts to endangered species. For example, in the aftermath of Hurricane Florence, at least a dozen carcasses of federally endangered Atlantic sturgeon (*Acipenser* sp.) were photographed by biologists and others in the Cape Fear watershed, sparking fears of severe impacts to the already-sparse population (Fig. 17.1f).

### 17.2.3 Mitigation of Hurricane Impacts and Enhanced Recovery

Given all this, what about protection of human infrastructure, water quality, inland ecosystems, etc.? The most obvious action is to severely decrease the input of greenhouse gases into the atmosphere to reduce warming and the frequency of large storms, but that is not going to undo what is already in motion. Because hurricanes are a natural phenomenon, can their environmental impacts be mitigated? We saw earlier that armoring river channels with dikes and levees only leads to problems displaced elsewhere.

Clearly, rebuilding of human habitats on river floodplains or low-lying barrier islands should be discouraged; such structures and uses make storm impacts worse. However, the availability of government-sponsored hazard insurance (supported by coastal lawmakers in the USA under pressure from builder's associations and wealthy individuals with beach homes) effectively enhances building in hazardous locales. Private insurers often will not insure such building. Regarding natural areas, some activities, such as fish restocking, are done to help such areas recover, at least to human-perceived standards. Under pressure from boaters, state agencies also remove woody debris including snags from creek and river channels (Table 17.3).

**Table 17.3** The recovery process timeline from hurricanes in river ecosystems. Note that exceptions can be found depending upon circumstances. (Compiled from Knott and Martore 1991; Van Dolah and Anderson 1991; Mallin et al. 2002; Paerl et al. 2001; Patrick et al. 2020; Tomasko et al. 2006; Burkholder et al. 2004; Balthis et al. 2006; Walker et al. 2020.)

- Recovery of water levels to normal seasonal levels: 2–8 weeks.
- Estuarine salinity levels returned to normal: 1–2 months.
- Recovery of DO to normal levels: 2–6 weeks.
- Decrease of nutrient concentrations to expected levels: 1–8 weeks.
- Decrease of fecal microbe concentrations to normal levels: 1–3 weeks.
- Recovery of the benthic community to normal: 2–6 months.
- Recovery of the fishery to normal: months to 2 years; but see next bullet.
- Enhancement of the fishery by agency stocking: stocking may occur from 6 months to two years after the storm. Note: impacts on rare or endangered species could last years to decades; they are not normally restocked.
- Removal of accumulated snags by state or federal agencies: weeks to months, depending upon available personnel and public pressure.

However, as discussed earlier (see Chapters 6 and 7), such snags and debris dams have a number of benefits for fish communities such as providing habitat and food.

One key is to avoid destruction of wetlands and their conversion to other uses. In fact, wetlands, especially coastal wetlands, have great value for hurricane protection (summarized in Costanza et al. 2008). Coastal wetlands perform several functions (Spalding et al. 2014), including decreasing the area of fetch (open water) for wind to form waves; increasing the drag on water motion; reducing the direct wind effect on the water surface; and reducing wave height. Wetlands directly absorb wave energy because aquatic vegetation baffles and reduces surges and waves, and reduces water flow velocities and shear stress. Vegetation helps maintain shallow depths, which also acts to reduce surges and wave heights; roots slow rates of erosion. A variety of wetland types are effective in reducing hurricane impacts. Mangroves and salt marshes are particularly useful in these functions, though submersed aquatic vegetation beds also provide protective services, as do shellfish beds and human-designed "living oyster" breakwaters (Spalding et al. 2014).

A large-scale statistical analysis of 19 coastal US states (Costanza et al. 2008) found that the loss of each hectare of coastal wetland corresponded to a median $5000 increase in storm damages (average $33,000). The annual value of coastal wetland depended on frequency of hurricane hits, wind speed, and coastal infrastructure, and ranged from $250/ha in Louisiana to $51,000/ha in New York, with a median of $3230/ha. The researchers found that value varied inversely with existing wetland area, but increased directly with value of coastal infrastructure (i.e. New York, Maryland, and Connecticut have few wetlands but a lot of valuable coastal infrastructure). As an example, Louisiana lost 480,000 ha of coastal wetlands prior to Hurricane Katrina through direct human activities such as wetland conversion or indirectly through preventing suspended sediments from reaching Gulf coastal marshes from the leveed rivers. The value of lost protective services was estimated as $1700/ha × 480,000 ha = $816 million. Note these figures do not include all the other wetland services (fish and shellfish production, nutrient absorption, recreation, etc.). Such estimates are not merely an academic exercise. An analysis of wetlands as storm risk buffers was commissioned by insurance giant Lloyd's (Narayan et al. 2016). They concluded that the presence of coastal wetlands in the northeast USA saved > $625 million in flood damages; the authors pointed out that the benefits of coastal wetlands conservation also accumulate upstream. Based on such coastal analyses, riverine wetlands, including floodplain forests, should provide some similar storm baffling effects up the river, but unfortunately floodplains are often compromised with land uses that decouple the floodplain from the river or outright enhance water pollution potential (see Chapters 10 and 13). Inland, where development already exists, resizing roadside culverts to enhance drainage from larger rainfall may help alleviate flooding. In coastal salt marshes the creation of runnels, which are largely hand-dug channels less than 30 cm deep and wide, are now being used to more rapidly drain standing waters from marsh interiors, improving vegetation growth (Besterman et al. 2022). This technique is hoped to allow time for more adaptation in salt and freshwater marshes in the face of encroaching sea level rise.

## 17.2.4 More Powerful Storms

Models indicate that climate change driven by increased greenhouse gas emissions will increase the frequency of the most destructive storms (Bender et al. 2010; Sobel et al. 2016), with indication that the maximum intensity of such storms will occur more frequently near land (Camargo and Wing 2021). The increase in sea surface temperatures in the tropical North Atlantic main hurricane development region will continue to drive increases in major hurricanes (Murakami et al. 2018). One of the areas expected to see an increase in the most powerful storms is the western Atlantic between 20 and 40°N latitude (Bender et al. 2010), which encompasses much of the US East Coast. Modeling is unclear as to whether or not climate change will bring more named storms in total, but the year 2020 truly stands out. According to NOAA (2020) the 2020 Atlantic season featured 30 named storms, the most on record for one season. There were 13 hurricanes of category 1 or more, the second highest number of hurricanes on record. NOAA (2020) noted that an average year has 12 named storms, of which 6 are hurricanes. In 2020 12 named tropical storms made landfall in the USA, some causing severe economic damage (Camargo and Wing 2021).

## 17.3 Climate Change

Human-induced climate change has become a major part of the political, industrial, and ecological conversation. Global warming, a major part of climate change, impacts river ecosystems in various ways, including increasing the strength of hurricanes, forcing changes in surface water or groundwater salinity and water availability, increasing trophic state, altering biotic species composition, and directly injuring cold stenotherms. The major factors accounting for climate-driven global warming are briefly discussed below.

### 17.3.1 Causes of Warming

Warming is being caused by excess greenhouse gases entering the Earth's atmosphere, especially carbon dioxide but also methane and nitrous oxide.

$CO_2$ increases are largely from industrial processes, especially fossil-fuel burning for power generation and cement production (see Chapter 12). Fires, both wildfires and controlled biomass burning, are a major contributor to atmospheric $CO_2$, accounting for as much as 50% of the $CO_2$ generated by fossil-fuel combustion (Bowman et al. 2009). In recent years global warming has contributed to the massive wildfires that have plagued the American West, Canada, Australia, and Siberia; large fires occurred in Greece as well in 2021. Such massive fires contribute large amounts of $CO_2$ to the atmosphere. The fires also directly pollute rivers with ash and other debris, changing water chemistry and causing massive fish kills. Finally, another byproduct of wild and controlled fire is aerosolized black carbon, or soot. As noted in Section 17.3.2, soot particles contribute to warming by absorbing solar energy and producing heat, and darkening snow and ice, leading to less reflection of sunlight.

Methane has the second strongest global warming potential (Ravishankara et al. 2009). Concentrations of this gas had been rising steadily until about 1999, leveled off for several years, and then began climbing again around 2007 (Nisbet et al. 2014). Sources of methane include natural wetland processes, thawing permafrost, the coal industry, biomass burning, and natural gas production, especially fracking (Nisbet et al. 2014). In terms of agriculture, much methane is generated by large-scale production of ruminant livestock, such as cattle, sheep, and goats (see Chapter 13). Ranking behind methane in global warming potential is nitrous oxide, which is also an important ozone-depleting substance (Ravishankara et al. 2009). It is produced naturally by bacterial reactions in soils and during fossil-fuel consumption, industrial processes, synthetic and organic crop fertilization, and biomass burning. Remember from Chapter 2 that nitrous oxide is also a byproduct of denitrification. Forests sequester $CO_2$ and work against warming, but following a few years of relative forest protection, Amazonian rainforest loss has rapidly accelerated in recent years, much of this illegal and caused by ranchers, loggers, and miners (Escobar 2020). Given all these (and other) sources of warming, what are the recognized and potential impacts to river ecosystems? Well, it's going to vary, depending upon region.

## 17.3.2 Impacts on Headwaters and Cascading Impacts

Many river systems have headwaters based in mountains. Downstream river flow variability depends (along with rainfall) upon snowpack melt at high altitudes, which is largely a function of temperature (Pederson et al. 2011). Thus, with persistent warmer temperatures, snowpacks cannot accumulate as much, and outright shrink. This leads to less river discharge downstream in higher-order streams, which can damage inland commercial shipping, impact hydropower production, cause water distribution conflict among political entities in arid regions, alter salinities in lower rivers, disrupt species composition, and impact river-based recreation. The same goes for glaciers; in some areas glacial meltwaters are used for irrigation and other purposes downstream, and shrinking glaciers contribute to water insecurity (Marzeion et al. 2014).

Recent warming of Alaska's Yukon River has led to severe heat stress impacting migrating Chinook salmon (von Biela et al. 2020). Impacts to salmon are not confined to the Yukon, however. In northern California and the Pacific Northwest salmon migrations are dependent upon, among other things, dependable sources of cold clear water to spawn in. When mountain snowfall and snowpacks are decreased due to climate warming, less runoff occurs and streams can become shallow and easily warmed to temperatures fatal to adult and young salmon (Service 2015). Warming is leading to glacier melt worldwide; whereas anthropogenic effects played a minor role in glacier melt in the early and middle twentieth century, since 1990 anthropogenic effects have been the major driver of glacier melt (Gardner et al. 2013). Melting of the Greenland and Antarctic ice sheets is also occurring; Greenland's ice sheet is melting at an accelerated pace, partially attributed to warming, but it has been reported (Kintisch 2017) that the melt has been exacerbated by a loss in albedo (reflection) due to accumulations of soot, algae, and bacteria on the ice—which increases the ice sheet's absorption of solar energy. As such, recent studies (summarized by Voosen 2021) show that the period 2014–2020 was the warmest on record to date, with melting ice sheets and glaciers causing sea level rise of 4.8 mm/year (and rising), with the increase since 1900 estimated to be about 20 cm. Climate change impacts the highest mountains on Earth—those in the Tibetan Plateau and nearby high Asian mountains. Research in these systems (Li et al. 2021) shown that climate warming and increased precipitation is melting glaciers and permafrost, which is increasing runoff as well as the associated suspended sediment loads to river basins far downstream. Remember that suspended sediments carry high concentrations of associated pollutants including nutrients and metals (see Chapter 1), thus raising pollutant levels in downstream reaches. Additionally, the increased suspended sediment loads fill reservoirs more rapidly, decreasing lifespans of hydropower systems (see Chapter 10) and degrading turbines (Li et al. 2021).

### 17.3.3 Changes in Rainfall

Rainfall changes due to climate change will vary from region to region, but some researchers predict hurricane-based rainfall to increase (Sobel et al. 2016; Sinha et al. 2017; Patricola and Wehner 2018; Maxwell et al. 2021). Based on known impacts of rain and runoff, we can expect increased N loading to rivers (see Chapter 2). Sinha et al. (2017) expect considerable increases in N loading in US areas such as the upper Mississippi watershed, the Great Lakes, and the northeast southward to the Mid-Atlantic Seaboard. On a global basis, those researchers also predict increased rain-based N loading in India and eastern China. With heavier rains one might also expect increased erosion from agricultural areas into river ecosystems; based on the strong associations between suspended solids and P (see Chapter 2) this will lead to increased P loading in some coastal regions. The result will be increased downstream eutrophication from increased N and P loading. Whereas turbidity may mute the eutrophication effect of nutrient loading in river channels (as discussed earlier regarding hurricane impacts), open downstream brackish and marine waters in sounds and bays will

experience algal blooms, including toxic and noxious taxa. Other impacts, as based on the 1993 Mississippi River flood (Rabalais et al. 1998), are likely to include stronger nearshore stratification and more severe bottom-water hypoxia in bays and sounds. Whereas increased diatoms may be good for coastal food chains, increases in cyanobacteria, dinoflagellates, and prymnesiophytes are likely to be problematic (Burkholder 1998).

A contributing factor to increased hurricane-driven rainfall is the translational speed, or how long it takes for a hurricane (and its rainfall) to exit an area after coming ashore. Using a study of tree rings representing a 300-year period, Maxwell et al. (2021) found that hurricane translational speeds have been slowing over time, resulting in larger extreme hurricane-related precipitation. From 1948 to 2018 hurricane rainfall was correlated with storm duration and negatively with translational speeds. Major hurricane-related precipitation events cited by the authors include Hurricanes Connie in 1955, Bertha and Fran in 1996, Floyd in 1999, Matthew in 2016, and Florence in 2018, some of which are discussed in Section 17.2.2.

Apart from more intense hurricanes, recent research (Huang and Swain 2022) has postulated that the state of California is likely to see one or more "megafloods" in the coming decades, with the chances enhanced by climate change. Using the massive California flooding of 1861–1862 as a benchmark, the authors used two separate models to assess how cool-season "atmospheric rivers" of moisture can lead to month-long rain events that will drop unprecedented amounts of rain in mountainous areas, which will then cause flooding in lowland areas such as the agriculturally rich Central Valley and population centers such as Sacramento, Los Angeles, and San Jose. The intense mountain runoff (both from rain and snowmelt) will cause highly destructive debris flows and major disruptions to emergency services, with staggering economic costs and large numbers of fatalities. To prepare, the authors suggest increased floodplain restoration and levee setbacks (discussed in Chapter 16), better informed reservoir operations, and revised and updated emergency services and evacuation plans. Such flooding occurred to a degree in early 2023.

## 17.3.4 Rising Sea Levels

In early stages of sea level rise, occurring in some areas at present, impacts on groundwater will be felt. Freshwater wells may turn brackish, for instance, with severe human health implications (Vineis and Khan 2012). Regarding sewage lines, expect increased problems of I&I as sanitary lines are submerged and leakages increase (Flood and Cahoon 2011; Cahoon and Hanke 2017). In areas served by septic systems, expect frequent flooding of such systems and release of wastewater (including BOD, nutrients, and fecal microbes) into upper groundwaters and surface waters (Humphrey et al. 2021).

How might rising sea levels impact riverine ecosystems? Well-established nontidal versus tidal freshwater reaches can be distinguished by differences in channel morphology, canopy influence and subsequent irradiance within the water column, and phytoplankton production dynamics (Ensign et al. 2012). Rising seas will push brackish estuarine waters farther upstream, altering riparian vegetation from freshwater marsh and riparian forest in favor of salt marsh vegetation and dead or "ghost" forests (see below). To what extent will a (possibly rather rapid) change in upstream tidal influence impact river channel morphology and survival of the riparian forest (and its canopy)? As waters spread out over land, impacts on riverine biota will vary according to the local terrain and salinity. Primary nursery areas will be submerged in some estuarine areas but expand into more upland terrain. Freshwaters moving well into forested areas could, according to theory, provide productive new habitat for many fish species in the "moving littoral" zone (from the flood-pulse concept). Reservoir ecology tells us that when new reservoirs are first flooded, there is a "trophic upsurge" from newly submerged terrain from both the photosynthetic and heterotrophic communities (see Chapter 11). However, things are likely to be more complicated by sea level rise. Recent work on coastal lowland forests (see Sacatelli et al. 2020 and references therein; also Ury et al. 2021) demonstrate that such freshwater-based forests on the US East Coast have been slowly dying and becoming what is now termed "ghost forests." Such forest death (and

conversion to marsh) comes about from chronic groundwater salinity increases (from sea level rise) that injure freshwater forests, coupled with acute major hurricane events that periodically douse the nearshore forest landscape with saltwater (storm surge) that may impact the forest for days or more. In such impacted coastal forests in South Carolina, researchers have determined that the salinization process has increased remineralization rates of N and P (Noe et al. 2013), making these nutrients bioavailable to aquatic algae and bacteria. Once such ghost forests are fully inundated from sea level rise it is unclear as to how favorable a fisheries habitat these changes will bring.

What about waters submerging former agricultural landscapes? Such submergence is likely to turn such landscapes into salt marsh or freshwater marsh, depending upon how far up the river such submergence persists. This might not be a positive thing for fish. Submerged agricultural soils may leach pesticides and herbicides into overlying waters, which are likely to bioconcentrate up the food chain from invertebrates into fish. Since agricultural fields are often heavily fertilized, N and especially P can be overly abundant in such soils (Cahoon and Ensign 2004); CAFO waste deposit fields will likewise be overloaded with nutrients. Soil-bound nutrients will remineralize into the overlying shallow waters; fertilized waters combined with abundant sunlight from the lack of forest canopy will lead to nuisance and toxic blooms of cyanobacteria in freshwater areas and dinoflagellates and prymnesiophytes in brackish waters, none of which will be good for food webs. In submerged areas of intense land use such as ports and industrial complexes there will be plenty of hard substrata to serve as habitat for fish and shellfish. However, such areas will contain legacy chemicals that will leach out of soils and building materials, including PAHs, metals, and especially petrochemicals, again likely biomagnifying into the food chains. There may be a lot of fish and shellfish, but you wouldn't want to eat them.

Sea level rise is already swallowing Pacific Islands, leading to population displacement. Some US cities like Miami, Florida, and Norfolk, Virginia are already getting hit hard by flooding on king tides, even on sunny days with no rain. The speed at which sea level rise impacts rivers and their watersheds will depend on multiple factors, including permafrost melting and methane release, continued ice sheet loss, and political directives from nations worldwide.

## 17.4 Summary

- Streams and rivers are normally resilient to flooding in that impacted biotic microbial, algal, invertebrate, and vertebrate communities rapidly (days to weeks) recolonize impacted areas from upstream, downstream, backwater, and hyporheic habitats.
- When rivers are impacted by human engineering efforts (armoring, dams) flood impacts can be displaced and enhanced laterally in unarmored locations and far downstream into estuarine waters, or constrained and funneled waters can overtop levees and cause immediate proximal flood damage to residential land and farmland.
- Flooding is a major consequence of rainfall from hurricanes (also known as typhoons or cyclones in non-Atlantic areas) along with wind and wave and tidal surge damage. Impacts in river systems can occur inland even hundreds of kilometers from the coast. Environmental impacts of such storms include floodplain forest damage, salinity changes, sewage and animal waste discharges, stream hypoxia and anoxia, large waterborne pulses of fecal microbes, chemical pollution, organism displacement, fish and invertebrate kills, and increases in human ailments.
- Recent modeling efforts predict climate change will bring an increase of the largest, most-destructive storms. Modeling is currently unclear regarding the increases of tropical storms in general, although recent years have shown record numbers of named storms.
- Climate change is predicted to produce more rainfall in several areas, including well-populated regions. Such rainfall is likely to bring enhanced riverine N loading, and subsequent increased eutrophication in lower rivers and estuaries.
- Warming is already strongly impacting snowpack accumulation and reducing riverine water supply in some areas, and warming is contributing to increased wildfires in the Western USA, Australia, and elsewhere, increasing $CO_2$ atmospheric loading and causing fish-killing pollution to rivers in fire-impacted areas.
- The impact of sea level rise on riverine fish and invertebrate communities is presently unclear, although slow death of coastal forests by increased groundwater and storm surge salinity has left "ghost forests" in formerly living forests. Flooding of formerly agricultural and livestock-rearing lands as well as residential and commercial and industrial areas is likely to cause leaching of N, P, herbicides, pesticides, metals, and organic contaminants into overlying shallow waters, impacting biotic communities.

## References

Avery, G.B. Jr., R.J. Kieber, J.D. Willey, G.C. Shank and R.F. Whitehead. 2004. Impact of hurricanes on the flux of rainwater and Cape Fear River water dissolved organic carbon to Long Bay, southeastern United States. *Global Biogeochemical Cycleas* 18:GB3015.

Balthis, W.L., J.L. Hyland and D.W. Bearden. 2006. Ecosystem responses to extreme natural events: impacts of three sequential hurricanes in fall 1999 on sediment quality and condition of benthic fauna in the Neuse River Estuary, North Carolina. *Environmental Monitoring and Assessment* 119:367–389.

Bender, M.A., T.R. Knutson, R.E. Tuleya, et al. 2010. Modeled impact of anthropogenic warming on the frequency of intense Atlantic hurricanes. *Science* 327:454–458.

Besterman, A.F., R.W. Jakuba, W. Ferguson, D. Brennan, J.E. Costa and L.A. Deegan. 2022. Buying time with "runnels" a climate adaptation tool for salt marshes. *Estuarine and Coasts* 45:1491–1501.

Bowman, D.M.J.S., J.K. Balch, P. Artaxo, et al. 2009. Fire in the Earth's system. *Science* 324:481–484.

Burkholder, J.M. 1998. Implications of harmful microalgal and heterotrophic dinoflagellates in management of sustainable marine fisheries. *Ecological Applications* 8:S37–S62.

Burkholder, J., D. Eggleston, H. Glasgow, et al. 2004. Comparative impacts of two major hurricane seasons on the Neuse River and western Pamlico Sound ecosystems. *Proceedings of the National Academy of Sciences of the United States of America* 101:9291–9296.

Cahoon, L.B. and S.H. Ensign 2004. Excessive soil phosphorus levels in eastern North Carolina: temporal and spatial distributions and relationships to land use. *Nutrient Cycling in Agroecosystems* 69:11–125.

Cahoon, L.B. and M.H. Hanke. 2017. Rainfall effects on inflow and infiltration in wastewater treatment systems in a coastal plain region. *Water Science & Technology* 75(7–8):1909–1921.

Camargo, S.J. and A.A. Wing. 2021. Increased tropical cyclone risk to coasts. *Science* 371:458–459.

Cornwall, W. 2021. Europe's deadly floods leave scientists stunned. *Science* 373:372–373.

Costanza, R., O. Perez-Maqueo, M. L. Martinez, P. Sutton, S.J. Anderson and K. Nulder. 2008. The value of coastal wetlands for hurricane protection. *AMBIO* 37:241–248.

Cox, A.H., M.J. Dowling, G.W. Loomis, S.E. Engelhart and J.A. Amador. 2020. Geospatial modeling suggests threats from stormy seas to Rhode Island's coastal septic systems. *Journal of Sustainable Water in the Built Environment* 6(3):04020012.

Dodds, W.K., K. Gido, M.R. Whiles, K.M. Fritz and W.J. Matthews. 2004. Life on the edge: the ecology of Great Plains prairie streams. *BioScience* 54:205–216.

Ensign, S.H., M.W. Doyle and M.F. Piehler. 2012. Tidal geomorphology affects phytoplankton at the transition from forested streams to tidal rivers. *Freshwater Biology* 57(10):2141–2155.

Escobar, H. 2020. Deforestation in the Brazilian Amazon is still rising sharply. *Science* 369:613.

Estuaries and Coasts (2006) Estuaries and Coasts Special Issue: *Hurricane Impacts on Coastal Ecosystems*, Volume 29, Number 6A, December 2006.

Fisher, S.G., L.J. Gray, N.B. Grimm and D.E. Busch. 1982. Temporal succession in a desert stream ecosystem following flash flooding. *Ecological Monographs* 52: 93–110.

Flood, J.F. and L.B. Cahoon. 2011. Risks to coastal wastewater collection systems from sea-level rise and climate change. *Journal of Coastal Research* 27:652–660.

Gardner, A.S., G. Moholdt, J.G. Cogley, et al. 2013. A reconciled estimate of glacier contributions to sea level rise: 2003–2009. Science 340:852–857.

Herrera-Garcia, G., G. Ezquerro, G. Tomás, et al. 2021. Mapping the global threat of land subsidence. *Science* 371:34–36.

Hu, C., F.E. Muller-Karger and P.W. Swarzenski. 2006. Hurricanes, submarine groundwater discharge, and Florida's red tides. *Geophysical Research Letters* 33: L11601.

Huang, X. and D.L. Swain. 2022. Climate change is increasing the risk of a California megaflood. *Science Advances* 8:eabq0995.

Humphrey, C.P., D. Dillane, G. Iverson and M. O'Driscoll. 2021. Water table dynamics beneath onsite wastewater systems in eastern North Carolina in response to Hurricane Florence. *Journal of Water and Climate Change* 12(5):2136–2146.

Kiaghadi, A. and H.S. Rifai. 2019. Physical, chemical, and microbial quality of floodwaters in Houston following Hurricane Harvey. *Environmental Science and Technology* 53:4832–4840.

Kintisch, E. 2017. Meltdown: as algae, detritus, and meltwater darken Greenland's ice, it is shrinking ever faster. *Science* 355:788–791.

Knott, D.M. and R.M. Martore. 1991. The short-term effects of Hurricane Hugo on fishes and decapod crustaceans in the Ashley River and adjacent marsh creeks, South Carolina. *Journal of Coastal Research* 8: 335–356.

Lapointe, B.E., L.W. Herren, D.D. Debartoli and M.A. Vogel. 2015. Evidence of sewage-driven eutrophication and harmful algal blooms in Florida's Indian River Lagoon. *Harmful Algae* 43:82–102.

Li, D., X. Lu, I. Overeem, et al. 2021. Exceptional increases in fluvial sediment fluxes in a warmer and wetter High Mountain Asia. *Science* 374:599–603.

Mallin, M.A. 2000. Impacts of industrial-scale swine and poultry production on rivers and estuaries. *American Scientist* 88:26–37.

Mallin, M.A. and C.A. Corbett. 2006. Multiple hurricanes and different coastal systems: how hurricane attributes determine the extent of environmental impacts. *Estuaries and Coasts* 29:1046–1061.

Mallin, M.A., M.H. Posey, M.R. McIver, D.C. Parsons, S.H. Ensign and T.D. Alphin. 2002. Impacts and recovery from multiple hurricanes in a Piedmont–Coastal Plain river system. *BioScience* 52:999–1010.

Marzeion, B., J.G. Cogley, K. Richter and D. Parkes. 2014. Attribution of global glacier mass loss to anthropogenic and natural causes. *Science* 345:919–921.

Maxwell, J.T., J.C. Bregy, S.M. Robeson, P.A. Knapp, P.T. Soulé and V. Trouet. 2021. Recent increases in tropical cyclone precipitation extremes over the US East Coast. *Proceedings of the National Academy of Sciences of the United States of America* 118 (41): e2105636118.

Minshall, G.W., K.W. Cummins, R.C. Peterson, et al. 1985. Developments in stream ecosystem theory. *Canadian Journal of Fisheries and Aquatic Sciences* 42:1045–1055.

Murakami, H., E. Levin, T.L. Delworth, R. Gudgel and P.C. Hsu. 2018. Dominant effect of relative tropical Atlantic warming on major hurricane occurrence. *Science* 362:794–799.

FLOODS, HURRICANES, AND CLIMATE CHANGE 395

Myers, M.F. and G.F. White. 1993. The challenge of the Mississippi flood. Environment 35(10):6–35.

Narayan, S., M.W. Beck, P. Wilson, et al. 2016. Coastal Wetlands and Flood Damage Reduction: Using Risk Industry-Based Models to Assess Natural Defenses in the Northeastern U.S.A. Lloyd's Tercentenary Research Foundation, London.

NCDEQ. 2018. North Carolina Division of Environmental Quality records.

Nisbet, E.G., E.J. Dlugokencky and P. Bousquet. 2014. Methane on the rise—again. Science 343:493–494.

NOAA. 2020. Record Breaking Atlantic Hurricane Season Draws to an end. News release, November 24. National Oceanic and Atmospheric Administration, US Department of Commerce, Washington, DC.

Noe, G.B., K.W. Krauss, R.G. Lockaby, W.H. Connor and C.R. Hupp. 2013. The effect of increasing salinity and forest mortality on soil nitrogen and phosphorus remineralization in tidal freshwater forested wetlands. Biogeochemistry 114:225–244.

Paerl, H.W., J.D. Bales, L.W. Ausely, et al. 2001. Ecosystem impacts of three sequential hurricanes (Dennis, Floyd and Irene) on the United States' largest lagoonal estuary, Pamlico Sound, NC. Proceedings of the National Academy of Sciences of the United States of America 98:5655–5660.

Parsons, M., C.A. McLoughlin, K.A. Kotschy, K.H. Rogers and M. Rountree. 2005. The effects of extreme floods on the biophysical heterogeneity of river landscapes. Frontiers in Ecology 3:487–494.

Patrick, C.J., L. Yeager, A.R. Armitage, et al. 2020. A system level analysis of coastal ecosystem responses to hurricane impacts. Estuaries and Coasts 43:943–959.

Patricola, C.M. and M.F. Wehner. 2018. Anthropogenic influences on major tropical cyclone events. Nature 563:339–346.

Pederson, G.T., S.T. Gray, C.A. Woodhouse, et al. 2011. The unusual nature of recent snowpack declines in the North American cordillera. Science 333:332–335.

Peierls, B.J., R.R. Christian and H.W. Paerl. 2003. Water quality and phytoplankton as indicators of hurricane impacts on a large estuarine ecosystem. Estuaries 26:5655–5660.

Penczak, T., G. Zieba, H. Koszalinski and A. Kruk. 2003. The importance of oxbow lakes for fish recruitment in a river system. Archiv für Hydrobiologie 158:267–281.

Rabalais, N.N., R.E. Turner, W.J. Wiseman Jr. and Q. Dortch. 1998. Consequences of the 1993 Mississippi River flood in the Gulf of Mexico. Regulated Rivers: Research and Management 14:161–177.

Ravishankara, A.R., J.S. Daniel and R.W. Portmann. 2009. Nitrous oxide (N$_2$O): the dominant ozone-depleting substance emitted in the 21st century. Science 326:123–125.

Sacatelli, R., R.G. Lathrop and M. Kaplan, 2020. Impacts of Climate Change on Coastal Forests in the Northeast US. Rutgers Climate Institute, Rutgers University, New Brunswick, NJ.

Service, R.F. 2015. Meager snow spell trouble ahead for salmon. Science 348:268–269.

Sinha, E., A.M. Michalak and V. Balaji. 2017. Eutrophication will increase during the 21st century as a result of precipitation changes. Science 357:405–408.

Sobel, A.H., S.J. Camargo, T.M. Hall, C-Y. Lee, M.K. Tippett and A.A. Wing. 2016. Human influence on tropical cyclone intensity. Science 353:242–246.

Spalding, M.D., S. Ruffo, C. Lacambra, et al. 2014. The role of ecosystems in coastal protections: adapting to climate change and coastal hazards. Ocean & Coastal Management 90:50–57.

Sparks, R.E., P.B. Bayley, S.L. Kohler and L.L. Osborne. 1990. Disturbance and recovery of large floodplain rivers. Environmental Management 14:699–709.

Tabb, T.C. and A.C. Jones. 1962. Effects of Hurricane Donna on the aquatic fauna of north Florida Bay. Transactions of the American Fisheries Society 91:375–378.

Tilmant, J.T., R.W. Curry, R. Jones, et al. 1994. Hurricane Andrew's effect on marine resources. BioScience 44:230–237.

Tomasko, D.A., C. Anastasiou and C. Kovach. 2006. Dissolved oxygen dynamics in Charlotte Harbor and its contributing watershed, in response to Hurricanes Charley and Frances and Jeanne—impacts and recovery. Estuaries and Coasts 29:932–938.

Tsui, M.T.-K., H. Uzun, A. Ruecker, et al. 2020. Concentration and isotopic composition of mercury in a blackwater river affected by extreme flooding events. Limnology and Oceanography 63:2158–2169.

Ury, E.A., X. Yang and J.P. Wright. 2021. Rapid deforestation of a coastal landscape driven by sea level rise and extreme events. Ecological Applications 31(5):e02339.

Van Dolah, R.F. and G.S. Anderson. 1991. Effects of Hurricane Hugo on salinity and dissolved oxygen conditions in the Charleston Harbor Estuary. Journal of Coastal Research 8:83–94.

Vineis, P. and A. Khan. 2012. Climate change-induced salinity threatens health. Science 338:1028–1029.

Von Biela, V.R., L. Bowen, S.D. McCormick, et al. 2020. Evidence of prevalent heat stress in Yukon River Chinook salmon. Canadian Journal of Fisheries and Aquatic Sciences 77:1878–1892.

Voosen, P. 2021. Global temperatures in 2020 tied record highs. Science 373:334–335.

Walker, L.M., P.A. Montagna, X. Hue and M.S. Wetz. 2020. Timescales and magnitude of water quality change in three Texas estuaries induced by passage of Hurricane Harvey. *Estuaries and Coasts* 44:960–971.

Wallace, J.B. 1990. Recovery of lotic macroinvertebrate communities from disturbance. *Environmental Management* 14:605–620.

Wang, S. and R. Toumi. 2021. Recent migration of tropical cyclones toward coasts. *Science* 371:514–517.

Yount, J.D. and G.J. Niemi. 1990. Recovery of lotic communities and ecosystems from disturbance—a narrative review of case studies. *Environmental Management* 14:547–569.

# Index